Metals Engineering

A Technical Guide

Metals Engineering

A Technical Guide

Leonard E. Samuels

Metals Park, Ohio 44073

Library of Congress Catalog Card Number: 87-73060

ISBN: 0-87170-315-7

SAN: 204-7586

Editorial and production coordination by
Carnes Publication Services, Inc.

Printed in the United States of America

To
Patricia

Acknowledgments

This book was written to an outline suggested by Timothy L. Gall, of the Reference Publications Department, ASM International. Without doubt, it would not have been started, and certainly would not have been finished, without his encouragement. I also acknowledge with gratitude the help and guidance received from many of my colleagues, particularly those at Materials Research Laboratories, and from my wife, Patricia, who learned to type so that she could convert my scrawl into a readable manuscript.

Apart from my colleagues at Materials Research Laboratories, the following have kindly granted permission to reproduce photographs which have not previously been published: Dr. M. Hatherly, Fig. 5.8(b) and 6.1; Professor I. J. Polmear, Monash University, Fig. 6.6(b), 6.6(c), 8.8(b), 11.8(a), and 11.8(b); Dr. A. S. Malin, University of New South Wales, Fig. 9.10; Dr. R. Hobbs, Broken Hill Proprietary Co. Ltd., Fig. 9.21; U.S. Steel Corporation on behalf of the late J. R. Vilella, Fig. 11.7.

Leonard E. Samuels

Foreword

As Dr. Samuels notes in his summary of Chapter 1, metals, central to the development of civilizations for over two millenia, are increasingly being challenged by alternative materials. Accordingly, engineers are now required to have a thorough understanding of *all* engineering materials in order to meet the manufacturing demands of today's economy. Nevertheless, the subject area of metals engineering has not grown any less complex, and the need to better understand the role of metals in engineering has not diminished; if anything, it has increased. Now, more than ever, a clear and concise overview of metals engineering is needed to guide the engineer in using metals effectively.

Metals Engineering: A Technical Guide meets this need. Dr. Samuels has written a book that provides an in-depth overview of metals technology without sacrificing the detailed information required of all materials specialists. Herein the reader is introduced to primary and secondary metallurgy: the disciplines concerned with the methods by which metals are extracted from their ores and processed into useful components. In doing so Dr. Samuels offers rich insight and clear explanations on how manufacturing methods influence the properties of a metal and, consequently, how the resulting metal component will behave when placed into service. The reader will learn how the choice of a metal can dramatically increase or decrease manufacturing costs, how long a metal will last when subjected to mechanical and chemical environments, and what factors might lead to its failure. To further aid the reader in understanding the relationship between the physical structure of a metal and its processing and use, attention is next given to the subjects of metallography; crystals, grains, and structures; solidification and casting; deformation and annealing; alloying; and heat treatment. In all, this book offers fundamental knowledge essential to making informed engineering decisions.

Dr. Samuels is eminently qualified to write this book. A native Australian, he has served as Director of Materials Research Laboratories, Australian Department of Defence, since 1980. From 1962 until 1980 he was Superintendent of the Metallurgy Division of MRL. An internationally recognized metallurgist and metallographer, Dr. Samuels has written, or coauthored, more than 75 papers in the scientific literature and is the author of four books — three of which, *Optical Microscopy of Carbon Steels*, the Third Edition of *Metallographic Polishing by Mechanical Methods*, and the present one, have been published by ASM. Among his numerous distinctions, Dr. Samuels is a former President of the Australian Institute of Metals, currently Chairman of the Australian National Committee on Manufacturing Engineering of the Institute of Engineers, and a Fellow of the Institution of Metallurgists and of ASM.

Metals Engineering: A Technical Guide should be on the required-reading list of all materials specialists.

TLG, Metals Park, OH

Contents

· 1 ·

The Role of Metals in Society

SUMMARY

The effective use of metals has been central to the development of civilizations for over two millenia. Nevertheless, metals are now coming under increasing competition from alternative materials, which increases the need for the development of a better understanding of their characteristics and how to use them effectively. The properties which make metals attractive must be appreciated as well as those which make them unattractive in particular circumstances. Likewise, the means by which their desirable properties can be obtained and their undesirable properties avoided must be understood. The production of usable metal from an ore mined from the earth involves a large number of distinct steps, and the term "metallurgy" is used to describe the broad technological discipline which covers all of these steps. However, the various groupings of these steps are backed by rather different branches of science, and so it is necessary to subdivide the discipline into primary metallurgy, which covers the steps necessary to produce molten refined metal from the ore, and secondary metallurgy, which encompasses the steps required to process this melt into usable material. As a further subdivision, physical metallurgy, with which we shall mostly be concerned, is taken to be a branch of secondary metallurgy which deals with the science and technology of the effects of composition, processing, and environment on the properties of metals and alloys. In this sense, a physical metallurgist can be described as a metals engineer who is part of the manufacturing engineering team.

The use of metals has always been a key factor in the development of the social systems of man. To realize this, one has only to recall the classification of early eras as the "stone age", the "bronze age", and the "iron age". At first, only those metals that were found naturally in the metallic state, such as gold and silver, were available for use. They were rare and were used almost exclusively for jewelry and ornamentation. Eventually, man learned how to prepare metals from the chemical compounds in which they occurred abundantly in nature. This led first to the copper-based *bronze age* and then to the *iron age*.

At first, however, metals were in short supply and hence expensive. They were used only in the most critical applications, especially for tools and armaments, and in privileged situations. Nevertheless, even in these limited roles they were often critical to the development of civilizations and societies. The ready availability of metals provided not only the wealth on which the development of great civilizations, including the Greek and Roman civilizations, was based, but also the tools which made possible the external manifestations of these civilizations. The architectural wonders of the ancient world could not have been constructed without effective metal tools. It is known, for example, that the Egyptians were carving massive statues out of monolithic blocks of granite prior to 1200 B.C. (Fig. 1.1), but there is no certain knowledge about the tools they used. It is known for certain, however, that they were manufacturing good steel chisels with preferentially hardened cutting edges before 700 B.C. At a less aesthetic, but nevertheless realistic, level, the flow and counterflow of

This statue is the oldest objet d'art on display at the Hearst Castle, San Simeon, California, where it is pointed out with pride to many hundreds of thousands of visitors each year. How many of these visitors pause to wonder what tools were used by the ancient sculptor? One suspects that something much harder than bronze or unalloyed iron would have been required. The matter is still one of controversy, but it is known that the Egyptians were using steel tools with hardened edges before 700 B.C.

Fig. 1.1. Statue of the Egyptian lion-headed goddess, Sekhmet, carved over 3000 years ago from a very hard granitic rock.

the development of weapons of offense and defense, often utilizing the most advanced metals technologies of the time, frequently played crucial roles in the rise and fall of societies. For example, it is said that a major reason for the success of the Roman legions was that they were equipped with short, thick swords with hardened steel edges (Fig. 1.2), whereas the swords of their opponents were of a soft iron and tended to bend during battle. The Roman swords were of quite complicated metallurgical construction which enabled them to have hard, effective cutting edges but tough cores which ensured that the blades did not break in battle (Fig. 1.2). Those societies which have had the best command of metals have been masters of their ages.

The wide use of metals became possible only when a number of engineering innovations were integrated together during the Industrial Revolution of the late 18th and early 19th centuries. Improved and large-scale methods of production reduced the cost of metals,* increased their availability, and improved their properties and reliability. This in turn made possible the widespread use of steam engines, machine tools, and large structures that were the most obvious artifacts of the revolution. Ever since, engineering and metallurgical developments have had a symbiotic relationship with one another. The practical implementation of an engineering innovation has often had to await the development of a new metallic alloy or a new metallurgical process to cope with new demands made by some critical component of the design. On the other hand, this type of definable need has often been the incentive for metallurgical developments. As good an example as any is the aircraft jet engine, the practical realization of which was ultimately dependent on the development of alloys for the blades in the turbine of the engine — alloys which can withstand for long periods of time a combination of high stress and high temperature in an aggressive environment. But there would have been no incentive to develop these alloys if the engineering development of the jet engine had not been pressing.

For most of the history of metallurgy, the metals and alloys used in industry, and the manufacturing processes to which they are subjected, have been the result of empirical developments — no doubt the result of a mixture of intuition, acute observation, and slow evaluation. Some of these developments, such as the all-important one of the hardening processes for iron alloys, are quite complicated. The hardening of steel implements was in wide use by at least the seventh century B.C., but it is only in the last 100 years, and more particularly in the last 50 years, that the developments have been placed on a reasonably sound scientific basis. Nevertheless, it has been chastening to find after metallurgy finally has become a technological science that often it has been very difficult indeed to improve significantly on the empirical developments of the past. Although there have been some outstanding exceptions,

---

*The price of steel was reduced tenfold during the second half of the 19th century.

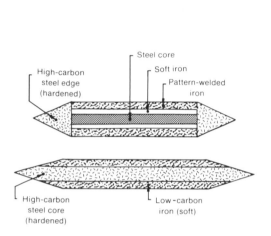

The construction of these swords varied; two typical examples are illustrated here. They characteristically had hardened edges and soft, tough shafts. ("Pattern-welded" means that several rods or bars were twisted together and then forged so as to weld them together.) The sketch of the soldier is a reconstruction based on material excavated from the ruins of Herculaneum, which was devastated with Pompeii by the eruption of Mount Vesuvius in 79 B.C. The material included the femur bone ghosted in the sketch. The soldier's broadsword is a prominent feature of his equipment. (Sources: left, C. Becker, *Arch. Eisenh.*, 1961, *32*, 661, and E. Schurmann and H. Schorer, *ibid.*, 1959, *30*, 127; right, *National Geographic*, 1984, *165* (5), 557.)

Fig. 1.2 The Roman soldiers were equipped with broadswords of a complicated metallurgical construction. These weapons gave them a marked technological advantage over their opponents.

modern developments have often been incremental and driven by a recognized engineering need to improve a particular property. What the development of a scientific base certainly has done has been to expedite considerably these developments by guiding them and by reducing the options that seem to be worth exploring. We have at least moved to a stage of what might be called enlightened empiricism. The scientific base has also enhanced immeasurably our understanding of problems that arise in the use of metals, and of how metals may be used more effectively.

This has certainly become a pressing need because, in recent times, metals have come under increasing competition from a range of other materials, notably the organic polymers known as plastics and various ceramics. The age of metals may well have passed through its zenith, but there would be few who would maintain that metals will not continue to constitute the infrastructure on which society is based. It is certainly likely that metals will be used more selectively in niches where their characteristics are unique and appropriate, and that greater demands will be made for them to meet stringent requirements in these niches. This will in turn require improvements in our understanding of the science of metals. It will also require that those who select metals for, and fabricate them into, engineering devices develop a better understanding of this science and the technology that is based on it. It is as an introduction to these mysteries that this book is designed.

WHAT IS A METAL?

Of the roughly 100 basic *elements** of which all matter is composed, about half are classified as metals. The distinction between a metal and a nonmetal is not always clear cut. The most basic definition centers around the type of bonding existing between the atoms of the element, and around the characteristics of certain of the electrons associated with these atoms (p. 223). In a more practical way, however, a metal can be defined as an element which has a particular package of properties.

Metals are *crystalline* when in the solid state (see Chapter 7) and, with few exceptions (e.g., mercury), are solid at ambient temperatures. They are good conductors of heat and electricity and are opaque to light. They usually have a comparatively high density. Many metals are *ductile* — that is, their shape can be changed permanently by the application of a force without breaking. The forces required to cause this deformation and those required finally to break or fracture a metal are comparatively high, although, as we shall see later, the fracture force is not nearly as high as would be expected from simple considerations of the forces required to tear apart the atoms of the metal.

One of the more significant of these characteristics from our point of view is that of crystallinity. We shall discuss this characteristic at some length in Chapter 7, but we need to develop here some understanding of this concept to help in the discussions which precede that chapter. A crystalline solid is one in which the constituent atoms are located in a regular three-dimensional array as if they were located at the corners of the squares of a three-dimensional chess board. The spacing of the atoms in the array is of the same order as the size of the atoms, the actual spacing being a characteristic of the particular metal. The

---

*Technical terms with which the reader may not be familiar are printed in italics the first time they appear. Many are defined in Appendix 1, "Glossary of Terms", p. 429.

directions of the axes of the array define the orientation of the crystal in space. The metals commonly used in engineering practice are composed of a large number of such crystals, called *grains*. In the most general case, the crystals of the various grains are randomly oriented in space. The grains are everywhere in intimate contact with one another and joined together on an atomic scale. The region at which they join is known as a *grain boundary*.

An absolutely pure metal (i.e., one composed of only one type of atom) has never been produced. Engineers would not be particularly interested in such a metal even if it were to be produced, because it would be soft and weak. The metals used commercially inevitably contain small amounts of one or more foreign elements, either metallic or nonmetallic. These foreign elements may be detrimental, they may be beneficial, or they may have no influence at all on a particular property. If disadvantageous, the foreign elements tend to be known as *impurities*. If advantageous, they tend to be known as *alloying elements*. Alloying elements are commonly added deliberately even in substantial amounts in engineering materials. The result is known as an *alloy*.

The distinction between the descriptors "metal" and "alloy" is not clear cut. The term "metal" may be used to encompass both a commercially pure metal and its alloys, and for convenience it will often be used in that sense throughout this book. Perhaps it can be said that the more deliberately an alloying addition has been made and the larger the amount of the addition, the more likely it is that the product will specifically be called an alloy. In any event, the chemical composition of a metal or an alloy must be known and controlled within certain limits if consistent performance is to be achieved in service. Thus chemical composition is the first of the two basic parameters that have to be taken into account when developing an understanding of the factors which determine the properties of metals and their alloys. We shall come to the second in a moment.

THE METALS THAT ARE IMPORTANT IN ENGINEERING

Of the 50 or so metallic elements, only a few are produced and used in large quantities in engineering practice. The most important by far is iron, on which are based the ubiquitous *steels* and *cast irons* (basically alloys of iron and carbon). They account for about 98% by weight of all metals produced. Next in importance for structural uses (that is, for structures that are expected to carry loads) are aluminum, copper, nickel, and titanium. Aluminum accounts for about 0.8% by weight of all metals produced, and copper about 0.7%, leaving only 0.5% for all other metals. As might be expected, the remainder are all used in rather special applications. For example, nickel alloys are used principally in corrosion- and heat-resistant applications, while titanium is used extensively in the aerospace industry because its alloys have good combinations of high strength and low density. Both nickel and titanium are used in high-cost, high-quality applications, and, indeed, it is their high cost that tends to restrict their application.

Metals are also used in significant quantities in nonstructural applications. Examples are:

Copper and aluminum: for conducting electricity
Iron and nickel: for magnets and magnetic materials
Tin, zinc, and lead: for corrosion resistance
Gold and platinum: for permanent attractive appearance and high value.

We cannot discuss these more esoteric properties here. Suffice it to say that a whole complex of properties in addition to structural strength is required of an alloy before it will be accepted into, and survive in, engineering practice. It may, for example, have to be strong and yet have reasonable corrosion resistance; it may have to be able to be fabricated by a particular process such as deep drawing, machining, or welding; it may have to be readily recyclable; and its cost and availability may be of critical importance.

These are all factors which are covered by the branch of metallurgical science and technology known as *physical metallurgy*, and this is the branch that is of most importance to engineers who use metals.

WHAT IS PHYSICAL METALLURGY?

The term "metallurgy" is an all-embracing one which may be used to cover anything having to do with the production, processing, or use of metals and alloys. We must therefore define the portion of this wide spectrum that we shall be considering — namely, physical metallurgy.

Only a few of the metallic elements are found in nature in the metallic form; examples are gold, silver, copper, and perhaps iron (in meteorites). Even so, none of these metals is now found in a form which is immediately useful. All useful sources of structural metals are present in the earth's crust as chemical compounds such as oxides, sulfides, and silicates. These chemical compounds are known as *minerals*. The metal-based minerals mostly are distributed widely and in very low concentrations, but occasionally they are found in reasonably high concentrations in *ore bodies*. Depending on the economics of the time, it may be profitable to extract, or *mine*, these ore bodies, an activity which is the province of the *mining engineer*.

The mineral or minerals of economic interest are still typically present in a mined ore only in quite low concentration. The important minerals have to be concentrated and separated from the unwanted minerals, a process known as *ore dressing* or *mineral processing*. The resultant concentrate is then reduced to the metal, frequently by a series of large-scale and complicated chemical processes, the over-all objective being to produce a metal of a purity that is acceptable for engineering use. These procedures are the province of the *process metallurgist* or *primary metallurgist*. Although it is to some extent a matter of definition, it can be said that the activities of the primary metallurgist cease when the refined metal has been produced in a liquid form.

Typically, the refined liquid metal is then solidified in a controlled manner. It may be solidified in a complex shape which is immediately useful (a *casting*). Alternatively, it may be solidified in a simple shape (an *ingot*) which subsequently can be reshaped by a variety of processes into a *semifabricated* product, such as a sheet or a bar. These semifabricated products are the feed stock of the manufacturing and construction industries. In effect, the metal after solidification is subjected to a series of deformation and thermal cycles which may be applied either separately or simultaneously. The crucial fact is that the deformation and thermal treatments change the internal structure of the metal as well as its external shape. Moreover, there is usually a causal relationship between these changes in structure and the all-important mechanical properties of the metal. The structure finally produced, as well as the chemical composition, determine these properties, and structure is often the more important of the two in this respect.

Physical metallurgy is the science and technology that studies these relationships between structure and properties and applies them to good effect. It is based on the determination and characterization of the internal structure. It then seeks to establish causal relationships between structure, together with chemical composition, and those properties that are important in the use of the metal in engineering practice, and then to develop ways of controlling composition and internal structure so as to optimize these properties. Finally, it seeks to establish criteria by which metals can be selected for use which have optimum properties for the particular application.

The last of these roles begins to overlap with those of design engineers, particularly mechanical and civil engineers. The physical metallurgist in these roles can be regarded as, and is often called, a *metals engineer* as a part of the engineering team. A team approach it certainly has to be, because the manipulation of metals by a physical metallurgist is of little significance in its own right. It only becomes significant when the result is used in an engineering product. On the other hand, a design engineer who selects a metal for use without consulting a metals engineer, either directly or indirectly, does so at his peril.

There is one bridge that certainly must be established between design and metals engineers. A designer starts by considering the loads that are to be applied to a body, which is never infinitely rigid. He then analyzes these loads into forces and deformations in individual elements inside the body. The metals engineer has to be able to understand the significance of these analyses, and to be able to relate them to properties of metals that he is able to determine. The design engineer has to appreciate the reverse process. This bridge is becoming increasingly important as greater demands are made for efficient designs, and as the complexities of the forces to which real structures are subjected are becoming better understood. A term commonly used to describe the bridge is *metallurgical mechanics*. It is certainly necessary to appreciate the significance of the mechanical properties that make metals attractive before the place that they occupy in the spectrum of engineering materials can be appreciated. Thus, we shall start by considering these matters. First, however, it is necessary to establish some background information about the techniques used to form and shape metals into useful engineering components.

FURTHER READING

R. Raymond, *Out of the Fiery Furnace: The Impact of Metals on the History of Mankind*, Pennsylvania State University Press, 1986.

R. F. Tylecote, *Metallurgy in Archaeology*, Edward Arnold, London, 1962.

R. F. Tylecote, *A History of Metallurgy*, The Metals Society, London, 1976.

· 2 ·

Processing Metals Into Useful Components

SUMMARY

There are several routes by which a metal may be processed into a useful form. One is by casting — that is, by solidifying molten metal directly into the desired shape, or at least one close to it. There are many ways by which this may be done, the optimum one depending on factors such as the melting point of the alloy, the dimensional precision and quality desired in the product, and the production rate required. The second main route, the making of wrought products, is one in which the molten metal is solidified into a large simple shape, called an ingot, which is then further shaped into a usable form by deforming it mechanically. The shaping may be done in a sequence of stages while the metal is either hot or cold and by a wide variety of processes. Examples are rolling, forging, extrusion, drawing, and spinning. A third, minor route is one in which a compacted powder is heated so as to weld, or sinter, together the particles of the powder. Each route and procedure has its own special characteristics, and each meets best some combination of production, metallurgical, and economic needs. Each depends to at least some extent on the properties of the metal being shaped, and each may have an influence, perhaps a subtle one, on the properties of the final product. Consequently, an understanding of these production routes is necessary for an appreciation of the characteristics of the product. Massive capital-intensive plants and specialized skills are required for many of the production processes. New procedures constantly are being devised to improve the product or the economics and to meet competition one with the other and with alternative materials.

In this chapter we shall discuss in a little more detail, but still in very broad outline, the means by which the refined molten metal that has been produced by primary metallurgy is converted into a form that can be used by a manufacturing or construction engineer. This is the field of *process metallurgy*. It is a field which is a mixture of production and metallurgical engineering, and which is a very capital-intensive activity.

Although an important objective of the processing sequence is to make a product as economically as possible, the methods used have to be compatible with the metallurgical characteristics of the particular alloy. Moreover, the product has to have assured properties and to be of appropriate quality in many respects. For example, it may be required to have a good surface finish, or to be free from harmful abnormalities such as surface cracks and folds. The attainment of these requirements is dependent on the proper manipulation of the mechanical and thermal cycles to which the material is subjected. Many subtleties of control of composition and of the various processing stages are involved, and many useful changes can be effected by the manipulation of these controls to produce what may seem to the uninitiated to be the same products but which to the user are significantly different. For example, the base steel for the tin-coated sheet (often called *tin plate*) used in the common food can is produced in many different grades. Some grades are more suitable for can bodies and some for can ends. Some are more suitable for containing particular products than others, depending on the corrosiveness of the product. Yet to you and me the various types of tin plate look to be the same.

Even in outline, the methods used for metal processing vary somewhat from metal to metal and from alloy to alloy, and we can scarcely deal with them all here. We shall confine most of our attention to the three structural metals that are produced in the largest quantities — namely, steel, aluminum, and copper. Moreover, an overwhelming proportion of these metals ends up in the form of plate or sheet (over 70% of all steel produced, for example, is made as *flat products*), and so our discussion naturally will be biased toward this type of product.

We shall first consider the processing of so-called *semifabricated* products. These are the sheets, bars, tubes, and wires that are sold to an enterprise which uses them to make a structure or some item which you as the consumer finally buy. We shall then consider processes which bring these semifabricated products closer to their final, useful forms. The metallurgist is much concerned with many of the production processes that are used in the final stages of manufacture, and at this final stage becomes part of a production engineering team.

There are several basic routes by which a molten metal may be taken through to a semifabricated product. The two main ones may be described as the *cast route* and the *wrought route*, and a third, minor one may be described as the *powder route*.

In the cast route, the metal is solidified into a shape which is usable or nearly usable as it stands. At worst, a little more processing, such as the machining of some regions, may be required. In the wrought route, the metal is solidified in a simple shape, which is then plastically deformed (see p. 66),

This is an example of the sophistication which foundry technology had reached in China 3000 years ago. The decorative pattern on the external surface of the bowl was cast on. Height, 17.2 cm (6.75 in.). (Reproduced by permission of the National Gallery of Victoria, Melbourne. Felton Bequest, 1969.)

Fig. 2.1. A liding (sacred cooking vessel) cast in bronze in the 11th century B.C.

either hot or cold, in a progression of stages which produces a desired shape. It is intriguing that the two great cultures of the world — namely, the Eastern culture originating in China and the Western culture originating in the Middle East and Europe — somehow or other developed their metalworking technologies each along a different route. The Chinese culture was based almost exclusively on the cast route. Casting technology, principally producing cast irons and the copper-tin alloys called bronzes, was developed to a very high level of perfection and sophistication (Fig. 2.1), a level which is only now beginning to be appreciated. But innovation stopped, and a major wrought technology was never developed. Early Western cultures, on the other hand, at first used castings (Fig. 2.2) but after the dawn of the iron age mainly followed the wrought route (Fig. 2.3). Some castings — principally bronzes — nevertheless continued to be used. The reasons for these differences in technological development are currently of great interest to technological and social archaeologists. Be that as it may, both routes need to be used in any complete manufacturing system, the more appropriate one to use in a particular application being chosen on the basis of cost, or fitness for purpose, or some compromise between the two.

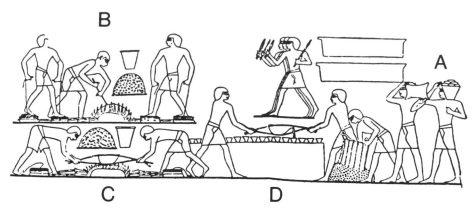

At A, two men bring supplies to the furnace, the leading one carrying a copper ingot which has the characteristic "oxhide" form used by early copper producers. The other man carries a basket of fuel. At B, the furnaceman tends his furnace, to which an air blast is supplied from foot bellows operated by two assistants. The crucible containing the now-molten metal is removed from the furnace at C. Two bowed green twigs are being used for this purpose and to carry the crucible to the vertical mold at D. Many such crucibles of metal would have been needed in close succession to produce a satisfactory casting of this size. (Source: P. E. Newberry, *The Life of Rekhmara*, Constable, London, 1900.)

Fig. 2.2. Set of Egyptian tomb paintings (reproduced here as tracings), dating from the second millennium B. C., depicting the casting of a pair of large temple doors and representing the earliest known record of metalworking in the Western world.

THE CAST ROUTE

A casting may be made directly from the molten metal produced by the primary metallurgists, but, more commonly, the refined metal is first cast into simple ingots which are of a size convenient for handling and transport to a casting workshop, or *foundry*. There they are remelted and, if necessary, the composition of the melt adjusted by adding alloying elements.

In either event, the molten metal is poured into a cavity which has been formed in a *mold*, the cavity in a pattern negative to the shape of the desired product. The cavity must actually be a little larger than the desired product to allow for the contraction that occurs during solidification (see p. 267). Various additions also have to be made to the mold cavity (Fig. 2.4). There must be a basin into which the molten metal can easily be poured, and from this basin channels (called *runners* or *gates* or *sprues*) must lead into the main cavity. *Risers* must be added at suitable positions to ensure that the shrinkage cavities which inevitably develop during solidification (see p. 268) are, as far as possible, not located in useful parts of the casting. There must also be means of ensuring that

Iron produced by early Western cultures was in the form of a spongy mass which was useless until it had been kneaded by forging it into a solid bar. This need opened the wrought route for fabricating metals, which soon became the dominant production method. In this illustration, the workman at the top is smelting ore, and those at the center are ready to knead his product by hand hammering, using a large stone as an anvil. The smith at the bottom is forming a sheet from the consolidated iron. He at least does not have to rely entirely on his own muscles, having the benefit of a tilt hammer driven by a water wheel. This illustration was included in *De Re Metallica* by Georgio Agricola, the first major work on mining and metallurgy published in Europe. This book was translated into English from the original Latin by Herbert Hoover and his wife. This was the Hoover who later became President of the United States and who practiced for a time as a mining engineer in Australia.

Fig. 2.3. Reproduction of a 1556 woodcut of German smiths using the basic techniques of ironworking that had by then been in use for over 2000 years and that continued to be used in principle until late in the 19th century.

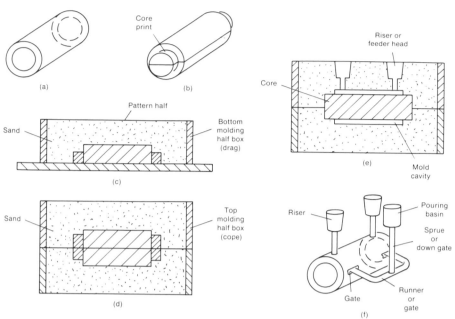

(a) The tube. (b) The pattern, which is split in half longitudinally, has protuberances (core prints) to mold depressions in which the core can be positioned. (c) A half molding box is set over one half of the pattern and the box is filled with sand, which is compacted around the pattern. (d) The filled half box is inverted. The second half of the pattern and the second half of the molding box are positioned and filled with sand, which is again consolidated. (e) The mold boxes are separated, the pattern removed, the separately molded core positioned, risers and feeds, etc., cut into the mold, and the two mold boxes reassembled. (f) The mold cavity is filled with molten metal. After solidification, the casting is removed from the mold, then appearing as in this sketch. The gates and risers finally have to be cut from the useful part of the casting.

Fig. 2.4. Steps in the production of a sand mold for casting of a tube.

the air originally in the mold cavity is displaced and not trapped within the solidifying metal. Also, parts of the main cavity may be filled by *cores* to produce cavities and holes in the final casting. The various appurtenances that as a result are attached to the casting after solidification have to be removed from the main casting after it has solidified and has been removed from the mold.

Casting provides a very direct and flexible manufacturing route, and is capable of producing a wide range of components. Small numbers of parts can be produced economically, often with minimal capital investment. Large production runs, on the other hand, can also be accommodated. Complex shapes can be produced, including shapes with internal cavities (e.g., the cylinder block of an automobile engine). Alloys which cannot be processed by the wrought route can be handled, but, on the other hand, not all useful alloys can be cast successfully. Indeed, special groups of alloys tend to have to be developed for castings and other groups for wrought products.

But, as with most things in this world, the advantages of casting are offset by penalties. Except for some special processes, dimensions cannot be controlled very well, and the surface finish obtained may not be all that is desired. More seriously, castings are heir to faults, both internal and external, that result from basic characteristics of the solidification process, which is discussed in Chapter 8. These faults may reduce mechanical strength and impair structural integrity and reliability.

Many of the potential faults can be avoided if sufficient skill is exercised by the foundry metallurgist. Nevertheless, castings which are stressed severely in service and in which a high degree of reliability is required may have to be inspected to ensure that objectionable abnormalities are not present (some of the inspection techniques that are used are discussed in Chapter 6, p. 203). This adds to the final cost. A technological specialty has been built up to assess the importance of any abnormalities that are discovered and to devise methods of modifying foundry procedures to prevent their formation in later production. Even so, the properties of castings tend to be (although not always are) inferior to and more variable than those of wrought products. This is partly because of the nature of alloys that are appropriate to use in castings, and partly because many castings are cooled comparatively slowly and at variable rates during and after solidification and thus have comparatively coarse structures (see Chapter 8). Nevertheless, by attention to detail, so-called *premium-quality* castings can be produced with guaranteed properties in the metal of the casting itself and with guaranteed freedom from significant internal defects. Such castings can, when the quality is needed, become competitive with wrought products.

Perhaps the public image of a foundry is one of stygian gloom, of dust and smoke and flame, and of hard manual labor. There are foundries that still fit this image, unfortunately, but a modern foundry can be as immaculate as any other type of production line. There is also perhaps an image among design engineers that the reliability of castings cannot be all that it should be. This too can be true, but again is not necessarily so. Quality in castings, however, is obtained only by the application of technological knowledge and skills of a high level. Skills of this nature are sometimes applied brilliantly but, as a consequence, quality castings cost more than any old casting and can be produced only in foundries where such high levels of skills and motivation do exist. These are facts which seem to escape many cost-conscious purchasers. Caveat emptor!

The preparation of the mold cavity is the crux of, and the key to, founding. A veritable legion of processes is now available for the purpose, but they all fall into two broad groups. In the first group, a new mold is made for each casting. Characteristically, each mold is made from a modified replica (called a *pattern*) of the desired part, the pattern being used repeatedly. In the second group, a *permanent mold* is formed in a block of metal and this mold is used repeatedly.

Sand Molding

A mold cavity can be hand sculptured in a compacted mass of sand, but more commonly the sand is compacted around a pattern of the object to be made. The basis of the technique employed is illustrated somewhat diagrammatically in Fig. 2.4. Note that the mold has to be split along some appropriate plane so that it can be opened and the pattern removed. More complicated parting planes than that illustrated may have to be used to achieve this. Even so, there are some geometric limitations to patterns that can be extracted successfully. For example, a pattern must have some *draft* at the parting plane to allow easy removal and certainly cannot contain re-entrants. Sprues, gates, and risers may have to be cut into the mold sand after the pattern has been removed, but some or all of them may be incorporated into the pattern.

Molding sands have to be carefully compounded. The sand has to hold together during molding and also while the mold is being filled with metal. But it must not then bake so hard that it either restrains too much the contraction of the solidifying mass of metal or cannot easily be broken away from the solidified mass. It must impart an acceptable finish to the surface of the casting. Basic sands consist of silica grains of an appropriate size and shape mixed with a small proportion of clay binder. More refractory base materials and other binding materials are also used in more advanced casting methods.

The preparation of a mold of the type illustrated in Fig. 2.4 is a manual operation requiring, incidentally, a good deal of technical skill. Manual techniques are still used in jobbing foundries when only small numbers of castings of a given type are required, but modified and increasingly more sophisticated techniques are used as the number of parts to be produced increases. Patterns are then made in more robust materials and may incorporate a number of parts into the one mold. The pattern halves may be mounted on plates corresponding to the parting plane of the mold, and may be designed specifically to facilitate mechanical handling during the molding sequence. This permits molding to be carried out by machines, and even by fully automated machines, as part of a production line. Other developments permit both surface finish and dimensional control to be improved.

Investment Molding

A different approach to the manufacture of molds that are used for one casting only has led to the development of *investment casting* processes (Fig. 2.5). A

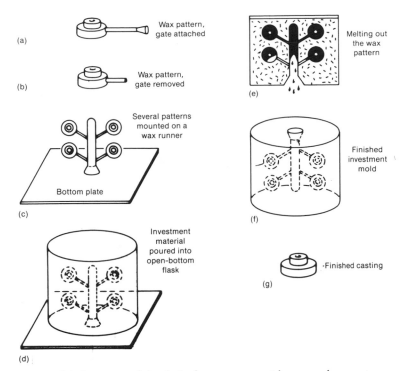

(a) A pattern of the desired component, with gate and sprue attached, is cast in wax using a conventional permanent mold. (b) The gate is removed from the pattern. (c) A number of patterns are attached to a separately cast wax runner and gate system. The assembly is attached to a bottom plate. (d) An open cylindrical flask is set on the bottom plate and the investment mold material is poured into the flask. (e) The mold is heated to melt the wax, which is poured out of the mold. The mold may be further heated to a higher temperature to burn out any remaining wax and to harden the mold material. (f) The mold is inverted, and is ready to be filled with molten metal. The mold may be preheated. (g) The finished casting is cut from its runner.

Fig. 2.5. Steps in the production of a mold by an investment process. Other methods of forming a mold around the wax patterns are also used, but the principle remains the same.

pattern is made in wax as a model of the component, with gates, etc. also made in wax and attached. This pattern can be made by injecting wax into a negative pattern by a simple process, and a number of these units can be joined to a system of gates and sprues which has been formed separately in wax. The mold material, often a ceramic, is formed around this wax model, but the mold this time can be made as a solid block because the whole assembly is heated to a temperature above the melting point of the wax and the wax poured out. The mold is subsequently heated to a high temperature to burn out the last traces of the wax and perhaps to harden the mold material.

The investment process was used to produce a complex turbine blade with good control of dimensions (left). Cooling passages can be incorporated in these castings, as illustrated by the sections of the blades at right. It is difficult and expensive to produce a similar blade by the wrought route, but this is done (without cooling passages) as a competitive process.

Fig. 2.6. Investment-cast turbine blades.

Components of considerable complexity can be made in this way, with re-entrants and holes and internal cavities if necessary (Fig. 2.6). If sufficient care and skill are exercised, excellent standards of surface finish and dimensional control can be achieved. This is one of the methods, for example, by which the turbine blades for jet engines are manufactured. These blades have a complex shape (Fig. 2.6) the dimensions of which have to be controlled closely. Reliability in the performance of these components is essential to the safe and economical use of modern aircraft.

The investment process, without the modern refinements that are aimed at the control of dimensions, has in fact been used almost since time immemorial by cultures as widely dispersed as those of the Sumerians of the Middle East and the Aztecs of the Americas. The process is known in this manifestation as the lost-wax process and was, and still is, used to produce jewelry and objects of art. The wax model in these applications is hand sculptured by an artist. Dentists also still use this technique to prepare bridgework and caps, the wax pattern being formed in a cavity cut into a tooth. If you have submitted to such a procedure, you will know how effective yet expensive a good casting can be.

Permanent Molds

An alternative process to those just described, which use disposable molds, is to machine the mold cavity into a block of metal (a *permanent mold*). The mold is made in a metal of higher melting point than the alloy to be cast and may include passages for water cooling. It can be used for many castings, but usually does not last forever because the thermal shock to which it is subjected every time it is filled eventually causes it to crack. Faster cooling rates are achieved than for sand casting, and hence finer structures are produced in the solidified metal. A better surface finish also is characteristically produced.

A permanent mold may be filled by pouring the metal in under gravity, as for a sand casting; in this instance, the process is known as *permanent mold casting*. Alternatively, the molten metal may be injected under pressure (Fig. 2.7), in which case the process is called *pressure die casting*, a term often contracted to *die casting*. Special casting machines are then required, but parts of complex shape, with thin-wall sections and good surface finishes, can be produced at high production rates.

The design of a permanent mold is a tricky business, particularly since the mold can be modified only with difficulty. The mold must be split so that the casting can be extracted without damage, and this can cause complications with irregularly shaped castings. The mold material is impervious to gases so that a smooth flow of metal which displaces the air contained in the cavity toward a suitably positioned outlet must be arranged. Precautions must be taken to ensure that local solidification of the metal does not occur too early and so block off the feeding of other portions of the mold cavity. Special steels have to be used in order for the mold to be able to withstand the casting temperatures and the shock imposed with each filling. In practice, this re-

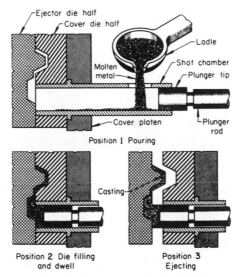

Fig. 2.7. Sketches illustrating the principles of casting in a permanent mold or die, the liquid metal here being injected into the die under pressure. A variety of casting-machine geometries and methods of injecting the metal are used in practice. (Source: *Metals Handbook Desk Edition*, **ASM, Metals Park, OH, 1985.)**

quirement imposes a limitation on the melting point of the alloy being cast, and thus limits the number of alloys to which the process can be applied. Zinc alloys, aluminum alloys, and copper alloys can be handled, with increasing difficulty in that order, but alloys with higher melting points cannot.

The permanent mold itself, and the equipment necessary to use it in the case of pressure die castings, are costly, and the expense usually is justified only when very large numbers of castings are to be produced. Nevertheless, die casting constitutes a major segment of the foundry industry. Your automobile and many of your domestic appliances abound in die castings, mostly in zinc alloys, although to a lesser extent now than a few years ago. Die castings have come under increasing competition from plastics. Plastics are molded in much the same way as die cast metals, but at lower temperatures. Simpler equipment therefore suffices. Plastics also are basically cheaper and will often do the job adequately. This is an example of the competition that constantly is occurring in the materials world. New materials are developed which find market niches in which they have adequate properties and economic advantages. Sometimes these developments open new markets; sometimes they displace old occupants.

THE WROUGHT ROUTE

The general path followed in the wrought route is to solidify the metal as a large mass which has a simple cross section and which is rather long in comparison with its lateral dimensions. This solidified mass is, again, called an *ingot*. The ingot is then plastically deformed (see p. 66) to reduce it to the desired cross section (e.g., a plate, a rectangle, a round, or a railroad rail, to name just a few), the reduction being accompanied by a corresponding increase in length.

The first stage of the reduction sequence typically is carried out hot, the metal initially being heated to a temperature as close to its melting point as is prudent. The force required to achieve the deformation is then at a minimum and stays low as working proceeds (cf., cold working, which we shall soon discuss). The temperature of the metal falls during the working operations, and the metal becomes more difficult to deform further. Eventually, a stage is reached where the metal must be reheated if working is to be continued. The heating-and-reduction sequence may be repeated several times.

However, there is a limit on what can be achieved by hot working, a limit set by the minimum thickness of section that can be produced and the surface finish and dimensional accuracy that can be achieved. For example, the plate for a ship's hull can be produced entirely by hot working, but not the thin sheet used in a food container. The final stages of reduction then have to be carried out cold, usually at ambient temperature. The metal has to be treated between hot and cold working to remove the surface oxide and scale produced during hot working. Any surface irregularities on the hot worked metal may also have to be removed so that they do not spawn extensive surface irregularities in the cold worked product.

The forces required to reduce most metals at ambient temperatures are much greater than those required during hot working. More robust equipment is required to achieve a given reduction, a requirement which sets limits on the dimensions of the stock that can be handled (for example, the width of a sheet). Moreover, the difference between hot and cold working goes even deeper because there is a fundamental metallurgical distinction between the two, which we shall discuss later (p. 317). Suffice it to say here that a metal becomes increasingly difficult to deform with increasing reduction during cold working; the yield stress of the material increases and it is said to *strain harden*, or *work harden* (see p. 68). This adds further to the limitations on the reductions that can be achieved in practice with given equipment. Strain hardening may thus require that the material be softened by a heat treatment (a *recrystallization annealing treatment*; see p. 308) at certain stages in a reduction sequence, stages at which further working has become too difficult. Note that during hot working this annealing process occurs concurrently with deformation (p. 317), and thus the problem does not arise.

Advantage can be taken, however, of the strain-hardening phenomenon. The product can be annealed after the final reduction, in which event its properties will be similar to those of a hot rolled product. On the other hand, its strength can be improved (up to doubled) by leaving it in the strain-hardened condition. This increase in strength is achieved at the expense of a loss in ductility. A range of properties can thus be achieved by controlling the reduction imparted after the last annealing treatment. These phenomena are discussed at more length in Chapter 9.

As already implied, these working processes, both hot and cold, do much more than change the shape of a block of cast metal. Cavities and other discontinuities that may have been present in the cast ingot are closed up and eliminated, provided that they haven't broken through to the surface of the ingot. The general structure of the material is refined. The variations in composition (segregation) in the cast structure are rearranged, generally to be present in a less harmful way. However, the structure does become directional. Any inhomogeneities that are present become elongated in a manner which exactly reflects the bulk changes in shape that occurred during working. One consequence of all of this is that the mechanical properties of wrought metals tend to be superior to, and more consistent than, those of corresponding cast alloys. Secondly, the properties of wrought products are usually more satisfactory in some directions than in others. One of the requirements desired of a working sequence usually is that things be arranged so that the direction in which the most satisfactory properties are obtained corresponds as closely as possible to that on which the major stresses to be imposed on the material in service. Nevertheless, the choice between cast and wrought products is not a clear-cut one in many instances. There are, of course, some shapes and sections and alloys that cannot be produced by the wrought route, and some that cannot be produced by the cast route. But the choice in intermediate areas is a matter of continuing debate. The answer depends on the relative economics and the standards attained by the two technologies, and this varies from time to time.

CASTING OF INGOTS

Whatever happens later, the wrought route has to start with the casting of a simple shape: an ingot. Ingots typically are made as large as possible, the optimum size being determined by the cost of, and demand for, the product. Ingots to be used as feed stock for large steel forgings may weigh up to 300 t (tonnes; see Appendix 2, p. 477), but this is exceptional. Ingots for rolled steel products commonly weigh 20 to 40 t. At the other extreme, ingots of precious metals may weigh only a few kilograms, but this is still large compared with castings that are likely to be made in these metals.

Static Casting

In any event, the general technique employed is to pour (*teem* is the jargon word used) the molten metal into a vertical open-end mold (Fig. 2.8). The mold is made of metal, usually a cast iron, and is tapered along its length so that the solidified ingot can be withdrawn easily from the mold (Fig. 2.8b and c), or vice versa (Fig. 2.8a and 2.9).

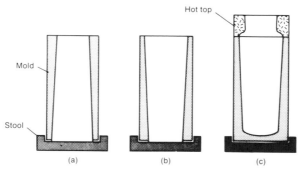

(a) An open-end-down mold. The mold is tapered so that it can, in this arrangement, be pulled off the solidified ingot, as illustrated in Fig. 2.9. (b) An open-end-up mold. The solidified ingot now has to be pulled out of the mold, which is a more difficult operation. A shorter pipe is, however, produced in the ingot. (c) An open-end-up mold with a rounded bottom, the open end being fitted with a hot top. This type of mold is used only for the highest-quality steels.

Fig. 2.8. Cross sections of typical molds used to cast ingots of steel.

Fig. 2.9. An open-end-down mold being withdrawn from a solidified steel ingot using an overhead crane. These ingots are rectangular in section and are the feed stock for rolling plate or strip. (Source: *The Making, Shaping and Treating of Steel*, 10th Ed., AISE, Pittsburgh, 1985.)

The processes which occur during solidification in molds of these types will be discussed in Chapter 8 (p. 264), where you will see that a number of characteristic but rather undesirable abnormalities are likely to form. The most inevitable of these is a conical cavity, called a *pipe cavity*, which develops at the open end of an ingot. The length of ingot containing this cavity, which may constitute up to 20% of the total ingot length, has to be cut off before working can proceed. A lesser length has to be removed from the base of the ingot to remove the defective portions discussed on p. 265. The cropped material usually can be recycled, but still represents a considerable economic waste. Other solidification abnormalities may also cause difficulties and losses during subsequent production stages. Moreover, ingot casting is a batch process, and batch processes are not ideal in large-scale production.

Continuous Casting

The limitations of batch casting of ingots that have just been mentioned have provided an incentive for the development of continuous methods of casting ingots.

It is convenient first to consider a semicontinuous process. The molten metal is poured into a shallow water-cooled jacket which has a false bottom

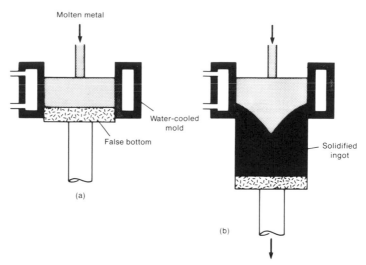

The molten metal is poured into a shallow water-cooled mold
which has a false bottom (a). The bottom is lowered at the rate at
which the metal solidifies, the liquid in the mold being topped up
continuously (b). The process is stopped when the lowering mechanism reaches the limit of its travel and the solidified ingot is removed. The process can then be restarted, with a different alloy
and a different mold section if desired.

Fig. 2.10. Semicontinuous casting of an ingot.

(Fig. 2.10a), the plan section of the jacket being the same as the cross section
desired in the ingot. The false bottom is lowered and melt is added to the trough
as the metal solidifies, the rates of both being adjusted to keep constant the
liquid level and the level of the solid–liquid interface (Fig. 2.10b). The net result
is that a solidified ingot is withdrawn into, perhaps, a pit. The process has to be
stopped when the base of the ingot reaches the base of the pit. The ingot is then
removed, and the process is restarted.

It is possible to make the process fully continuous by modifying it in one of
several ways, but these modifications involve only the manner in which the
ingot is handled after it has solidified and not the principles involved. For example, in one arrangement the emerging ingot is cut at regular intervals and
the cut lengths are removed from the path of the casting train (Fig. 2.11a). In
another arrangement, the emerging ingot is bent around gently so that it
emerges horizontally from the casting machine (Fig. 2.11a). The length of the
ingot that can be produced then is limited only by the rate at which liquid metal
can be supplied to the equipment. The ingot can, of course, subsequently be cut
into lengths if so desired. It is also possible with metals of lower melting points
(e.g., copper, aluminum, and cast iron) to cast directly a continuous ingot that
emerges horizontally from a bath of molten metal (Fig. 2.11b).

Losses due to the formation of pipe cavities are greatly reduced in semicon-

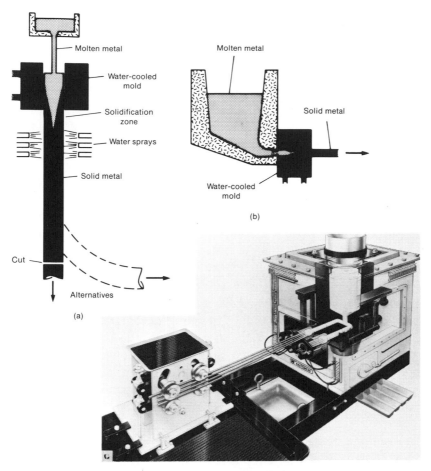

In (a), the ingot is solidified vertically in the same way as for semicontinuous casting, but the emerging ingot either is cut at regular intervals and the solidified length removed from the casting train, or is bent gently to emerge horizontally. In the latter event, the emerging ingot can either be cut into lengths or be further processed continuously. Units of these types produce hundreds of tonnes of steel per hour and occupy large special-purpose buildings.

In (b), the ingot emerges horizontally from the mold in which it has solidified. This particular type of unit can be adapted to quite low production rates, one such unit being illustrated in the cut-away drawing in (c) [Source for (c): Rautomead Ltd.]. It is shown producing eight strands of copper alloy rod for wiredrawing stock and has a production rate of tens to hundreds of kilograms per hour. It occupies a space only several metres square. Continuous-casting units are now available to cover the whole range of metal-production needs.

Fig. 2.11. Continuous casting of ingots.

tinuous casting and are virtually eliminated in continuous casting. The final product is more uniform and generally of higher quality than corresponding statically cast ingots. This is because of the intrinsic nature of these processes and because they can be more closely controlled. In fact, semicontinuous casting was pioneered by the aluminum industry to overcome the severe production and quality problems encountered with statically cast ingots of highly alloyed aluminum. Continuous casting was subsequently extended to the copper and steel industries either to improve productivity or to produce higher-quality grades of material.

An additional production advantage is that sections can be produced which are much smaller and of greater length than those which can sensibly be produced in batch cast ingots. The need for preliminary breakdown stages in primary rolling mills, which we shall soon discuss and which are so capital-intensive, is thereby eliminated. Advantage was first taken of this development by setting up small so-called mini steel mills, mills which are much smaller than the conventional plants in which breakdown rolling mills can be justified. Mini mills can be based on local supplies of steel scrap in areas remote from integrated steel mills and can be designed to supply local markets for simple products such as reinforcing bars for concrete structures. Developments of continuous-casting technology soon made it apparent, however, that it could be applied with considerable economic and quality advantages to steel plants of conventional size. An increasing number of new plants producing slabs for plate and strip or billets for bars are being based on the new technology, and old plants are being converted. In the case of flat products, it will soon be possible to cast continuously slabs which are so thin that the use of a secondary mill (see p. 31) as well as a breakdown mill is obviated. It may also soon be possible to obviate a hot strip mill (see p. 32), the cast strip being fed directly to a continuous cold strip mill. The price of survival even in an industry as old as the steel industry is the need to keep up with the times. These new casting plants are enormous in size and are a triumph of mechanical and metallurgical engineering.

At the other end of the scale, units with quite low production rates have been developed with which small to medium-size companies can produce their own semifinished shapes (Fig. 2.11c). They are particularly suitable for the more costly metals, such as copper and the semiprecious and precious metals. The semifinished products can be used directly as feed stock for finishing operations such as wiredrawing (p. 35) and cold rolling (p. 35).

HOT WORKING OF INGOTS

Assume now that a statically cast ingot has been produced and that it is to be hot worked. The ingot may be allowed to cool out and then be reheated for hot working. This is a common practice in the aluminum and copper industries.

However, in the steel industry, where high temperatures have to be used and large quantities of material have to be handled, the hot ingot, wherever possible, is placed in a heated refractory chamber (a *soaking pit*). Its temperature is allowed to equalize and to adjust to the correct hot working temperature. This saves energy. Even though the metal when heated to an appropriate temperature is in its most plastic condition, massive equipment is still required for the first stages of hot working. Large pieces of metal have to be handled and the objective is to achieve large reductions. There are three main ways in which the deformation is carried out — namely, by forging, by rolling, and by extrusion.

Forging

Ingots are forged by squeezing the metal between two anvils, one of which is actuated by a large press (Fig. 2.12). The faces of the anvils may be flat or may be recessed with a vee-shape or concave depression. In any event, reduction is achieved by deforming, in steps, localized volumes of the workpiece, and the shape finally produced is entirely at the control of, and one might say is deter-

Fig. 2.12. A large ingot being forged in an open-die hydraulic press. This photograph gives an impression of the massive equipment needed in the steel industry. (Source: Same as for Fig. 2.9.)

mined entirely by, the skill of the forge operators. Holes may be punched and the forging may be bent, but even so only comparatively simple shapes can be produced and then only with modest control of dimensions. Nevertheless, forgings for many important components are produced in this way, including some very large components such as rotors for power-generating turbines and propeller shafts for ships.

This process is called *open-die forging*. *Closed-die forging*, which is used to produce more closely shaped forgings, will be discussed later (p. 40).

Rolling

In ingot rolling, the hot ingot is again squeezed, but this time by forcing it through a gap between two counterrotating rollers (Fig. 2.13). The gap is

The ingot is the bright mass arrowed at lower right. Only one of the pair of rolls can be seen, the other one being located beneath the table of rollers on which the ingot is traversed. The direction of rotation of the rolls can be reversed so that the ingot can be traversed backward and forward through the rolls, the gap between the two being reduced progressively after each pass. (Source: Same as for Fig. 2.9.)

Fig. 2.13. A steel ingot being rolled in a typical large primary mill.

smaller than the thickness of the ingot so that the ingot is reduced in thickness and increased in length, but increased little in width. The process is repeated, the gap width being reduced with each pass, until the desired reduction has been achieved or the material has cooled too much to be rolled further.

An ingot with a rectangular section rolled in this way would produce a rectangular *slab* of about the same width, although the width of the slab may be increased by cross rolling during several of the early passes. A square-section ingot can be rolled successfully on its two perpendicular faces to produce a square or slightly rectangular shape, which is known as a *bloom* in large sizes or as a *billet* in smaller sizes.

The equipment in which this first stage of breaking down an ingot is performed is called a *primary rolling mill*. It is massive and very expensive (Fig. 2.13). The product may need to be further worked in a smaller *secondary rolling mill* (Fig. 2.14). For flat products (plate, sheet, and strip), the reduction of a slab is continued using cylindrical rolls, as in Fig. 2.14. This may be done in a single rolling mill in which the roll gap is reduced after progressive passes. Alternatively, a number of mills may be set up in series, the roll gap of each mill or *stand*

Four rolls are used in this mill, only the upper two of which can be seen in the photograph. The central pair are the working rolls. The larger-diameter outer pair support the working rolls and so reduce the extent to which they bend during a working pass. This is also a reversing mill which handles plates of finite length, one at a time. (Source: Same as for Fig. 2.9.)

Fig. 2.14. The product of a primary mill is usually processed further in a secondary mill. The mill illustrated here is producing plate and so uses plain cylindrical rolls.

Billets from a primary mill enter from the left and pass through
five mill stands in succession. The emerging strip is cooled by the
water sprays over the exit table at right, and is then picked up by
an automatic coiler (not shown in this photograph). Sets of spare
rolls are lined up at the right of each mill stand. (Source: Same as
for Fig. 2.9.)

Fig. 2.15. A continuous hot strip mill.

being adjusted to be smaller than that of the preceding stand. A long strip is
thus produced, strip which can be coiled and handled subsequently as a continuous strip in the next stages of fabrication, or be cut into sheets of any required
length. These *continuous* strip mills (see Fig. 2.15) are now common in the steel
industry for producing thin, flat products, and are spectacular operations.
Think about it for a moment and you will realize that the strip emerges with
increasing speed from each pass, so that the roll speeds have to be adjusted
accordingly in progression. The strip typically emerges finally at speeds of 30
to 50 km/h (20 to 30 mph) and then is picked up by a coiling device. Imagine the
mess if the red hot strip got out of control, as, fortunately, it only rarely does.
Moreover, the reductions and temperatures and the final cooling rate have to
be carefully controlled because all have significant influence on the structure
and properties of the strip that is produced. A continuous strip mill is a masterpiece of engineering and metallurgical design and control. Virtually all of the
steel sheets used in automobile bodies and in food containers are rolled as an
intermediate reduction stage on mills of this type. It would scarcely be practicable to meet the modern market for these items without recourse to high-
production units such as continuous mills.

The rolls used to produce wide sheet and strip may look rigid enough but
inevitably deform (bend) elastically under the rolling loads (see p. 63). The

sheet produced consequently is thicker at its center than at its edge, an undesirable state of affairs which has to be kept within bounds. This problem is compounded by the fact that it is desirable for mechanical reasons to use rolls which have as small a diameter as possible. The dilemma is resolved by using small-diameter working rolls and supporting them by large-diameter backup rolls. This is called a *four-high mill*. The stands in continuous strip mills are invariably four-high mills.

Sections other than simple plate or sheet have to be rolled in secondary mills through rolls in which pairs of grooves have been cut. This includes round bars, the structural shapes used in buildings and bridges, and railroad rails. The dimensions and shape of the gap formed by each successive pair of grooves is changed so that the section of the feedstock is altered progressively until the last pair forms the desired section (Fig. 2.16). It is a skilled business designing a sequence of roll gaps to produce a perfectly filled final section in the most effective manner. It is possible to produce simple sections, such as rounds, on continuous mills which have a number of stands in line. More complex sections, however, may have to be produced on single-stand mills, although each mill may have three rolls (a *three-high mill*) on top of one another (as illustrated in Fig. 2.16), so that the section can be passed backward and forward to ease handling.

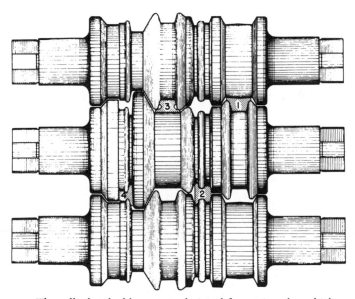

The rolls sketched here were designed for use in a three-high nonreversing mill to produce railroad rails. A square billet is passed backward and forward through grooves 1, 2, 3, and 4 in succession. (Source: Same as for Fig. 2.9.)

Fig. 2.16. Grooved rolls used in secondary mills to produce complex shapes.

Extrusion

The extrusion process starts with either a round ingot or a prerolled billet which has been heated to an appropriate hot working temperature and placed in an open-ended cylinder (Fig. 2.17). The ram of a large hydraulic press projects through one end of this cylinder, and the other end is closed by a plate in which a central hole has been machined. When the press ram advances, the hot plastic metal is extruded out of the hole in the die plate like toothpaste out of its tube, the extrusion emerging with the same section as the hole in the die. Complicated sections, including tubes, can be made. You can obtain an appreciation of the range of complicated shapes that can be produced by noting the wide variety of sections used in domestic metal window and sliding door sets.

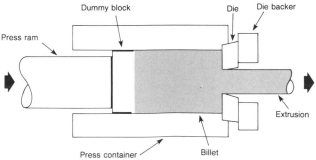

The press is surrounded by much ancillary equipment required to service the activity. The heart of the press is circled, the working components of which are depicted in the accompanying sketch. A press of this type operates semicontinuously. It has to be opened when the extrusion of a billet has been completed and a new billet loaded. The die can be changed to suit the required section. (Source of photo: Commalco Ltd.)

Fig. 2.17. A horizontal extrusion press used to produce aluminum alloy bars and sections.

All are produced by extrusion, and are possible only because of the versatility of this process.

There are two main limitations to the extrusion process. The first is imposed by the high pressures that are required to force even hot metal through a die hole. Massive presses are required (Fig. 2.17), and there is a limit on the reductions that can be achieved and the sizes of sections that can be produced. The second limitation concerns the availability of die materials which can withstand adequately the high temperatures and stresses imposed. This limits the temperatures that can be used and in turn the alloys that can be extruded. Thriving industries are based on the extrusion of metals of lower melting points such as lead, aluminum, and copper. The extrusion of steel, however, is difficult. It can be accomplished in a limited way by surrounding the extrusion billet with a layer of woven glass. The glass extrudes with the product and protects the die material, although only to a limited extent.

Cold Rolling

The production of thin strip starts with the output from a hot strip mill, a product which usually has to be pickled to remove the layer of surface oxide (scale) that is inevitably produced during hot rolling. The strip is then rolled in mills with cylindrical rolls similar in principle to those used in hot rolling. The surfaces of the rolls have to have a good finish and to be flooded with lubricant, both being needed to produce a smooth finish on the final product. Sheet may be reduced in a single-stand rolling mill, but increasingly steel strip is being rolled in continuous multistand mills not too dissimilar in appearance from continuous hot strip mills. The stands are invariably four-high mills to ensure maximum uniformity in thickness across the strip. Even this may not be good enough. Mills in which the working rolls are supported by clusters of backup rolls (clusters of as many as nine) are then used, but can be operated only as single stands. The product may be shipped to the fabricator in coil form or it may be cut into lengths of sheet.

Sheet and strip may be annealed at intermediate stages of reduction or after final reduction. In this event, the heating is likely to be carried out in an atmosphere the composition of which is controlled to ensure that the surface finish is not impaired by oxidation and scaling.

Cold Drawing

The second important method of cold working is known as *drawing*. It is used principally to manufacture wires, which are also ubiquitously important for making useful items such as electric power generation and distribution cables, springs, and pins and needles. The process is again simple in principle, but not so simple in practice when large quantities have to be produced economically and to high quality standards. The feed stock is usually either a hot rolled bar or rod. Increasingly with some metals, however, a continuously cast bar is pulled through a die plate in which a hole has been cut, the section of the hole being smaller than that of the feed stock (see sketch in Fig. 2.18). The stock that emerges has, give or take a little, the same section and dimensions as those of the die hole. The stock is drawn through a succession of dies until a wire of the desired dimensions has been produced. Intermediate annealing heat treatments may again be necessary at certain stages.

Wires ranging from large to very small in section can be produced in a range of materials with good surface finish, provided that the die hole is of the right shape and finish and the lubrication is adequate. The wire can be annealed after finishing; this is done, for example, if it has to be bent sharply in subsequent production or if maximum electrical conductivity is required. It may, on

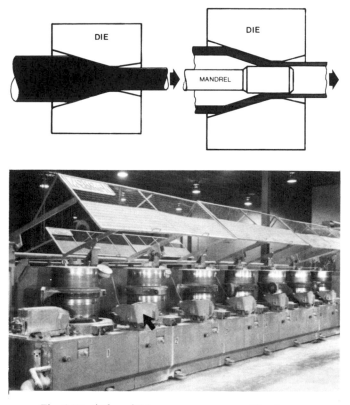

Fig. 2.18. (Above) Die arrangements used for drawing wire (left) and tube (right). (Below) A modern multistand drawing machine producing steel wire in seven steps, or drafts. Arrow shows location of one die block. The wire is wound on intermediate vertical drums which pull it through each die and take up any slack. (Source: Same as for Fig. 2.9.)

the other hand, be produced with various degrees of enhanced strength resulting from strain hardening during drawing (see p. 68). The wires used in steel wire cables are strengthened in this way.

The principles of wiredrawing were first developed in medieval times, when men sitting on swings provided leg power to pull the wire through a single die hole length by length. Chain-mail armor was an important application for such wire. Now, power-driven multidie machines produce wire at very high speeds (see photo in Fig. 2.18). The material used in the die then is important, because the die wears and eventually has to be replaced. Tungsten carbide and, in smaller sizes, diamond are the two materials now widely used, these being the two hardest materials that can be obtained in suitable forms.

Although most wire is produced principally in round sections, other simple sections (e.g., squares, rectangles, and hexagons) are produced in limited amounts. Products other than wire are also cold drawn as a finishing stage. Bars may be reduced a little to improve control of size and shape, to improve surface finish, and to enhance some properties. Tubes may be finished by cold drawing, for similar reasons, by being drawn over a mandrel (see sketch in Fig. 2.18).

Metal is pulled through the die under tensile forces during drawing, and you might wonder when you have read Chapter 3 why it does not fracture in the manner that it would if it were subjected to a simple tensile test (see p. 68). In fact it will, much to the consternation of the production team, if the reduction sequence during drawing is not adjusted correctly. The trick is to limit the reduction at any stage such that the stress required to pull the material through the die does not exceed its fracture strength, the fracture strength here being that of the newly drawn and strain hardened material, which is the part that is pulled. The total required reduction must therefore be obtained in a series of judiciously arranged steps. Moreover, the process is very sensitive to the presence of any abnormalities in the material, which reduce its tensile strength to a level below that expected at any stage. Fractures do then occur even with a well-designed die sequence, in which event production has to be interrupted while the string of dies is rethreaded. High-quality feed stock consequently is a premium in cold drawing practice.

THE POWDER ROUTE

Although this is a minor route, it survives and is even expanding because it fills a few niches in the spectrum of engineering production methods. For example, parts with complicated shapes can be produced close to final size, thus eliminating such expensive machining. Alloys whose melting points are too high for convenient handling by the cast-wrought route can be produced economically. Alloys of compositions which could not possibly be made by the cast-wrought route can be made by simply mixing various powders to make up the required mean composition. If desired, a product can be made with a controlled amount of evenly distributed internal porosity.

The powders required can be produced in a variety of ways, but most tend to be expensive. This factor very much determines the economic viability of the powder route. The powder, often with a little binder added, is compressed in a die which has a cavity that is a negative pattern of the shape desired. Pressing may be carried out either cold or at a somewhat elevated temperature, but in either event a rather fragile compact is produced but one that can be handled. This green compact is heated for a period of time at a high temperature approaching the melting point of the alloy; this heating is carried out in a controlled atmosphere to ensure that the powder does not become oxidized or otherwise contaminated. The individual particles *sinter* together during heating. The atoms at the surface of the particles move about and interchange at points of contact so that bridges are formed between particles. An evening out of composition may also occur by the exchange of atoms between particles of different composition. Cavities are present between the particles in a green compact, and these cavities shrink as the particles meld together. They may even be, but usually are not, completely eliminated. The bulk compact shrinks as the cavities are eliminated, and allowance has to be made for this in the die in which the powder is compacted. On the other hand, advantage can be taken of these cavities by arranging to leave some in the compact deliberately. The pores may be filled with a lubricant, for example, to make an effective bearing. The bearings for small electric motors are made in this way.

We have so far described *solid-phase sintering*. There are also processes in which a liquid phase forms at the sintering temperature (*liquid-phase sintering*). An important example, although the product is not strictly a metal, is found in the metal carbides that are now so widely used as tools for machining metals and in wiredrawing dies. Tungsten carbide powder (other carbides such as those of titanium and niobium may also be added) is mixed with a small amount of a nickel or cobalt powder; the nickel or cobalt melts at the sintering temperature, wets the carbide particles, reacts with them to some extent, and binds them together after cooling.

FINAL FABRICATION

We have now seen how metals are produced as basic shapes, such as bars and sheets and castings, shapes which are handed over to the manufacturing or construction engineers for fabrication into usable products. Some of the processes used in this final step make major metallurgical contributions. The efficacy of the production process may depend significantly on the metallurgical characteristics of the metal being processed. Also, the metallurgical characteristics of the metal may be affected for good or bad by the production process. A wide range of processes are used in everyday production. These processes are extensively classified and discussed in many publications, some of which are listed under "Further Reading" at the end of this chapter. Here, however, we will only be able to outline a limited number of the more basic of these processes.

Forging

The hot forging that we shall now discuss is similar in principle to the forging of ingots discussed earlier (p. 29). The feed stock now, however, is a wrought bar, and the aim is to produce shapes close to those of the final parts.

This type of forging may also be carried out with open dies, the methods used to drive the forging hammer ranging from the muscles of a blacksmith (Fig. 2.3) through power-driven hammers (Fig. 2.19a) to large hydraulic presses (Fig. 2.19b). The shape of the product, then, is determined to a large

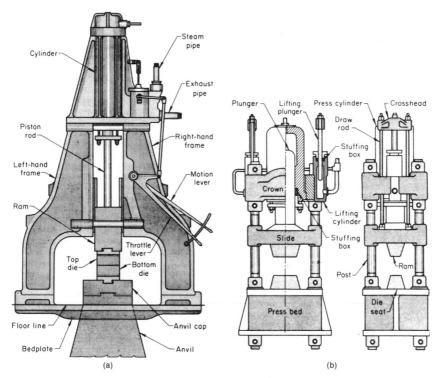

(a) A steam-driven power hammer used for open-die forging. Hammers of this type may also be driven by compressed air. This is the modern equivalent of the tilt hammer illustrated in Fig. 2.3. (b) A hydraulic press used for closed-die forging. (Source: Same as for Fig. 2.7.)

Fig. 2.19. Examples of forging hammers and presses.

degree by the skills of the hammer operator. At best, shapes of only limited complexity and dimensional accuracy can be produced, but much forging is still carried out in this way in general engineering and jobbing shops.

The forging of more complicated shapes typically is carried out in *closed dies*. A pair of matching die blocks is used in each of which a negative impression of one half of the intended part is machined (Fig. 2.20). A number of impressions may be included in the one die, as in Fig. 2.20, so that the final shape can be developed in a number of steps (Fig. 2.21). Some of these preforming steps may also be carried out in other forging devices, as for the series illustrated in Fig. 2.21. It is necessary to arrange for the metal to spill out into a gutter surrounding the main die cavity to ensure that this cavity fills completely and the *flash* so produced then has to be trimmed off as a final stage. The die halves are brought together by a variety of hammers or presses which are operated mechanically, pneumatically, or hydraulically and which are of the general type illustrated in Fig. 2.19.

Complicated and irregular shapes with reasonable dimensional control can be produced by closed-die forging, so reducing the amount of machining needed to produce an engineering component. Examples include crankshafts for automobile engines, spanners, hammers, blades for steam turbines and jet engines, and undercarriage components for aircraft.

Several other basic forging techniques are used to produce controlled shapes. The shank of a bar may be gripped so that a projecting end can be hit axially by a die. The end is then *upset* by the die to produce a bulb of controlled shape. Inlet and exhaust valves for automobile engines are produced in this way, where it is a much more economical and metallurgically more satisfactory process than that of machining the stem down from a large bar. Another

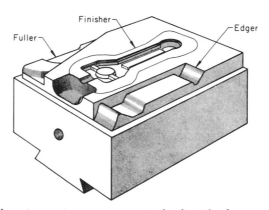

Three impressions are present in the die. The first, at right, rough-forms the blank; the second, at left, forms the edge profile; and the third, at center, develops the final shape. (Source: *Metals Handbook*, 8th Ed., Vol 5, ASM, Metals Park, OH, 1970.)

Fig. 2.20. One half of a multi-impression die used for closed-die forging, the die shown here being used for forging an automobile connecting rod.

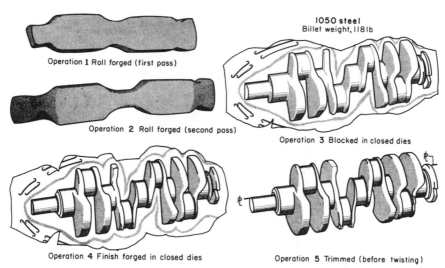

Operation 1 Roll forged (first pass)

Operation 2 Roll forged (second pass)

1050 steel
Billet weight, 118 lb

Operation 3 Blocked in closed dies

Operation 4 Finish forged in closed dies

Operation 5 Trimmed (before twisting)

The final shape has been developed by several preliminary stages which preform the bar feed stock, and then in two stages in a closed die. The forging flash is trimmed off, and the shaft finally is twisted to produce the appropriate offset in the cranks. (Source: Same as for Fig. 2.20.)

Fig. 2.21. Stages in a typical closed-die forging sequence producing a crankshaft.

common procedure is to pierce a hole in a disk and to roll the annulus circumferentially to produce a thinner ring of larger diameter (e.g., the rim tire for a railroad wheel).

A limited amount of forging is also carried out cold, even in stiff materials such as steel. The heads of bolts are often formed by a cold upsetting operation. Shallow patterns are also raised on sheet by stamping a blank between two appropriately engraved dies in a manner analogous to closed-die forging. Coins and the embossed handles for table flatware are produced in this way.

As mentioned earlier, the objective of these forging processes can be more than simply producing a particular shape economically. An additional objective may be arrangement of the grain structure in the material so that it is elongated and, as far as practicable, aligned everywhere parallel to the major system of tensile stresses imposed on the component in service. For example, the forging sequence for the crane hook illustrated in Fig. 2.22 was designed to produce a grain structure that flowed smoothly around the bearing journals into the adjoining flanges. The grain flow in a hook cut from a plate would go transversely across the jaw, and the jaw would then be more liable to failure in service. Likewise, the grain flow in an automobile engine valve would run straight across the head if it were machined from bar. Experience shows that this is a most undesirable state of affairs from several points of view. The grain flow desirably follows the contour of the head when the head is formed by upsetting.

Etching has developed flow lines, which delineate bands of segregated material that contain above-average numbers of nonmetallic inclusions. The material is weaker in tension when
stressed across these lines than when stressed parallel to them. The
forging process has been designed to ensure that the flow lines are
everywhere parallel to the surface of the hook. They would have
extended straight across the highly stressed bend in the jaw if the
hook had been cut from plate, and the hook would then have been
more likely to fail in service. Safety is a prime requirement in
lifting gear, hence forged hooks are used even though they are
more costly to manufacture. (Source: Same as for Fig. 2.7.)

**Fig. 2.22. An etched section of a forged steel crane
hook.**

Sheet Forming

Most sheet is formed into a contoured shape before use, whether it be a sauce-
pan or a panel for an automobile body. Cutting a shape out of a flat sheet is the
simplest process used, and this is frequently done with a punch and die set in a
blanking press to produce outlines, or in a *piercing press* to cut holes. Next, the
sheet may be bent around a former, or, at the next higher level of sophistication, pressed into a shape (*press forming*) imposed by a punch and a die in a
manner somewhat analogous to closed-die forging. A succession of dies may be
used to develop progressively a complicated shape, and this can be done in a set

of dies in which strip enters at one end and finished components pour out at the other.

A variety of types of stresses are imposed on the metal during press forming. One of the most critical requirements, however, is that the sheet should be able to be bent around a small radius. The outside surface of the sheet is then stretched, and it must be able to withstand this stretching without fracturing. The ductility (see p. 73) of the sheet material in the direction in which it is stretched then becomes a critical material property. The surface quality of the sheet also becomes important.

More severe processes are needed to produce deeply recessed parts. *Spinning* is one that is applicable to shapes with circular symmetry. A circular blank of sheet is spun in a lathe and forced around a mandrel of pre-established shape that spins with it (Fig. 2.23). Tubes can also be shaped in this way. The spinning is frequently carried out manually, which requires a degree of skill to wheedle the blank around the mandrel without fracturing or wrinkling it. Powered tools are necessary, however, to spin thicker sheet and plate.

Drawing is a more controlled process which can produce parts that are irregular in section. It is also a process suited to closely controlled mass production. A sheet blank is held against the surfaces of a die in which a hole has been cut (Fig. 2.24A), the shape of this hole being that required of the outer surface of the wall of the product. A punch is made to descend through the die hole, and the shape of the nose of this punch is made to be that required in the base of the product. There is a gap between the vertical portions of the punch and the die hole. The blank is at first bent over and wrapped around the nose of the punch as it descends, as shown in Fig. 2.24B(b). The outer portions of the blank then move radially toward the die hole and flow around the die radius, and finally proceed into the die cavity through the gap between the punch and die, as

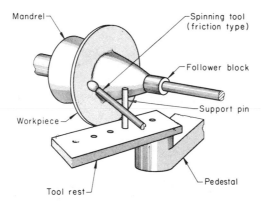

Fig. 2.23. Technique used to form a cup from sheet by spinning. The precut disk is eased around a shaped mandrel as they rotated together at high speed. (Source: *Metals Handbook*, **8th Ed., Vol 4, ASM, Metals Park, OH, 1969.)**

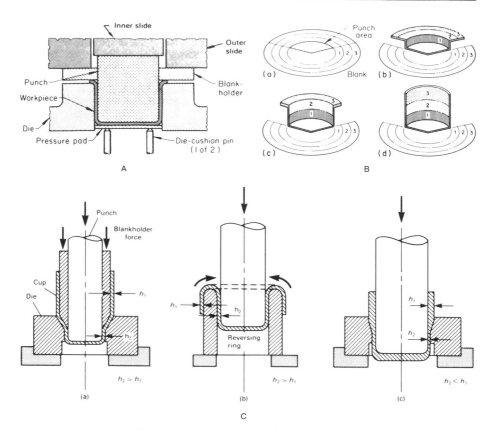

Part A illustrates the manner in which a blank is held between a blankholder and a die while it is pushed by a punch through a correspondingly shaped hole in the die.

Part B illustrates stages in the flow of the metal during the drawing of a cylindrical cup from a circular disk. The metal in the outer rings of the blank has to thicken as it flows toward the punch, and then to stretch as it flows around the die radius into the walls of the cup.

Part C illustrates several ways by which a cup may be redrawn to produce a longer cup with a thinner wall: (a) straight redrawing; (b) reverse redrawing; and (c) ironing.

Fig. 2.24. Principles involved in the deep drawing of sheet. (Source: Same as for Fig. 2.7.)

shown in Fig. 2.24B(c). The sheet usually maintains approximately its original thickness as it passes through the die gap. The pressing ends up in close contact with the punch, and is stripped from the punch as a last step, as shown in Fig. 2.24B(d). A flat blank is thus converted into a deep closed-end container without substantial change in wall thickness. The primary cup may be drawn further, perhaps with intermediate annealing heat treatments, in a variety of ways (Fig. 2.24C) to produce a longer pressing of smaller cross section. This may be carried out in continuous presses. Although the distinction is arbitrary, the process is called *deep drawing* when the cup is deep compared with its diameter and when some thinning occurs in the wall of the cup. Simple cups, complex cup-shape objects such as cartridge cases for ammunition, and irregularly shaped boxes such as oil sumps for automobile engines are drawn in the above way, among a very wide variety of objects.

The apparent mechanical simplicity of the drawing process is deceptive from the point of view of the metal. Much is required of the sheet. It first has to thicken as it moves toward the die hole, because its area is reduced, but then has to thin down as it flows around the die radius (Fig. 2.24B). The thinning occurs either because the sheet is stretched by the tensile stresses imposed at the die radius or because it is drawn through the gap between the die and punch. Individual elements of the blank have to change shape in different and complex ways, as can be visualized from Fig. 2.24B. This asks much of the sheet material, and special qualities are required of sheet intended for deep drawing, particularly for a drawing operation of any severity. The absence of surface nicks and cracks, adequate uniformity of mechanical properties in all directions in the plane of the sheet, and good ductility in all directions become important. A small grain size is also required to ensure a good surface finish in the drawn part (see p. 311).

Welding

One of the great metallurgical advances that has occurred in recent decades has been the development of techniques for producing high-strength joints between metals — hopefully, joints as strong as the metal itself. Along with this has gone the development of alloys with which full advantage can be taken of the joining techniques.

The joining of metal parts by bolting and riveting has been used since the earliest times. In general, however, such joints are bulky and do not permit the full potential strength of the metal to be utilized because, if for no other reason, many holes have to be drilled in the metal to make the joint. Metals are also joined by melting an alloy of low melting point in such a way that it wets the two surfaces concerned and then solidifies between them. Soldering is an example. Although this technique is very useful, the joints produced are comparatively weak. An intimate metal-to-metal bridge must be made to produce a joint of high strength, particularly if the joint is to be stressed in tension.

Blacksmiths have been making joints of this nature in iron and steel since the inception of the iron age. They heated the two surfaces to be joined in a forge to a red heat and then hammered them together. This process works because the oxide which coats hot iron melts at a comparatively low temperature. The iron is heated to above the melting temperature of the oxide, and the joint surfaces are slid past one another a little during hammering so that the molten oxide layer is broken up. The uncontaminated surfaces thereby brought together are bonded by atoms on the mating surfaces joining in what is now called a *diffusion bond*. Globules of oxide usually are left along the joint plane so that a joint of only moderate strength is produced. This principle can also be used to join other metals, but rather more sophisticated procedures are then required. Uncontaminated surfaces have to be brought into close contact under pressure at a temperature at which easy diffusion can occur (see p. 235). This can be achieved simply with noble metals, such as gold, which never have oxide layers on their surfaces; the surfaces are simply placed in contact and heated. The goldsmiths of the Pharaohs were very adept at using this technique in 3000 B.C. It is more difficult to perform, however, with the more common metals, but techniques for doing so have been developed in recent years. These techniques involve heating of the joint-to-be under pressure in a vacuum, which requires elaborate equipment. Diffusion bonding is used in production, however, particularly by the aerospace industry, to join high-strength alloys that could not be joined by any other means.

Many and various welding processes are now available, but perhaps the most significant developments have been in *fusion welding*. In fusion welding processes, surface layers of the two metal halves to be joined (usually of the same composition) are melted by a source of intense heat, such as an oxyacetylene flame or an electric arc of some type. A suitable gaseous atmosphere or a layer of flux is maintained in order to purge away surface contaminants from the pool of molten metal and to protect it from further oxidation. A *filter rod* of the same or similar composition as the parent metal may also be melted simultaneously with the weld faces so that a pool of molten metal forms and fills the gap between the two faces (Fig. 2.25). The molten pool solidifies as the heat source is made to progress along the joint line, forming a bridge of cast metal (Fig. 2.26). A joint that is as strong, or almost as strong, as the parent metal can be produced, although this requires attention to a number of details.

Most of you will have seen someone joining sheet metal in this way using an oxyacetylene torch as the heat source. You may also have seen someone carrying out an electric-arc weld, striking an arc between a rod of metal, which is coated with a flux, and the workpiece. The metallic core of the rod melts and is consumed in the process. The coating also melts to form a protective flux cover over the pool of metal that is formed under the arc. More advanced methods of arc welding feed the wire and flux into the arc region mechanically. Other methods strike the arc between a nonconsumable electrode and the workpiece and feed the filler wire into the heated region using a gas shield to protect the hot and molten metal. Still others work on the same general principle but use a consumable wire electrode. Some of these welding processes rely very much on the skill of the operator, but many of the advanced processes can be closely controlled and automated.

The welding conditions have to be just right to produce a weld that does not contain holes, cracks, or discontinuities that might impair its strength. This is

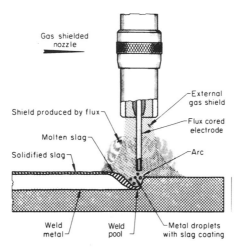

Gas shielded nozzle

Shield produced by flux

Molten slag

Solidified slag

External gas shield

Flux cored electrode

Arc

Weld metal

Weld pool

Metal droplets with slag coating

Heat is supplied to the joint area, here by an arc struck between an electrode and the workpiece. The electrode is a wire which is fed into the weld zone. The wire melts, as does some of the parent plate, to supply the weld metal. The wire has a core of flux powder which also melts to form a protective cover over the molten weld pool. Further protection is provided by a shield of inert gas. The electrode is moved from left to right in synchronism with the solidification of the weld pool.

Fig. 2.25. Principle of one method of fusion welding. (Source: Same as for Fig. 2.7.)

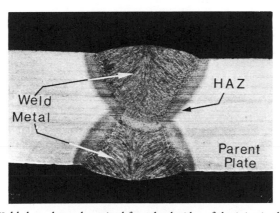

Weld Metal

HAZ

Parent Plate

Welds have been deposited from both sides of the joint in this instance. The darker-etching bands in the parent plates immediately adjacent to the welds are zones whose structure has been affected by a heating and cooling cycle as the weld pool passes by. Optical macrograph. Magnification, 1×.

Fig. 2.26. A section through a fusion weld joint, etched to distinguish the central area of cast weld metal which bridges the two plates being joined.

not easy to do in what, after all, is a casting process (Chapter 8). Critical welds routinely have to be nondestructively inspected for such abnormalities (see p. 203). The composition of the weld metal also has to be right if its intrinsic strength is to match that of the parent metal; it must not be either much stronger or much weaker. The parent metal in a zone adjacent to the weld is heated to a temperature close to its melting point, and this zone, called the *heat-affected zone* (HAZ), is then cooled rapidly (Fig. 2.26). Structures of poor strength or toughness may form if the alloy composition and welding conditions are not properly selected.

So welding has become an advanced specialist technology. Even so, not all alloys are candidates for fabrication by welding. In fact, alloys which have good mechanical strength commensurate with acceptable welding characteristics have had to be developed specially for the common base metals. The most important of these weldable alloys undoubtedly are the weldable grades of steel (see p. 94). The many welded bridges, ships, oil storage tanks, and pressure vessels that are such a feature of modern technology depend for their effectiveness on these steels and on modern methods of welding them. Much success has also been achieved in developing weldable aluminum alloys of comparatively high strength. Of course, simpler and cheaper steels and aluminum alloys of lower strengths can also be welded, and this is done extensively as a useful production technology. To use automobiles as an example again, the various panels of the body are joined together by a series of localized welds called *spot welds*. These welds are made by passing an electric current locally through the two sheets. A small nugget of metal melts at the interface, because of the high electrical resistance at this point, and this nugget, when solidified, forms a bridge between the two sheets.

Machining

The final stage of metal processing, particularly when close control of dimensions is required, is characteristically the removal of some of the metal by machining. Irrespective of whether it is the drilling of a hole, the turning of a bar in a lathe, or the grinding of a surface that is involved, the principle of material removal is the same. The sharp edge of a comparatively hard material is forced into the surface of the metal and then moved past it. A layer of metal, called a *machining chip*, is peeled off the surface. This seems to be a simple enough mechanical process and you might well ask what it has to do with metallurgy.

Actually, the mechanics of the separation of a chip is by no means well understood. What is sure is that complex deformation processes involving large and localized strains in the metal workpiece are involved (Fig. 2.27). The plastic behavior of the metal is consequently an important factor in the total system, as well as the geometry of the cutting tool and machine parameters such as feed and speed. Both the energy that has to be expended to remove the

The tool is the triangular shape at left. The rumpling of the surface of the steel as it approaches the tool and flows into the chip has been caused by severe plastic deformation during separation of the chip. The properties of the workpiece material have a major influence on the magnitude and distribution of this deformation which, in turn, affect the whole process of chip formation. Scanning electron micrograph. Magnification, 600×.

Fig. 2.27. Side face of a chip being machined from a steel workpiece.

chip and the surface finish that is produced on the component are affected by metal characteristics, some metals and alloys being much more difficult to machine than others on either or both of these counts. Moreover, the structure and properties of the workpiece material have an influence on the rate at which the cutting tool wears, which is important to the economics of production. Above all, perhaps, the production engineer seeks uniformity in the machining characteristics of the work materials which he has to use, particularly in these days of computer-aided and automated manufacturing methods. This is a problem that is squarely in the hands of the metals producer, and requires an understanding of what really are the characteristics that confer a particular set of machining characteristics on a metal. Much progress has been made, but the problem cannot be said to have been fully resolved.

Machining is an expensive process. It is also a wasteful one, since all too often more than half of the metal that enters the many machine shops of the world ends up as scrap in the form of machining chips. This provides a strong motivation for the development of methods of fabrication which produce components whose shapes and dimensions approach as closely as possible the net shape of the final products (*near-net-shape manufacturing*). Fabrication by powder metallurgy methods, by precision casting, and by precision closed-die forging are examples.

HEAT TREATMENT

Heat treatment (subjecting a metal to a controlled cycle of heating and cooling) is, as we noted in Chapter 1, one of the principle techniques available to a metallurgist for varying and controlling the properties of a metal. Heat treatment may be carried out between processing stages to enhance the processing characteristics in subsequent stages of fabrication; annealing between the stages of a cold rolling sequence is an example. It may be carried out as either a late or a final processing stage to produce improved properties in the final component; a hardening heat treatment of a steel is an example. Almost always the objective is to produce a specific internal structure in the metal, a structure that is known to be associated with desirable properties. This is, perhaps, the processing stage to which the science and technology of physical metallurgy is applied with greatest effect.

The Heating Stage

The key to successful heat treatment is the selection of the temperature to which the metal is heated, the temperature being chosen to produce a particular structural change. There is a wide range of possibilities. It may be a temperature at which recrystallization of a cold worked structure occurs (see p. 306). It may be one at which one of the constituents present dissolves in another, as in the solution treatment of aluminum alloys (p. 370). It may be a temperature at which a complete change in the internal structure occurs, as in heating for the hardening of steels (p. 388). Or it may be a temperature which modifies a structure produced by an earlier thermal cycle, as in the tempering of steels (p. 408) or the precipitation hardening of aluminum alloys (p. 371). The rate at which the piece of metal is heated to the selected temperature is usually not of particular importance, but the time for which it is held at that temperature is. The time must be long enough to ensure that the intended temperature is reached throughout the bulk of the piece, and then for any further period necessary to ensure that the desired structural change occurs to completion (p. 358). The former usually is the more important of the two.

So, in principle, all that is needed to carry out a heat treatment is a thermally insulated enclosure (called a *furnace*) which can be heated to the desired temperature and into which the pieces of metal being heat treated can be placed and left until they are judged to have reached the temperature of the enclosure throughout (Fig. 2.28). However, sophistications build up as the complexity of the heat treatment increases and the production requirements vary.

The first need is a means of reliably and accurately measuring and controlling the temperature of the enclosure. In the "good old days", a metalsmith judged the temperature of the component being heat treated from its color, at

The furnace consists of an open-end steel container lined with insulating bricks. The open end can be closed by lowering a vertical sliding door, which also is lined with insulating bricks. This particular furnace is heated by radiant electrical resistors, which are the horizontal short black strips that can be seen on the inner sidewall of the furnace enclosure. The furnace is equipped with a roll-top table in front of the charging door to assist handling of the workpieces.

Fig. 2.28. A typical box-type heat treating furnace.

least when the temperature was high enough for a distinguishable red color to develop. With experience, he did fairly well but not nearly well enough for modern requirements. Nowadays, temperature is measured by a sensor, most frequently a *thermocouple*. This consists of a joined pair of wires of two special dissimilar alloys, the joint being placed at the sensing point. An electromotive force is developed which is proportional to temperature; this electromotive force is measured and read out in terms of a temperature scale. Moreover, the electromotive force can be made to operate a control device which turns on or off the heat source to the enclosure when the temperature rises above or falls below prescribed limits. Such a device can also be programmed to heat or cool the enclosure at a controlled rate. There are, however, many pitfalls in measuring temperature accurately and reliably. *Pyrometry* is a technology of crucial importance to metallurgy and is perhaps too often neglected more than it should be.

A second, associated need is that the enclosure must be adequately uniform in temperature. Otherwise, components placed in some parts of the enclosure will not attain the temperature that the pyrometer records, and the heat treater won't know the temperature that it actually did reach. This is a function of the design of the furnace, of the arrangement of its insulating and heating units. Sometimes it is even necessary to incorporate a circulation fan in the furnace enclosure to obtain satisfactory temperature uniformity, particularly in furnaces operating at low temperatures (Fig. 2.29). This trick has recently been adapted to domestic cooking ovens.

The atmosphere within the enclosure is, at its simplest, that of the air from the surrounding atmosphere. This often is acceptable, but there are situations where it is not. For example, the oxygen in air reacts with most metals at heat treatment temperatures (see Chapter 5), producing a thick surface layer of oxide known as a *scale*. This may occur to an unacceptable extent. Oxidation on steel surfaces may also remove preferentially from the adjoining surface layers some of the critical alloying element of the metal; the most important example of this is the removal of carbon from the surface of steel (the steel is then said to be *decarburized*). Some metals (e.g., titanium), on the other hand, absorb gases from an air atmosphere to the detriment of their properties.

When such problems become serious, the ordinary atmosphere has to be flushed out of the furnace enclosure and replaced by one which is neutral to the surface of the metal at the heat treatment temperature. Equipment to provide such a controlled atmosphere has to be added to the furnace complex, and its

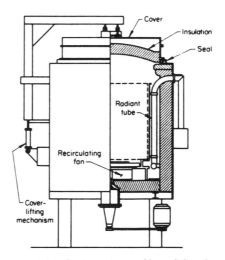

This furnace is circular in section and has a lid at the top so that workpieces can be lowered into a pit-shape enclosure. This particular furnace is heated by burning a gaseous fuel within radiant tubes. The fan circulates the atmosphere within the enclosure to improve the uniformity of temperature throughout the workpiece region. (Source: *Carburizing and Carbonitriding*, ASM, Metals Park, OH, 1977.)

Fig. 2.29. A pit-type heat treating furnace fitted with a recirculation fan.

operation has to be understood and controlled if it is to be really effective. Alternatively, the atmosphere may, if the furnace container is suitably designed, be pumped out and heating carried out in a vacuum.

On the other hand, advantage can be taken of the possibility of reaction between the furnace atmosphere and the surface of a metal. For example, steel can be heat treated in an atmosphere the composition of which is adjusted to add carbon to the surface layers (*carburize* the surface). Likewise, nitrogen can be added in a beneficial way to the surface of steel. These additions to the surface layers allow the production of components which have hard, wear-resistant surfaces but strong, tough cores, which are very advantageous in some engineering applications.

We have so far considered only what might be called a box furnace — that is, a furnace with a simply shaped container, with a lid or door, into which the component can be placed for heating and then removed when it has reached the desired temperature (Fig. 2.28 and 2.29). Furnaces of this type are flexible and are used widely for heat treating components in batches, often in small numbers although the individual components may be large. The economy of heat treatment of large numbers of similar components, however, may require a furnace through which the components can be progressed continuously (on a moving belt, for example).

A typical heat treatment furnace is heated either by electrical resistance elements, as in Fig. 2.28, or by the combustion of a liquid or gaseous fuel. In the latter case, the combustion may be made to occur in a tube separate from the main furnace chamber (as in Fig. 2.29) so that the products of combustion and the furnace atmosphere do not mix. The two can then be controlled independently. The choice between electrical and combustion heating depends on a number of economic and technological factors.

A large number of other methods of heating are used to meet special needs. Baths of molten salt are one example; they heat rapidly and provide a comparatively simple way of heating small components in a neutral environment, particularly at high temperatures. Heating by electrical induction in the component itself (*induction heating*) is another; the heated region can then be confined to the surface layers and even to localized areas if the frequency of the induction current and the shape of the induction coil are chosen correctly. Surface layers can also be heated locally by applying an intense source of heat such as a combustion flame, a laser beam, or, in an evacuated container, a focused beam of electrons. Such heating methods are costly but have advantages which in some applications justify their cost.

The Cooling Stage

Our later studies of the structural changes that occur in metals will indicate that there are many occasions where the rate at which a metal is cooled from a heat treatment temperature is of critical importance. As distinct from control of temperature, only a limited amount of control can be exercised over the cooling rates that are achievable in heat treatment practice. This is true, first, because only a limited number of cooling media are available that will extract heat from the workpiece at different rates, and secondly, because the cooling rate achieved in the workpiece by a given medium depends on the thermal characteristics of the workpiece (its surface area, section thickness, temperature, and thermal conductivity).

Low cooling rates are achieved by cooling the workpiece in the furnace, either by decreasing the temperature of the furnace at a controlled rate or by simply turning off the heat supply to the furnace. Somewhat higher cooling rates are achieved by removing the workpiece from the furnace and allowing it to cool in air. The highest cooling rates are achieved by quenching the workpiece in a liquid, with the cooling rate being determined by the nature of the liquid. Quenching in water, brine, or caustic solutions gives the fastest cooling rates; the cooling rate may in fact then be too fast on occasions because large thermal stresses can be induced which cause fractures or cracks to develop in the workpiece. This is a common problem, for example, when quenching steels to harden them. Oils or, more recently, certain water-base polymer solutions can be used to achieve somewhat lower cooling rates. Molten salts, blasts of gas, or fog mists may be used to achieve even lower cooling rates. The operation and maintenance of a quenching bath can be quite complicated. The bath usually has to be agitated, and the quenching medium may have to be filtered and cooled in a recirculating system. The quench bath has to be located close to the heating furnace and arranged so that components can be transferred from one to the other without serious loss of temperature. This too may cause some design problems.

We have a complex piece of production equipment by the time that our basic heated enclosure has been equipped with temperature-control equipment, mechanisms to progress the workpieces continuously through the furnace enclosure, generators to supply a controlled atmosphere to the enclosure, and a quench tank with all its appurtenances, all requiring close attention and control and perhaps controlled by a microprocessor. This is a far cry indeed from the blacksmith who heated, say, a steel tool in an open coke fire, judged its temperature by eye, and quenched it by immersing it in a cask of water.

FURTHER READING

E. P. DeGarmo, *Materials and Processes in Manufacturing*, Macmillan, New York, 1979.

S. Kalpakjian, *Manufacturing Processes for Engineering Materials*, Addison-Wesley, Reading, MA, 1984.

Metals Handbook Desk Edition, H. E. Boyer and T. L. Gall, eds., American Society for Metals, Metals Park, OH, 1985.

The Making, Shaping and Treating of Steel, H. E. McCannon, ed., Association of Iron and
    Steel Engineers, Pittsburgh, 1984.

T. Altan, S. Oh, and H. Gegel, *Metal Forming: Fundamentals and Applications*, American
    Society for Metals, Metals Park, OH, 1983.

H. E. Boyer, *Practical Heat Treatment*, American Society for Metals, Metals Park, OH,
    1984.

Metals Handbook, 9th Ed., Vol 4, *Heat Treating*, American Society for Metals, Metals
    Park, OH, 1981.

· 3 ·

The Basic Mechanical Behavior of Metals

SUMMARY

Engineers typically select metals in the first instance on the basis of the manner in which they support an externally applied load, mostly a steadily applied tensile load. Specifically, they are interested in the manner in which the shape of the metal changes (strain) with unit applied load (stress) and with how well failure by fracture is resisted. Several quite distinct properties have to be distinguished here. The first is the extent to which the material expands or contracts recoverably under comparatively small stresses. This property of elastic stiffness is an unchangeable characteristic of the element on which the alloy is based and has to be taken as is. The second property is the stress that can be withstood without the metal changing shape permanently, or plastically. This property, commonly called the yield strength, establishes the maximum stress at which the material can be used in most structures, and can be varied considerably by metallurgical means. Indeed, this is one of the main objectives of physical metallurgy. The third characteristic is the ease with which and the extent to which the material can be deformed at stresses beyond the yield strength. This is of interest particularly when the material has to be shaped by deformation. Finally, the stress that can be withstood without fracture (the fracture strength) needs to be known and, perhaps of greater importance, the manner in which fracture occurs when it does. Users, above all, do not want a structure to fail catastrophically and without warning. These characteristics depend on a number of metallurgical factors but also are determined by the geometry of the component and the nature of the stress system. An understanding of the factors which determine whether a structural material fails in a brittle manner instead of first stretching in a discernable way consequently is of fundamental importance to both those who make metals and those who use them. Poor fracture toughness, as it is called, can set a limit on the degree of strengthening that can safely be achieved in a number of important alloys, including steels. However, improvements in metal manufacturing procedures and in design concepts have greatly alleviated this problem. Moreover, the possibility of obtaining an acceptable level of fracture toughness, in comparison with other available structural materials, is the very reason for the dominance of metals in structures which are stressed in tension. These various mechanical properties have to be characterized in comparatively simple laboratory tests, but such tests necessarily have their limitations and so have to be interpreted carefully and correlated with service experience.

Metals are selected for use in engineering structures when, and only when, they have some property or combination of properties that is especially appropriate to meet a specific need. They may, for example, be used because they most readily provide the required electrical conductivity, or the required corrosion resistance, or the required strength, or simply because they can be formed most economically into the required shape.

These properties are of three general types:

1. *Chemical properties* pertain to the chemical reactivity of the metal in various environments. Included are resistance to corrosion by liquids and gases, and the ease with which protective or decorative layers can be attached to a surface. These properties are largely determined by the bonding forces between the constituent atoms.
2. *Physical properties* are those properties not defined as mechanical properties that are determined by the properties of the constituent atoms. They include density, electrical and thermal conductivity, and thermal expansion.
3. *Mechanical properties* are the properties which are concerned with behavior when forces are applied.

Chemical properties require and deserve separate treatment, and will not be considered at any length here (but see Chapter 5). Most physical properties are *structure-insensitive*, or nearly enough so: that is, they cannot be altered by varying the structure of the metal, although they may be sensitive to chemical composition. Consequently, they are not susceptible to the wiles of a metallurgist. Quantitative data on physical properties can be found in many handbooks, such as the *ASM Metals Reference Book* listed under "Further Reading" at the end of this chapter. Most mechanical properties, on the other hand, are *structure-sensitive*, and the control of mechanical properties by the control of structures is, as we have implied earlier, central to physical metallurgy. We shall in the present chapter be concerned with the mechanical behavior of metals under simple stresses applied at room temperature. Some other important modes of mechanical behavior will be discussed in Chapter 4.

The behavior of metals under tensile stresses is of particular importance in this context because it is under these circumstances that metals are notably superior to alternative structural materials, such as concrete, stone, and ceramics. Concrete and stone, for example, are excellent structural materials if precautions are taken to ensure that they are stressed only in compression. Disaster inevitably follows when they are placed in tension. Disaster can follow too when metals are placed in tension (see p. 83), but this is not inevitable. This is what design and metals engineering are all about — namely, to ensure that metals are used to their utmost without disaster occurring. Indeed, a designer's first objective is to ensure that a component or structure will be able to withstand, without collapsing or breaking, the forces that are applied to it in service. Above all, the designer must ensure that the component does not fail without giving due warning; catastrophic failure too often results in catastrophic loss of human life.

There are also other limiting parameters in the mechanical behavior of materials. Even if proper precautions are taken to ensure that a structure does not fail, it still must not deflect so much that either its functions are impaired or the confidence of its users is undermined. For example, excessive deflection of the wing of an aircraft not only would affect its aerodynamic performance but might also influence you as a discerning observer to patronize a different type of aircraft. Aircraft wings always deflect; the question is how much. One of the requirements of a design is to ensure that these types of deflections are kept within acceptable limits.

A design engineer has first to determine the type (tension, compression, or shear) and then the magnitude of the forces that will be applied throughout a structure under the loads that it is expected to bear in service. In spite of some recent major improvements in methods and in the use of computers, this is a difficult and uncertain task in a structure of any complexity, and structures have a habit of becoming increasingly more complex. Examples abound where, even after the most careful analysis, structures fail at positions that nobody expected. The problem that the design and metals engineers consequently share together is that of selecting a material that will adequately withstand the predicted stresses, and hopefully be forgiving enough to cope with the uncertainty of these predictions. Many other parameters, such as cost, availability, and ease of fabrication, will of course also have to be considered.

The second problem that design and metals engineers share is that of devising means of determining parameters that adequately characterize the mechanical behavior of a metal. At the highest level, they must produce quantitative data that properly describe the way in which the metal will behave under the stress systems actually encountered in service. At the lowest level, they may merely have to provide assurance that a particular batch of material is up to the standard expected of it — but, it should be noted, up to standard in those characteristics that will be important in service. Moreover, the parameters will have to be determined in laboratory-scale tests which can be carried out in a reasonably short time and which are as simple and as cheap as possible.

Those concerned with the processing and fabrication of metals look at mechanical properties from a somewhat different viewpoint. They frequently want to deform metals, to change their shape in a predictable way, and to be able to do this as easily and as cheaply as possible. They certainly also want to be able to do this without the material breaking up, and without introducing abnormalities which might adversely affect subsequent service performance. They need test parameters which characterize these aspects of the behavior of the metal.

Those who process metals are also interested in the behavior of metals in operations such as machining and welding. These processes may seem to be simple enough to the layman, but actually they are extremely complex at the level of sophistication at which they are now carried out. These processes may be critically dependent on specific material characteristics, characteristics that are different for each process. Successful control of these processes consequently depends on these characteristics being identified, and on simple tests being devised to control them.

This is a tall order by any standard. All simplified tests require careful interpretation, and some of them require very careful interpretation indeed. Some well-established tests are downright misleading. Our purpose in this chapter will be to develop an understanding of the more important tests used to characterize mechanical properties and, in particular, to develop an understanding of what these tests really mean and how the data derived from them should be used.

STRESS AND STRAIN: WHAT THEY MEAN TO AN ENGINEER

"Stress" and "strain" are terms which in everyday usage tend to be regarded as synonymous, often being used as vague descriptors of the pressures of life. To an engineer, each has a precise meaning, and these meanings are different. The difference seems to present some intellectual difficulties to nonengineers (even the Webster's and Oxford dictionaries have it wrong), but you must develop a clear understanding of these terms if you are to get to first base in developing a feel for the importance of the mechanical properties of materials and the place that metals have in the hierarchy of structural materials.

Stress is simply the force that happens to be acting in a given region of a body divided by the area over which the force acts:

$$\text{stress}\,(s) \;=\; \frac{\text{force}\,(f)}{\text{area}\,(a)} \tag{Eq. 1}$$

If the stress acts normal to the area, it is called a *normal stress*. If it acts parallel to the area, it is called a *shear stress*. The magnitude of a stress is expressed in units of force per area, and a variety of units, such as pounds per square inch (psi), thousands of pounds per square inch (ksi), tons per square inch, or kilograms force per square millimetre (kgf/mm²), are commonly used. By international agreement, however, the preferred unit is the SI (Système Internationale)* unit which is expressed in terms of a unit of force known as the newton (1.0 newton = 0.102 kilogram force = 0.225 pound force†), and we shall use this unit exclusively. Stress usually is most conveniently expressed in meganewtons per square metre (MN/m²), called megapascals (MPa).

Strain, on the other hand, describes the change in shape that occurs in the metal when the force is applied. A bar of metal stretches when it is pulled — that is, when a *tensile stress* is applied. The extension might be too small for you to notice when the stress is small, but it can be measured nevertheless. Strain is defined as the amount of stretch per unit length‡:

$$\text{strain (e)} \; = \; \frac{\text{change in length } (\ell)}{\text{original length (L)}} \qquad \text{(Eq. 2)}$$

Strain is not expressed in any unit but as a dimensionless number. For example, if a bar 10 cm long were stretched 1 cm, it would be said to have been strained 10% (tensile). If it were squeezed 1 cm, it would be said to have been strained 10% (compressive). By convention, compressive stresses are said to be positive and tensile stresses to be negative.

Stress and strain are thus quite different things, like chalk and cheese — but not quite like chalk and cheese, because they are related. You can't have one without the other. We must now explore this relationship between stress and strain.

BEHAVIOR IN TENSION

The Tensile Test

The simplest way of establishing the relationship between stress and strain is to carry out a so-called *tensile test*. A test piece, preferably one that is circular in section and shaped as sketched in Fig. 3.1, is machined from the metal. The purpose of this test-piece design is to ensure that the stress and the strain are reasonably uniform throughout the central *gage length* portion of length ℓ. They can then be determined simply as defined above.

---

*See Appendix 2 (p. 477).

†See Appendix 3, p. 485, for detailed information on conversions.

‡Strain defined in this way is known as *engineering strain*. *True strain* or *natural strain* is more likely to be used in scientific circles. This is the natural logarithm of the ratio of length at the moment to the original length. Unless qualified to the contrary, strain can be taken to be engineering strain.

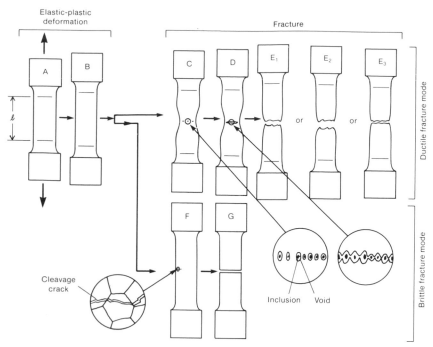

A – The undeformed test piece. B – Test piece strained in the plastic range. C – Necking down has commenced; inset illustrates development of a layer of internal voids in the neck region. D – An internal fissure, shown in more detail in the inset, has developed across the neck region. E_1 – The ring of material between the internal fissure and the external surface has necked down to zero, resulting in final fracture of the test piece in the form of two cups. E_2 – The remaining ring has torn along a conical surface on which a sheet of voids has formed; a cup is formed on one side of the fracture and a cone on the other. E_3 – The fissure has extended to the surface, producing planar fracture surfaces. F – A cleavage crack has developed on specific planes in each grain. G – Cleavage crack has been extended by further cleavage to form a planar fracture.

Fig. 3.1. Diagrammatic illustrations of the ductile (A B C D E) and brittle (A B F G) modes by which metals deform and then fracture when subjected to an increasing tensile stress (test pieces shown in longitudinal section).

The two shoulder sections of the test piece are gripped and pulled apart at a steady rate. This is done in a special machine in which the force required to stretch the test piece can be measured and recorded continuously throughout the test. The *nominal* (tensile) *stress* is then calculated as the force divided by the original section area. The extension of a *gage length* marked on the central reduced section of the test piece can also be measured throughout the test, and the tensile strain can be calculated as a percentage increase in the original gage

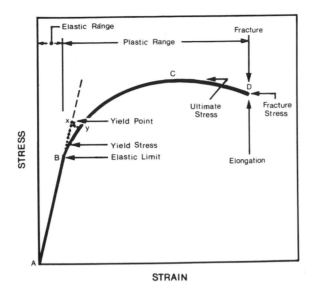

The solid line is the typical curve. The dotted line labeled "Yield point" is a modification found in certain special circumstances. The salient features of these curves are discussed in the text. NOTE: The relative magnitude of the elastic strain is greatly exaggerated in this stress-strain curve, and in those sketched in Fig. 3.3, 3.5, and 3.6, for clarity of illustration. The elastic portion of the curve (AB) should at this scale be nearly parallel with the stress axis.

Fig. 3.2. Diagrammatic illustration of a characteristic stress-strain curve for a metal stressed in tension.

length. Thus, nominal stress can be plotted against engineering strain as a *stress-strain curve*, a generalized example of a stress-strain curve for a ductile material being given in Fig. 3.2.

In a sense, the test piece that we have just described is a structure in itself, even if a simple one. It can also be taken to represent a volume element in a complex structure throughout which volume the stress is reasonably uniform.

Elastic Deformation

At small stresses and strains, and up to a certain limit which we shall discuss later, stress is linearly proportional to strain (section A-B of the stress-strain curve in Fig. 3.2); moreover, the original gage length is recovered when the stress is removed (Fig. 3.3A) — that is, there is no permanent change in shape.[*] This is called *elastic deformation*. The relationship at this stage is as follows:

---

[*]These statements may not be absolutely true, but are so to at least a good enough approximation for many purposes.

$$\frac{s}{e} = E \qquad\qquad (\text{Eq. 3})$$

where E is a constant, was established by Robert Hooke as long ago as 1678 and hence is known as Hooke's Law. Hooke, however, expressed the law in terms of the force applied to, and the change in shape of, a complete structure, because the generalized concepts of stress and strain had not been thought out at that time. Hooke also did not appreciate the significance of the proportionately constant E. It was Robert Young who, in about 1800, realized that E was a characteristic constant of a particular material, and this constant has been known as *Young's modulus* ever since. Strictly speaking, E should be called the *elastic modulus*, but the term "Young's modulus" is widely used for the modulus under tensile and compressive stresses.

Young's modulus is a structure-insensitive property. It is, in fact, directly related to the forces that bond atoms together (see p. 284). This means that all simple iron alloys (steels and cast irons), for example, have the same value of E. All simple copper alloys have the same value of E too, but this is a different value from that for iron alloys. It also means that the modulus cannot be altered by a metallurgist by any trick of alloying or thermal or mechanical treatment. Values of E for some common metals at ambient temperature are given in Table 3.1. Note that the units for E are those of stress. The values decrease with increasing temperature — about 5% per 100 °C (180 °F) for steels — and more rapidly so when the melting point is approached.

What is the importance of this elastic behavior and of Young's modulus? First, Hooke's law indicates that all structures deflect when they are loaded and, within certain limits as we shall soon see, recover when they are unloaded. The only question at issue is how much they deflect. This is where Young's modulus comes in. It is the indicator of how large the deflection will be when a given load is applied; materials with comparatively small values of Young's modulus deflect comparatively large amounts, and vice versa. Putting this another way, the modulus determines the stiffness, or conversely the springiness, of a structure.

A structure, such as an airplane wing or a spring, may deflect too much for safety or comfort under the loads that it is expected to bear in service, even if these loads are far below those which would cause it to break or collapse. In this event, there are only two possible corrective measures available: either the design has to be changed to reduce the stress to which the material is subjected, or the material has to be changed to one which has a higher Young's modulus. It will do no good at all to use a higher-strength alloy of the same metal. In fact, this may worsen things because the designer may be tempted to take advantage of the higher strength to lighten the structure, thereby increasing the stress and, consequently, the deflection. This is an example of a situation in which intuition and a lack of real understanding of a phenomenon might result in embarrassment.

Table 3.1. Elastic Moduli and Tensile Strengths of Some Representative Metals and Nonmetals

Material	Elastic modulus, GPa Absolute	Specific	Tensile strength(a), MPa
Metals			
Pure metals:			
Aluminum	70	26	70
Beryllium	295	160	290
Copper	130	14	250
Iron	210	27	420
Magnesium	45	26	180
Molybdenum	270	25	700
Nickel	220	25	340
Titanium	120	26	300
Alloys:			
Aluminum	70	26	80 to 400
Copper	130	14	250 to 750
Iron	210	27	400 to 2000
Titanium	120	26	750 to 1500
Nonmetals			
Artificial:			
Plastics	1.4	1.5	15 to 70
Concrete	17	7	4
Glass	70	28	150
Kevlar fibers(b)	130	90	2700
Carbon fibers	410	180	2000
Boron fibers	750	330	3100
Natural:			
Granite	30	11	20
Wood(c)(d)	13	25	100
Bone(d)	21	46	140
Sapphire	420	105	. . .
Diamond	1200	340	. . .

(a) Indicative values only. (b) Kevlar is the trade name of a product manufactured by E. I. du Pont de Nemours & Co. (c) Spruce, along the grain. (d) Wood and bone have very complex structures.

Difficulties may arise with either of the above methods of stiffening a structure. It may not be possible to reduce the stress without making an unacceptable sacrifice in the form of an increase in the weight of the structure. After all, the whole idea of using a metal is to take as much advantage as possible of its strength. Moreover, it may not be possible to find a metal that has a higher value of Young's modulus, particularly when costs are taken into account. It so happens that steel, the most common structural material, has about the highest Young's modulus of all metals (Table 3.1). The exceptions are all exotic and rare metals (such as beryllium, molybdenum, and ruthenium) the use of which would be considered only for the most specialized applications. This is fortunate in one way, because it adds to the relative benefits of using this comparatively cheap high-strength material, but it does mean that there is nowhere to go with metals when a steel structure is excessively springy. An excessively springy aluminum structure, on the other hand, can be improved by using titanium (if the extra costs are acceptable) or steel (if the extra weight is acceptable). In fact, improvement with metals is really possible only if weight is not a consideration.

A value known as the specific Young's modulus (the modulus divided by the mass density of the material) can be used to give a measure of this problem of improving stiffness without increasing weight, because it so happens that the specific moduli of most common metals are much the same (Table 3.1). They are, nevertheless, high compared with most other commercial structural materials (Table 3.1). Some natural materials do quite well on the same count, but they have comparatively low strength and some are really complex structures rather than materials as such (e.g., wood and bone). So we have to look for entirely new classes of material to get out of this springiness dilemma. Some possibilities have emerged in recent years, examples of particular importance being fibers of carbon and boron and the polymer known by its Du Pont trade name of Kevlar. Fibers of carbon, prepared in a particular way, have a modulus nearly four times that of steel and a specific modulus over ten times as good. Boron fibers are even better yet. Both these fibers can be incorporated into a matrix of either a plastic or a metal, resulting in a composite which is exceedingly stiff. But they are expensive. Nevertheless, there has been strong interest in recent years in these *composite materials* for use in aerospace structures and other important items such as tennis racquets and golf clubs. Kevlar is not quite as good but is much less expensive. Moreover, it can be woven into products as disparate as sails for racing yachts and bulletproof vests, giving products which perform outstandingly in service. All of these materials also have very high strength. As interesting and important as these materials may be, we cannot pursue them here. This book is about metals only.

Plastic Deformation

Sooner or later, but certainly by a strain of 1%, the linear proportionality between stress and strain breaks down in a gross manner. Two things then happen in a simple tensile test. First, the strain increases more rapidly with increasing stress than it did in the elastic range, increasingly so as the stress increases (section BC in Fig. 3.2). Secondly, the gage length does not return completely to its original length when the stress is removed (B in Fig. 3.1; Fig. 3.3). The material changes its shape permanently, and is said to have been *plastically deformed*. The plastic strains soon become much larger than those which can be achieved in the elastic range; the sketch in Fig. 3.2 is somewhat distorted in this respect in that the magnitude of the elastic strain is exaggerated for clarity.

The stress at which plastic deformation commences (B in Fig. 3.2) is referred to as the *elastic limit* of the material. It is reached after an elastic strain of between 0.25 and 1.0% in metals, but at a strain of less than 0.5% in those that are most commonly used. Elastic limit is, however, a somewhat imprecise concept because the elastic and plastic ranges shade into one another. The greater the sensitivity with which strain is measured, the lower the elastic limit seems to be; and the softer and the more ductile the metal, the more likely it is that this is so. A more reproducible characterizing parameter is the stress that is required to cause a certain permanent strain, usually a strain of either 0.1 or 0.2%, as indicated by the value e_p in the sketch in Fig. 3.3B, which will be explained soon. This is called the *proof stress* in Europe, and usually the *yield stress* or *yield strength* in the United States.* Yield strength values quoted throughout this book are those of the 0.2% yield stress. In the latter usage, care must be taken to distinguish between yield stress and yield point (see p. 68). Nevertheless, the general concept of an elastic-to-plastic transition and of an elastic limit is a useful one. Elastic limit and yield stress are very much structure-sensitive properties, and for this reason they are wide open to metallurgical manipulation.

The elastic-plastic transition is an important one in practice. Both the designer and the user of a structure are likely to be distressed if the structure sags noticeably and stays that way when the service loads are applied, even if it doesn't actually break. So care needs to be taken to ensure that the expected service stresses stay well within the elastic limit (or appropriate yield stress) of the material that has been selected for use. Herein lies the advantage of using materials of high strength, although when saying "high strength" we actually mean high elastic limit, or proof stress, or yield stress, whichever term you prefer. Higher stresses can then be applied without the material deforming permanently, even if it does deform as much as ever elastically. It is in fact this strength figure that should be used in design calculations instead of the tensile strength as defined later.

As an example, a spring made from a low-strength steel might deform permanently and so become useless after only a small load has been applied and only a small deflection has been achieved. A spring of the same dimensions but made from a steel of higher strength would have the same stiffness but could be used to carry much higher loads and to sustain much larger deflections without distorting permanently. Springs in practice tend to be made from the highest-strength steel that one thinks one can get away with. What one can get away with is determined by the sensitivity to catastrophic fracture in the presence of surface imperfections, a matter which we shall discuss soon (see p. 83).

The metallurgist who has to form metals looks at these matters in a different light. He wants to change shape permanently, and the yield stress defines the minimum stress that he has to apply to achieve this end. Young's modulus is important here because it gives a measure of how much the dimension of the

---

*The term "flow stress" is also used in scientific work. This is the true stress required to continue deformation beyond a nominated strain.

formed part will change (will *spring back*) when the forming forces are re-moved.

Before leaving this section we need to note one form of anomalous behavior at the elastic-plastic transition. When this anomaly is present, the metal extends elastically to rather larger strains than usual, and then yields suddenly to release the additional elastic strain, as indicated by the dotted line in Fig. 3.2. Plastic straining thereafter occurs in the usual way. The stress at which this sudden yielding occurs is called the *yield point*, not to be confused with yield stress. Two values of yield point may be distinguishable — namely, the *upper yield point* (x in Fig. 3.2) and the *lower yield point* (y in Fig. 3.2). The lower yield point is the more significant one. Although this phenomenon occurs in only a few alloys, it so happens that one group of them is industrially important. This group is the low-carbon steels, which may exhibit this phenomenon when in some conditions but not in others (p. 328).

Strain Hardening

The shape of section B–C of the stress-strain curve sketched in Fig. 3.2 implies that the yield stress increases with increasing plastic strain. To understand this point more clearly, imagine that a test piece has been stressed to a little beyond the elastic limit, as sketched in Fig. 3.3(b). The test piece would recover elastically with the same Young's modulus as during the original application of stress if the stress were then removed. It would re-extend elastically, again with the same Young's modulus, if a stress were reapplied. Now, however, a stress rather higher than the original elastic limit would have to be applied to cause plastic yielding (Fig. 3.3b). This type of cycle can be repeated any number of times, as indicated for a third cycle in Fig. 3.3(c). The net result is that the yield stress has increased with increasing amounts of plastic strain. This is called *strain hardening* or, sometimes, *work hardening*. It is one of the basic methods by which metals can be strengthened (see Chapter 9). The rate of strain hardening determines the slope of the B–C section of the stress-strain curve in Fig. 3.2: the higher the rate of strain hardening, the steeper the slope (compare the two curves in Fig. 3.5a).

The sketches in Fig. 3.3 illustrate the point that the strain achieved permanently by the application of a given stress is the total strain produced by the application of the stress minus the elastic recovery that occurs when the stress is removed (e_p in Fig. 3.3b). These sketches also illustrate the manner in which yield stress is defined. It is the stress for which e_p is 0.1 or 0.2%, whichever the case may be.

Fracture in Tension

The gage length of our simple test piece has to this stage elongated uniformly (B in Fig. 3.1). Actually, the diameter of the gage length has also contracted,

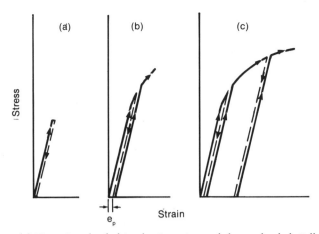

(a) Test piece loaded in elastic region and then unloaded. Full recovery of the elastic strain occurs. (b) Test piece first loaded beyond the elastic limit and then unloaded. A permanent strain (e_p) now remains after the elastic component of strain has recovered. When then loaded again, the test piece first strains elastically with the same elastic modulus as before but plastic deformation eventually resumes. (c) Test piece taken through a second cycle of loading and unloading. A higher stress than during the first cycle is required for plastic deformation to resume. This is because the metal has strain hardened.

Fig. 3.3. Portions of stress-strain curves for a metal subjected to cycles of loading and unloading. Elastic components for loading and unloading should be coincident, but are shown separated for clarity.

the ratio of the elastic component of this contraction to the elastic elongation being known as *Poisson's ratio*. This ratio is also a structure-insensitive property, and is between $1/2$ and $1/3$ for metals.

Let us consider first the behavior of what can be described as *ductile* materials, ductility being a concept that we shall define more precisely later. At a certain stage, the elongation begins to be concentrated in one localized region of the gage length, as also does the contraction in its diameter. The test piece is said to begin to *neck down* (C in Fig. 3.1). The detailed reason for necking need not concern us here, but in general terms deformation begins to concentrate in narrow planar zones inclined to the tension axis instead of being distributed uniformly throughout the gage length. Instabilities of this nature inevitably take over once they start, the strain thereafter being concentrated in the regions where the instability was first initiated. The force that the test piece can sustain then starts to decrease when the rate of decrease of the area of the neck exceeds the rate of strain hardening in the neck region. The nominal maximum stress* that the test piece sustains (C in Fig. 3.2) is termed the *ultimate (tensile)*

---

*The *true* stress — that is, the applied force per unit of section area at the time — is higher than the nominal stress once the test piece begins to neck down. The nominal stress is usually of the greater engineering significance.

strength. The necking-down process continues until, in a theoretically pure metal, no section is left. The test piece then separates, or fractures. The stress at which this occurs is called the *fracture stress* or the *tensile strength* (D in Fig. 3.2). In engineering tests, it is again calculated as the failure load divided by the original section area of the test piece.

Real metals do not neck down to zero section. No matter how carefully made, they contain numerous small particles of foreign matter and constituents of various types. Voids develop around these particles as the metal elongates and stretches away from the particles, and this becomes significant at about the stage at which the neck starts to form. A sheet of voids then develops internally within the neck, the sheet being normal to the stress axis (C in Fig. 3.1). The regions between the voids then neck down individually to zero section, and the sheet of voids develops into an internal fissure across the neck (D in Fig. 3.1). The final stage during which the fissure extends to the outer surface to complete the fracture varies. In some materials, the ring of material around the fissure necks down completely, producing two fracture surfaces shaped like flat-bottom cups (E_1 in Fig. 3.1). In other materials, a second system of voids develops on the surface of a cone extending from the edge of the fissure to the external surface. Separation occurs on this conical surface; one fracture surface is shaped like a flat-bottom cup and the other like a truncated cone (E_2 in Fig. 3.1); a cone side with a small cup-type portion is illustrated in Fig. 3.4a). These first two types are known as *cup* and *cup-and-cone* fractures. In still other materials, the original fissure may extend to the surface, producing essentially planar fractures (E_3 in Fig. 3.1). The fracture type varies with the metal, its purity and inclusion content, and the temperature, but is of rather secondary importance. The details of fracture shape also vary in pieces of different shapes, but the principles remain the same.

In any event, fracture surfaces of all of these types have a dull matte appearance to the eye. When looked at more closely they can be seen to be composed of a multitude of small dimples centered around a small foreign particle (Fig. 3.4b). Such particles are the void-nucleating particles referred to above, and the rims of the dimples are the edges formed when the intermediate regions necked down to zero section. The dimples formed under pure tensile stresses are roughly circular (as in Fig. 3.4b), and those formed by tearing are cusp shaped with the cusp pointing in the direction from which the fracture started. This is simply a result of the geometry of the separation process. It is this mechanism of fracture by void formation and coalescence that is the basic characteristic of one of the boundary modes by which metals fracture — the most ductile mode.

Fracture occurs by a very different *cleavage* mechanism in the other boundary mode, which exemplifies the most brittle type of fracture. Fracture in this case occurs soon after yielding without significant necking down, developing a planar fracture (Fig. 3.4c). The individual grains within the metal (see p. 240)

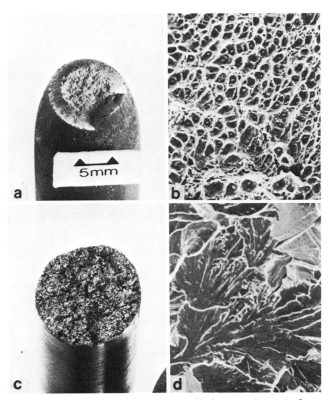

(a) A ductile, fibrous fracture in steel. Photograph; magnification, 1×. (b) A fracture similar to (a) but viewed at a higher magnification in a scanning electron microscope. The fracture surface is composed of contiguous dimples each centered on an inclusion that was present in the alloy. Scanning electron micrograph; magnification, 500×.

(c) A brittle cleavage fracture in steel. Photograph; magnification, 1×. (d) A fracture similar to (c) but viewed at a higher magnification in a scanning electron microscope. The fracture surface is composed of flat facets with steps between them. This type of fracture is sometimes called quasicleavage because of these steps. Pure cleavage would occur on an atomically flat plane. Scanning electron micrograph; magnification, 350×.

Fig. 3.4. Appearance of representative fracture surfaces obtained in metals stressed in tension.

split, or cleave, apart on a particular crystal plane (F in Fig. 3.1). The cleavage spreads from grain to grain, a continuous fracture path is soon developed, and the material parts (G in Fig. 3.1; Fig. 3.4c). Very brittle materials cleave across single crystal planes, absorbing comparatively little energy in the process. But in most metals the fracture jumps occasionally from one plane to an adjoining parallel plane. This introduces steps into the fracture surface, steps which have to be dragged along with the main cleavage fracture. These steps are, for example, indicated by the radiating lines in Fig. 3.4(d). This is sometimes called *quasicleavage* and absorbs a little more energy during propagation than pure cleavage. So, in practice, some energy is required to propagate cleavage-type fractures in metals, but never nearly as much as for ductile fractures.

The tendency for fracture to occur by cleavage is determined by the crystal structure of the metal and the temperature. Metals which have hexagonal or more complex crystal structures typically fracture by cleavage below a certain temperature which usually is well above room temperature. Those with body-centered cubic structures may have the same temperature-sensitive characteristics, the transition temperature varying with their alloy content and metallurgical condition. Mixtures of the two types of fracture may also occur. Iron is an important example of a body-centered cubic metal in which this range of behaviors is possible. Metals with a face-centered cubic crystal structure always fracture by a ductile mode irrespective of the temperature.

Metal structures do not behave in the ductile manner that we have come to expect of them when cleavage fracture intervenes, as we shall discuss later in this chapter. Moreover, tensile strength as measured in a simple tensile test is likely to be lowered. For example, the tensile strength measured in the test illustrated in Fig. 3.2 would be lowered if brittle fracture intervened at B than if the material had continued to deform and had fractured in a ductile mode at D. Intermediate cases between these two extremes are also possible when mixtures of brittle and ductile fracture occur.

Another basic parameter which might affect the values of tensile strength, measured as above, for materials which have similar yield strengths is the rate of strain hardening. The sketches in Fig. 3.5(a) illustrate that alloy A, which has a higher rate of strain hardening than alloy B, has a higher tensile strength even though the yield strengths of the two alloys are the same. A proviso is that premature failure by some other mechanism does not intervene. The warning is clear that the ratings of alloys by tensile strength may not be truly indicative of their ratings by yield strength.

Tensile strength is, in fact, commonly used to rate the "strengths" of materials (typical values for some metals and for a few nonmetallic structural materials are listed in Table 3.1). But it has severe limitations in this role which need to be kept in mind. As we have just seen, it is not necessarily a reliable indicator of yield strength, which is a property of greater fundamental significance. It is not necessarily an indicator of the stress sustained at fracture either, because it is calculated on the basis of the original section area of the test piece and not that immediately prior to fracture. Moreover, very rarely indeed do structures fail in service by being torn apart as a test piece is in a tensile test. As we shall see later, service failures commonly involve phenomena such as fatigue, creep, and corrosion, and the ratings for resistance to failure by these modes typically are quite different from those assessed by simple tensile tests. We shall soon see, moreover, that even the static stress that a structure can withstand frequently is determined by the behavior of the material in the presence of a notch, a condition which is not duplicated at all in a standard tensile test. Admittedly, there are general relationships between tensile strength and some of the properties that are important in service (see, for example, Fig. 4.4, p. 109), but they are empirical and of limited applicability at best. Tensile strengths sort strong sheep from weak goats but do not grade various types of either sheep or goats very well.

On the other hand, measurements of tensile strength undoubtedly have a place in quality control. A user has cause for concern if a particular batch of an alloy which is required to have a certain composition and to have received a certain heat treatment does not have a tensile strength within the range that experience has indicated that it should. The advantage of this parameter is that it is one of the simplest mechanical properties to measure. Only the cross-sectional area of the gage length of the test piece and the load that the test piece sustains at fracture have to be measured.

Ductility

Two other pieces of information can be obtained after a tensile-test piece has been fractured and the pieces reassembled in contact:

1. The elongation that has occurred along a gage length which was marked on the virgin test piece, measured as a percentage ($\Delta l/l \times 100\%$ in Fig. 3.1). This is called the *elongation* (Fig. 3.2) or, more strictly, the *elongation to fracture*.
2. The ratio of the cross-sectional area of the fracture surface to that of the original gage length, expressed as a percentage. This value is called the *reduction in area*.

These values are taken to be measures of the ductility of the metal.

"Ductility" is a general term used to describe the ease with which a metal can be deformed plastically without cracking or fracturing. We shall see throughout this book that many — perhaps too many — of the terms used in physical metallurgy have acquired a variety of meanings, but probably no term is more slippery than "ductility". Different engineers see ductility in different ways depending on what they require of a metal. To a structural or mechanical engineer, ductility is taken to be the opposite of brittleness and to mean the ability of the metal to yield a little in a structure rather than to fail catastrophically. It is true that metals which have large elongation values in tensile tests are generally more forgiving in this way than those which do not, but a more realistic assessment of their likely behavior has to be based on the concept of *toughness* (p. 83). To a process metallurgist, on the other hand, ductility is taken to indicate the ease with which a metal can be shaped or formed without fracturing or cracking. But a multitude of factors determine this, not the least of

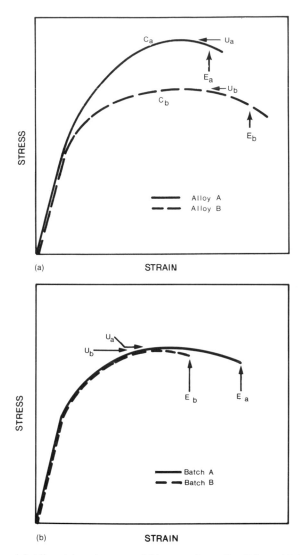

(a) Alloy A has the same yield strength as alloy B but a higher rate of work hardening, as signified by the steeper slope of the stress-strain curve for alloy A in the section labeled B-C in Fig. 3.2. The result is a higher value of tensile strength (cf. U_a and U_b) and a lower value of elongation to fracture (cf. E_a and E_b). (b) The materials represented here are two batches of the same alloy. They have the same yield strength and the same rate of work hardening, but batch B contains more inclusions than batch A. Consequently, the test piece from batch B starts to neck down earlier and fractures earlier than that from batch A. The elongation value is reduced (cf. E_a and E_b), but the tensile strength is not greatly affected (cf. U_a and U_b).

Fig. 3.5. Diagrammatic representations of the effects of several materials parameters on stress-strain relationships. (*Continued on facing page*.)

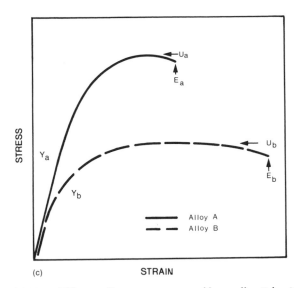

(c) Two different alloys are represented here, alloy B having a lower yield stress (cf. Y_a and Y_b) and a lower rate of strain hardening than alloy A. Alloy B consequently has a lower tensile strength (cf. U_a and U_b) but a higher elongation value (cf. E_a and E_b) than alloy A.

Fig. 3.5 (continued).

which is the nature of the forming process involved. So the relevance of information obtained in a tensile test depends on how closely the conditions of the forming operations are duplicated in the test. Finally, a quality-control engineer uses ductility as an indicator of the quality of a batch of material, but may not obtain a clear indication of what quality factor is involved. So we need to recognize the factors which determine the "ductility" values obtained in a tensile test.

Consideration of the series of sketches in Fig. 3.1 will make you realize that the measured elongation value is made up of two parts. The first is the uniform elongation which occurs before necking starts (Fig 3.1A and B) and which is intrinsic to the alloy concerned. The second is the elongation which occurs during the development of the neck and of the subsequent final fracture (Fig. 3.1C to E), and which is determined by a number of other factors, some of which are directly related to the properties of the material and some of which are not.

One of the most important of these factors is the presence of phases or inclusions which initiate the voids discussed in connection with Fig. 3.1. Some of them may be characteristics of the alloy but some may be present more or less accidentally. For example, pieces of slag and refractory and other extraneous matter almost always get into commercially made melts of metal, and some of them remain trapped in the metal when it solidifies (see Chapter 8). These particles are known as *nonmetallic inclusions*, and their volume, size, shape, and distribution often vary considerably with the skill and care exercised, and the costs incurred, during the preparation of the metal.

The influence of these phases and inclusions is illustrated schematically in Fig. 3.5(b). Stress-strain curves are sketched for test pieces taken from two batches (A and B) of an alloy. The two batches are supposed to have different inclusion contents (B more than A) but to be identical otherwise. The test piece from batch B commenced to neck down significantly earlier than that from Batch A, as indicated by the earlier drop in nominal stress. This occurred because the void sheet developed in the neck earlier and more rapidly. Fracture also occurred earlier for the same reason, so there was less elongation in the neck, which is manifested by the smaller elongation between the points of maximum stress and final fracture. The net result is that E_b is smaller than E_a, in this instance solely because of an incidental difference in inclusion content. In other instances, however, the difference could be due to the nature of the phases present or to differences in the volume fractions of the phases present, differences arranged deliberately by varying the composition of the alloy. Only further investigation would sort out which.

Note, incidentally, that the effect illustrated in Fig. 3.5(b) has not greatly affected the value of the tensile strength. This is because the stress-strain curve was fairly flat in the region where the neck developed, a phenomenon typical of metals.

Another factor which has a significant effect on the elongation value is the geometry of the gage length of the test piece. For example, a longer test piece of the same cross section would exhibit a smaller elongation value because the proportionate contribution of the necking elongation would be smaller. Engineers and metallurgists consequently have agreed on test-piece gage-length geometries that make it possible for the results of their tests to be compared reliably. These test-piece geometries have been published by various national standards institutions.

In spite of all this, the elongation values measured in a tensile test do contain a component which is indicative of the ease with which a metal can be expected to deform. Some alloys give much larger elongation values than others, even allowing for the effects that we have just been discussing, and can fairly be said to have greater ductility. For example, alloy A whose characteristics are illustrated schematically in Fig. 3.5(c) undoubtedly is intrinsically more ductile than the stronger alloy B. At the other extreme, very brittle alloys (i.e., alloys which fracture close to B in Fig. 3.2) have small and perhaps even unmeasurable elongation values.

One intrinsic factor affecting the elongation value that we can easily understand here is that of strain hardening. Suppose that an alloy has in the annealed condition the stress-strain curve in tension represented by the continu-

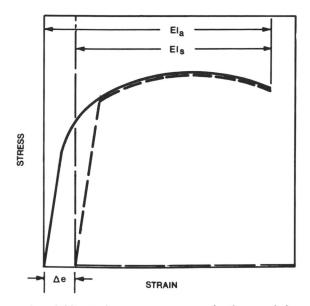

The solid line is the stress-strain curve for the annealed material. The broken line is for the same material which has been subjected to a strain Δe and then tested. The elongation value is then $E\ell_s$ instead of $E\ell_a$. The stress axis for the strain-hardened material has been displaced to the right in this sketch to make the point clearer.

Fig. 3.6. Diagrammatic representation of the effect of prior strain hardening on the elongation value obtained in a tensile test.

ous line in Fig. 3.6. Then suppose that the material was deformed plastically in tension by the strain Δe. For the reasons outlined in the discussion of Fig. 3.3, the stress-strain curve of the deformed material would be that of the broken line sketched in Fig. 3.6. The elongation now has the smaller value of $E\ell_s$ instead of $E\ell_a$. This would also have been so if the material had been plastically strained in compression (see below). Strain hardening reduces ductility in the sense that it reduces the capacity of the material to be deformed further in tension. But it does not necessarily reduce its ability to be deformed further in compression.

Much the same things as those mentioned above can be said about the reduction-in-area values measured in a tensile test, although the necking process has a more dominant influence on the value measured. Consequently, there are circumstances where reduction in area is used to give a better indication of likely performance in practice than could be obtained from elongation data. Nevertheless, both values have to be interpreted with caution; the circumstances under which they are to be used have to be related to those under which they were determined.

BEHAVIOR IN COMPRESSION

A compressive stress is a normal stress in which the force is applied to the action plane in a direction opposite to that for a tensile stress. It squeezes instead of stretches. By convention, it is called a *positive normal stress*. The behavior of a metal in compression at low stresses is the same as its behavior in tension, in that the material first undergoes linear elastic deformation with the same Young's modulus as for tension,* and then deforms plastically at the same yield stress as for tension. The rate of work hardening during the early stages of plastic deformation is also the same. The maximum strain that can be achieved before fracture occurs is different and larger, however, because the mode of the failure that terminates plastic deformation is different. The failure mode now depends largely on the geometry of the structure or test piece.

A test piece of a plastic material which is thin in the compression direction in comparison with its lateral dimensions can be strained large amounts without fracturing, much larger amounts than for tension because the deformation instabilities that cause necking down and so initiate fracture in tension do not develop. Strains in excess of 99% reduction in thickness are possible, and can be achieved in practical operations such as cold rolling. When the length of the test piece is about the same as its width, failure typically occurs by shearing along a plane aligned at approximately 45° to the compression direction. In ductile materials, this occurs by the same void-coalescence mechanism as for tensile fracture. The dimples on the fracture surface are elliptical, the long axes of the ellipses being in the direction in which shear tearing occurred. The strain to fracture typically is larger than for tensile straining. In brittle materials, shearing occurs in a cleavage mode, usually after a much smaller strain.

Slender columns (length more than five to ten times the thickness) when loaded in compression tend to buckle elastically, after which the distorted structure crumples and crushes. Resistance to elastic buckling then becomes the stress-limiting factor, and it is determined by the cross-sectional dimensions of the column as well as the elastic modulus of the material of construction. The relationship is described by a number of basic theorems of structural analysis, but, roughly, thicker sections are much stiffer than thinner ones, by a power relationship of as much as three or four, depending on the geometrical shape of the section. So we now have a situation where materials with a low mass density do have an advantage not indicated by the earlier discussion of simple tensile or compressive stressing. Remember that we indicated there (p. 66) that nature has decreed that the advantage of low density is about counterbalanced by reduced elastic modulus, all metals having about the same specific elastic

---

*There are a few exceptions with materials which contain internal discontinuities. For example, gray irons contain flakes of graphite, and the cavities in which they are contained close up when a compressive load is first applied, resulting in anomalous elastic behavior.

Table 3.2. Comparison of the Weights of Rectangular Beams of Equal Width and Thickness

Metal	Relative Weight
Beryllium	2.2
Iron	10
Copper	13
Titanium	7.0
Aluminum	4.9
Magnesium	3.8
Nickel	11

modulus. But thin columns in compression can be made stiffer by increasing their thickness with less sacrifice in weight for materials with lower density because the advantage accrues faster than the loss in elastic modulus (Table 3.2). Many engineering structures contain long thin elements which are loaded in compression (thin struts and columns and panels and plates), and the behavior of these elements may set the limit to the behavior of the structure as a whole. Beams are an important example, one surface being stressed in tension and the other in compression, and many structures contain beams. An aircraft wing is in effect a box-shaped beam with thin upper-surface panels loaded in compression in flight. Herein lies the basic advantage of using low-density materials in their construction.

BEHAVIOR IN SHEAR

That a structure can be strained in either tension or compression is pretty obvious, but there is a third basic method of straining which is perhaps not quite so evident. This is straining by *shear,* which occurs when a test piece or an element of a structure is subjected to opposing forces that are not in alignment, which is in fact the general case in practice. This is illustrated diagrammatically in Fig. 3.7, which also indicates the type of strain distortion that occurs.

The shear stress is calculated, as before, as the force per unit area over which it acts, this time the force being parallel to that plane. Shearing strain is best characterized, however, by a change in angle (θ in Fig. 3.7). The shear angle usually is expressed in radians* because this is a simple number; strain, then, as for tension and compression, has no units. Shear stress varies with shear strain in much the same way as for tension and compression, with both elastic and plastic regions occurring. However, the value of the elastic modulus is different, and this time is called the *shear modulus* or the *modulus of rigidity.* This

---

*A radian is the angle at the center of a circle subtended by an arc whose length is equal to the radius. It is about 57°.

UNDEFORMED   SHEARED

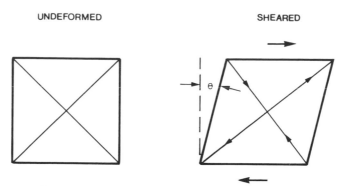

A shear strain applied in the direction indicated by the arrows distorts a square into a parallelogram. The distortion angle (θ), in radians, is used as a measure of the shear strain. Note that one diagonal of the square has lengthened and the other has shortened. Thus a shear strain can be resolved into a tensile and a compressive strain, respectively, in these directions, and vice versa.

Fig. 3.7. Sketches illustrating the concept of shear strain.

modulus is again structure-insensitive. The mode of failure is also different in shear, being determined by the tensile component of strain (see below). Hence the terminal stages of the stress-strain curve are different from those for tension alone.

A little further thought makes it apparent that a shear strain can be regarded as a tension or a compression strain or both, acting at 45°. For example, one diagonal of the sheared block sketched in Fig. 3.7 is extended and the other one is compressed. It follows in reverse that a shear stress acts at 45° to every tension and compression stress. The stress in a volume element of a structure is usually most conveniently analyzed in terms of normal stresses (tension or compression) and a shear stress. We have, in fact, noted earlier that the fracture of a test piece stressed in tension or compression frequently involves macroscopic shear on planes aligned at close to 45° to the stress axis. This is the plane of maximum shear stress. Moreover, we shall see later (p. 287) that all plastic deformations occur by shearing on a microscopic scale.

HARDNESS

"Hardness" is another one of those vague terms that have different meanings to different people and in different contexts. This term can be used to mean high strength (hard) as opposed to low strength (soft). It can mean resistance to scratching. It can mean resistance to wear. We shall consider here the context in which it means resistance to indentation by a harder substance in what is in effect a compression test. Hardness is then a measure of mechanical properties.

Hardness is determined quantitatively in tests in which a blunt indenter of known geometry and made from a considerably harder material than the test material (often it is made of diamond) is pressed into the surface of the test material under a known load. The load is made large enough to produce a permanent impression, or indentation. Some dimension of this impression is measured after the load is removed, and from this a *hardness number* is deduced by a standard procedure. The principles of two common types of indentation hardness tests are illustrated in Fig. 3.8.

The *Brinell* test uses a spherical indenter, and the *Vickers* test uses a pyrami-

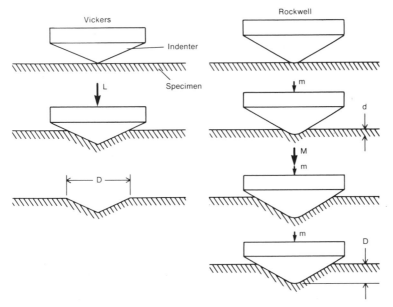

In the Vickers test, a pyramidal indenter is forced into the surface of the specimen by applying a load (L). The load is removed after a prescribed time, the indenter is withdrawn from the permanent indentation that has been made in the surface, and the diagonal (D) of the indentation is measured; from this measurement the surface area of the indentation is calculated. The Vickers hardness number is defined as the applied load divided by this area, so this hardness value is in units of force. The Brinell test is carried out similarly, except that a spherical indenter is used.

In the Rockwell test, a conical indenter with a spherical tip is first embedded into the surface of the specimen by applying a minor load (m). An additional major load (M) is then applied, and removed after a prescribed time. The depth (D – d) is measured and is related to an empirical scale of numbers. Different values of M have to be associated with different hardness scales, as do indenters of other shapes.

Fig. 3.8. Sketches illustrating the principles of two common methods of carrying out an indentation hardness test.

dal indenter the shape of which approximates that portion of a sphere which is active in the Brinell test. In both, the projected dimensions of the indentation are measured (Fig. 3.8), and a hardness number* is calculated by a standard procedure. This number is effectively the compressive stress that was needed to plastically deform the material a sufficient amount to produce the indentation. To a first approximation, therefore, the hardness number is a measure of the yield stress of the material in compression. Note, however, that a volume of material beneath the indentation is strain hardened, so that to a better approximation the hardness number is a measure of the yield stress of the material after it has been strain hardened a little. Experience indicates that a strain of 8% compression is about the right equivalent to use. So the hardness number is, if you like, a measure of the 8% yield strength in compression. Strains of this magnitude can be achieved even in quite brittle materials because the volume of material compressed is highly constrained by the rest of the workpiece. Thus some information can be obtained easily about the mechanical properties of materials that could scarcely be tested in any other way.

The other indentation test commonly used is the *Rockwell* test,† which employs a blunt cone with a spherical point as the indenter and measures the depth of the indentation after part of the indenting load has been removed (Fig. 3.8). This test is much easier to carry out than a Brinell or a Vickers test and consequently is popular, particularly for routine testing. The Rockwell hardness scale is, however, an arbitrary one, and several unrelated scales are required to cover the range of hardnesses that are encountered in metallurgical practice.

Hardness tests are all simpler to carry out than tensile tests, and the equipment required is less elaborate and less costly. As a result, much use is made of them. They are very useful in a comparative sense, but sometimes too much is made of them. Hardness numbers frequently are used to derive an estimate of the tensile strength of a metal, for which purpose conversion tables are available (Appendix 3, p. 485). Even ignoring the limited significance of tensile strength itself, it should by now be apparent that such a correlation will be valid only for materials which have the same relationship between points B and C in Fig. 3.2 (more strictly, between point B and the 8% yield stress). Conversion charts should therefore be used with caution, and only for the type of alloy for which they were empirically developed. Hardness values also are often used to assess wear resistance. This usage also needs to be approached with caution, because correlations between hardness and wear resistance are indirect at best.

---

*Referred to as HB for the Brinell test and as HV for the Vickers test.

†Rockwell hardness numbers are referred to as RC numbers when determined in a Rockwell C Scale test, RB numbers when determined in a Rockwell B Scale test, etc.

TOUGHNESS

Until about the middle of the present century, engineers and metallurgists relied almost entirely on information obtained from simple mechanical tests of the type that we have just been discussing to rate metals and their alloys. They played safe by attempting to ensure that the calculated design stress never exceeded some fraction of the yield strength or, more likely, the tensile strength. This fraction was called a *factor of safety* or, by the more cynical, a factor of ignorance. Indeed a factor of ignorance often was involved, in the sense that this factor attempted to allow for one important characteristic of the behavior of metals under tensile stresses which was not understood at the time.

The events that probably drew the most attention to this gap in knowledge were the failures of the "Liberty" ships that occurred during World War II. These ships were of a very basic design and were fabricated entirely by welding, itself something of an innovation at the time. They were produced in large numbers. Perhaps not too much was expected of these ships, but they were at least expected to hold together in one piece. So there was great consternation when some broke in two at sea, and even more consternation when one broke in two alongside its fitting-out wharf. All of this occurred in spite of checks that the design stresses were much smaller than the tensile and yield strengths of the plates used in the hulls, as they are normally determined. Actually, serious failures occurred in only some ten ships out of the several thousand that were built, but these failures were spectacular enough to stir the engineering world.

To make matters worse, once the existence of the problem was recognized it was soon realized that it was not confined to ships or to steel structures. Moreover, it was not new. A search through engineering records indicated that many major steel structures had failed suddenly and unexpectedly for as long as steel had been used in quantity. Many types of structures — bridges, pressure vessels (Fig. 3.9), pipe lines, and the structural skeletons of buildings, to name a few — had failed through the years. One of the more mind-boggling examples was a large tank containing molasses, the failure of which resulted in the drowning of a number of unfortunate bystanders. Moreover, structures continue to fail to this day, although, to the credit of engineers and metallurgists, with greatly reduced frequency. Most of these failures have occurred in steel structures, but some have also occurred in high-strength aluminum and titanium alloys. The failure of the cabin of the first commercial jet aircraft, the De Havilland "Comet", was one of these. In retrospect, the two features that made such failures more common and more spectacular for a period of time were the use of welding in place of riveting as the preferred method of fabrication and the use of alloys of increasingly higher strengths.

In any event, it became apparent that a factor of considerable ignorance

A brittle crack was initiated at a welding defect in the region indicated by the arrow and branched as it propagated through the main wall of the vessel. The vessel consequently broke up into a number of large fragments, some of which were propelled for a considerable distance. (Source: The Welding Institute, England.)

Fig. 3.9. A large welded pressure vessel which failed catastrophically during a hydraulic proof test.

was seriously inhibiting engineering developments. The hunt started in earnest and, after a few false starts, a critical mechanical property of materials was identified to which little attention had been paid in the past. This property — called *toughness* or, more specifically, *fracture toughness* — was defined qualitatively as the resistance to the propagation of cracks or similar sharp discontinuities. The current view of the importance of this property is well put by Gordon in his book *The New Science of Strong Materials*, listed under "Further Reading" at the end of this chapter:

> "The worst sin of an engineering material is not lack of strength or lack of stiffness, desirable as these properties are, but lack of toughness."

This view is based on the premise that there is nothing worse than for a structure to break suddenly and unexpectedly without having given the user due warning that disaster was impending. It is also based on an acceptance of the reality that a real structure must be assumed to contain cracklike discontinuities.

Something About Stress Concentrators

The question at issue is whether the cracklike discontinuities likely to be present in a structure will propagate further when service stresses are applied. In attempting to answer this question, it is necessary first to realize that the stress at the tip of such a discontinuity is much higher than the average stress in the surrounding regions of the structure. The higher stress is present throughout only a small local volume of material at the root of the discontinuity, but this may be all that is required to cause a crack to propagate. In fact, not even something as sharp as a crack is required. Any irregularity, such as a hole, a step, a notch, or a scratch, raises the stress locally by a factor much larger than you might imagine.

A formula that gives an adequate indication of the stress-concentration factor (C) at such an irregularity is:

$$C = 1 + 2\frac{L}{r} \qquad\qquad (\text{Eq. 4})$$

where L is the length of the crack, notch, or re-entrant feature, and r is the radius of its tip. This formula indicates that the stress at the edge of a round hole (r = L) is three times higher than the average stress, which would take care of most of a designer's factor of safety. A re-entrant with a sharp corner, in which case r would be much smaller than L, would concentrate the stress even more. The ultimate in sharp cracks would be one for which r was about the radius of an atom; in this instance the crack would have a stress-concentration factor approaching infinity.

Thus it does not require much of a stress concentrator to develop very large stresses in a small volume of material at the root of the concentrator. The stress at the root might then locally exceed the fracture stress of the material even when the average stress applied is quite small.

Fracture Energy

The wonder is, you might now conclude, that a metal (or any other material for that matter) holds together at all when a stress concentrator of any consequence is present. The reason why metals do hold together is that energy has to be supplied to form the new surfaces that arise when a crack increases its length. Consequently, a crack cannot drive itself through a structure unless an adequate supply of energy is available.

It was A. A. Griffith, an English aircraft engineer, who at a young age realized this in 1920. As with Sorby's metallographic innovations, described in Chapter 6, it took some 30 years for the significance of Griffith's work to be fully recognized, but this should not be blamed entirely on the conservatism of engineers, although they are conservative enough. Concepts of this nature have to fall into place, and perhaps have to be supported by several other supplementary concepts and to be needed to cope with pressing technological problems, before they are likely to be translated from scientific curiosities, however brilliant, to the body of widely used knowledge.

We now need to consider from whence comes the energy needed to drive a crack across a structure. Everyone realizes that mechanical energy is stored in a spiral spring when it is wound up, and that this energy can be released when required to drive things such as clocks and toys (at least not so long ago they were used for such purposes). What may not be so apparent is that this energy is stored because the material of the spring is elastically deformed when the spring is wound up. The ribbon of steel that constitutes the spiral spring is bent to a smaller radius when the spring is tightened, which means that some elements are extended. Energy is required to do this, just as it is required to extend the simple tensile-test piece discussed on pp. 61 to 68. Energy exists in the universe in many forms, but the type that we are dealing with here is known as mechanical energy. It is, for example, the type of energy that you have to expend when you lift a weight. Mechanical energy is defined simply as force multiplied by the distance through which the force acts. Several units of energy are in common usage to confuse us all: foot-pounds, ergs, and calories are in wide use, to mention just a few. The SI unit of energy, which we shall use, is the joule (J) (Appendix 3, p. 485). One joule is the amount of work done when a force of one newton acts through a distance of one metre.

The energy per unit volume of material (g) expended when the gage length of a simple tensile-test piece is extended is given by the area subtended by the appropriate portion of the stress-strain curve (Fig. 3.2). Thus:

$$g = 1/2(s \cdot e) \qquad\qquad (Eq. 5)$$

This energy is stored as *strain energy* in the material and is released when the stress is released, as happens when the material fractures.* In fact, the strain energy in a volume of material between a crack tip and the free surface of the material will be released when the crack extends. The energy released will thus be larger the longer the crack.

So we have a source of energy to drive a crack forward. We now need to know how much energy will be required to drive the crack; this amounts to how much energy is needed to form the new surfaces that are created in the process. It is implicit in this argument that surfaces of solids have energy associated with them. It is perhaps easiest to see that this is so by considering first a

---

*You will be well aware of this if you have seen a steel rope break. The strain energy is then released as kinetic energy, which causes the ends of the rope to whip about with considerable speed and danger.

liquid. We can imagine easily enough the surface-tension forces in a liquid. They are the forces that, for example, counterbalance the slight pressure of air within a bubble of water. We all know that this pressure has to be increased to increase the size of the bubble — that is, that work has to be done (energy has to be supplied) to increase the surface area of the liquid film of the bubble. The quantity known as the surface energy of the liquid is the energy that has to be supplied per unit increase in area. This can be measured quantitatively in a number of ways. This concept can be extended to a bubble inside a liquid, including a liquid metal. Thus the surface energy of a liquid metal can be measured in essentially the same way as for any other liquid. The values for metals turn out to be large — ten to twenty times those for the liquids that we normally encounter.

The next step is one that you will have to accept on faith, but only on the faith that physicists can make reasonable interpretations of their experiments which indicate that the atomic structure of a liquid does not change much when the liquid solidifies (see p. 250). From this, it is reasonable to conclude that the surface energy of a solid is not very different from that which the material had when it was molten. This, as we have already seen, can be measured.

So we have reached the point where we know the energy required to increase the length of a crack, we know how much energy is available, and we know that the amount of this energy required per unit increase in crack length will be larger the greater the initial length of the crack. Remember that our basic premise is that, if the crack is to propagate, the energy released when the crack increases its area must exceed the energy needed to create two surfaces of the same area. This follows no matter how high the stress at the crack tip. Griffith saw that it also follows that this energy balance will not be met until the crack exceeds a certain critical length (a_c). Cracks shorter than the critical length will stay put, no matter how sharp they are and how effective they are as stress concentrators, because not enough energy would be released should they lengthen. They are not a cause for concern no matter how fearsome they may appear to be. Cracks longer than the critical length, however, will take off unstably, and are a cause for great concern no matter how innocuous they may appear to be. This is known as the Griffith criterion for fracture.

The sixty-four dollar question is the value of the critical crack length in a particular material under a particular applied stress. It will be difficult, if not impossible, to build a safe structure that has to operate in tension if the critical length is small. The risk is then high that a crack could be present whose length is greater than the critical length, and the probability of detecting such a crack, even with good inspection methods, will be small. On the other hand, there can be confidence in the safety of a structure when the critical length is large, because a crack approaching critical length is not likely to be present and the probability of detecting it is high.

Griffith was able to devise the following simple formula for estimating the critical length:

$$a_c = \frac{2GE}{\pi s^2} \qquad \text{(Eq. 6)}$$

where a_c is critical length in metres, G is surface energy in J/m^2, E is Young's modulus, and s is nominal local stress in N/m^2. However, this formula indicates that a_c would be very small in all materials. The value would be about 1 μm (40 μin.) for most materials, including metals. A crack of this length would not be visible to the unaided eye. In fact, quite sophisticated methods of detection would be necessary to reveal it. All structures are likely to contain defects larger than this critical size, which implies that no material would be safe to stress in tension and that metals would not be much better in this respect than, say, glass. They are in fact a lot better.

The fallacy arises from the assumption that the only energy required to produce a fracture is that required to make the new surfaces. This assumption implies that the material behaves purely elastically right up to fracture and that no energy is absorbed in the fracturing process. We have seen earlier that this is not the case for metals (see p. 68) and that, moreover, the energy absorbed varies over quite a wide range depending on the metal and the mode of failure.* Thus, we have to modify Eq. 6 to:

$$a_c = \frac{2W_f E}{\pi s^2} \qquad \text{(Eq. 7)}$$

where W_f is the *work of fracture*, still in units of J/m^2.

The difference between W_f and G is the factor that distinguishes between what empirical experience classifies as "brittle" and "tough" materials. W_f is only five to ten times larger than G in a material such as glass. The critical crack size then is still only 5 to 10 μm (200 to 400 μin.) at the stress at which ordinary glass fractures in tension. Although rather special techniques are required, numerous cracks of about this size can be detected in the surface of commercial glasses, cracks which are the result of accidental and unavoidable damage caused by, for example, small pieces of dust rubbing across the surface. The strength of glass is reduced even further if a larger crack is introduced into the surface, which is what a glass cutter does. On the other hand, glass rods which are completely free from surface cracks can be prepared by very careful laboratory procedures. The tensile strength is then very high — perhaps a thousand times larger than normal. But one has only to breathe on the surface of

---

*In fact, the area underneath the stress-strain curve of a tensile test (see Fig. 3.2) is a measure of the work that has to be done to fracture the metal.

such a rod to introduce cracks that return its tensile strength to normal values. The same can be said of all those structural materials that experience has taught engineers not to use in tension. Their small critical crack length, a consequence of their poor toughness, is the reason why they cannot be used safely.

On the other hand, W_f may be between ten thousand and one million times higher than G for a ductile metal. This is because, as we saw earlier (p. 68), the propagation of fracture in highly ductile metals requires that many small volumes of material be extended and necked down, which requires that much work be done. The critical crack length is then hundreds of millimeters, perhaps even a metre or so, at the stresses that these metals are expected to withstand. This crack length far exceeds the length of any flaw that is likely to be introduced during manufacture. It also characterizes a flaw that can be expected easily to be detected by available methods of nondestructive inspection (p. 203). A structure which has a critical crack length of this order of magnitude can fairly be said to be a safe structure, and the material to be a safe structural material. *Herein lies the key to the reason for the dominance of metals as structural materials which have to be stressed in tension.*

However, we also mentioned earlier (p. 70) that some metals, including irons and steels, do not always fracture in this highly energy-absorbing ductile manner. Some fail in a quasicleavage manner which requires a considerably smaller amount of work to be done during fracture, the amount varying with the number of irregularities developed during the process of cleavage. Fractures may also occur which involve different amounts of the two basic fracture modes. Thus some metals can be expected to be much tougher than others, and the toughness of some can be expected to vary over a range. Moreover, it follows that "brittleness" and "toughness" are not absolute concepts but have to be regarded as describing vaguely the two ends of a spectrum of behaviors. A relatively brittle material is one which has a comparatively small work of fracture and hence a short critical crack length even at small stresses. A relatively tough material is one which has a comparatively large work of fracture and hence a long critical crack length even at high stresses. Clearly, an engineer needs to be able to quantify these concepts before he can usefully apply them to design or to the selection of structural materials.

Unstable Propagation of Cracks

However, it is worth considering first what happens when a crack of critical length does start to propagate. We saw in Eq. 4 that the stress-concentration factor at the tip of a crack is determined by the ratio of the length of the crack to the radius of its tip. In most materials, the tip radius remains fairly constant, so the stress-concentration factor increases when the crack lengthens. This makes matters worse, because a crack once started continues to propagate and at increasing speed. It in fact soon reaches the theoretical maximum speed, which is about one-third of the speed of sound in the material concerned (the speed of sound in steel is about 5,000 m/s, or 11,000 mph). The running crack also emits elastic waves which are propagated through the material at the speed of sound and which are likely to be reflected in a complex manner from various discontinuities in the structure. These reflected waves may arrive back in time to interfere with the propagating crack tip, causing it to branch and rebranch. A structure consequently is likely to break up into many fragments, as was the case for the structure illustrated in Fig. 3.9.

In any event, the crack will continue to run until either a major discontinuity is reached (e.g., the edge of a plate) or the driving stress is relieved. For the former reason, crack-stopping discontinuities may be deliberately built into a structure. They may also, advertently or inadvertently, be built into the structure of the material. Laminations in a plate, for example, tend to hold up the progression of cracks running in a direction perpendicular to the laminations. The pattern welding used in Roman swords (Fig. 1.2) inadvertently served this purpose.

The absence of crack-stopping discontinuities is the reason why integral all-welded structures are more likely to fail completely and spectacularly than riveted structures. Thus the cracks in the Liberty ships, once started, could progress right around the hull unless they ran into something such as a porthole. Riveted structures are not proof against the propagation of running cracks, but the cracks usually stop at the edge of the plate in which they were initiated at a riveted joint. They usually do so, for example, in riveted ships. This, incidentally, is no reason for abandoning welding as a method of construction. Its advantages are too great. It is reason, however, for ensuring that the materials used in welded structures are tough enough to prevent this type of failure from occurring.

Measuring Fracture Toughness

Now we will return to the need to quantify fracture toughness. It has been difficult to devise quantitative measures of fracture toughness which can be used to establish sound design and inspection procedures. The task is still incomplete, particularly for alloys with high toughness. A range of tests has had to be developed to cover the range of toughness and design requirements met with in practice.

The simplest types of tests can be applied to materials for which the plastic component of fracture is comparatively small. These will be materials, such as high-strength steels, that fracture soon after point B in Fig. 3.2 is passed in a simple tensile test. Putting it another way, they will be materials that deform essentially in a linearly elastic manner. The analysis of this situation is called *linear elastic fracture mechanics* (LEFM). The Griffith criterion can, under these circumstances, be shown to be equivalent to an instability which occurs when a critical stress distribution is developed. A stress criterion of this nature is more useful to an engineer than the Griffith energy approach, because engineers are used to calculating stresses. In the most important case of fracture by simple tearing, the stress at the crack tip can be described by a simple *stress-intensity factor*, K^*:

$$K = s(\pi a)^{1/2} \qquad \text{(Eq. 8)}$$

where s is the gross stress applied to the body, as in Eq. 1, and a is the crack length. The crack becomes unstable when a critical stress intensity (K_c) is reached ahead of the crack:

$$K_c = Y \cdot s_f \cdot a_c^{1/2} \qquad \text{(Eq. 9)}$$

where s_f and a_c are the fracture stress and critical crack size, respectively, and Y is a constant of the component geometry and the loading system. The critical stress-intensity factor (K_c) is a quantitative measure of toughness for the particular mode of failure. Those in the fracture mechanics world quote K_c in units of either $MPa \cdot m^{1/2}$ or $ksi \sqrt{in.}$, which are strange units to the uninitiated but useful to the initiated.

One type of specimen, known as a compact tension specimen, which is used to determine K_c is illustrated in Fig. 3.10(a). The specimen is machined to standardized dimensions, particular care being taken with the preparation of the notch. It is then subjected to cyclic loading to develop a sharp fatigue crack at the root of the notch. The length of the fatigue crack is measured, and a tensile stress is applied in the direction indicated and increased at a controlled high rate until unstable fracture is initiated. A value of s_f for a particular value of a is thereby measured. In turn, values of K_c and of a_c for other values of s can be calculated by using Eq. 9. This test is obviously much more difficult and expensive to perform than the simple tensile tests described earlier.

A different approach is necessary, however, for materials in which significant amounts of plastic deformation occur before fracture. This is so for most medium- and low-strength alloys. The approach then necessary is called *general yielding fracture mechanics* (GYFM). The aim of GYFM is to provide an alternative measurement to K_c which still identifies the onset of unstable fracture. This parameter needs to be relatable quantitatively to K_c so that use can still be made of the quantitative analyses that are possible by applying LEFM (e.g., Eq. 9). One approach that has been developed involves measuring the displacement of the faces of the tip of the crack in a specimen such as that illustrated in Fig. 3.10(a). The displacement at the onset of unstable fracture is called the *crack-*

---

*Strictly, this applies only to a tearing type of fracture which, by convention, is known as mode I fracture. The stress-intensity factor is then known as K_{I_c}.

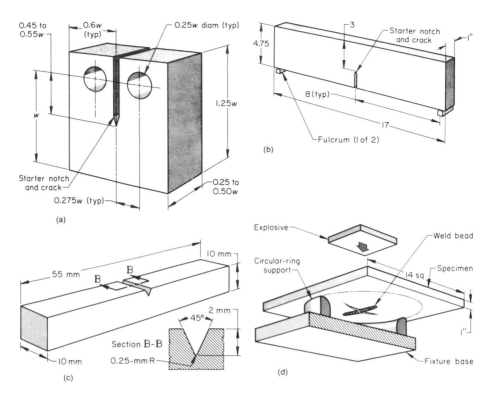

(a) Compact tension specimen for determining plane-strain fracture toughness with a tearing mode of failure. A fatigue crack is produced at the bottom of the machined notch, and the specimen is then fractured in tension by applying a load (horizontally here) through the loading holes. (b) Specimen used for a dynamic tear test. A crack with a sharp tip is produced in the specimen by making a brittle electron-beam weld at the base of the notch or by pressing it with a knife edge. The specimen is fractured in bending with the notch in tension. (c) Charpy V-notch specimen, one of a range of similar specimens commonly used to evaluate notch toughness in a simple test. The specimen is fractured by impact, and the energy required to produce failure is thus determined. (d) Explosion bulge method of determining nil-ductility temperature.

From *Metals Handbook*, 8th Ed., Vol 10, American Society for Metals, Metals Park, OH, 1975.

Fig. 3.10. Examples of the types of specimens used to determine quantitative values of fracture toughness for metals. (Dimensions are in inches except where otherwise indicated.)

opening displacement (COD). The COD value, d, can be related to the stress-intensity factor, K, by:

$$ds_f = \text{constant} \left(\frac{K^2}{E} \right) \qquad \text{(Eq. 10)}$$

Uncertainties arise in the experimental determination of all of these values, as well as in ascribing a value to the constant (its value is, however, not too far from unity). Nevertheless, this approach has proved to be a useful one, particularly for assessing weldable grades of structural steel.

Although complex and expensive tests of the types that we have just described are needed to obtain basic information on fracture toughness, they are not always necessary for the practical application of the concepts. A number of simple and comparatively inexpensive tests have been developed which are effective in all stages of design, quality control, and inspection. Certainly, these tests have to be calibrated in terms of basic fracture-toughness parameters, and the calibration in each case has only a limited range of applicability. Their limitations have to be recognized and understood so that a test appropriate to the application can be chosen and then applied correctly. They can, nevertheless, be applied very effectively.

These simpler tests include impact tests carried out on notched bars (Fig. 3.10c), the impact energy required to fracture the bar at a known temperature being measured (the Charpy test). Another test (the Dynamic Tear Test) uses a plate specimen which has a notch in one edge and which is bent at a high strain rate with the notched edge in tension (Fig. 3.10b); the load required to cause catastrophic failure at a particular temperature is measured. Another test uses an as-received plate on the surface of which a bead of brittle weld metal has been deposited, with the weld metal being notched for good measure. Explosive charges are detonated against the plate to produce a bulge with the notch at the tension surface (Fig. 3.10d). The lowest temperature at which the flaw does not propagate into brittle catastrophic failure is measured (the Nil-Ductility Transition Test). Each of these tests may have to be carried out at several temperatures to obtain a guide to the temperature sensitivity of the toughness of the material. Each has a particular field of usefulness, depending on the end requirement. The explosion bulge test, for example, is of particular interest when testing steel plate for the hulls of naval vessels because it simulates the underwater explosions to which they may be subjected in service.

Crack Growth Before Unstable Fracture

Structures may contain either cracks or other stress concentrators that initially are much smaller than the critical size. However, these stress concentrators may grow to critical size if they are subjected to cyclic stresses in service by a process known as fatigue (see p. 104) or by other mechanisms such as stress corrosion (see p. 158). A major factor determining the safe life of the structure then is the rate of growth of such a fatigue crack, which may be determined by material parameters other than those which determine fracture toughness itself. It may therefore be a second parameter that has to be considered when assessing the resistance of a structure to catastrophic failure.

The interplay between these two factors is well illustrated by modern gun tubes, which are manufactured from steels of a higher strength than those formerly used. A gun tube is subjected to a large stress cycle each time the gun is fired. Fatigue cracks may be initiated at the bore surface and grow radially through the tube wall and longitudinally along the tube as increasing numbers of rounds are fired. The rate of growth of the cracks determines the number of rounds that can be fired before failure occurs by one of two possible modes. The first possibility, if the fracture toughness of the steel is high enough, is that the crack extends completely through the wall of the tube before it reaches critical size. The gun will then no longer function, but the consequences will be no more serious than that. This can be described as *fail safe*. If the steel has inadequate fracture toughness, on the other hand, the crack may reach critical size first. The tube will then fracture explosively when the next round is fired, usually with very serious consequences. Ships have been lost as a result.

The Fracture Toughness of Engineering Alloys

Very many of the metals and alloys commonly in use have more than adequate fracture toughness, having critical crack lengths approaching a metre, as mentioned earlier. With them, the possibility of unstable crack propagation scarcely arises. There are alloys, however, for which it does. They include some steels, aluminum alloys, and titanium alloys that have been developed to high strength levels. It is generally true that, for a particular type of alloy, fracture toughness decreases when strength is increased, but the correlation fortunately is not a strict one. Metallurgists have discovered ways of producing alloys which have good fracture toughness as well as high strength, but perhaps not beyond a certain limit.

For example, the original attempts to develop weldable steels with high strengths were based on increasing first the carbon content, and then both the carbon and manganese contents, of the steel. Virtually the only limitation was that the weldability should be preserved. Strength was certainly increased, but toughness was reduced drastically to the extent that the critical crack length was reduced to only a few millimetres at the stresses which the steels were now expected to withstand (curves 1 and 2 in Fig. 3.11). These steels were unsafe. Serious service failures occurred, and the steels had to be abandoned. New generations of steels were then developed in succession, taking into account the newly developed concepts concerning fracture toughness. The increase in

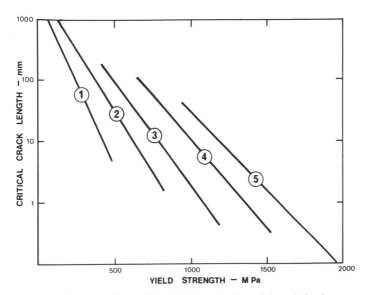

Toughness is indicated here by the critical crack length for fracture when the steel is stressed at two-thirds of its yield stress. Although represented here by discrete curves, the characteristics of the various types of steel shade into one another in reality. Generalized descriptions of these steels are as follows: 1 – plain carbon mild steels; 2 – carbon-manganese steels (pearlitic structures); 3 – special low-carbon alloys (fine grain size, advanced steelmaking practices, martensitic structures); 4 – special low-carbon alloys (martensitic structures, quenched and tempered); 5 – special low-carbon alloys (fine grain size, advanced steelmaking practices, martensitic structures). The meanings of terms such as "pearlitic" and "martensitic" will be discussed in Chapter 11.

Fig. 3.11. Variation with yield stress of the toughness of some successive generations of weldable steels.

strength was obtained by different methods. The carbon content was kept low, but other alloying elements, such as vanadium and niobium, were added in small amounts. The grain size was kept small and the steelmaking practices were improved. The steels were heat treated carefully after rolling. All of this added to costs, but steels were produced with greatly improved toughness for a given strength (curves 3 to 5, Fig. 3.11). The critical crack length at useful stresses was restored to tens of millimetres. But there is still a limit on the increase in strength that can be safely obtained (curve 5, Fig. 3.11).

The safety mark was also overstepped for awhile with ultrahigh-strength steels and also with high-strength aluminum alloys. For example, a military aircraft was built using a very high-strength steel in a critical wing component. This steel turned out to have a critical crack length of a millimetre or less at the required operating stresses and temperatures. A wing fell off in flight before the magnitude of the problem was fully realized. The component cannot be

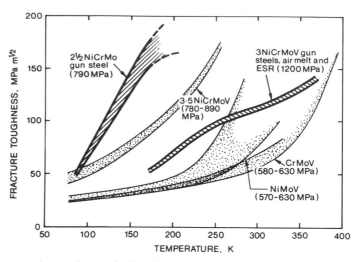

The toughness of all steels decreases with temperature. The decrease occurs over a narrow temperature range in four cases, and moreover over a temperature range above or close to ambient temperature in two cases. Note also that steels with approximately the same yield stress may have significantly different toughness-temperature characteristics. The characteristic curves of the tougher steels are located more toward the left in this diagram.

Fig. 3.12. Variation with temperature of the fracture toughness of five high-strength steels. The yield strength of each steel and the major alloying elements present are indicated.

replaced, certainly not with a steel of lower strength. Consequently, the only recourse open is to inspect the component regularly for these very small cracks. This aircraft has become something of an inspector's nightmare.

Substantial advances have also been made in improving the fracture toughness of these ultrahigh-strength steels (Fig. 3.12) and high-strength aluminum alloys. Adjustments have been made in composition, thermal and mechanical treatments, and processing methods, the objective usually being to produce finer and more homogeneous structures in the final product. Substantial improvements have also been made by ensuring that the rate of growth of cracks during cyclical loading is reduced (Fig. 3.13). Nevertheless, there is a barrier which has proved to be difficult to penetrate. Engineers have had to back off a little from using alloys as high in strength as those they were wont to employ a few years ago in order to obtain acceptable fracture toughness.

Several other problems arise with these commonly used steels. The first is that the fracture toughness decreases with temperature, often quite markedly so over a small temperature range (Fig. 3.12). There is a transition from quite tough behavior above a certain temperature range to quite brittle behavior below that temperature range. Worse still, the critical temperature range can

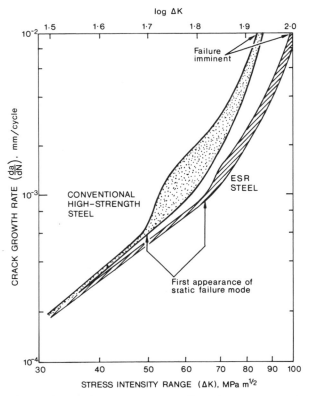

The crack-growth rate increases rapidly with an increase in the magnitude of the cyclical stress, but may differ considerably for different steels, even for those of similar yield strength. Improvements in the quality of the steel resulting from improved steelmaking practices have a major effect of this type. The electroslag refining (ESR) process is an example.

Fig. 3.13. Variation of the rate of growth of fatigue cracks with stress intensity for a number of high-strength steels.

be around ambient temperature, and can vary significantly from batch to batch. Consequently, a structure could be quite safe when the weather is warm but fail unexpectedly during a cold snap. A welded bridge that was, at the time, the pride of my city performed very well during its first summer, but then winter came. A girder fractured spectacularly when a heavy truck drove over the bridge early one cold morning. Not all of the girders broke, fortunately for the truck driver, because the batches of steel used were not equally susceptible to the brittle mode of failure. But many were found to be on the verge of failure, and expensive repairs to the whole structure were deemed necessary. A second requirement when high-strength steels are being developed, therefore, is that the brittle-to-ductile transition temperature should be below the likely range of operating temperatures. Considerable improvements have been achieved in this respect too, as the curves in Fig. 3.12 testify.

The toughness of these steels also typically is adversely affected by increases in strain rate, hence the emphasis in carrying out fracture-toughness tests at high strain rates. Ideally, a steel should be assessed at a strain rate at least equal to that to which it will be subjected in service.

The toughness of titanium alloys and some other metals is also temperature-sensitive. The fracture toughness of aluminum alloys, on the other hand, is not affected by either temperature or strain rate. The toughness of some special types of steels is not sensitive to temperature or strain rate either. The so-called precipitation-hardened austenitic steels (p. 382) are examples, but they are expensive and so their use is confined to special applications.

The susceptibility of fracture toughness to variation with temperature and strain rate is structure-sensitive. In the cases just mentioned, this susceptibility is sensitive to crystal structure (p. 224), but other structural factors can also be important. Grain size (p. 240) is particularly important, reduction of grain size being the one factor that improves both yield strength and fracture toughness simultaneously. The presence of brittle constituents, such as carbides in steels and the intermetallic constituents in aluminum alloys, adversely affects toughness. Nonmetallic inclusions can have similar effects. Reducing the volume content of these constituents, reducing the size of the particles in which they are present, changing the shape of the particles to a more rounded one, and improving the homogeneity with which they are distributed all improve fracture toughness. Some of these improvements can be achieved by adjustments in composition and heat treatment, but many have been achieved by improving processing methods and processing control. It follows that variations in toughness in a particular alloy are possible from source to source, and among different batches from a single source. It is common practice, therefore, to require certification of a manufacturing source, and to institute quality control on individual batches, when toughness is critically important.

Another consequence of the sensitivity of toughness to structure is that the toughness of an alloy may vary with the thickness of the product and be directional in that product. This occurs because the hot rolling schedule used in producing, say, a thick plate is different from that used for a thin plate. This affects adversely the inevitable variations in composition and structure throughout the plate (see p. 258) and so adversely affects toughness. Moreover, the toughness of thick plate in a direction across its thickness is likely to be inferior to that in a direction parallel to its surface. Structures made from very thick plates, such as those used in the platforms for deep-sea oil-drilling rigs, are just the ones that might be stressed in the through-thickness direction, and account must be taken of this when steels are selected for these critical welded structures. However, modern steelmaking practices which reduce segregation and control the numbers, sizes, and shapes of nonmetallic inclusions provide steels which can safely be used in such applications.

FURTHER READING

ASM Metals Reference Book, 2nd Ed., T. L. Gall, ed., American Society for Metals, Metals Park, OH, 1983.

J. E. Gordon, *The New Science of Strong Materials, or Why You Don't Fall through the Floor*, 2nd Ed., Pelican, New York, 1976.

J. E. Gordon, *Structures, or Why Things Don't Fall Down*, Pelican, New York, 1978.

G. E. Dieter, *Mechanical Metallurgy*, McGraw-Hill, New York, 1961.

D. Broek, *Elementary Fracture Mechanics*, Martinus Nijhoff, Netherlands, 1986.

· 4 ·

The Behavior of Metals
in Some Important
Mechanical Environments

SUMMARY

Metal components fail in service by a number of mechanisms involving the application of stresses. These are in addition to those discussed in the preceding chapter. The main ones are known as fatigue, creep, and wear.

Fatigue may occur when a tensile stress, which may be small compared with the tensile fracture strength, is applied repeatedly for a sufficiently large number of times. This situation arises frequently in, for example, rotating machinery. A crack starts at the surface of the component and grows until the section remaining can no longer support a single application of the stress. Sudden failure then occurs. Fatigue life (number of stress cycles withstood before fracture) is determined, for a given material, principally by the magnitude of the cycling stress, which may be concentrated considerably by irregularities at the surface of the component, such as notches, cracks, and even machining marks. It is also determined by several of the metallurgical characteristics of the material in a way which may not be directly related to their influence on yield or tensile strength. This is another factor which may limit the degree of strengthening that can usefully be achieved in a given alloy. Only modest success has been achieved by metallurgists in improving the resistance of alloys to the initiation of fatigue cracks, although considerable success has been achieved in reducing their rate of growth. The task of combating fatigue lies largely in the hands of the designer.

Creep occurs when a comparatively small stress is applied constantly for a long time at a high enough temperature, a set of circumstances which may exist in a component required to operate at elevated temperatures. The component gradually extends (when the stress is tensile), and cracks develop in the stressed region and grow until fracture eventuates. The dimensional changes may be enough to render the component unserviceable. The temperature at which creep becomes significant is related roughly to the melting point of the alloy, but considerable improvements have been made by metallurgical means in some alloys by increasing the acceptable operating temperatures and reducing creep deformation rates.

Wear can occur whenever surfaces move past one another in contact. Material is removed from one or both surfaces progressively, and the component may eventually be rendered unserviceable due to dimensional changes or deterioration in surface finish. Catastrophic failures are rare. The mechanisms of wear are various and not well understood. Hence wear has to be combated largely by empirical means, for which purpose a considerable body of information is available.

We discussed in the preceding chapter the basic behavior of metals when a stress is applied, but we considered only the immediate reaction of the metal to a single application of stress. Moreover, we considered only the behavior at ambient temperatures in air, although this was not explicitly stated. Even so, we saw that some metals have a failing in that they may fracture in an unexpectedly brittle manner. But this is only the beginning of the story of how metals may fail in unexpected ways — behavior which follows because they are expected to continue to perform their functions for long periods of time and under much more complex stress systems and environments than those which were considered in Chapter 3.

Hard experience has shown that there are quite a number of circumstances under which the performance of a metal component deteriorates in service even to the extent of complete failure. Some of these circumstances are mechanical and some are chemical in origin, and some are due to a combination of these two factors. We shall consider circumstances of the first type in the present chapter and those of the last two types in Chapter 5. The three most important factors of mechanical origin that we shall consider here are: *fatigue*, which is caused by the application of cyclical stresses; *creep*, which is caused by the application of a stress for a long time at relatively high temperatures; and *wear*, which is caused by interactions between two surfaces when they are made to rub against one another.

As well as having adequate properties of the types discussed in Chapter 3, before it can be accepted into engineering practice, a metal or alloy may be required to be proof against one or more of these modes of failure. At least it must be sufficient proof against such failure to ensure that a component has an economically acceptable service life. Moreover, impending failure must be predictable with sufficient reliability to give a reasonable hope that it will be recognized before a disastrous failure does occur. Metals, unlike diamonds, are not necessarily expected to last forever, but they are expected not to fail catastrophically.

Different alloys vary considerably in their resistance to one or more of these failure modes and, moreover, in a manner which more often than not is unrelated to their short-term mechanical properties. The selection of an alloy for use in particular circumstances consequently has to include consideration of its resistance to failure by any of the modes that are judged to be possible in the particular application. The anticipation of all possibilities and the accurate assessment of the magnitudes of their importance are not easy, particularly in more complex situations. Needless to say, therefore, design and metals engineers don't always get it right the first time, and failures do occur in service. It can fairly be said that the development of advanced engineering devices often proceeds by a series of progressive approximations, or partly by trial and error if you prefer to put it that way. An unanticipated problem may arise after a period of service; this problem may be investigated and solved, but a new problem may then be revealed elsewhere. And so the cycle goes on until a failure-proof device has been developed. By then, likely as not, the performance demands on the device will have been increased and the cycle may have to be restarted.

The proper investigation of a service failure when one does occur is thus crucial to engineering development. The mechanism of failure has to be identified and the design or materials parameters responsible have to be established before sound remedies can be prescribed. Failure analysis consequently has become an important facet of metals engineering, and has to be backed by a thorough understanding of the possible failure mechanisms. Studies of specific failure modes have become specialty disciplines in themselves, and all we can do here is to provide a basic introduction to the understanding of the more important ones.

You should rest assured, however, that engineers do not just sit back waiting for failures to occur. Key components of critical structures are inspected, often at regular intervals, for the appearance of cracks or other features which might lead eventually to failure. This is particularly likely once the possibility of failure has been recognized. A number of techniques of nondestructive testing have been developed for this purpose (p. 104), but the detection of an offensive crack or similar discontinuity or *flaw* is only the first step. It is not economically sensible, and it may not even be practicable, to remove from service a component in which any sort of flaw has been detected. The detection of the flaw has to be followed by an evaluation to decide whether or not it is really likely to lead to a problem in service. If this is judged to be so, then it is classified as a *defect*, perhaps with a qualifying adjective to indicate the order of its seriousness. The final step is to decide whether the component has to be removed from service and either repaired or replaced, hopefully by a modified, more-failure-proof component. All of this is a multidisciplinary activity involving a high level of professional sophistication in which the metals engineer has, or should have, a significant role.

Nor should you take a pessimistic view of the usefulness of metals from these discussions of their failings. All structural materials have failings of one sort or another. The important thing in engineering practice is that these failings be recognized, understood, and combated where necessary and possible.

BEHAVIOR UNDER CYCLICAL STRESSES: FATIGUE

The meaning of the word "fatigue" in common usage is "weariness after exertion". The operators of the early railway systems found that their wheel axles all too often seemed to become weary after exertion. They broke transversely adjacent to a wheel after some use and in a most unexpected way. Operators of rotating machinery, which was at the same time coming into wide use, had similar experiences. They called this phenomenon *fatigue*.

It was said once that fatigue fracture occurred because the metal had "crystallized" in use, a misconception which persists to this day even among some who should know better. As we shall see in Chapter 7 (p. 221), metals are always crystalline. Indeed, the pioneer metallographer Sorby deduced this from his investigations of the structure of iron in the 1860's (p. 186) and said specifically that the concept of crystallization during fatigue was false. At about the same time, moreover, the basic facts about the fatigue phenomenon were beginning to be established by Wöhler, a railway engineer in Germany, who showed that a distinct mechanical characteristic of metals was involved.

A railroad axle is loaded as a simple beam when it is stationary; tensile stresses are induced in the lower regions of the shaft between the bearing and the wheel, and compressive stresses are induced in the diametrically opposite regions. The stress at any point on the axle surface consequently changes from tension to compression each time the axle rotates. The material of an axle thus is subjected to cyclical stresses varying in a simple sinusoidal manner about zero stress, and it was this cycling of the stresses that was soon established as being responsible for fatigue failure.

Many structures are subjected to cyclical stresses in service. Sometimes the cycles are regular, as in axles and shafts, but sometimes they are complex and irregular in both magnitude and frequency. For example, the stresses in the undersurface of an aircraft wing go through a simple cycle from compression to tension each time the aircraft takes off; before takeoff the wing supports its own weight only and sags at the tip; after takeoff it supports the weight of the fuselage and deflects upward at the tip. But during flight the wing strikes air turbulences and bends and twists as the aircraft undergoes various maneuvers. This imposes additional, variable, and complex cyclical stresses on the wing surfaces about a positive mean stress that is imposed by the wing's supporting of the fuselage, and this is different from a simple rotating shaft in which the stress cycles about zero mean stress. The presence of residual stresses (pp. 302 and 361) has the same consequence.

The Basic Characteristics of Fatigue

The story of the development of a fatigue fracture starts with the initiation of a microscopically small crack, which almost invariably occurs at a point right at the surface* in a region where the cyclical stresses are highest. The crack then grows inward with increasing numbers of stress cycles, the direction of growth being determined by the stress system. It grows in a manner essentially perpendicular to the maximum tensile stress range. The area of the section that remains intact in the component eventually becomes too small to support another application of stress, which is now applied in the presence of a severe stress concentrator (p. 85). The component then fractures suddenly. This might occur under apparently normal operating conditions, perhaps to everyone's surprise and consternation, but doom has been impending for some time.

The two regions of different fracture modes — namely, fatigue-crack growth and final overload fracture — can usually be recognized on a fatigue-fracture surface (Fig. 4.1a). The region of fatigue-crack growth may be slightly darkened, but more significantly is characteristically covered with conchoidal markings radiating from the point of crack initiation (Fig. 4.1a). These markings are the results of irregularities in the progression of the crack; the crack path may deviate slightly once in a while or crack growth may slow down or speed up occasionally, causing irregularities in the fracture path. These markings are sometimes known colloquially as *beach marks*. Closer examination at high magnification may detect a second system of finer *striations* on the fracture surface (Fig. 4.1b). Each of these striations marks a step in the advancement of the fatigue crack, one striation for each cycle of stress. The appearance of the zone of final fracture varies, but has the characteristics of one of the modes of failure that occur in the particular material under a static overload in the presence of a severe stress concentrator (p. 83). It can be distinguished from the fatigue zone on this basis.

Fatigue cracks also exhibit some distinguishing characteristics when examined in section by the techniques of optical microscopy outlined in Chapter 6 (p. 184). It may be apparent that the initiation of the crack has been influenced by the presence of surface metallurgical abnormalities which act as stress concentrators (Fig. 4.2), either alone or supplementing a macroscopic stress concentrator (as in Fig. 4.2). The cracks usually can also be seen to extend in a generally straight transgranular path but in detail in a series of differently oriented straight segments, one segment in each crystal grain (Fig. 4.2; see also p. 321).

---

*Exceptionally, fatigue cracks may initiate below the surface when comparatively severe stress concentrators are present below the surface or when unusual stress conditions produce peak stresses there.

(a) A fatigue fracture developed in the cheek of a crankshaft from a diesel engine. The fatigue crack was initiated at the point indicated by the arrow and progressed across the cheek in the direction indicated by the arrow. The conchoidal markings centered about the point of initiation are characteristic of fatigue and are the result of variations in the progression of the fatigue crack. Magnification, ¼× (approx). From D. J. Wulpi, *Understanding How Components Fail*, American Society for Metals, Metals Park, OH, 1985.

(b) The fine striations shown in this photograph can often be seen on a fatigue fracture surface when it is examined at high magnification. Each line marks a step in the progression of the crack during a single stress cycle. This example is a fatigued laboratory test piece in which the magnitude of the cycling stress was changed from time to time. The variations in the spacing of the striations correlate with these variations in stress. This is evidence that each striation marks the advance of the crack during a single stress cycle. Replica electron micrograph; 2975×. From *Metals Handbook*, 8th Ed., Vol. 9, American Society for Metals, Metals Park, OH, 1974.

Fig. 4.1. Appearance of the surface of typical fatigue fractures.

Fatigue Properties of Metals

It is difficult to reproduce in the laboratory the complexities of the stress cycles to which many real components are subjected in service, although attempts are made to do so when the structure or component is sufficiently important. It is

The cracks were initiated at a stress-concentrating fillet in the component, the stress concentration having been enhanced by the presence of surface cracks (dark networks) formed at the fillet surface during heat treatment. Thus both mechanical and metallurgical stress concentrators were present. The cracks have propagated in a series of short, nearly parallel steps, but follow a general path which is parallel with the fillet radius. Optical micrograph; magnification, 750×.

Fig. 4.2. Section through fatigue cracks in a steel component.

usual, as a simplified laboratory representation, to investigate the characteristics of a cylindrical test piece of the type sketched in Fig. 4.3, the test piece being loaded as a beam and rotated at constant speed. The surface layer of the test piece is then subjected to a sinusoidal stress cycle about zero mean stress, which is the simplest stress cycle possible. If the test piece is suitably shaped, as sketched in Fig. 4.3, and is smoothly finished so that it contains no significant stress concentrators, the position of the maximum stress is known and its magnitude can be calculated. This again is a considerable simpification of real life. The story of the development of a fatigue crack given above can nevertheless be confirmed in an experiment with such a test piece, which is what Wöhler did in his pioneering investigations. Some quantitative data can also be obtained on the fatigue behavior of the material from which the test pieces were machined.

The number of stress cycles required to cause a crack to initiate and the

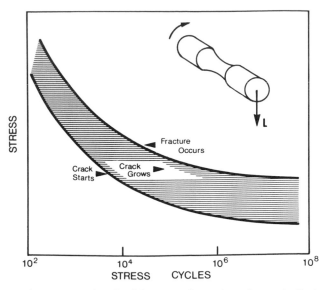

The inset is a sketch of the type of test piece that typically is used to obtain this information. Note that the number of cycles has to be plotted on a logarithmic scale.

Fig. 4.3. Diagrammatic representation of the variation with stress (S) of the number of stress cycles (N) required first to initiate a crack and then to cause final fracture by fatigue.

number required to cause final fracture typically are found to vary with the magnitude of the cyclical stress in the manner illustrated diagrammatically in Fig. 4.3. This is called an *S–N* curve, which is short for stress (S) versus number of stress cycles (N). At comparatively high stresses, a crack is initiated after a comparatively small number of cycles and the crack grows rapidly with increasing numbers of cycles. Consequently, final fracture also occurs after a small total number of cycles. At comparatively low stresses, on the other hand, many more cycles are required to initiate a crack (nevertheless, a crack is initiated eventually even at the smallest stresses). Moreover, the number of stress cycles required for the crack to grow to a sufficient length to cause final fracture is also greater. From this it follows that the total number of cycles required to cause failure increases considerably. There is also now a longer period between crack initiation and final failure which, in its practical equivalent, gives an inspector a better chance of discovering the crack before a catastrophic failure occurs.

The parameter commonly used to characterize the fatigue behavior of a metal is the stress that it can withstand for a specified number of stress cycles; 10^7 cycles is a number frequently adopted. This is called the *endurance limit*, and it is always only a fraction (at most one-half, and frequently less) of the tensile strength of the metal (Fig. 4.4). Thus it is easy to see why a component might

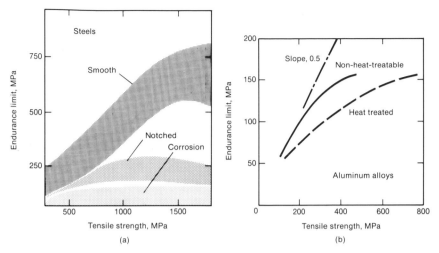

There is wide scatter in these relationships, as indicated in (a), but the endurance limit of steels measured with smooth polished test pieces in air is about one-half the tensile strength. The endurance limit is greatly reduced, however, when a notch is introduced into the test piece, particularly with steels of medium-to-high tensile strength; the endurance limit then does not vary much with strength. The endurance limit is reduced even further, notch or no notch, in a corrosive environment, tap water in this instance. The endurance limit then varies very little with tensile strength. The endurance limit of aluminum alloys is improved with increasing tensile strength to an even smaller degree than that for steels (b). This is particularly true for heat treated aluminum alloys of higher strengths, very little improvement being obtained at strengths near the higher end of the range. Adapted from (a) *Prevention of Fatigue of Metals*, Battelle Memorial Institute, Wiley, New York, 1941, and (b) P. C. Varley, *The Technology of Aluminum and Its Alloys*, Newnes-Butterworths, London, 1970.

Fig. 4.4. Variation of endurance limit with tensile strength for typical structural steels (a) and aluminum alloys (b).

fail in fatigue even when the designer's calculations have indicated that it can withstand a single application of the service stress with plenty to spare. Hence the surprise and consternation when the wheel axles began to fail in early railway vehicles.

There are exceptions to the general rule that metals have only an endurance limit as just defined. Some have a genuine *fatigue limit* in that they do not fail no matter how numerous the stress cycles if the cycling stress is less than a certain value. The S–N curve then becomes parallel to the N axis in Fig. 4.3, and this usually occurs after about 10^6 to 10^7 cycles. It so happens that steels are one of these exceptions, which is another one of their advantages. Titanium alloys and a few aluminum alloys also have this characteristic. Under optimum conditions, the fatigue strength of a steel is about one-half of its tensile strength, but never more than this (Fig. 4.4).

The Importance of Stress Concentrators

Experiments using a simple rotating-beam test piece can also demonstrate the significance of one other important factor in fatigue — namely, the role of stress concentrators. The fatigue or endurance limit found when a notch is machined in the neck area of a test piece of the type sketched in Fig. 4.3 can be much lower than that of an unnotched specimen. The more severe the notch, the greater the reduction in endurance limit is likely to be. The general magnitude of this effect for a range of steels can be seen by comparing the upper two curves in Fig. 4.4(a). Note that the deleterious effect of a notch increases with the strength of the steel and that any advantage of increasing the yield or tensile strength is soon lost. Similar effects are exhibited by all common alloys, with some alloys being more sensitive to these effects than others. Practical experience has shown that fatigue cracks very commonly are initiated at stress concentrators such as holes, changes in section, machining marks, cracks, or like metallurgical surface flaw. An example of where a combination of stress concentrators of design and manufacturing origins has been responsible for the initiation of a fatigue crack is shown in Fig. 4.2.

This marked effect of stress concentrators is not exactly a surprise once the magnitude of the stress-concentration factors of even seemingly innocuous notches (p. 85), and their influence on the energy required to produce final fracture (p. 87), are recognized. Nevertheless, this effect still is sometimes overlooked during design and manufacture. The elimination of, or at least the reduction of the severity of, stress concentrators is perhaps the single most important thing that can be done to proof a component against fatigue. It is better to do this than to attempt to use alloys of excessive strength, as discussed above in connection with Fig. 4.4(a). It is better, likewise, to reduce the severity of a stress concentrator at which a service fatigue crack has been initiated when attempting to cure a fatigue problem than to substitute an alloy of higher strength.

Fatigue in Real Life

The curves sketched in Fig. 4.3 might be taken to imply that the number of stress cycles at which a fatigue crack initiates and at which failure occurs can be predicted precisely when the magnitude of the cycling stress is known. This is not so, even allowing for uncertainties in estimating the actual stress level. Wide scatter is found in practice for both types of values under apparently similar conditions. Fatigue is by its very nature a statistical phenomenon. Fatigue cracks are initiated at microscopically small points on a surface, so that any minute localized feature which causes the stress to be a little higher than usual, or the strength of the material to be a little lower than usual, constitutes a weak spot at which crack initiation becomes easier than usual. This is so even when no major stress concentrators are present. There is a random aspect associated with the locations of these weak spots and with their degrees of weakness. An S–N curve of the type sketched in Fig. 4.3 consequently has to be associated with a particular probability of failure, and so do estimates of the fatigue lives of engineering components.

Actually, data obtained in small-scale laboratory testing are difficult to apply quantitatively to real engineering components. There are too many uncertainties in the scaling-up process — uncertainties about how well are known the actual values of the mean and cyclical stresses, the magnitudes of the stress-concentration factors, and the conditions of the metal at likely crack-initiating points. There are even problems of geometric scaling, large components having lower endurance limits than small laboratory test pieces or scale models, possibly because of the higher probability of a very weak spot being present at a critical position. This is one reason why designers, particularly in the aircraft industry, may go to the considerable trouble and expense of carrying out fatigue tests on full-scale structures, even attempting to reproduce the expected service stress spectrum as closely as possible. This is, of course, not always economically feasible, in which event small-scale tests backed by service experience must suffice.

Controlling Fatigue

Considering the complexity and uncertainty of these effects, it is perhaps not altogether surprising that fatigue failures still do occur rather regularly. Fatigue is one of the commonest methods by which metals fail in service even to this day. Sometimes a fatigue failure is the result of the designer's failure to take heed of simple lessons of the past, such as that of the importance of stress concentrators. Sometimes it occurs because the manufacturer has introduced stress-concentrating flaws and has failed to carry out adequate inspection to detect these flaws. Sometimes, however, it is impracticable or beyond the wit of man with his present state of knowledge to design and manufacture an absolutely fatigue-proof component for a particular complex application. Those who have to operate the component then must take precautions to avert a catastrophic failure. Two courses of action are open to them once the sensitivity of the component to fatigue has been discovered.

First, a limit can be set on the life of the component, at which stage it is removed from service and either repaired or replaced — preferably by a modified component with improved fatigue resistance. This life generally has to be determined from experience and a conservative value chosen, with more conservative values being set for critical components for which only a low probability of failure is acceptable. This procedure can be wasteful of costly components. Secondly, the offending component can be inspected for cracks at appropriate intervals, for which purpose a range of techniques is available (p.

203). The conservative approach then is to remove the component from service when a crack is first detected, but this too may be wasteful. A crack does not actually make a component unsafe until it has grown to a certain size (p. 93), and it may be possible to estimate with reasonable accuracy, from a mixture of theory and experience, what this maximum safe size is. In this event, the growth of a crack can be monitored and the component removed from service only when it nears an unsafe state. This is the more technologically advanced and the more economical approach.

Metallurgists have, of course, been trying for some time to develop alloys which have improved fatigue resistance as well as improved yield strength with the objective of alleviating these problems. They have, it must be admitted, not been very successful in developing alloys which have improved resistance to the initiation of fatigue cracks. Many of the structures which have been devised to strengthen metals (see Chapter 10) unfortunately break down when the material is subjected to alternating stresses. Precipitation hardening (p. 367), for example, may increase yield strength considerably but endurance limit only a little, an unfortunate consequence of which is that the majority of high-strength aluminum alloys, which are strengthened by precipitation harding, do not have commensurately improved fatigue resistance (Fig. 4.4b). The phenomena involved in crack initiation are complex and not yet fully understood, and it is this, perhaps, that most of all has frustrated attainment of the sought-after improvements.

On the other hand, metallurgists have been quite successful in modifying high-strength alloys to reduce the rate of growth of a fatigue crack once it has been initiated. They have also been successful in modifying these alloys so that larger cracks can be tolerated with safety. Both of these features were discussed earlier (p. 94). The net result of these improvements is that the gaps between the initiation and failure lines of S–N curves (Fig. 4.3) has been widened considerably, and this has great practical advantages. First of all, the total life of a component is increased. Secondly, the chance that a crack will be detected before fracture occurs is improved because the crack is present for a longer time. Thirdly, the practicability of the procedure of assessing serviceability by monitoring the growth of the crack is enhanced.

BEHAVIOR AT HIGH TEMPERATURES: HOT SHORTNESS AND CREEP

Stresses Applied for a Short Time

We have so far discussed only the behavior of metals at room temperature, and we are now to consider how this behavior may be modified at high temperatures, considering first the short-term consequences. In a most general way, both yield and tensile strengths decrease with increasing temperature, as illus-

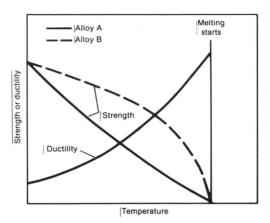

Fig. 4.5A. Diagrammatic representation of the typical variation with temperature of the strength and ductility of metals.

trated diagrammatically in Fig. 4.5A, both becoming zero at the temperature at which the metal starts to melt. More quantitative information for three important groups of alloys is given in Fig. 4.5B, the comparison there being based on specific yield strength and indicating the magnitude of the temperature at which the alloy types can no longer be used effectively. Aluminum alloys, for example, become ineffective at temperatures above 200 °C whereas titanium alloys retain their strength at temperatures well above this. The changes are

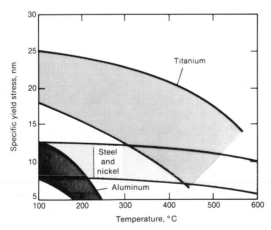

Specific yield strength is the 0.2% yield strength divided by mass density, and compares the weight effectiveness of the alloys. The effectiveness of aluminum alloys is lost at operating temperatures of 200 to 250 °C, of titanium alloys at 500 to 600 °C, and steels and nickel alloys at somewhat higher temperatures again. (Adapted from Fig. 1.3, p. 7, in I. J. Polmear, *Light Alloys: Metallurgy of the Light Metals*, Edward Arnold, London, 1981.)

Fig. 4.5B. Variation of the specific 0.2% yield strength with temperature for typical aluminum alloys, titanium alloys, and steels and nickel alloys.

basically due to changes in fundamental properties, but they can be augmented by destruction of the microstructures which have been used to strengthen the alloys. The characteristics of some of the more important of these strengthening structures will be discussed later.

Ductility typically increases with increase in temperature, but only until melting commences, at which point it drops rapidly to zero. There may be perturbations in this general behavior when secondary phenomena intervene, but the result is that metals stressed in tension at temperatures approaching their melting point deform easily and eventually fracture at comparatively small stresses but with considerable ductility. The reduction of area at fracture in tension may approach 100% (Fig. 4.6). In compression, deformation also occurs at correspondingly small yield stresses and deformation to increasingly larger strains become possible. This is one of the main reasons why metals are easier to deform and shape at high temperatures when they are being fabricated.

But the story changes drastically when the temperature actually reaches the melting point, or perhaps even a temperature close to the melting point. Melting always commences at the grain boundaries (p. 241), and only a very small amount of melting is necessary to form a thin film of molten material around the boundaries. Both strength and ductility are then reduced to zero in either tension (Fig. 4.5A) or compression. Theoretically, this drop in strength and ductility can be expected to occur at a temperature very close to the melting point in a pure metal or at one close to the solidus temperature (p. 255) in an alloy. But in practice, the presence of impurities may reduce the temperature at which melting starts to a lower value than expected. The deterioration in properties will then occur at a correspondingly lower temperature, perhaps a good deal lower than expected. In any event, an intercrystalline crack or fracture is likely to develop when the metal is strained even a small amount while at a temperature which would normally be quite safe. As examples, fractures may

The steam pressure caused a local bubble to form in the tube at the overheated area, and a tensile fracture eventually developed in the bubble. At the high temperature, the steel necked down to a section of almost zero thickness before fracture occurred. Macrograph; magnification, 2X.

Fig. 4.6. Longitudinal section of a boiler tube which overheated locally in service.

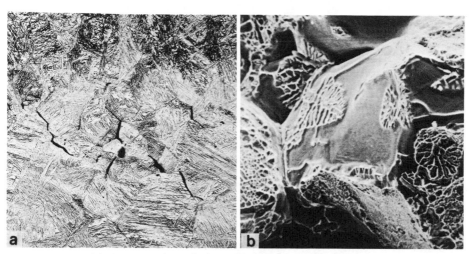

(a) A section through the weld showing that intercrystalline cracks are present in the parent metal in a region immediately adjacent to the weld. This region has been heated to a temperature close to, but certainly below, the nominal melting point of the steel. Optical micrograph; magnification, 75×.

(b) The surface of one of these cracks after it has been opened up. The areas containing the fern-shape phase were molten at the welding temperature, forming thin films of liquid around grains of otherwise solid parent metal, so allowing the grains to separate easily. The ferns can be identified as sulfides by microprobe analysis (p. 199), indicating that the hot shortness is due to the presence in the steel of excessive amounts of sulfur as an impurity. The ferns are dendrites (p. 252) of manganese sulfide formed when the molten films subsequently solidified. Scanning electron micrograph; magnification, 500×.

Fig. 4.7. Intercrystalline cracks developed adjacent to a weld in an alloy steel due to hot shortness in the steel.

develop during hot rolling when quite normal practices are followed, and cracks may develop in the heat-affected zone adjacent to a weld where the metal has been heated close to, but still below, the melting point (Fig. 4.7). This phenomenon is known as *hot shortness*. This term may be applied to alloy types which lose their strength and ductility at such low temperatures that they cannot be hot worked safely at any reasonable temperature. It may be applied also to an individual batch of an alloy in which the effect has been induced by a particular impurity, as in the example illustrated in Fig. 4.7. The regions that melt to cause hot shortness commonly are so thin that advanced investigational techniques may be required to detect them and to identify the offending impurity (Fig. 4.7b). Once this has been done, however, the metallurgist knows what to avoid in the future and how to diagnose the cause of hot shortness if and when it does occur.

The rate at which strength deteriorates with temperature varies with different metals and with different alloys of one metal. Alloys have been developed in which the strength holds up better than usual with increases in temperature, as illustrated diagrammatically in Fig. 4.5B, with the intent that they be used at high temperatures. But often the long-term behavior under stress that we shall soon discuss then comes to be of even greater importance. Alloys for such applications must also have good oxidation resistance at the intended operating temperature (p. 170). Otherwise, they would waste away by oxidation, and the advantage of their high strength at temperature would soon be lost.

Stresses Applied for a Long Time (Creep)

It was implied in Chapter 4 that a metal strains a certain amount when a stress is applied and that no further straining occurs thereafter. This is true enough for most metals at room temperatures, but is not so for any metal at a sufficiently high temperature. The metal then continues to strain under constant stress, and this time-dependent deformation under an applied stress is called *creep*.

"High temperature" is a relative term in this context. Roughly, creep becomes significant at temperatures above about one-half the melting point of the particular metal, the temperature being measured in kelvins, or K (°C + 273). By "significant", we mean here that the mechanical strength of the material has become limited by creep rather than by normal yielding. Thus creep is significant even at room temperature in metals such as lead, which has a low melting point. The result could be seen until a recent restoration in St. Paul's Cathedral, London, where the lead damp course beneath the supporting pillars extended several centimetres beyond the masonry; the lead had extruded out by creep over the centuries under the compressive dead load of the pillars. Creep in aluminum alloys, on the other hand, does not start to become significant until temperatures of about 200 °C (390 °F) are reached. Temperatures above about 500 °C (1020 °F) are required for ordinary structural steels, and temperatures above 900 °C (1650 °F) for some special nickel alloys, to choose a few examples. Note also from these examples that the relationship with melting point thus is not a direct one but is at best only a guide (the melting point of nickel is about the same as that of iron). Many factors are involved, and the creep-limiting temperature has to be established for each individual alloy. Moreover, as we shall soon see, the phenomena that occur depend on the magnitude of the stress as well as on temperature. Low stress at high temperature is roughly equivalent to high stress at low temperature.

Basic Characteristics of Creep

The creep characteristics of a metal in tension, to which we shall confine our attention from now on, can be described by measuring the increase in strain that occurs with time at constant stress and temperature. The sketch in Fig. 4.8

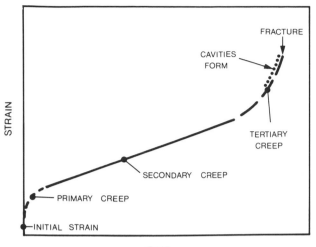

Fig. 4.8. Diagrammatic illustration of the variation of creep strain with time at constant stress and temperature. The different regimes of creep are drawn with different line values.

illustrates diagrammatically the type of relationship which typically is found and which involves four regimes of behavior. The first is an initial elastic, and perhaps also a plastic, extension that occurs when the stress is first applied; this is the normal elastic-plastic yielding. Next comes a period of *primary creep* during which additional strain occurs with time, first at a comparatively high rate but then at a steadily decreasing rate until a regime of steady-state *secondary creep* is entered. The creep rate is constant and at a minimum during this period of secondary creep, a period which extends over most of the creep life. Eventually, however, the creep rate accelerates again during a period of *tertiary creep* that typically terminates in fracture, provided that the stress is not relieved by the increased strain.*

A family of strain-time curves covering a range of stresses and temperatures is necessary to characterize fully a metal in creep. Part of such a series covering a range of stresses at one temperature is shown in Fig. 4.9. As is typical, increasing the stress has increased the creep rate during the secondary creep stage, and hence also the total strain that occurred in a given time. Up to a point, the total strain to fracture also has increased. The time to fracture has decreased. Full characterization of this nature obviously requires the accumulation of large amounts of data, which are costly to obtain. In practice, therefore, data on the time to fracture and total strain to fracture (*stress-rupture ductil-*

---

*A metal may not, however, exhibit all three of these stages of creep. The absence of a primary creep stage is not uncommon at high stress or high temperature. Sometimes there is no tertiary creep stage, particularly in cast alloys. In the latter event, fracture occurs abruptly after only a small strain.

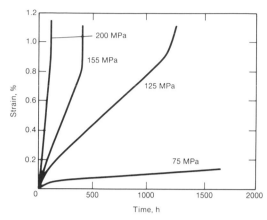

Fig. 4.9. Creep curves for a low-alloy steel at four stress levels at 600 °C (1110 °F).

ity) over an appropriate range of stresses and temperatures usually must suffice. The metal is then characterized by the variation of one of these parameters with one of the others, or with some combination of the others; one combination commonly used is illustrated in Fig. 4.10. The parameters that are selected in a particular case, however, will depend on the requirements of the designer. A designer may want to know, for example, the conditions which ensure that the creep strain will not exceed a certain value (typically 1%) or, on the other hand, may want to know those conditions that will ensure a certain minimum life without fracture; it is this type of information that is provided in Fig. 4.10.

The mode of final fracture during creep also depends on both stress and temperature, and varies from ductile to brittle. A transition between the two general types of fracture occurs at a fairly definite temperature, called the *equicohesive temperature*, which is a characteristic of the strain rate for a particular metal. In general, lower creep rates, longer rupture times, or higher temperatures promote brittle types of fracture.

Ductile fractures develop by the normal process of void formation and growth (p. 68), and so are transcrystalline. They are associated with large fracture strains. Brittle fractures occur with little necking down at fracture (Fig. 4.11) and occur along the grain boundaries of the metal. They occur by unique mechanisms under circumstances where deformation phenomena in the region of the grain boundaries dominate over those occurring within the grains (the reverse is true at low temperatures; see p. 296). One mechanism involves the formation of chains of small cavities along grain boundaries, principally on boundaries which are inclined to the axis of tensile stress (Fig. 4.12a). The cavities grow and join up to separate the grains, at which stage they constitute an intercrystalline separation. These cracks grow in turn and some join up to form a continuous fracture path. This is sometimes called cavitation creep. In the other important mechanism, wedge-shape cracks initiate along the grain

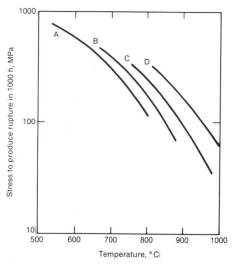

The characteristics of a series of successively improved alloys are illustrated here. All alloys are heat treated, and the heat treatment schedule has been optimized for each composition. A – The original alloy, containing 20% chromium and small amounts of titanium, aluminum, and carbon. B – Similar to A, but with 20% of the nickel replaced by cobalt. C – Titanium and aluminum contents increased; molybdenum added; chromium content reduced. D – Titanium and aluminum contents increased further. (Adapted from *Nimonic: Structures of Nimonic Alloys*, Henry Wiggin & Co., Ltd., Hereford, England, 1971.)

Fig. 4.10. Variation with temperature of the stress to creep rupture for several nickel-chromium superalloys used in the hot-end components of aircraft jet engines.

boundaries at points where the boundaries meet obstacles (e.g., other grain boundaries, second phases, or surfaces; see Fig. 4.1). The cracks then tend to grow along grain boundaries which are approximately normal to the applied stress. This type of fracture tends to occur at lower creep temperatures and higher stress levels.

Controlling Creep

Metallurgists have been highly successful in developing alloys with improved creep resistance, difficult though the task has been. Various methods are available to strengthen metals (Chapter 10, p. 323), but the strengthening structures are not always stable at elevated temperatures. So strengthening methods which are adequately stable have to be chosen from among them. The alloy must also be designed to resist oxidation and corrosion in the intended working environment, which can be very aggressive, (e.g., combustion gases). Often, also, an adequate degree of fracture toughness both at room temperature and at

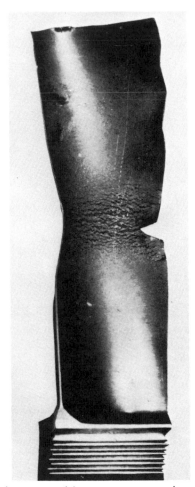

These blades run at red-hot temperatures and are subjected to large radial stresses due to centrifugal forces. Note the elongation of the blade and the band of intergranular cracks in the regions adjoining the developing fracture. Photograph; shown here at 75% of actual size. From *Metals Handbook*, 8th Ed., Vol. 10, *Failure Analysis and Prevention*, American Society for Metals, Metals Park, OH, 1975.

Fig. 4.11. A creep failure in a turbine blade from an aircraft jet engine.

operating temperature is required, and perhaps also adequate fatigue resistance at the operating temperature.

A good example of the successful development of creep-resistant alloys is found in the nickel alloys containing about 20% chromium and, as originally developed, small amounts of titanium and carbon. These alloys are used extensively in the turbine blades of aircraft jet engines. Indeed, the development of alloys of this type was the key to the success of the first fully opera-

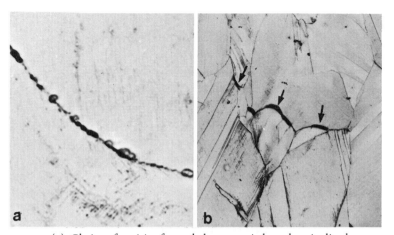

(a) Chains of cavities formed along a grain boundary inclined to the strain axis. (b) Cracks developed along grain boundaries aligned approximately normal to the strain axis. The cracks were initiated mostly at points where three grain boundaries met. Optical photomicrographs of brass strained in creep in a vertical direction. (a) 2000×; (b) 750×.

Fig. 4.12. Illustrations of the two mechanisms by which intergranular creep fractures develop.

tional jet engine designed by Whittle* in the early 1940's. The original alloy has since been much improved (Fig. 4.10), the motive being that one of the main parameters that determines the performance of a jet engine is the maximum temperature at which the components of the turbine can operate without creeping excessively. Modifications to the original alloy have enabled them to be used at progressively higher temperatures. These improvements in creep resistance were made by adjusting both composition and heat treatment. Even further improvements have been made by an approach which recognizes that grain boundaries, particularly those which are aligned normal to the major stress axis, are a weakness in creep. Techniques have been developed for casting blades in which all of the grain boundaries are aligned parallel to the length of the blade, which is parallel with the major stress axis. The ultimate achievement of this approach has been to cast blades which have no grain boundaries at all — that is, blades which are single crystals (p. 240). This is done as a production routine and is a very clever piece of technology. It is an example of how even an esoteric laboratory technique can be converted into an engineering reality when the needs and rewards are great enough.

Thus metallurgists have made a major contribution to the outstanding improvements that have been achieved in aircraft jet engines since Whittle's first

---

*A jet engine developed simultaneously in Germany actually flew in an operational service aircraft before Whittle's engine, but the materials in this engine were so inferior that the engine could operate for only a few hours.

efforts. The large fuel-efficient passenger aircraft that we now take so much for granted would not be possible without this contribution. It has, of course, not been the only contributing factor, but it has been an essential one. An example which is more mundane but perhaps more important to civilization lies within the domestic electric lamp. The light from these lamps is emitted from a coil of fine tungsten wire which is mounted within an evacuated glass bulb, the wire being heated to a high temperature by the passage of an electric current. The higher the temperature, the brighter the light and the higher the efficiency of the device. Pure tungsten sags excessively by creep under its own weight at desirable operating temperatures. Thus the commercial success of this light source, which changed life styles so dramatically, depended on the development of a tungsten alloy which had adequate creep resistance at an effective operating temperature. This was an achievement of a distinguished American metallurgist, Zay Jeffries.

Fatigue at High Temperatures

The direct effect of temperature on endurance or fatigue limit is marked and detrimental, the precise relationship varying widely depending on the alloy and temperature. Moreover, the superimposition of cyclical stresses on a steady stress in the creep range greatly reduces the creep life compared with that expected from the creep conditions alone. The cyclical stresses can result from either vibrations or stresses applied externally in the normal way. They can also result, however, from stresses induced by the thermal cycle itself. A component may be heated or cooled unevenly or may expand unevenly during heating, because it has a nonsymmetrical section. The resulting distortion may induce stresses of a significant magnitude.

Hot fatigue cracks characteristically initiate at a single nucleus at the surface of a component and extend inward, as do normal fatigue cracks. These are characteristics by which they may be distinguished from creep cracks. The surface of the crack usually oxidizes, in which event the extent of the oxidation is greater in the surface regions of the crack than toward its root because the surface regions have been exposed to the oxidizing atmosphere for a longer period. Hot fatigue cracks sometimes grow along an intercrystalline path, but most commonly they grow along a transcrystalline path in the same manner as normal fatigue cracks.

WEAR

The metal surfaces of many engineering components come into contact with other solids while the two are moving past one another. Shafts running in bearings and shovels digging in soil are examples. The surface of the component may be consumed gradually in the process, and it is then said to undergo *wear.* More formally, wear can be defined as the undesired removal of material from contacting surfaces by an essentially mechanical action. Simple though this definition may be, it covers in practice a multitude of mechanical actions which are difficult to identify. Hence wear is difficult to predict and to correct. A saving grace, perhaps, is that wear is a gradual, and usually a slow, process which rarely results in catastrophic failure. The performance of a wearing component typically only deteriorates until the component no longer performs its function as effectively as it should and so has to be repaired or replaced. Nevertheless, wear ranks high among the reasons for component replacement in industry. It is of considerable economic importance in an advanced economy, particularly since many wearing components are precision machined and are costly.

There are so many different conditions under which wear can take place and so many possible mechanisms of wear that it is difficult to develop a classification system on which to base rational discussion. We shall therefore discuss only two broad situations: wear between two contacting metal surfaces, and wear between a metal surface and the small hard particles known as *abrasives.*

Wear Caused by Metal-to-Metal Contact

The area over which two metal surfaces actually make contact is much smaller than it would seem to be. Real surfaces, no matter how smooth they may appear to the eye, are quite rough on a microscopic scale. The topography of a machined surface, for example, always consists of an irregular series of crests and troughs, as illustrated in a rather exaggerated way in section in the sketches in Fig. 4.13. Actual contact consequently is made only at comparatively few asperity pairs and over a small area of each asperity at that (Fig. 4.13). The forces applied through each contacting asperity thus are always quite large

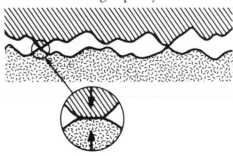

The inset illustrates that the contacting asperities are flattened when even a small force is applied between the two surfaces. This is because the real area of contact is so small that the unit pressure at the rare contacting points is bound to be high enough to cause local plastic yielding. In reality, contacting asperities are more widely spaced than shown here.

Fig. 4.13. Schematic representation illustrating that two surfaces make contact only at a few asperities.

irrespective of the value of the macroscopic forces applied between the two surfaces. Consequently, the asperity is likely to flatten by plastic deformation (inset in Fig. 4.13). The wear processes occur at these contacting asperities, and it is from them that material is removed as wear debris. However, it seems that many mechanisms of material removal are possible.

The earliest suggestion was that a proportion of the flattened contacting asperities weld together when relative movement starts. Further relative movement would then require that this bond be broken, and perhaps, it is suggested, the break does not occur at the weld plane but within one of the adjoining asperities (Fig. 4.14a). A fragment would then be removed from one surface, which would become the wearing surface, and transferred to another. The ease with which a weld bond could be made and an asperity could fracture become important parameters determining wear rate by this mechanism. This concept of adhesive wear is, however, at best a hypothesis and encounters a number of difficulties on close analysis. It certainly can only be regarded as being one of several possibilities.

A number of alternative mechanisms have also been proposed, each of which implies somewhat different controlling parameters. One of them proposes that plates of material are delaminated from one surface by a process of subsurface fracture, the development of which requires repeated rubbing against a contacting asperity (Fig. 4.14b). Another suggests that a hard asperity may shear off a weaker one if it has a suitable shape and if they meet flank to flank (Fig. 4.14c). Yet another suggests that the one asperity removes the protective oxide film from the other, thus allowing progressive chemical removal of material from an asperity which happens to be repeatedly denuded (Fig. 4.14d).

None of these hypotheses is quantitatively predictive, although they all allow some guidelines to be developed, which include the following. Smooth surfaces are preferable because more asperity contacts will be made and so the pressure and shearing force on each will be smaller; moreover, the asperities will be more obtusely shaped and hence less easily sheared off. Hard surfaces are preferable because their contacting asperities will deform less and will also be less easily broken off. The presence of thin oxide layers and the like on the metal surface may be desirable in some circumstances because they could deter adhesion by inhibiting actual metal-to-metal contact. In other circumstances, such as when a chemical-removal mechanism is operating, they can be detrimental. Mating pairs of metals which are chemically compatible might be undesirable because the probability of adhesion would be increased.

Wear Caused by Abrasives

Over half of the wear experienced in industry is caused by the action of abrasives. The abrasive may be intrinsic to the operational system, such as in agriculture or mining where the abrasive particles are contained in the soil or ore.

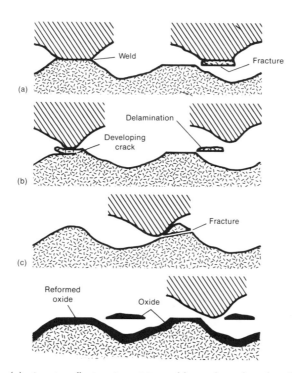

(a) *Asperity adhesion.* Asperities weld together when they first come into contact (left). In some cases at least, fracture to allow further relative movement between the two surfaces occurs not along this weld but within one of the asperities (right). A fragment is thereby removed from the latter asperity, and this fragment may later be released as a wear particle. (b) *Delamination.* A subsurface crack develops within an asperity after repeated rubbing contacts (left). The crack grows, eventually causing a plate-shape fragment to detach (right). (c) *Asperity shear.* A stronger asperity shears the top off a weaker asperity by pressure on its flank. This can occur, however, only if the asperities have a particular shape. (d) *Oxide removal.* One contacting asperity removes the protective oxide layer from the crest of another (right). The bared asperity reoxidizes (left) and so gradually wastes away after repeated contacts.

Fig. 4.14. Schematic representations of some mechanisms of metal-to-metal wear.

Alternatively, it may be present inadvertently, such as when a lubricating oil becomes contaminated with gritty particles.

An abrasive is composed of small particles, often called *grits*, which are irregular in shape and so contain a number of sharp points. They are usually thought of as being very hard, much harder than a metal. They need, in fact, to be two or three times harder than the metal if they are to indent into the metal surface without being deformed too much themselves. We also usually think of them in terms of nonmetallic materials, such as quartz and aluminum oxide, but fragments of a hard metal could act as an abrasive in relation to a soft metal.

Wear fragments from a hard steel, for example, would be more than hard enough to abrade an aluminum alloy.

In any event, three modes of interaction between an abrasive grit and a metal surface are possible. In the first, the grits are sandwiched between, and roll between, two surfaces between which a force is applied. This is called *three-body abrasion* (Fig. 4.15a). The edges and corners of the grits then indent into the metal surfaces, most deeply into the softer of the two, producing a series of pits as the grits tumble across the surfaces. This imparts a dull matte appearance to the surface but removes little material from a ductile metal unless its surface has become embrittled by repeated previous indentations. On the other hand, tumbling grits remove material rapidly from brittle materials, or even from brittle phases in an alloy, because cracks form around the indentations and

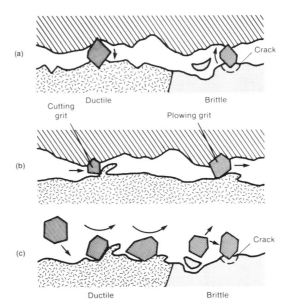

(a) *Three-body abrasion.* The abrasive grits tumble between the surfaces. They produce indentations in ductile material (left), removing little or no material. The indenting grits may, however, develop cracking in brittle material, removing an irregular chip from the surface (right). (b) *Two-body abrasion.* The abrasive grits embed in one surface and sweep through the other one. Depending on the geometry of the contacting point, either a micro-machining chip is cut out of the surface (left) or a groove is plowed (right). The latter is the more common event in practice. (c) *Erosion.* The abrasive grits are projected onto the surface. Again depending on the geometry of the contacting point and on its angle of incidence, either a micromachining chip is cut or a groove is plowed in a ductile material (left). Cracks and irregular chips may, however, be produced in a brittle material by all contacting points (right).

Fig. 4.15. Schematic representations of some mechanisms of abrasive wear.

small, irregular chips are broken out of the surface as a result (Fig. 4.15a). The wear rate is then determined by a diversity of factors such as the force applied between the two surfaces, the number of sandwiched grits which transmit this load, the shape of the grit points, and the hardness and brittleness of the wearing material.

The second possibility is that the grits become embedded in one of a pair of contacting surfaces (the softer of the two) and hence are able to sweep across the second surface while held in a fixed position, somewhat as would the grits of an abrasive paper. This mode is known as *two-body abrasion* (Fig. 4.15b). Grooves are then produced in the second, harder surface, the volume of any groove depending on the shape of the contacting grit point, the force applied through this point, and the hardness of the metal. The first factor determines the section of the groove and the last two its depth (cf. hardness tests, p. 80). This volume is not necessarily removed from the surface, however, and fortunately so, because otherwise the wear rate would be tremendously high. Material is removed only if the shape of the grit point, considered as a machining tool, is within certain limits, and only a minority of points are likely to have the required shape in practice. Wear under these circumstances is effectively a micromachining process (Fig. 4.15b). Nevertheless, all contacting points, whether they machine out a chip or not, produce grooves in the surface and so the worn surface has a shiny, grooved appearance. The wear rate for a given applied force again depends partly on the hardness of the metal, which determines the volume of the grooves being produced, and partly on a range of factors such as the number of grits that are in contact at any one time, the shape of these grit points, and an ill-defined property which determines how easily a micromachining chip can be separated from the wearing surface.

A third possibility is for the abrasive grits to be projected at speed against the wearing surface (Fig. 4.15c), as might occur when gritty particles are carried in a stream of gas or liquid. This is termed *erosion* — or, more strictly, *abrasive erosion* (cf. erosion-corrosion, p. 154). An irregular chip can then be broken out of the surface of a brittle material or a ductile material which has been embrittled by repeated impacts, in the same way as for three-body abrasion. A micromachining chip can also be cut out of the surface when the kinetic energy of the grit, its angle of incidence, and the shape of its point are appropriate for it to act as a machining tool. Wear rates can be very high. A matte surface is produced.

All of the mechanisms mentioned so far assume that the abrasive grit is several times harder than the wearing metal. Perhaps the commonest abrasive encountered in practice is quartz, which is the gritty constituent of many sands and soils. Quartz is relatively soft for a mineral. It is little harder than many tool steels, and certainly is no harder than the phases in the special alloys that are used to combat erosive wear. Different mechanisms of material removal altogether are then involved. Multiple passes of grits over the one point appear to be required to break small fragments out of the hard phases.

Controlling Wear

All in all, very complex phenomena are involved in all wearing systems, and, moreover, it would be difficult to define and measure the controlling parameters in a particular system. It is not surprising, therefore, that the control of wear still has to proceed largely by empirical methods. Wear systems are always many-faceted systems of which the metal is only one component. Thus wear resistance is not a unique materials parameter and cannot be predicted directly from other basic mechanical, physical, or chemical properties. In the ultimate, data accumulated from practical experience with specific types of wear systems have to be used to rate the wear resistance of materials but are applicable only to that particular type of wear mechanism. Laboratory tests have been devised to give relative ratings, these tests attempting to duplicate particular types of wear systems. Useful and necessary though these tests may be, their reliability depends entirely on how closely they reproduce the essential elements of a particular practical system. These elements are not always known with certainty, nor necessarily are those of the laboratory test, and so such tests are at best useful only as guides.

On the other hand, the fact that wear involves an integrated system means that it can be protected at a number of points. Many metal-to-metal systems can be protected by using a so-called lubricant which is designed to maintain a film of foreign material between the two moving components, thus inhibiting actual contact between the metals. The foreign material may take the form of long-chain organic molecules which are adsorbed on the two surfaces and which slide easily over one another. Lubricating greases have these characteristics. The lubricant, on the other hand, may take the form of a discrete but thin film of liquid which is maintained between the two surfaces in some way; hydrodynamic effects due to the very relative motion of the two surfaces may be important here. Most lubricating oils work in this way. Your automobile would run for only a few minutes if these lubricating techniques were not used and were not so effective. Nevertheless, a lubricating layer can break down on occasion, particularly during start-up, and this is when some wear can occur even in the best-protected system. For this reason, further defenses often need to be built into a system by selecting appropriate combinations of metals to run against each other as satisfactorily as possible.

Much experience has been built up to guide the selection of desirable combinations of contacting materials. Many special *bearing metals* have been developed, for example, in which to run rotating shafts, and some of these alloys have effective self-lubricating properties. Other methods have been developed to improve the surface characteristics of one of the mating pair when less-than-optimum combinations have to be used. Thus methods have been developed to harden the surface layers of components which have had to be made from softer metals to withstand the stresses and shocks of service. Methods have also been developed to deposit on critical surfaces layers of materials which either help to retain a lubricant film or have self-lubricating properties.

Abrasive wear can be combated in oil-lubricated systems by recirculating and filtering the oil, with the filter hopefully removing the abrasive grits. Lubrication systems in automobile engines are so protected. The elimination of abrasives from a wear system is, however, not always possible. Excavating equipment used in earth moving and mining are important examples; pipelines carrying mineral slurries are another. Special wear-resistant alloys have been developed for these applications, alloys which typically contain large areas of a very hard phase dispensed in a more ductile matrix. These alloys usually are somewhat brittle, but, where this is a serious disadvantage, some can be deposited as a hard facing on a tougher backing (a process known as *hard facing*).

FURTHER READING

Metals Handbook, 9th Ed., Vol 11, *Failure Analysis and Prevention*, American Society for Metals, Metals Park, OH, 1986.

D. J. Wulpi, *Understanding How Components Fail*, American Society for Metals, Metals Park, OH, 1985.

H. O. Fuchs and R. I. Stephens, *Metal Fatigue in Engineering*, John Wiley, New York, 1980.

J. E. Dorn, *Mechanical Behavior of Materials at Elevated Temperature*, J. E. Dorn, ed., McGraw-Hill, New York, 1961.

Wear Control Handbook, M. B. Peterson and W. O. Winter, eds., American Society of Mechanical Engineers, New York, 1980.

· 5 ·

The Behavior of Metals in Some Important Chemical Environments

SUMMARY

The metals used in engineering practice, with the exception of the noble precious metals such as gold and platinum, intrinsically are chemically unstable, because it is energetically favorable for them to revert to a chemical compound, such as an oxide, of the type in which they are found in nature. They are said to corrode when this reversion occurs by a chemical process, most typically in practice by a reaction with oxygen in the presence of water. Metals normally are surrounded by an environment of this nature. Corrosion thus in principle cannot be avoided, the only question at issue being the rate at which it does occur. Corrosion in aqueous environments occurs by an electrochemical process that is understood well enough for sound analyses to be developed of the factors which control the rate of general corrosion of various metals and the means by which it might be reduced to an acceptable level in practice. Unfortunately, however, metals sometimes corrode in a nonuniform manner not predicted by such analyses. Pits, crevices, or cracks may develop, and even selective leaching of alloying elements may occur. Moreover, there may be an adverse interaction between corrosion and stress. The phenomena known as stress corrosion, hydrogen embrittlement, and corrosion fatigue are examples. Only a limited understanding has been developed of these complex phenomena, and they can be more difficult to control than general corrosion, especially in certain types of alloys. The formation of an oxide layer in a water-free atmosphere at a comparatively high temperature is also a form of electrochemical corrosion. A good understanding of this process has been developed also, and of the factors which determine the reaction rate. This enables much to be explained about the oxidation of metals, the method by which it can be controlled, and how to develop oxidation-resistant alloys.

The occurrence of oxidation corrosion usually is made apparent by the formation of a layer of oxides, or a scale, on the surface. The material then wastes away. There are cases, however, where certain alloying elements may oxidize selectively in a surface layer of considerable depth beneath a scale. This is called internal oxidation, and its occurrence is not so obvious as are other forms of corrosion. The material remaining in the affected layer is depleted of the alloying elements concerned, and the properties of the alloy in this layer consequently are changed.

The metals used in engineering practice, with the exception of rare and precious metals such as gold and platinum, are not in a stable condition. The stable state in which they had existed for eons in nature was that of a chemical compound, and energy had to be supplied to extract them from these compounds. A metal is thus in a higher-energy state than the mineral from which it came, and the laws of nature, as codified by the laws of thermodynamics, say that a consequence is that a metal will revert to the lower-energy state of a chemical compound if it is able to do so. It is said to *corrode* when this occurs by a chemical process during exposure to a natural environment. The questions of practical importance are the conditions under which this corrosion can occur, and the rate at which it occurs when it does.

In practice, metals are surrounded by many environments which provide conditions that are appropriate for corrosion to occur. For example, oxygen is almost invariably present to permit reversion to an oxide. So the wonder is that metals last long enough in a metallic condition to be useful, let alone for the thousands of years that copper alloys, and even irons and steels in some circumstances, are known to have lasted. These metals last because the reversion rate is slow, sometimes even vanishingly slow. Nevertheless, the struggle against corrosion is a constant and never-ending one which, as every user of metals knows, is not always won. So corrosion joins with the phenomena discussed in Chapter 4 as a major cause of the replacement, and sometimes of the failure, of metal components.

Because the rate at which the reversion reaction occurs determines whether metals corrode and the rates at which they do so, we have to develop some understanding of these reactions. Two situations can be distinguished conveniently for this purpose. The first is one in which a liquid is present, the liquid of importance being water. This can be called *liquid corrosion*. The second is one in which no liquid is present, typified by the exposure of metals to oxidizing atmospheres at elevated temperatures. This can be called *gas corrosion*, but more commonly is described as *oxidation*.

LIQUID CORROSION — UNIFORM

Basic Electrochemical Mechanisms

As we shall soon see, the corrosion of a metal in the presence of a liquid is an electrochemical process, and so the corrosion can occur only in the presence of a liquid in which the metal can form *ions*. All metals can ionize in water, but not in many other liquids; this is why water plays such a central role in corrosion. The electrochemical nature of corrosion can at its simplest be illustrated by the attack of zinc by a dilute solution of hydrochloric acid in water. A vigorous reaction occurs in which the zinc is dissolved to form zinc chloride and hydrogen, the zinc chloride staying in solution and the hydrogen being evolved as a gas. This reaction is described in simple chemistry books by the equation:

$$Zn\ (s)\ +\ 2HC\ell\ (aq)\ \rightarrow\ ZnC\ell_2\ (aq)\ +\ H\ (g) \qquad \text{(Eq. 1)}$$

or more basically as:

$$Zn\ (s)\ +\ 2H^+\ (aq)\ \rightarrow\ Zn^{2+}\ (aq)\ +\ H_2\ (g) \qquad \text{(Eq. 2)}$$

where the (+) superscripts designate positively charged ions, and where (s), (aq), and (g) indicate a solid, a solution in water, and a gas. In detail, however, this reaction involves a number of discrete steps which are illustrated schematically in Fig. 5.1(a). These steps are:

1. An atom at the surface of the zinc releases two electrons:

$$Zn\ \rightarrow\ Zn^{2+}\ +\ 2\ electrons\ (e) \qquad \text{(Eq. 3)}$$

   and the zinc ion produced diffuses into the bulk of the solution — that is, a zinc atom dissolves and is removed from the metal (the zinc corrodes).
2. Hydrogen ions present in the bulk of the solution diffuse to the surface of the zinc where they react with the electrons released by the ionization of the zinc. This can occur at the point of ionization of the zinc or at a more remote point to which the released electrons have traveled within the metal.
3. These electrons neutralize the charges on the hydrogen ions to form molecules of hydrogen gas:

$$2H^+\ +\ 2e\ \rightarrow\ H_2 \qquad \text{(Eq. 4)}$$

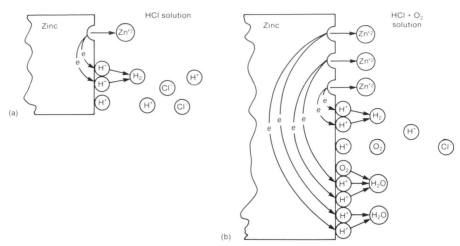

Fig. 5.1. Diagrammatic illustrations of the electrochemical reactions occurring during the corrosion of zinc in air-free (a) and aerated (b) dilute hydrochloric acid solutions.

4. A bubble of hydrogen gas is nucleated at the zinc surface and grows until it breaks free.

Step 1 (Eq. 3) is defined in chemistry as being an *oxidation reaction*, and step 3 (Eq. 4) as being a *reduction reaction*. Any reaction which can be divided into two (or more) partial reactions of oxidation and reduction is defined as being an *electrochemical reaction*. In the corrosion context, the oxidation reaction is called an *anodic reaction*, and this is the corrosion-causing reaction. The reduction reaction is called a *cathodic reaction*. An important point is that the two reactions must occur simultaneously and at the same rate during corrosion. Otherwise, the metal would spontaneously become electrically charged, which is impossible. Thus, if one of the reactions is slowed down, then the other has to slow down to match, and so the corrosion rate is reduced (and vice versa). The fact that one or the other of these reactions can proceed only at a finite rate is the reason why metals corrode at a finite rate, and in turn is the reason why they can be used at all.

Thus all measures of controlling corrosion, either those which occur naturally or those which are implemented by man, are based on the concept of slowing down one of the reactions, hopefully to a rate of zero. It is, in practice, safest to attempt to do this by controlling the cathodic reaction. The penalty for achieving only partial control then is only that the corrosion is reduced instead of being eliminated, and this is better than nothing. On the other hand, the penalty for the breakdown of anodic control even over only small parts of the metal surface is for corrosion to be concentrated at those parts, with the likelihood of the development of a serious pit. Corrosion-control measures need to be assessed from this point of view.

Basic Factors Affecting Cathodic Reactions

Cathodic reactions other than the one described by Eq. 4 are possible, the specific one depending on whether the solution is acidic or alkaline. The one most commonly encountered in acidic systems involves the oxygen reduction reaction illustrated schematically in Fig. 5.1(b) for the dissolution of zinc. The basic reaction is:

$$O_2 + 4H^+ + 4e \rightarrow H_2O \qquad \text{(Eq. 5)}$$

which is shown in Fig. 5.1(b) as occurring in addition to the hydrogen evolution reaction. The extra reaction increases the total rate of the reduction reaction and in consequence the corrosion rate. Thus the corrosion rate is increased by the presence of oxygen or any other oxidizer in an acidic solution. Certain metallic impurities in the solution also cause additional cathodic reactions, increasing the corrosion rate further, but we need not go into them. The point is that corrosion rate can be increased by a number of factors which speed up the reduction reaction.

The oxygen reduction reaction that occurs in neutral or alkaline solutions is different. It is:

$$O_2 + 2H_2O + 4e \rightarrow 4OH^- \qquad \text{(Eq. 6)}$$

This is, in fact, the cathodic reaction that occurs during the corrosion of most metals in oxygenated waters, which characteristically are nearly neutral. (Nevertheless, acidic conditions are increasingly being encountered in polluted industrial atmospheres.) The hydrogen reactions (Eq. 4 and 5) obviously cannot occur in distinctly alkaline solutions because no hydrogen ions are present, and the oxygen reduction reaction (Eq. 6) cannot occur either when no oxygen is present. There are then no cathodic reactions available. Corrosion of metals consequently cannot occur in deoxygenated alkaline solutions.

Basic Corrosion Phenomena in Practice

The anodic reaction in every corrosion reaction is the oxidation of a metal to its ion. This can be written in the general form

$$M \rightarrow M^{+n} + ne \qquad \text{(Eq. 7)}$$

where n is the valence of the metal. One specific example was given in Eq. 3 above. Other examples are:

$$Cu \rightarrow Cu^{+2} + 2e \qquad \text{(Eq. 8)}$$

$$Fe \rightarrow Fe^{+2} + 2e \qquad\qquad (Eq.\ 9)$$

$$Al \rightarrow Al^{+3} + 3e \ . \qquad\qquad (Eq.\ 10)$$

Let us take iron as the first example, considering first its corrosion in a nearly neutral solution. The anodic reaction occurs as for Eq. 9 above, and the cathodic reaction as for Eq. 6, the product being ferrous hydroxide, viz.:

$$2Fe + 2H_2O + O_2 \rightarrow 2Fe^{+2} + 4OH^- \rightarrow 2Fe(OH_2) \quad (Eq.\ 11)$$

Ferrous hydroxide [$Fe(OH)_2$] is insoluble in water and so is precipitated out of solution at the metal surface. Ferrous hydroxide, however, is unstable and soon oxidizes to ferric hydroxide [$Fe(OH)_3$], which product is the destructive, unsightly, and familiar *rust*. But this additional reaction does not affect the principles that we are discussing. Rust is porous and nonadherent and does nothing to isolate the metal surface from the environment (cf. aluminum, which is discussed below). So rusting proceeds unabated until all the iron is wasted away.

Now let us see how the rusting of iron can be reduced by controlling the cathodic reaction. Iron cannot, and does not, rust in the absence of either oxygen or water because the cathodic reaction cannot then occur. This solves the popular mystery of the famous Iron Pillar of Delhi which shows little signs of rust after more than 1500 years of exposure in the open. Its survival is due to the almost complete absence of moisture in the atmosphere at the site and not to any undiscovered secrets of ancient ironmakers in producing corrosion-resistant iron. Iron does not corrode in oxygen-free water either. Thus the recently discovered wreck of the Titanic clearly has suffered little if any rusting after more than 70 years of exposure to seawater at the bottom of the Atlantic Ocean. The wreck is resting at a depth of about 4000 m (13,000 ft), where there is no oxygen. It would have rusted away years ago if it had run into a surface reef instead of an iceberg in mid-ocean. Iron does not corrode in oxygen-free alkaline solutions either, because, as we noted earlier, a cathodic reaction is not available. This is why the steel tubes in steam boilers do not rust away rapidly, as you might expect if you thought about it for a moment. Boiler feed waters are treated to maintain alkaline, oxygen-free conditions, and very effective procedures for doing this are available.

Unfortunately, however, moisture and oxygen abound in the environments to which irons and steels are commonly exposed, and these environments are typically nearly neutral. Engineers consequently, in the general situation, have to seek ways of keeping rusting within reasonable bounds. The method of doing this with which you will probably be most familiar is to attempt to seal off the surface from the environment by coating it with the adherent, nonconductive material known as paint. Paints are indeed effective, but only for as long as they remain adherent, continuous, and impervious to moisture. Normal corrosion also occurs at any breaks in a paint coating, such as those resulting from mechanical damage. The price of this passive method of corrosion control consequently is constant maintenance. Other, more active control methods are available and will be discussed in following sections.

The problems encountered with the corrosion of iron are increased because the corrosion product is not protective. This is not always the case, an extreme example being aluminum. All who are interested in the conservation of energy know by now that it takes a lot of energy to reduce aluminum oxide to metallic aluminum. On these grounds, as previously explained, aluminum would be expected to corrode very readily. We all know, however, that it has good corrosion resistance under common conditions of usage; it is much better in this respect than iron, which is intrinsically a less active metal. The reason is that the product of corrosion this time is aluminum oxide (alumina), which forms a continuous adherent layer over the corroding surface and which provides an effective physical barrier that prevents further corrosion. This is a form of anodic control but is effective because the oxide layer is self-forming and self-healing.

The protective alumina layer is very stable in neutral and many acid solutions, in all of which aluminum has good corrosion resistance. This layer is, however, attacked by alkalis, and so aluminum is not resistant to corrosion in alkaline solutions. But as a bonus on the positive side, the alumina layer can be produced artificially in a thickened form by an electrolytic treatment known as *anodizing*. The thickened layer enhances the corrosion protection and has the further advantage that it is hard and abrasion resistant. Moreover, it can be colored in several ways to improve the appearance of the final product. Aluminum alloy architectural sections are now commonly treated in this way, usually to produce a pleasing brown color.

Titanium is another intrinsically reactive metal which in practice has excellent corrosion resistance in many environments, notably seawater and other chloride-containing solutions, because it forms a protective surface layer naturally. A layer of titanium dioxide forms in this instance, and this layer too can be enhanced by anodizing.

Copper, on the other hand, is a metal that we can use as an example of a comparatively noble metal. The hydrogen cathodic reaction (Eq. 3) then usually cannot occur in acidic solutions, but the oxidation reaction (Eq. 5) can. The result is that copper does not corrode in acids unless oxygen or an oxidizing reagent is present. The hydroxide-producing reaction (Eq. 6) can occur in nearly neutral or alkaline solutions, but only very slowly. Copper alloys consequently are quite resistant to corrosion in such solutions. Copper alloy artifacts are known to have survived for thousands of years with little corrosion damage (see, for example, Fig. 2.1, p. 13). Indeed weathered copper alloy objects are usually covered with a pleasing greenish film, called a *patina*. The formation of a patina is due to a complex series of reactions, in addition to the main corrosion reaction. These reactions are caused by the presence of minor impurities in the environment, sulfur being the one principally responsible. Sulfur in the environment, incidentally, is also responsible for the less-than-pleasing black tarnish that all too frequently forms on silver, an even more noble metal. The presence of ammonia in an alkaline solution may also cause corrosion of copper, in spite of what we have just said, but this and the immediately preceding examples are exceptions to the main story and are of only specialized interest. There are some other exceptions to the basic story which are of considerable general importance, and so we shall pursue them in the following sections.

Polarization

It is sometimes possible to slow down either the oxidation reaction or the reduction reaction by slowing down one of the processes associated with them in the list on p. 133. The system is then said to be *polarized*. The term *activation polarization* is used to describe a situation in which corrosion ends up being controlled by the rate of a reaction which occurs at the metal/electrolyte interface (e.g., by reducing the rate at which hydrogen bubbles form). The term *concentration polarization* is used where corrosion is controlled by the rate of diffusion in the bulk electrolyte (e.g., by reducing the rate at which hydrogen ions can diffuse in the corroding solution). Environmental variables have different effects on corrosion rate depending on whether or not they have an influence on the polarization reaction if it is controlling the corrosion sequence. For example, agitating the corrosive medium increases the rate of diffusion in the liquid medium and so increases the corrosion rate if concentration polarization happens to be in control. But it does not do so if activation polarization happens to be in control. This is a matter of some practical importance and an indication of the number of factors that can affect the corrosion rate.

Passivity

A phenomenon known as *passivity* is of even greater importance than polarization. Essentially, it is a desirable loss of chemical activity exhibited by certain metals and alloys in moderately to strongly oxidizing conditions. The metals concerned then become essentially inert under the particular conditions and act as though they were noble metals. The metals that are most likely to exhibit this kind of behavior are among those which are commonly used in engineering practice and include nickel, silicon, chromium, and titanium. The addition of one of these elements in an appropriate amount as an alloying element may also confer passivity on another base metal. Important examples of this are the so-called stainless steels (which might better be referred to as corrosion-resistant steels). The key alloying element in stainless steels is chromium, which confers an excellent degree of passivity on iron in many environments when it is present in amounts greater than about 12%. The passivating characteristics of the elements concerned are frequently the reason why they were chosen as industrial metals in the first place even though the basic reason may not have been understood. Zinc, tin, and cadmium also exhibit passivity under more limited conditions.

Passivity is difficult to define as a phenomenon but can be described by characterizing the metal by the type of behavior illustrated diagrammatically in Fig. 5.2. Normally, the corrosion rate progressively increases rapidly with increases in the oxidizing power of the corroding solution, as indicated by line AB in Fig. 5.2 and its dotted extension. But for a metal which passivates, the corrosion rate decreases to a low value when the oxidizing power reaches a certain level (BC in Fig. 5.2). It then stays low over a range of oxidizing powers (CD in Fig. 5.2), but eventually increases again (DE in Fig. 5.2). The metal is said to be passive over the range of conditions for which the corrosion rate is low (CD in Fig. 5.2).

The reduction in corrosion rate at the active-passive transition is often large. It can be as much as a thousand or even a million times. The precise causes of these transitions are not completely understood, but in general terms they can be ascribed to the formation of either a surface film or a protective barrier which is stable over a considerable range of oxidizing powers but which eventually breaks down in very strongly oxidizing solutions. In this sense, passivity is a case of activation polarization. It is also a form of the less desirable anodic form of corrosion control, but is effective because the control-

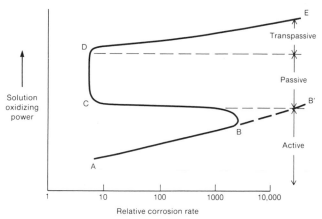

The corrosion rate of normal metals increases steadily with oxidizing power, as indicated by the portion AB of the main curve and its dotted extension BB′. For a metal which passivates, however, the corrosion rate drops remarkably when a certain oxidizing power is exceeded, as indicated by BC of the full curve. The corrosion rate then remains at a low value for a range of oxidizing powers (CD) but eventually increases (DE).

Fig. 5.2. Diagrammatic illustration of the variation in corrosion rate with the oxidizing power of the corrosive liquid for metals which passivate.

ling barrier is self-forming and self-healing. Nevertheless, trouble is possible if passivation doesn't quite work, because accelerated local corrosion then occurs at points where the passivation breaks down.

In any event, the existence of the passivation phenomenon is often the reason why metals do survive so well in industrial environments that are sufficiently oxidizing.

Galvanic Effects

A difference in electrical potential develops between many pairs of dissimilar metals when they are immersed in a conductive solution. This difference in potential causes electrons (a current) to flow between the two in the external circuit if they are electrically connected. This is the effect upon which electrical batteries depend. The difference in potential also encourages the movement of ions in the conductive solution from one metal toward the other; a consequence is that the corrosion of one metal is increased (step 1 in the list on p. 133 is speeded up) and that of the other metal is decreased (step 1 in the list on p. 133 is slowed down). The metal for which the corrosion rate is increased is said to become *anodic*, and the metal for which it is decreased is said to become *cathodic*. This is called *galvanic corrosion* in the former case and *galvanic protection* in the latter. These effects are direct results of the electrochemical nature of corrosion. They are named after Aloisio Galvani, Professor of Anatomy at Bologna, Italy, who observed the phenomenon concerned as long ago as 1789, although he did not correctly diagnose its cause.

The driving force for galvanic corrosion — the electrical potential developed between the two metals — varies considerably with different pairs of metals and different liquid media. The likely magnitude of galvanic effects in corrosion varies accordingly. The galvanic series listed in Table 5.1 gives an indication of the relationships between common engineering alloys exposed to a common corroding medium (unpolluted seawater). The more widely that a pair of metals is spaced in this table, the more marked the galvanic effect is likely to be, the corrosion rate of a metal low in the table being increased by contact with a metal higher in the table. Galvanic corrosion is not likely at all between bracketed pairs of metals. Note that a metal may appear at different positions in the table depending on whether its surface is active or passive. However, a table of this type can be used only as a guide, confirmatory corrosion tests being desirable — particularly in unfamiliar circumstances.

Combinations of dissimilar metals which are electrically connected are used commonly in engineering structures, including many which are subjected to corrosive environments. Among these are cases where the corrosion rate of one of the metals is increased to a severe extent. Sometimes this is due to an error in materials selection, such as when iron fittings were used in the first

**Table 5.1. Galvanic Series of Some Commercial Metals
and Alloys in Seawater(a)**

Noble, or cathodic	Platinum
	Gold
	Graphite
	Titanium
	Silver
	Stainless steel, austenitic, P (18% Cr, 8% Ni, low C)
	Stainless steel, ferritic, P (10 – 30% Cr, high C)
	⌈Nickel-chromium-iron alloy, P (80% Ni, 13% Cr, 7% Fe)
	⌊Nickel, P
	Silver solder
	⌈Nickel-copper alloy (70% Ni, 30% Cu)
	│Copper-nickel alloys (60 – 90% Cu)
	│Copper-tin bronzes
	⌊Copper-zinc brasses
	⌈Nickel-chromium-iron alloy, A (80% Ni, 13% Cr, 7% Fe)
	⌊Nickel, A
	Tin
	Lead
	Lead-tin solders
	Stainless steel, austenitic, A (18% Cr, 8% Ni, low C)
	Stainless steel, ferritic, A (10 – 30% Cr, high C)
	⌈Cast iron
	⌊Steel
	Aluminum alloys, precipitation hardened
	Cadmium
	Aluminum, commercially pure
	Zinc
Active, or anodic	Magnesium and magnesium alloys

(a) P indicates passive condition; A indicates active condition.

wooden sailing vessel whose hull was sheathed with copper to resist the ravages of the Toredo shipworm. A British naval frigate was so fitted in 1763, before Galvani's discovery, and was docked after some service to assess the effectiveness of the sheathing. The sheathing had indeed been effective in deterring the Toredo worm, but the rudder was about to fall off the hull because its iron hinges had all but corroded away — an early example of accelerated galvanic corrosion due to an incorrect choice of materials. This error was certainly excusable at the time. The British Navy solved the problem by the obvious strategem of using copper alloy fasteners, including nails, although the cause of the problem was not of course recognized at the time. Errors of this type are still made (Fig. 5.2A) but now are perhaps less excusable. It is a factor the possibility of which a designer needs always to keep in mind.

The components of the frame are anodized aluminum alloy extrusions joined at the corners by internal steel fixing pieces. Aluminum is anodic to steel and has corroded completely through from the inside in a region where it has contacted the steel fitting. The steel (large arrow), on the other hand, is virtually uncorroded. The aluminum elsewhere is not seriously corroded, although a few corrosion pits (small arrows) have developed, particularly on horizontal surfaces.

Fig. 5.2A. An example of accelerated galvanic corrosion in an aluminum alloy architectural fitting exposed to a marine environment.

The use of a combination of metals which is unfortunate from this point of view is, however, sometimes unavoidable. The propellers for modern steel-hulled ships are commonly cast from copper alloys because the particular alloys used are outstandingly the most suitable for the purpose from many other points of view. The propellers do cause accelerated corrosion of the adjacent stern regions of the hull, and either this has to be accepted or suitable precautions have to be taken to combat the problem. We shall soon discuss methods of doing this.

The coatings applied to metals to improve their corrosion resistance, their appearance, or some other surface characteristic constitute an important group of cases where inappropriate combinations of metals from this point of view may have to be used to meet other criteria. Many of the electroplated coatings used on steel are in this category, an example being the nickel coatings which are widely used to protect steel components from atmospheric corrosion and for decorative purposes. (The nickel is covered by a layer of chromium to provide a permanently attractive appearance and the coating system is known as *chrome plating*; but it is the nickel that provides the basic corrosion protection.) Nickel has good corrosion resistance under these circumstances and performs its protective role well, provided that the coating is continuous. However, nickel is cathodic to steel and so tends to stimulate corrosion of the underlying anodic steel at any discontinuities which might be present in the coating (Fig. 5.3a); even quite fine pores suffice. The corrosion then extends as a pit into the steel (Fig. 5.3a), a pit which might well penetrate the section thickness. The pit also extends preferentially along the nickel/steel interface (Fig. 5.3a), which may cause the coating to peel off. The long-term effectiveness of this protective system consequently depends on the nickel deposition process being designed and controlled to ensure that the deposit is adequately continuous. This, among other things, involves depositing a certain minimum thickness of nickel, and the cost-conscious will want to have the minimum possible thickness. There is a balance between cost and quality. This phenomenon is an example of the pitfalls of anodic control of corrosion.

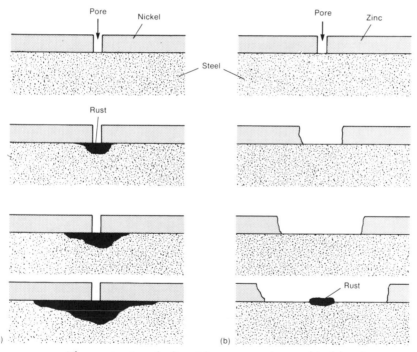

(a)　　(b)

The coating is cathodic with respect to the metal in (a) and anodic in (b). Accelerated corrosion occurs at the discontinuity in the first case, developing a pit. The coating is consumed in the second case, but the base metal continues to be protected for as long as some coating remains in the neighborhood.

Fig. 5.3. Diagrammatic illustrations of the effects on corrosion of a discontinuity in a protective coating on a metal.

The relative areas of the cathode and anode have an important influence on these corrosion phenomena. In the copper-sheathed sailing ship, the area of iron was small compared with that of copper; consequently, the galvanic current density was large at the iron anodes and so their corrosion rate was considerably increased. This is true also for a pore in a nickel coating. The ratio of anode to cathode area is large in a steel-hulled ship fitted with a copper alloy propeller, and so the corrosion of the iron is not so markedly increased.

Galvanic effects can also contribute to corrosion on a microscopical scale when phases or constituents (p. 154) are present which are anodic to the matrix metal. Local galvanic cells may then be set up wherever particles of the phase outcrop at the surface and the general corrosion rate is thereby increased. For example, commercially pure aluminum contains very small particles of phases due to the presence of impurities, iron, and silicon, and these phases are anodic to aluminum. The consequence is that the general corrosion resistance of commercially pure aluminum is much inferior to that of high-purity grades which contain little of these phases. Moreover, corrosion pits tend to develop in commercially pure grades, such as the pits you may have noticed on the inner surface of old domestic cookware, and this pitting is probably initiated by groups of the offending phase particles. Similar effects are possible in other alloys. Often the alloy can be designed to ensure that deleterious phase particles are not present, but perhaps only in special grades at higher costs. Again taking aluminum as an example, the impurities responsible are inevitably present in material smelted by standard methods. Special, more costly smelting procedures are necessary to produce the high-purity grades which have improved corrosion resistance.

Cathodic Protection

The galvanic effect can, however, be used beneficially to reduce or even eliminate corrosion by using it to effect cathodic control. Let us go back to the discussion of Fig. 5.1 as an aid to understanding the principles involved. The diffusion of zinc ions into solution in this illustration (part of step 2 in the list on p. 133) constitutes a flow of electrical current between the corroding metal and the solution. Corrosion will be reduced if this current, or flow of ions, is reduced to zero. Thus a current of any magnitude imposed in the opposite direction will provide some corrosion protection, and a reverse current exceeding a critical value will prevent corrosion altogether. This critical current is a characteristic of the corrosion system but has to be determined empirically.

The protective current can, in practice, be obtained from either one of two sources. The first is to use the galvanic effect in reverse by connecting the metal which needs to be protected to a more active metal, such as one that is listed considerably below it in Table 5.1 (*sacrificial protection*, this is called). The second is to connect the metal and a more noble metal to the negative and positive terminals, respectively, of an external electrical power supply (*impressed current protection*). In either event, the metal being protected is the cathode in the system, and hence the procedure is known as *cathodic protection*.

Coating steel with zinc is an example of the application of sacrificial protection. The steel is coated either by immersing it in molten zinc (hot dipping) or by electroplating. It is then known as *galvanized steel* (little did Galvani realize how widely his name was to be commemorated, and by engineers at that). The corrosion resistance of zinc itself is not particularly good, and zinc would not be used as a coating for its corrosion resistance alone. But zinc is strongly anodic to steel (Table 5.1) and so protects the steel at any point where the steel is exposed by discontinuities in the coating (edges, scratches, etc). The zinc corrodes sacrificially — i.e., the zinc layer is consumed while it is protecting the steel (Fig. 5.3b). The area of bared steel is progressively enlarged, but the steel does not begin to rust until the zinc becomes too remote to be protective, which occurs when the current density developed by the galvanic couple falls locally below the critical value. The protection is also lost, of course, when the zinc coating is removed by normal corrosion of the zinc itself, and so the original thickness of the coating is important on this count also. Other types of coatings can be used to protect steel; a zinc-iron alloy (*galvannealed coatings*), aluminum-zinc alloys (*zincalume coatings*), and electroplated cadmium are three that are used extensively. Each has its advantages and disadvantages. Cathodic protective coatings can also be devised for other metals, using Table 5.1 as a guide. For example, a coating of aluminum is used on aluminum alloy sheet, the composite being known as *alclad sheet*. The cladding provides cathodic as well as general protection.

All of this is the reverse of the situation with nickel coatings on steel and illustrates the principle that cathodic methods of protection are to be preferred to anodic methods.

The same galvanic type of protection can be achieved by using, for steel, a separate block of zinc (or magnesium or aluminum) connected by an electrical conductor to the steel. Both the zinc and the steel have to be immersed in the corroding medium to complete the electrical circuit, and the protection extends for only a finite distance from one of these *sacrificial anodes* — namely, the distance over which the critical protective current is maintained. So anodes have to be spaced at appropriate intervals. The anodes are consumed gradually and thus have to be replaced from time to time. Steel structures which are immersed in the sea commonly are protected in this way, including the hulls of ships, in which the effect is used to overpower the adverse effects of copper alloy propellers mentioned earlier. This technique is also used to protect buried pipelines and large water-storage tanks.

The second method of obtaining cathodic protection is, as we have already mentioned, to impress a current from an external source, an inert electrode and the metal that is to be protected being connected to a source of direct current in the appropriate sense. Again, a critical current density has to be exceeded for full protection to be achieved. Such a system can be controlled more precisely than a sacrificial system but has to be controlled carefully (see section on hydrogen embrittlement below). It is particularly useful in circumstances where the regular replacement of sacrificial electrodes is inconvenient.

Both types of cathodic protection systems can be very effective, but a good deal of specialized technology is required for their effective and economical operation. Special alloys are required to ensure that the anodes remain active (they tend to become sealed off by corrosion products) and that sacrificial anodes corrode evenly. Impressed currents have to be of the correct magnitude. Many other design details as well determine whether full protection is achieved economically.

There is an unfortunate obverse side to the coin of impressed current protection because stray direct currents from many sources abound in the ground beneath modern built-up communities. These currents, if of the wrong sense, can cause accelerated corrosion of buried steel structures. Designers of systems which are likely to produce stray currents, such as electrified railway systems, consequently may be required to arrange their designs to reduce stray currents generated by the system to the minimum. Designers of buried structures often have to install protection systems to cope with stray currents.

LIQUID CORROSION — NONUNIFORM

It has been implicit in the discussion to this stage that material is removed uniformly from a corroding surface. Corrosion does most commonly occur in this way, allowing that more severe general corrosion is to be expected in areas of a structure which happen to be exposed to more severely corrosive conditions than others. Steel piling used in wharfs and piers is an example of the latter situation. The corrosion of piling is always most severe in the splash zone at and above high-tide level because this zone is alternately wetted and dried and is subjected to variable aeration.

General corrosion is the easiest situation to cope with. The lives of structures can be predicted from established data, and nothing occurs that should not really be expected. However, highly localized corrosion does occur under some circumstances and can be more insidious and serious. It may result in premature failure, and even in unexpected failure, because its presence and extent may not be easy to discover. We shall now discuss a number of the more important modes of localized corrosion.

Pitting

The accelerated local corrosion which can occur at discontinuities in coatings on metals in certain combinations has already been discussed. These are special although important cases, but there are more general situations where corrosion leads to the development of localized depressions in a surface. The depression is called a *pit* when it is deep with respect to its surface diameter. Pitting is one of the most destructive and insidious forms of corrosion. It can cause failure by perforation of thin sections (Fig. 5.4b) after the dissolution of only a small absolute amount of material. Moreover, the pit may be difficult to detect be-

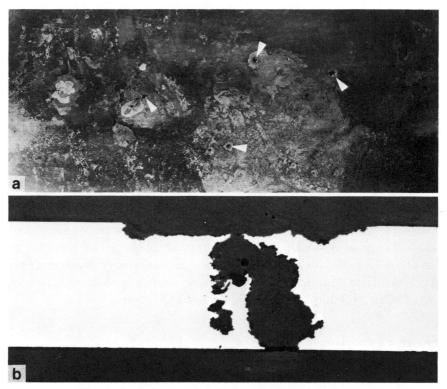

The pits have developed at the centers of patches of general corrosion, indicated by arrows in (a). They have penetrated the wall thickness of the tank, as can be seen in the section through a typical pit shown in (b). The inner surface of the tank is at the top of this photomicrograph. The corrosion occurred because the fuel had become contaminated with water and a mat of fungus had grown across the tank surface. (a) Photograph; 1×. (b) Optical micrograph; 100×.

Fig. 5.4. An example of pitting corrosion developed at the bottom inner surface of an aluminum alloy fuel tank.

cause of its small diameter and because its mouth may be covered with corrosion product (Fig. 5.4a). Worse still, pits commonly develop at awkward positions, such as the inside bottom surface of a container, where they are next to impossible to detect. Vehicular fuel tanks are an example in which leaking fuel may be the first indication that pits have developed in the bottom inner surface of the tank. This was so with the tank illustrated in Fig. 5.4, and constitutes a potentially dangerous situation.

Pitting corrosion can be regarded as an intermediate case where the material is balanced between general corrosion and complete corrosion resistance. The pits usually initiate only after an extended period of exposure, perhaps due simply to random local variations in corrosion rates. However, they grow rapidly once initiated because they establish conditions within themselves which accelerate the corrosion reaction. The corrosive ions concentrate in the direction of gravity and so it is found that pits mostly develop in upward-facing horizontal surfaces and grow downward. This is why they tend to form on the bottom inner surfaces of tanks, as mentioned earlier, and on horizontal surfaces more rapidly than on vertical surfaces (Fig. 5.2A).

The tendency for corrosion pits to develop varies considerably with both the environment and the metal. Commercially pure grades of aluminum are, for example, particularly susceptible, but not high-purity grades. You probably have noted the pits that often develop at the inner *bottom* surface of aluminum cooking utensils. Most pitting is caused by the presence of chloride and chloride-containing ions in the corroding solution and occurs under stagnant conditions. It may be prevented by the addition of chromate or silicate ions; some commercial corrosion inhibitors contain these types of ions, but they must be added with care because pitting is accelerated when the ions are present in insufficient concentrations. Many alloys are susceptible to pitting corrosion, stainless steel alloys as a class being notoriously susceptible.

Crevice Corrosion

Intense localized corrosion frequently occurs at the edges of and within crevices and adjacent to shielded areas on a surface. This may occur even when no noticeable general corrosion has occurred (Fig. 5.5). Lap joints, surface deposits and encrustations (sand, dirt, corrosion products), holes, and the interfaces under bolt and rivet heads are common susceptible areas. The crevice must be wide enough to permit entry of the corrosive solution but narrow enough to maintain a stagnant zone in which the corrosion products can concentrate. Fibrous gaskets and absorbent deposits are particularly deleterious because they draw in solution and then maintain a stagnant zone.

Crevice corrosion is related to pitting corrosion, the initiating pit having in effect already been provided by the surface geometry of the component. Thus only the corrosive conditions necessary for pit growth are needed. Conse-

Plastic tubing had been wound around the austenitic stainless steel tube section at the position indicated by the circle. The unit had then been immersed in seawater. Marked localized corrosion (arrows) occurred at the crevice formed between the tubing and the metal surface. The remainder of the stainless steel is uncorroded.

Fig. 5.5. An example of crevice corrosion.

quently, the attack can occur in a wide range of media. Nevertheless, crevice corrosion, like pitting, usually is most intense in media containing chloride ions. A long initiation period is often required before the attack commences, and the attack, once started, proceeds at an ever-increasing rate. Metals and alloys which depend on oxide films or passive layers for their corrosion resistance are particularly susceptible because these films are destroyed by the high concentrations of chloride and hydrogen ions that develop in crevices. Stainless steels, aluminum alloys, and titanium alloys thus are highly susceptible.

Intergranular Corrosion

Highly localized attack occurs along the grain boundaries (p. 241) of some alloys. The grains may even become separated (Fig. 5.6), in which event the strength of the material is destroyed. Grains may even fall out of the surface. This *intergranular* (or *intercrystalline*) *corrosion* is another of the insidious forms of corrosion which can cause serious loss of strength after the dissolution of only a very small amount of material and perhaps with little or no external indication that corrosion has occurred at all. An exception to the latter is when the grains

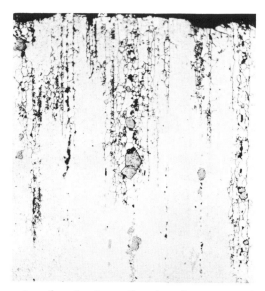

A section of an aluminum alloy aircraft wing spar which had been exposed only to the environments normally encountered in service. Corrosion has penetrated completely around the grain boundaries to a considerable depth. The material would have no strength in these regions. Optical micrograph; 250×.

Fig. 5.6. An example of intergranular corrosion.

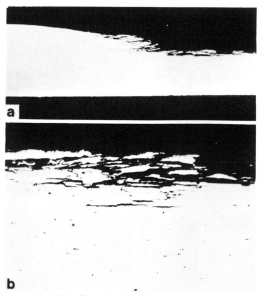

The grains in this alloy are slab-shaped with the major boundaries parallel to the surface which has corroded. Intergranular corrosion has caused grains to flake off the surface in layers; that is, to exfoliate. Optical micrographs: (a) 10×; (b) 500×.

Fig. 5.6A. Exfoliation corrosion in an aluminum alloy.

are slab-shaped with the elongated grain boundaries parallel to a corroding surface. This type of grain structure is commonly found, for example, in aluminum alloys. The corrosion products formed along the elongated grain boundaries during intergranular corrosion force the grains apart in parallel layers (Fig. 5.6A). The affected surface layers flake off in a visible way. This is called *exfoliation corrosion*.

Fortunately, only comparatively few alloy types are susceptible to intercrystalline corrosion and even then, for the most part, only when the alloy is in a particular state in which an abnormal structure has been developed at the grain boundaries. There are two possible types of sensitive structures. In the first, a thin film of a phase or a segregate which can corrode preferentially is present at the grain boundaries (Fig. 5.7a). The phase concerned can result from the presence of a particular impurity or alloying element or can be caused by a thermal treatment. The sensitivity to intercrystalline corrosion of a number of aluminum alloys, such as the one illustrated in Fig. 5.6, seems to be due to this type of effect. The second possibility is that the alloying element that confers the corrosion resistance on the alloy is depleted in zones adjoining the grain boundaries during formation of the phase. The alloy-depleted zones then corrode rather than the grain boundaries themselves (Fig. 5.7b), but the end result is the same.

The corrosion-resistant austenitic stainless steels are examples of alloys susceptible to the second type of effect. The corrosion resistance of these alloys depends on the presence of at least 12% chromium in the alloy, and the chromium has to be present at this concentration at every point in the surface.

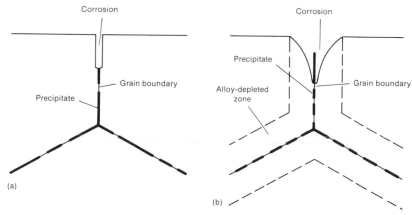

In (a), precipitates are present at the grain boundaries and the precipitates corrode preferentially. In (b), the formation of a precipitate at the grain boundaries drains a critical alloying element from the adjoining regions in the grains. It is the alloy-depleted bands that corrode preferentially in this case.

Fig. 5.7. Sketches illustrating diagrammatically two mechanisms by which intercrystalline corrosion can occur.

Austenitic stainless steels contain at least 18% chromium but usually also contain small amounts of carbon. The carbon can combine with the chromium to form a chromium carbide, and it may do this under circumstances where the carbides form at the grain boundaries of the alloy. In this event, the chromium comes from narrow adjoining zones in the grains, and the chromium content in these depleted zones can easily fall below the critical 12% level. The depleted zones may then be corroded rapidly by a medium which the alloy normally would resist (Fig. 5.7b).

Manufacturers of stainless steels take care to ensure that the material they supply does not have a susceptible structure, but the structure may still be developed if the material is subsequently heated within a certain *sensitizing range* of temperatures (500 to 800 °C, or 930 to 1470 °F). These alloys commonly are welded, and it is inevitable that some zone adjacent to a weld will be heated to temperatures in the sensitizing range and so become susceptible to intercrystalline corrosion. The integrity of an expensive structure may then be jeopardized by the corrosion of a very small amount of its material in narrow bands adjacent to the welds. This phenomenon is called *weld decay*. More general intergranular corrosion can occur, of course, if the material is not properly treated in the first place or if it is subsequently heated as a whole in the sensitizing temperature range by error. Several techniques for alleviating this problem have been developed. One is to reduce the carbon content of the steel to such low level that insufficient carbon is present to remove significant amounts of chromium from where it is needed in the alloy. The second remedy is to add small amounts of elements (principally titanium and niobium) which form carbides at the sensitizing temperature in preference to chromium. The carbon is soaked up by these elements before chromium carbides get a chance to form. The chromium is then left where it can continue to protect the steel.

Selective Leaching

One of the elements in a solid-solution alloy (p. 228) may be removed preferentially during corrosion, and this is referred to as *selective leaching.* Actually, both elements of the solid solution may dissolve during the corrosion process, one of them then being redeposited on the corroding surface.

This phenomenon can occur in a number of alloys. The important example in practice, however, is the selective removal of zinc from copper-zinc alloys (brasses) containing more than about 15% zinc (Fig. 5.8). The corrosion process is then known as *dezincification.* Copper is left at the corrosion site as a spongy mass with no mechanical strength. The dezincification may progress from the surface either as plugs (Fig. 5.8a) or as a uniform layer (Fig. 5.8b). In brasses of very high zinc content in which two phases are present in the microstructure (p. 232), the dezincification occurs preferentially in one of the two phases (actually, in the phase of higher zinc content, the so-called beta phase) and penetra-

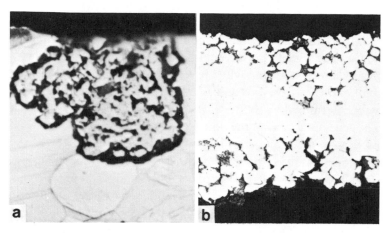

The phenomenon in this type of alloy is called dezincification because it is the zinc of the alloy that is removed selectively during corrosion. Dezincification can occur by the development of a localized plug (a) or by the selective corrosion of one of the constituent phases (b).

Part (a) shows a section of a Cu-30Zn alloy tube which has dezincified in a plug. The material removed from the plug has been replaced by a spongy mass of copper which has no strength. Optical micrograph; 500✕. Part (b) (source: M. Hatherly) shows a section of a sheet of a Cu-40Zn alloy. This section has not been etched to reveal the microstructure of the alloy, but it consisted of a mixture of alpha and beta phases similar to that for the material illustrated in Fig. 11.5(b) (p. 355). Here, however, the areas of beta phase are visible in the two surface layers because the phase has been selectively dezincified. Optical micrograph; 100✕.

Fig. 5.8. Examples of selective leaching during the corrosion of copper-zinc brasses.

tion is consequently accelerated. An example is illustrated in Fig. 5.8(b). In all cases, mechanical strength is lost in proportion to the extent of penetration, a deterioration which is unlikely to be apparent from external appearances. A hint may be obtained from a change in surface color from the goldish hue of brass to the reddish hue of copper, but this is scarcely a strong indicator. Selective leaching may also occur in some other types of copper alloys, when it is usually known by a term indicating the element which has been leached selectively (e.g., dealuminification for the loss of aluminum from a copper-aluminum alloy). In all alloys, however, the phenomenon occurs only in specific corrosive media, the characteristics of which are difficult to define. For example, the high-zinc brasses commonly used in domestic plumbing fittings are susceptible to dezincification in mains water, but fortunately only in the water from a few local regions. If the use of brass fittings is banned in your area, that is probably why.

The susceptibility of brasses to dezincification can in some cases be adequately suppressed by adding certain alloying elements in minor amounts. Tin in amounts of about 1% and arsenic, antimony, or phosphorus in amounts of 0.02% to 0.05% are effective in brass containing 30% zinc. This is an alloy that is widely used because of a good combination of properties, including corrosion resistance. There was a time when it was used in the condenser tubes of the main engines of naval vessels and the sea-readiness of the vessels was then effectively determined by the state of the steam condensers of their main engines. The condenser tubes at the time were made from a straight 30% Zn brass, which is very prone to plug dezincification of the type illustrated in Fig. 5.8. Dezincification was likely to penetrate the walls of the tubes, causing them to leak. This all too often occurred unexpectedly, and without warning, after a short service life. The discovery of the effectiveness of the simple addition of tin (the additions of arsenic, antimony, and phosphorus came later) largely solved the problem. The common name appropriately given to the resistant alloy was, and still is, admiralty metal. This alloy is still used widely in marine condensers, although high-performance naval condensers now use an even better copper-nickel alloy.

A phenomenon which involves a different mechanism but which can still be called selective leaching occurs in an alloy which contains a phase (p. 230) that is markedly different chemically and electrochemically from its matrix. The gray cast irons which are used so commonly and which contain flakes of graphite as a microconstituent are examples, graphite being strongly cathodic to iron (Table 5.1). The consequence is that accelerated corrosion occurs in the iron at the iron/graphite interface under quite mildly corrosive conditions (Fig. 5.9). This corrosion penetrates deeply because the graphite flakes are interconnected in three dimensions. The phenomenon is often called *graphitization*, but this is another one of the unsatisfactory terms in metallurgical usage. Graphitization is better used to describe the formation of graphite in iron. *Graphite corrosion* is the better term to use for the phenomenon just discussed.

Erosion-Corrosion

The rate of corrosive attack is significantly accelerated by rapid relative movement between the corrosive fluid and the metal surface. This phenomenon of *erosion-corrosion* is characterized by the development of deep smooth grooves, waves, and rounded holes in the corroding areas, the features being aligned along the local direction of liquid flow (Fig. 5.10). The attack can be rapid, resulting in failure of a component in a much shorter time than would have been expected from experience with static conditions.

The accelerated attack occurs because the moving liquid removes the films which are protecting the surface from corrosion, and is further enhanced because reaction products are swept away from the corrosion site. Since most metals depend on a surface film of some sort for their corrosion resistance, most are susceptible to erosion-corrosion. Metals which are easily damaged or worn mechanically (e.g., copper alloys) are particularly susceptible.

This iron has been immersed in seawater and corrosion has occurred preferentially in the matrix at the interface with the graphite flakes (light gray). Corrosion has consequently extended along the flakes, which are interconnected, much more rapidly than would otherwise have occurred. Corrosion by selective leaching in this manner can occur quite rapidly, leaving a spongy black mass with no strength. Optical micrograph; 100×.

Fig. 5.9. An example of selective leaching of the iron matrix in a gray cast iron.

This phenomenon must be distinguished from abrasive erosion (p. 127). Erosion-corrosion occurs by a chemical process in which metallic ions are taken into solution, whereas abrasive erosion occurs by mechanical processes in which discrete pieces of metal are physically removed from the surface. Nevertheless, the presence of abrasive solids in a corrosive medium may aggravate erosion-corrosion by facilitating the removal of protective films.

It is the local relative velocity of the corrosive medium that determines whether erosion-corrosion occurs at a particular point. Thus turbulences in the flow of the corrosive liquid are of major importance, and erosion-corrosion frequently occurs at geometric irregularities in a surface which disturb the laminar flow of the corrosive liquid (e.g., at ledges, crevices, deposits, and bends). An extreme case is where the conditions of the fluid flow result in a high-velocity jet impinging on the metal surface (*impingement corrosion*). A pipe,

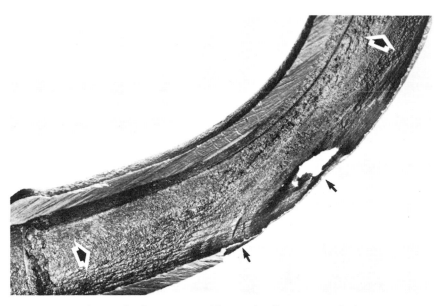

The pipe has been sectioned longitudinally to expose the inner surface. The white arrows indicate the direction of flow of the water. The black arrows indicate the area at which the pipe has been penetrated. A considerable surrounding area of the pipe wall has also been wasted from the inner surface. Photograph; $\frac{1}{3}\times$.

Fig. 5.10. An example of impingement corrosion at the inner surface of a bend in a steel pipe which carried a stream of mains water.

for example, may be penetrated at one point in a bend without corrosion occurring elsewhere.

A special case of erosion-corrosion, called *cavitation damage*, is caused by the formation and collapse of vapor bubbles in the corrosive liquid at the metal surface. These bubbles form when the conditions induced by the fluid flow produce regions of low pressure, low enough for the liquid to boil at the temperature of the liquid (i.e., for the local pressure to fall below the vapor pressure of the liquid). The bubbles collapse explosively, producing shock waves which remove the protective film on the metal surface at the sites of the bubbles. Corrosion of the bared metal then occurs until the protective film is re-formed, initiating a pit. The embryonic pit encourages the formation of another bubble at the same site, and repetition of the process results in the development of a deep hole. Marine propellers, hydraulic turbines, and pump impellers are examples of components susceptible to this type of damage. Severe damage and even penetration can occur after only a short period of operation (Fig. 5.11). Incidentally, the cavitation usually also induces vibrations in a rotating system. Unsatisfactory hydrodynamic design normally is the primary cause of cavitation, and then correcting the design is the only sure way of eliminating the problem. But the severity of the effect can sometimes be re-

This copper-alloy marine propeller was in service in a work boat for only a short period. Severe cavitation damage has occurred in a band toward the blade tip, and the blade has even been penetrated at some points.

Fig. 5.11. An example of cavitation damage.

duced by using a more corrosion-resistant alloy, by improving the surface finish of the susceptible surfaces, or by coating the susceptible surfaces with a resistant material.

CONTROLLING CORROSION

We noted early in this chapter that, thermodynamically, it is a wonder that metals do not corrode away rapidly in the environments to which they are ordinarily exposed. We have seen, moreover, that the phenomena which prevent this are numerous, and vary with different metal/environment systems. Consequently, corrosion control may occasionally be balanced somewhat on a knife edge, with apparently small differences in the system causing considerable differences in service behavior. Nevertheless, a large body of knowledge has been built up which, if interpreted on the basis of the guiding principles that are available, usually enables a satisfactory alloy to be selected to cope with a given environment. This body of knowledge also contains guidance on what to do and what not to do when integrating metals into an operational system and using that system.

The first step in corrosion control, therefore, depends on this knowledge being used by engineers who select and use metals. Many fewer corrosion problems would arise in service if this were done more frequently. Admittedly, there are situations where the system is too complex to be dealt with simply. There are even situations where the actual exposure environment cannot adequately be predicted. There are others where one of the nonuniform modes of corrosion may unexpectedly be induced. Specialist advice may then be needed, but there are still few situations where corrosion cannot be overcome or at least alleviated. Needless to say, as with other failure modes, any serious service failure by corrosion needs to be competently investigated to establish the true cause or causes of failure. Only then can remedial measures be selected which have a good chance of success.

CORROSION AND MECHANICAL STRESS IN COMBINATION

Several of the mechanical properties of metals discussed in Chapter 3 can be adversely affected when the metal is exposed to a corrosive environment, perhaps only a mildly corrosive one, while being simultaneously stressed. These are called *environment-sensitive mechanical properties*. The consequences can be unexpected and serious, and include some of the more common mechanisms by which metals actually fail in service. The most important of the phenomena are known as *stress corrosion, hydrogen embrittlement*, and *corrosion fatigue*. These three phenomena will now be discussed in turn.

Stress Corrosion

A stress sustained over a period of time while a metal is exposed to certain environments may cause cracks to initiate and to propagate under conditions where neither the stress nor the environment acting alone would have caused cracking. The propagation of these *stress-corrosion cracks* may eventually lead to structural failure or at least to ineffective performance of a component (Fig. 5.12). Stress corrosion constitutes a mechanism of service failure which ranks with fatigue in its frequency in highly stressed components. There may be no evidence of impending failure, including no indication of general corrosion, other than the presence of fine cracks. There are similarities here with fatigue failure (p. 104) although the circumstances under which, and the mechanisms by which, the two types of failure occur are quite different.

Stress-corrosion cracks develop only when the stress is tensile in nature and when its magnitude exceeds a threshold value (Fig. 5.13), a value which is near the yield strength (p. 67) of the metal. The rate of crack propagation thereafter increases rapidly with increasing stress (Fig. 5.13). The required stress may, however, be the sum of two components: first and most obvious, the externally applied stress; and second and more insidious, residual stresses induced during

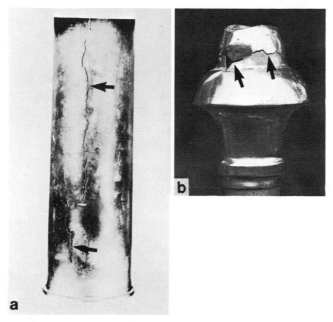

(a) An ammunition cartridge case which has been used and then stored. Hoop stresses introduced into the case during firing were responsible for the development of the cracking, the corroding medium most likely being normal traces of ammonia in the storage atmosphere. Photograph; ½×.

(b) A spun ornamental knob. Residual stresses induced during spinning were responsible for the development of the cracking. The component had not been effectively stress relieved after spinning. Photograph; 1×.

Fig. 5.12. Stress–corrosion cracks in Cu-30Zn brass components.

manufacture and assembly. The manufacturing stresses may be induced during fabrication processes (such as drawing or spinning), heat treatments involving quenching, or finishing operations (such as machining or electroplating). Assembly stresses may be induced by procedures such as the press fitting of interference-fit bushings and the distortion of components to align bolt and rivet holes. It is the maximum local value of the stress that matters. Hence the presence of stress concentrators is detrimental, as it is for fatigue.

Residual stresses frequently are the cause of unexpected and apparently spontaneous failure by stress corrosion because they can be high enough to cause failure by themselves. Failure can then occur during storage of components or in ornamental usages (Fig. 5.12). Residual stresses were the cause of cracking in the classic case in which the existence of stress corrosion as a phenomenon was first recognized. This was the splitting of the cartridge cases of rifle ammunition stored by the British Army in India during the 19th century, splitting which did not occur when the same ammunition was stored in England. Because the cracking occurred mostly during certain climatic seasons of

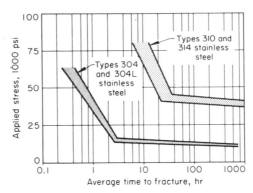

Types 310 and 314 stainless steels contain more chromium and nickel, two critical alloying elements, than type 304. Note that in each case failure does not occur unless the stress exceeds a critical value, and that this value is larger for the more highly alloyed steels. The time to failure decreases quite rapidly with increasing stress above the critical value, but is always longer for the more highly alloyed steels.

Fig. 5.13. Effect of stress on the time for complete failure by stress corrosion for two grades of stainless steel. Time to fracture here is a measure of the rate of crack growth.

the year, the phenomenon was called *season cracking*. This is a term which is still used occasionally but which is perhaps best avoided. The residual stresses in these cartridge cases were induced during deep drawing of the case and during crimping of the mouth of the case to attach the bullet.

Once initiated, stress-corrosion cracks extend laterally and grow into the underlying section in a plane which is normal to the principal tensile stress. They tend to branch extensively in the process (Fig. 5.14 and 5.15), but there are occasional exceptions to this. Only a limited number of separate cracks are likely to initiate and then tend to be confined to a limited area of a component. This is in contrast to the more widespread and random nature of intercrystalline corrosion (cf. Fig. 5.6 and 5.14).

The cracking can develop in pure metals, but only under unusual conditions. It is more characteristic of alloys, particularly those containing large amounts of alloying elements. Perhaps the most notable examples, in the sense that service failures are most commonly encountered in these metals, are high-strength aluminum alloys, high-zinc copper alloys (brasses), mild steels, high-strength steels, and austenitic stainless steels. Even for such alloys, susceptibility varies significantly with temperature, composition, and metallurgical structure in complicated ways. The susceptibility of brasses, for example, increases with the content of the important alloying element zinc, but that of austenitic stainless steels decreases with the contents of the important alloying elements chromium and nickel (Fig. 5.13). The susceptibility of austenitic stainless steels is also sensitive to the accidental presence of nitrogen; it is only those

This is a section through one of the cracks in the component illustrated in Fig. 5.12(a). Optical micrograph; 100×.

Fig. 5.14. An example of an intergranular stress-corrosion crack.

stainless steels containing more than about 0.05% nitrogen that are likely to crack.

Grain size is always important, a small grain size being favorable. Elongated grain shapes are unfavorable in one respect in that the material then has reduced resistance when stressed in a direction perpendicular to the major axis of the grains. This situation arises in practice with thick aluminum alloy plate and extrusions in which the grains can be shaped like pancakes that are greatly elongated in the direction of working. Special heat treatments can be used to overcome this deficiency.

The environments which induce stress-corrosion cracking are specific to particular metals, and only a limited range of environments can cause cracking in any one metal. Some examples of cracking environments are listed in Table 5.2, most of them being at worst only mildly corrosive in a general sense. The presence of oxygen is important in most of them. Many of these environments are likely to be encountered in everyday usage, and some are virtually impossible to avoid. Moisture containing chlorides that will cause cracking of aluminum alloys is a case in point, because moisture and traces of chlorides are ubiquitous. Traces of ammonia that can cause cracking of brasses are frequently present in the atmosphere due to the decomposition of organic matter and the presence of animal waste products. This is why the cracking of the cartridge cases in India was seasonal. Organic matter decomposes more rapidly, producing higher concentrations of ammonia in the local atmosphere, under some climatic conditions than under others, and certainly more rapidly in the hot climate of India than in the cold climate of England.

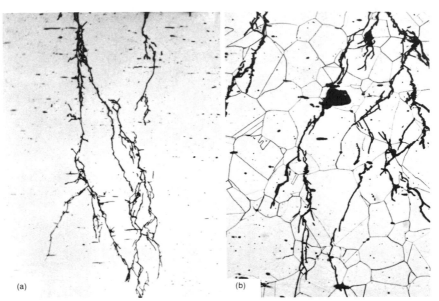

(a)   (b)

Sections through cracks in austenitic stainless steels. The branching characteristics of stress-corrosion cracking is well illustrated in (a). The photomicrograph in (b) illustrates that the cracks propagate in a straight line across each grain but in different directions in each. This is because they propagate along a specific crystal plane in each grain. Optical micrographs: (a) 100×; (b) 150×. From *Metals Handbook*, 8th Ed., Vol 7, *Atlas of Microstructures of Industrial Alloys*, American Society for Metals, Metals Park, OH, 1972.

Fig. 5.15. Examples of transgranular stress-corrosion cracks.

Other causative environments are met with only under special circumstances, but circumstances that nevertheless are encountered in applications for which an alloy might have been selected. Moreover, the environment to which susceptible regions of a structure is actually exposed is not always what it might seem. For example, stress-corrosion cracking is a serious potential cause of disastrous failure in steel boiler drums, yet the concentration of caustic or nitrate in feed water is never anywhere near that needed to cause stress-corrosion cracking. These failures characteristically occur near faying surfaces where small leaks permit soluble salts to build up high local concentrations. Hence it is established practice to make special additions to boiler feed water to inhibit this type of failure.

Stress-corrosion cracks initiate only at a surface — and only at a surface that is exposed to the damaging environment. Once initiated, they propagate laterally into the section thickness. Eventually, one of two things happens. Either the crack reaches a critical length and initiates catastrophic failure (p. 89), or the section is penetrated and the component distorts, or leaks, or be-

**Table 5.2. Environments That Cause Stress-Corrosion
Cracking in Some Common Alloys**

Alloy	Environment
Aluminum alloys	Chloride-containing solutions, including contaminated water vapor
Copper alloys	Ammonia and amines for high-zinc brasses; range of solutions for other specific alloys
Gold alloys(a)	Chlorides, particularly ferric chloride; ammonium hydroxide; nitric acid
Nickel alloys	Hot caustics
Low-carbon steels	Hot concentrated caustic and nitrate solutions; anhydrous ammonia
Corrosion-resistant steels(b)	Hot chloride solutions; caustics; saline solutions
Titanium(c)	Fuming nitric acid; seawater; methanol – hydrochloric acid mixtures

(a) Alloys containing less than 67% gold.
(b) Austenitic nickel-chromium steels.
(c) A pre-existing crack of some type usually must be present as an initiator.

comes ineffective in some other way. The rate of growth of the cracks therefore can become important in determining service life. The growth rate varies over a very wide range, from 10^{-1} to 10^{-9} m/s, depending on the alloy, the environment, and the stress.

The cracks may be either intergranular (Fig. 5.14) or transgranular (Fig. 5.15). These two types of cracks propagate by different mechanisms. Intergranular cracking, which is the more common, occurs by a dissolution-dependent mechanism. Protective films at the root of a crack rupture under the local stress while the sides of the crack remain protected. Preferential anodic dissolution consequently occurs at the crack tip, and the crack thereby lengthens. The small area of the anode compared with that of the cathode ensures rapid corrosion at the anodic area. Transgranular cracks, which are less common but which include the industrially important cases of austenitic stainless steels and admiralty metal, occur by environmentally induced cleavage. The crack advances in a series of discrete steps by brittle cleavage fracture on a characteristic crystal plane, the cleavage being induced by the environment in a manner which is not yet understood. The progression of the cleavage crack is interrupted each time it encounters a grain boundary, because fracture has to be initiated anew on the appropriate plane in the new grain. Grain boundaries do not have such an influence on the progression of intergranular cracks.

Nevertheless, both mechanisms imply that grain size and shape should influence the ease of crack propagation. The distance that an intergranular crack has to travel is longer the smaller the grain size, and is also longer when the crack is traveling perpendicular to elongated grains than when it is traveling parallel to them. The progression of a transgranular crack is interrupted more often by grain boundaries the smaller the grain width in the direction of crack propagation.

The reasons why cracks initiate in the first place, and why they initiate either at a grain boundary or at the trace of a particular crystal plane, whichever the case may be, are, however, not yet understood. Crack initiation is certainly much more difficult in some alloys than in others and plays a more important role in determining total life. Titanium alloys, for example, are very resistant to stress-corrosion cracking in salt water when they are crack free, but stress-corrosion cracks may propagate comparatively readily once a crack has been initiated or if a sharp crack of any other type is present. Engineers never like to base their decisions, however, on an assumption that a structure will be crack-free.

The mechanisms of stress corrosion are so imperfectly understood that it is only to be expected that methods of controlling the phenomenon are either general or empirical in nature. A further complicating factor is that either intercrystalline or transgranular cracking can occur in the one alloy, often even in the same environment. The existence of the two basically different and competing cracking processes severely complicates efforts to develop resistant alloys and conditions. One mode might be cured only to open up the possibility of cracking by the other mode.

The simplest way to prevent stress-corrosion cracking is, in principle, to lower the stress below the threshold value. This can be achieved by reducing the applied stress, but there is a limit to how far this can be taken in structures which are designed to be stress bearing. The second possibility is to reduce internal stresses; this, in any event, is always desirable. Precautions can be taken to keep the stresses induced during manufacture and assembly to the minimum, but often they cannot be reduced to anywhere near zero. A considerable reduction in stress level can often be achieved, however, by a stress-relieving annealing heat treatment carried out at a low temperature (p. 307) after the last possible fabrication step. The component illustrated in Fig. 5.10(b) would not have developed stress-corrosion cracks if it had been adequately stress relieved in this way after spinning. All brass pressings and spinnings should be adequately stress relieved. Eliminating the active component of the environment is another possibility, as are the use of inhibitors, protective coatings, or cathodic protection. But the adoption of these approaches is not always practicable and may not be fully effective. Finally, if all else fails, the alloy in use may have to be replaced with one which does not stress corrode in the particular environment.

The high-strength aluminum alloys used widely in aircraft and other lightweight structures illustrate the type of dilemma that can arise. These alloys stress corrode in an environment which can scarcely be avoided with certainty: water and chlorides are everywhere. Standard corrosion protection measures often cannot be implemented, and high stresses are applied to structures made from these alloys (which is the reason for using them). The heat treatments to which these alloys have to be subjected to achieve their strength (p. 376) also may introduce internal stresses. So the specter of stress-corrosion failure is always present around these alloys. Moreover, ever since these alloys were developed, metallurgists have constantly been making endeavors to improve their strength. Unfortunately, this has typically been accompanied by a deterioration in resistance to stress corrosion, and so parallel endeavors have had to be undertaken to improve stress-corrosion resistance as well (cf. problems with fracture toughness; see p. 94). Only limited success has been achieved here. Aluminum alloys have, for example, been introduced into service which have turned out to be too susceptible to stress-corrosion cracking and have had to be withdrawn. Improvements have been made to some of the more acceptable types of aluminum alloys by subtle adjustments to composition and heat treatment, but there is still debate as to whether even the improved alloys are sufficiently reliable in practice and whether it wouldn't be more effective to settle for alloys which are highly resistant to stress corrosion even if they are a little weaker.

The saga just described also highlights another serious problem which arises when attempts are made to develop stress-corrosion-resistant alloys and which is general to the field. This is the problem of devising short-term laboratory tests which fully and reliably predict performance in service. Many of the unsatisfactory aluminum alloys just referred to exhibited reasonable stress-corrosion behavior in accepted laboratory tests and trials. Their deficiencies, particularly in lack of complete reliability, were revealed only by the hard experience of service.

Nevertheless, there are many situations for which experience has been accumulated and reliable methods of preventing failure by stress corrosion have been thoroughly established. For example, the fact that brasses must be given a stress-relieving heat treatment after drawing or spinning to avoid failures of the type illustrated in Fig. 5.12, and the details of the required heat treatment and of the tests needed to confirm that the treatment has been successful, are all well established. But the implementation of these precautions requires knowledge and equipment, and costs money. Most service failures occur when one of these factors is missing.

Hydrogen Embrittlement

Hydrogen can enter into a metal from the environment in a number of ways, and it then diffuses rapidly throughout the material, even at room temperature. The result is that concentrations of hydrogen much higher than normal are built up in the metal. This can have a number of consequences, probably the most important to a user being delayed fracture under a sustaining load by a phenomenon called *hydrogen embrittlement*. Only a comparatively few metals are susceptible to this phenomenon, but prominent among them are the high-strength steels. Steels having tensile strengths above about 1000 MPa (145 ksi) are susceptible, and thereafter sensitivity increases markedly with increasing tensile strength. We shall confine our attention to such steels.

The methods by which the hydrogen can be introduced into a steel include corrosion in aqueous environments, improperly controlled cathodic protection treatments, acid pickling, electroplating, and even exposure to hydrogen gas — particularly at high pressures. All are situations where atomic hydrogen is formed at the surface by either an electrochemical or a catalytic reaction. The initiation of a crack requires the application of a stress above a threshold level, a stress which can again be the sum of internal and externally applied stresses. There is a time delay before failure occurs, the time being shorter the higher the stress and the hydrogen content. Thus a component having high internal stresses may crack soon after the hydrogen is introduced or after some days or months of delay in storage, depending on the stress level and the hydrogen content. Cracks consequently may appear, apparently mysteriously, after components have been inspected and passed as being crack-free. On the other hand, cracking and failure of components having a comparatively low level of internal stresses may occur only after the service stress is applied, perhaps after some delay. Fracture occurs with little evidence of ductility (Fig. 5.16a), and the fracture surface is bright. It is usually intergranular with respect to the austenitic grain boundaries (p. 384) of the steel (Fig. 5.16b), but sometimes various amounts of cleavage fracture may be present. Although there are obvious similarities with stress-corrosion cracking, there are also significant differences in detailed behavior. The mechanisms of cracking are probably different in the two cases, but there is at present no agreed view on the mechanism of hydrogen-embrittlement cracking.

High-strength steel bolts, particularly when electroplated (usually with cadmium or zinc) for corrosion protection, are among the components in which failure by hydrogen embrittlement is likely to be encountered. They can be used to exemplify the principles of the precautions that have to be taken to avoid failure by this mechanism. Bolts of the type under consideration are heat treated to strengths in the susceptible range and are stressed beyond their yield strength when correctly tightened. Consequently, the only course of preventive action open is to keep the hydrogen content down to an acceptable level. The bolts typically are pickled in acid after heat treatment, and this is likely to introduce a considerable amount of hydrogen, although the amount can be reduced by adding inhibitors to the pickling bath. The hydrogen that remains needs to be expelled, however, and this can be done by baking at a relatively low temperature (150 to 250 °C, or 300 to 480 °F) for several hours. Additional hydrogen is likely to be introduced if the bolts are subsequently electroplated, although this too can be limited by proper choice of the plating bath and conditions. Nevertheless, the new hydrogen has to be removed by another baking treatment. But this is now more difficult because the plated coating constitutes

The bolt fractured during tightening at a much lower stress than expected and with little indication of ductility, as shown in (a). The fracture surface is shown at a higher magnification in (b). The fracture is intergranular, but otherwise is featureless. (a) Photograph; 1×. (b) Scanning electron micrograph; 400×.

Fig. 5.16. An example of hydrogen embrittlement in a high-strength steel bolt.

a barrier to the egress of the hydrogen. Careful control of the plating and baking conditions is necessary for success. Thus, much attention to detail is necessary to produce bolts which are proof against failure by hydrogen embrittlement.

Combating hydrogen embrittlement is more difficult when the hydrogen is introduced during service (e.g., during corrosion or cathodic protection). If the ingress of hydrogen cannot be prevented by adjustments to the environment (and this is often the case), the only completely effective preventive measure remaining is to reduce the strength level of the steel to below the susceptibility level. The problem is increased when hydrogen sulfide is present, even when it is present in the small amounts that can result from modern industrial atmospheres. It is even further increased when high concentrations of hydrogen sulfide are present, as in some untreated natural gases and crude oils. Hydrogen sulfide reacts with iron, atomic hydrogen being one of the reaction products. This has caused problems in pipelines carrying high-sulfur gases and crude oils, because the material in the heat-affected zones adjacent to welds can be in the susceptible strength range.

Corrosion Fatigue

The behavior of metals in fatigue (p. 104) is affected seriously in the presence of a corrosive environment, failure at a given cyclical stress occurring after many fewer stress cycles than would be required in a more inert environment. Another consequence is that even those alloys which have definite fatigue limits (p. 109) no longer do so. The presence of a particular environment is not required for the deterioration in properties, as it is for stress-corrosion cracking. The sole requirement is that the environment be sufficiently corrosive, although there is not necessarily a direct correlation between general corrosiveness and effect on corrosion fatigue. The effect does not occur, on the other hand, if the corrosion and cyclical stressing occur successively or alternately but only if they occur simultaneously.

The deterioration in fatigue properties is usually assessed by comparisons with those obtained in air — which, however, is not necessarily a completely inert environment. Many metals have a higher endurance limit in vacuo than in air. Nevertheless, even neglecting this, the deterioration in endurance limit is considerable (Fig. 4.4, p. 109). The corrosion endurance limit ranges from about 50% to 10% of the limit in air. Moreover, for a particular alloy type the corrosion endurance limit is independent of metallurgical structure and so shows little correlation with strength. Thus the endurance limit of steels under even mildly corrosive conditions is much less than that in air and does not increase at all with an increase in the tensile strength of the steel (Fig. 4.4a, p. 109). The combination of corrosion with a cycling stress destroys completely the benefits of all efforts made to improve the strength of steels, as assessed by static mechanical tests.

Corrosion-fatigue cracks always initiate at a surface which is exposed to the active environment and proportionately occur at a much earlier stage in total life than do normal fatigue cracks. Characteristically, many cracks are initiated over an area of generally high stress (Fig. 5.17a) instead of one or two cracks in the region of the very highest stress, as in normal fatigue. Initiation is sometimes preceded by the development of a corrosion pit, but this is not essential. In any event, the mechanism of initiation is not understood. It is likely that several mechanisms are possible and that the fastest one in the particular circumstances is the one that dominates.

The mechanism of propagation is not understood either. The cracks normally propagate on a transgranular path, although in a few instances (e.g., copper alloys) they are intergranular. They tend to branch more than normal fatigue cracks, and to be wider and to be filled with corrosion product (Fig. 5.17b; cf. Fig. 4.2, p. 107). Total life in normal fatigue is determined solely by the number of stress cycles undergone and so is independent of the frequency of cycling. This is not so for corrosion fatigue, because lower frequencies expose the material to corrosion for longer times per stress cycle. The total number of cycles to failure consequently decreases with the frequency of the stress cycle.

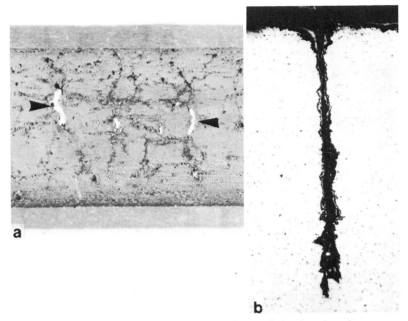

Numerous circumferential cracks (arrows) developed at the inner surface of the tube, some of them penetrating the wall thickness (a). A longitudinal section of the tube through a crack shows that the crack is comparatively wide and lined with corrosion product (b). Compare with the normal fatigue crack shown in Fig. 4.2 (p. 107). The cracking was caused by alternate expansion and contraction of the tube which induced cycling longitudinal tensile stresses. (a) Photograph; 1×. (b) Optical micrograph; 75×.

Fig. 5.17. An example of corrosion fatigue in an economizer tube from a steam boiler.

Normal fatigue is a serious enough problem in engineering structures. Corrosion fatigue obviously is a much more serious one, even more so because it is so difficult to combat. Failure prevention can be approached on one or more of the following lines: reduce the cyclic stress; eliminate or protect against the corrosive environment; use a more resistant material.

Reducing the cyclical stress to a level below the endurance limit in corrosion fatigue may require that an impractically small working stress be used. Nevertheless, the mitigation of stress concentrators where they have had an influence is still likely to be helpful, as it is for normal fatigue. It may not be possible, either, to eliminate the corrosive environment: the shaft of a pump handling a corrosive liquid simply has to operate in a corrosive environment. If it is not possible, protection against the environment has to be considered. Organic or metallic coatings are often effective, although coatings which provide anodic control must be absolutely continuous at vulnerable surfaces and must not be damaged in service in these regions. Coatings which provide cathodic control are, as usual, likely to be more reliable. Cathodic protection is probably the most widespread preventative system in use. Corrosion inhibitors may also be helpful in closed systems, such as recirculating water systems. The final option is to change the alloy used, either in composition or structure, in the exposed surface layers at least. The new alloy will, in general terms, have to be one with improved corrosion resistance and not just one with improved strength. The acceptability of this solution may depend on economic considerations.

GAS CORROSION: OXIDATION

We used the term *oxidation* earlier in this chapter to describe an electron-producing reaction which is one component of an electrochemical reaction. This term is also used to describe the reaction between a metal and a gas containing oxygen to produce a solid oxide compound of the metal. The oxygen (most commonly) comes directly from the surrounding air, but can also originate from reaction with other gases, such as carbon dioxide and water vapor. The terms *tarnishing* and *scaling* are also used to describe this phenomenon.

A thin oxide layer often is invisible, but sometimes it is just discernible, in which event it may be called a *tarnish*. A tarnish may impair surface appearance, which can be important enough, but even more important surface properties may also be affected. For example, a tarnish layer prevents the surface from being wetted by a molten metal, as is needed in a soldering or brazing operation. This is why fluxes have to be used in these operations, the fluxes being compounded to dissolve the oxide layer. On the other hand, thin oxide layers can be advantageous, such as when they inhibit further oxidation. As indicated elsewhere in this chapter, many "stainless" alloys depend on this effect.

Thicker oxide layers are often easily discernible as separate layers on the surface, and this is when they tend to be known as *scales*. Scaling represents an economic loss of material which has been wasted during processing. Moreover, a scale may have to be removed at some additional cost before processing can proceed. Rapid scaling also may set a limit on the temperature at which an alloy can be used effectively. An alloy may have adequate mechanical and physical properties at the temperature concerned but oxidize away so rapidly that it does not have a useful life.

Basic Mechanisms of Oxidation

All common metals and alloys react with oxygen at room temperature, but they do so more readily at elevated temperatures. The reason for this is the same as for corrosion in aqueous solutions — namely, that metals are thermodynamically unstable. So again the questions of practical importance are why metals last for as long as they do without oxidizing excessively and how they can be made to last as long as possible.

The formation of a metal oxide* can be described by the general equation:

$$aM + \frac{b}{2} O_2 \rightarrow M_a O_b \qquad \text{(Eq. 12)}$$

This might appear to be the simplest possible type of chemical reaction, but it only describes the end result, or equilibrium result, that nature would like to achieve. As for liquid corrosion, the reaction has to proceed by a number of separate steps in sequence, and the slowest of these steps determines the rate at which equilibrium is actually approached — that is, the kinetics of the reaction.

Let us follow the steps of a characteristic oxidation sequence, starting with a hypothetical surface which is absolutely clean and free from oxide and which is then exposed to an oxygen-containing gas. The first oxygen molecules to land on the surface are adsorbed and dissociate into atoms. These atoms are at first physically, and then chemically, bonded to metal atoms exposed at the surface, eventually to begin to form discrete particles of metal oxide. The oxide particles are nucleated at certain favorable isolated sites (e.g., small surface steps, exposed crystal defects, and impurity particles) and then grow laterally until they contact one another, and the whole surface eventually becomes covered with a layer of oxide. This film can be assumed for the moment to be adherent, continuous, and uniform in thickness. Subsequent growth of the oxide layer requires that atoms of metal at the metal/oxide interface migrate, or diffuse, through the pre-existing oxide layer to some place where they can meet an oxygen ion (Fig. 5.18). To do this, the metal atoms first have to ionize (lose electrons). The electrons released by the ionization of the metal atoms must also migrate through the oxide — this time to the oxide/oxygen interface, where they provide the charge necessary to ionize additional adsorbed oxygen atoms.

Thus reaction (Eq. 12) has to be considered as occurring by the following three steps in sequence:

$$M \rightarrow M^{+2} + 2e \qquad \text{(Eq. 13)}$$

$$1/2 \, O_2 + 2e \rightarrow O^{-2} \qquad \text{(Eq. 14)}$$

$$M^{+2} + O^{-2} \rightarrow MO \qquad \text{(Eq. 15)}$$

This sequence is in all respects similar to an electrochemical corrosion reaction, as defined earlier. The reaction given in Eq. 13 is an oxidation reaction, the

---

*Actually, in a few cases, a solid oxide may not form at all because the oxygen dissolves in the metal. In a few other cases, the oxide melts or is volatile above a certain temperature; the consequences of this will be described later in the section on catastrophic oxidation.

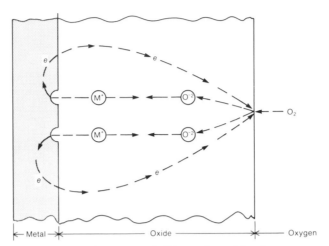

**Fig. 5.18. Diagrammatic illustration of the electro-
chemical reactions that occur during the oxidation
of a metal in oxygen gas.**

same as Eq. 3 above, and the reaction given in Eq. 14 is a reduction reaction, the
same as Eq. 4 above. Herein lies the justification for classifying the oxidation
phenomenon as a corrosion reaction, the oxide layer playing the same role as
the electrolyte solution plays during corrosion in a liquid. The oxidation and
reduction reactions must again proceed at the same rate if the impossibility of
the system becoming charged is to be avoided. It also follows, again, that the
over-all reaction rate is limited to that of the slowest of the steps involved. In
gas corrosion, it is the diffusion of one or another of the ions required to enable
the reaction in Eq. 15 to proceed that almost invariably is the rate-controlling
step.

Clearly, therefore, we need to understand a little about diffusion in the
solid state if we are to understand the kinetics of the oxidation of metals. Diffusion will be discussed in more detail in a subsequent chapter (p. 235), where it
will be shown that the presence of vacancies in the crystal lattice is essential for
diffusion to occur in a crystalline solid. So it will suffice here to say that metal
oxides are crystalline solids, that the metal and oxygen atoms are present in the
crystal lattice (p. 223) as ions, and that the vacant sites in the oxide lattice that
are necessary for diffusion to occur (p. 235) arise in the following way. The
compositions of metal oxides usually are represented as conforming to strict
chemical formulas, such as Cu_2O for cuprous oxide; but typically they don't.
The oxides of most common metals are deficient in metal, the actual composition of a typical cuprous oxide, for example, being about $Cu_{1.8}O$. The ideal
situation is called *stoichiometry* and the actuality is called *nonstoichiometry*. A consequence of nonstoichiometry is that not all of those sites in the crystal lattice of
the oxide that are available for metal ions to occupy are actually occupied.
That is, there are vacant metal-ion sites in the crystal lattice of the oxide, and

the concentration of vacant sites depends on the degree of nonstoichiometry of the oxide. On the other hand, there can be no vacant oxygen-ion sites. Thus metal ions can diffuse through the oxide layer but oxygen ions cannot do so. The metal ions consequently have to diffuse right out to the oxygen/oxide interface to meet oxygen ions so as to form the additional molecules of oxide. The oxide layer grows at its outer surface and not at the metal surface, as you perhaps might have expected.

The deficiency of ions in the crystal lattice of the oxide is also responsible for the electrical conductivity of the oxide, but we need not go into that. More strictly, the oxides are semiconductors in the same sense that silicon used in transistors and integrated circuits is a semiconductor.

Even though an adequate concentration of vacant ionic sites may be available in metal oxides, the diffusion of ions is still vanishingly slow in most of them at room temperature. This is the basic reason why metals do not oxidize away before our eyes. It is the slow kinetics of the oxidation reaction that staves off the fate decreed by thermodynamics. The diffusion rate increases significantly with increases in temperature (p. 236), approximately doubling for each 10 °C (18 °F), and the oxidation rate typically increases correspondingly. Under all circumstances, however, the rate of growth of the scale can be expected to slow down as the scale thickens because the distance over which diffusion has to occur increases. The rate of growth of a scale does in practice decrease progressively with time until some other phenomenon, such as the development of cracks in the scale, intervenes.

Other things being equal, the diffusion rate is determined by the concentration of ionic vacancies in the oxide, and this varies with the degree of nonstoichiometry of the oxide. Hence the variation in the oxidation resistance of different metals. A metal such as aluminum has much better oxidation resistance at a comparable homologous temperature than, say, copper even though it is a less noble metal. This is because aluminum oxide is nearly stoichiometric whereas the oxide of copper is not. The vacancy concentration in the oxide can also be affected by alloying elements. Some alloying elements block up metal-ion vacancies in the oxide of the parent metal when they enter the scale and so reduce the oxidation rate. Some increase the vacancy concentration and so increase the oxidation rate. Rules are available which help to predict this behavior. Some elements are very effective in this role even when present in only small amounts. Rare-earth elements, such as cerium, are examples of good vacancy blockers.

Deviations From the Basic Mechanism

There are many factors, however, which complicate this basic model. First of all, the scale may be composed of a number of layers of different oxides because the metal concerned forms several stable oxides at the exposure temperature.

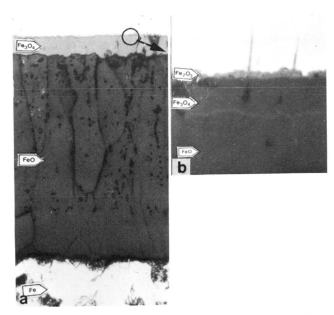

Photomicrograph in (a) shows a section of the scale formed on steel heated in air at 800 °C (1470 °F) and cooled slowly. Two of the layers in this scale, the compositions of which are labeled, can be seen here, but the third (Fe_3O_4) is so thin that it cannot be seen at this magnification. The scale has been etched to show that it is composed of columnar grains and that one layer contains chains of small cavities (dark spots). Optical micrograph; 1000×. Photomicrograph in (b) was taken at a higher magnification (2000×) to show the thin surface layer of Fe_3O_4.

Fig. 5.19. An example of a multilayer scale formed during oxidation.

Iron is an example of this (Fig. 5.19), forming three oxides of increasing oxygen content (nominal compositions: FeO, Fe_3O_4, and Fe_2O_3). The stable oxides generally form in sequence in a scale with the most oxygen-rich oxide at the outer surface. The relative thickness of each oxide is determined by the relative rates of ionic diffusion. An oxide in which diffusion is faster can be thicker because the migrating metal ion can travel over a longer distance in a unit of time. The oxides actually present in a scale on such a metal may vary with the exposure conditions and thermal cycle, in which event the characteristics of the scale change accordingly, and this is sometimes of considerable importance in practice. Other phases may also be introduced because of the oxides produced from the different elements of an alloy. They may react with the base-metal oxide to produce a type of compound known as a spinel. For example, the oxides of nickel and chromium formed on nickel-chromium alloys react to form $NiCr_2O_4$, which is more effective in reducing diffusion through the scale than would be a mixture of nickel and chromium oxides.

Another complication arises because cavities, or pores, often form in a scale toward the metal/oxide interface, particularly when the scale becomes thick (Fig. 5.20). This probably results from the agglomeration of the very vacancies that are essential for diffusion in, and growth of, the scale. On the one hand the pores act as barriers to diffusion in the scale, and on the other hand they provide a source of oxygen close to the metal interface because they fill up with oxygen due to decomposition of oxide at their surfaces. In any event, rate-controlling processes other than solid-state diffusion take over when large volumes of pores have formed. Cracks may also develop in the scale, and these cracks tend to expose unprotected metal to the atmosphere. The oxidation rate then increases suddenly until a protective layer is re-established. Breakaways in oxidation rate of this nature tend to occur successively as new generations of cracks form to the detriment of over-all oxidation resistance.

The formation of cracks is important in practice, and occurs for two possible reasons. The first is that the volume of the oxide is never the same as that of the unit of metal from which it originated. If it is larger, which is the case with all engineering metals, compressive stresses develop in the scale, which is then likely to rumple and crack eventually. If it is smaller, tensile stresses develop in the scale, which then cracks even more readily. The scale then may not even be able to cover the metal surface completely, and this is accompanied by particu-

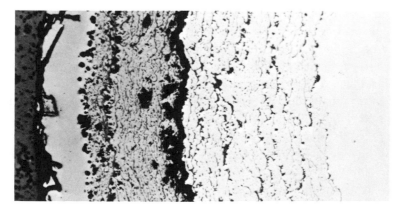

This is a section of the scale formed on a complex nickel-silicon-aluminum-chromium alloy heated in air at 1000 °C (1830 °F). Large amounts of porosity have developed at certain levels in the scale. The alloy has internally oxidized to a considerable depth beneath the scale, the elements which have oxidized being aluminum and manganese. Small particles of their oxides have been precipitated in networks which correspond to the grain boundaries of the alloy. The composition of the alloy remaining in this layer consequently is depleted of these elements. Optical micrograph; 200×.

Fig. 5.20. An example of the development of porosity in an oxidation scale, and of internal oxidation accompanying scale formation.

larly poor oxidation resistance (fortunately, this occurs with only a few unusual metals). Sodium is an example which, as every chemistry student knows, disappears before your eyes, perhaps even violently, if exposed to air. Experience indicates that good protective scales form only when the volume ratio (the *Pilling and Bedworth Ratio*, it is called) is reasonably close to, but greater than, unity. The second source of cracking is differential expansion of the scale and the metal during thermal cycling. The coefficients of expansion of oxide and metal need to be nearly equal to reduce the risk of this type of spalling to the minimum. The likelihood of the occurrence of either of these types of cracking is reduced if the scale has good plasticity. So many factors determine the effectiveness of a scale in inhibiting oxidation.

A final complicating factor is an important one which can be turned to practical advantage. The presence of an alloying element in a metal can result in the formation in the scale of a layer of an oxide of that alloying element. This layer can be in addition to the layers of parent-metal oxide or can be a replacement for it. An important point is that the alloy oxide may be more nearly stoichiometric than the parent-metal oxide, in which event it will be the more protective of the two. Selective oxidation of an element which forms a nearly stoichiometric oxide can in fact reduce the oxidation rate of the whole system, perhaps to a very low value. This *selective oxidation* of an alloying element can occur, however, only when the alloying element is less noble than the parent metal and when several other criteria are met. An important one is that the alloying element has to be present in a concentration exceeding a critical value.

Most of the special heat-resistant alloys used in engineering practice depend on this phenomenon, and this includes the common iron-chromium and nickel-chromium alloys. Let us take nickel-chromium alloys as an example. The oxidation rate of these alloys at first increases slightly with the addition of chromium (Fig. 5.21). This is because chromium at these dilute concentrations simply enters the nickel oxide scale, and the characteristics of chromium are such that the concentration of metal-ion vacancies in nickel oxide is increased. This is detrimental (see above). However, the oxidation rate begins to decrease rapidly when a certain chromium concentration is exceeded (Fig. 5.21), and this is the stage at which a layer of chromium oxide (Cr_2O_3) begins to form in the scale. The oxidation rate then falls rapidly and attains a small and fairly constant value when a somewhat higher chromium concentration is reached (Fig. 5.21). By then, the scale is composed entirely of Cr_2O_3 and is very thin, much thinner than the scale formed under comparable exposure conditions on alloys of low chromium content (Fig. 5.22). The critical chromium concentration for the onset of the decrease in oxidation rate is about 10 wt %, and that for the attainment of minimum oxidation rate is something approaching 20 wt % (Fig. 5.21). Somewhat higher concentrations are required to achieve equal effects at higher temperatures. Most commercial heat-resistant alloys contain over 20 wt % chromium.

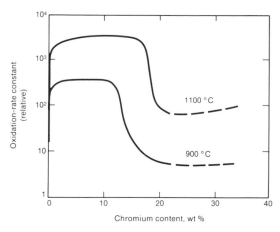

Fig. 5.21. Variation of the oxidation rate of nickel-chromium alloys with chromium content. The alloys were heated in oxygen at the temperatures indicated.

We have so far confined our discussion to the case where the metal oxide is deficient in metal ions. There are a few metals whose oxides are deficient in oxygen. Niobium is one. Oxygen ions can then diffuse through the scale but metal ions cannot do so, and so the scale grows at the metal/scale interface. Scales which grow in this way invariably are very porous, and consequently afford the metal poor protection. The oxidation resistance of niobium is so poor that its use at high temperatures is virtually precluded even though it has some other attractive characteristics from this point of view.

Internal Oxidation

We have just seen that a minor alloying element may oxidize selectively in the external scale when present in a sufficient concentration. Such alloying elements, when present in less than this threshold concentration, may still oxidize preferentially but then do so in a layer immediately beneath the surface scale (or beneath the surface in a few cases where no scale forms). The result is the formation of an *internally oxidized* layer, as it is called, in which many small particles of oxides of the alloying elements are precipitated (Fig. 5.20 and 5.22a). The metallic material remaining becomes depleted in the alloying elements that are internally oxidized.

Internal oxidation is possible whenever the alloying element is less noble than the parent metal of the alloy, and many common engineering alloys are in this category. This includes alloys of silver, copper, nickel, and iron containing aluminum, zinc, silicon, and chromium. However, several additional criteria have to be met before internal oxidation actually does occur. First, oxygen has to get into the surface layers of the metal. It follows that oxygen has to be

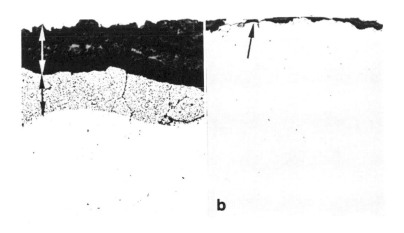

(a) An alloy containing 9.3% chromium. The white arrows delineate a nickel oxide scale layer and the black arrows an internally oxidized layer containing precipitates of a chromium oxide. Much of the section has wasted away and even more has been affected by internal oxidation. Optical micrograph; 500×.

(b) An alloy containing 14.2% chromium. The surface layer (arrow) is very thin and is composed essentially of a chromium oxide. Very little of the section has wasted away and none has been affected by internal oxidation. Optical micrograph; 500×.

Fig. 5.22. An example of the effects of chromium on scale formation on, and internal oxidation of, nickel-chromium alloys heated in air for 800 h at 1200 °C (2190 °F).

available at the metal/oxide interface, usually from pores in the adjoining scale, and has to be soluble in the parent metal. Oxygen also must be able to diffuse in the parent metal, and to diffuse faster than the minor, less-noble alloying element so that an oxygen concentration gradient can be established by diffusion. It is this diffused oxygen that causes the internal oxidation. Diffusion characteristically occurs faster along grain boundaries than through the body of a crystal grain (p. 242); hence the oxygen concentration in the diffused layer is greatest in the region of the grain boundaries, and the internal oxidation precipitates tend to be concentrated there (Fig. 5.20). Finally, as we have already noted, internal oxidation occurs only in comparatively dilute alloys: specifically, only when the alloy concentration is less than that at which selective oxidation and external scale formation occur. There is then a transition from internal oxidation to external scale formation. To continue to use nickel-chromium alloys as an example, alloys containing 10% chromium or less form an external scale based on a nickel oxide. They also internally oxidize, precipitating fine particles of Cr_2O_3 (Fig. 5.22a). Alloys containing 15% chromium or more not only oxidize at a greatly reduced rate because a diffusion-inhibiting external layer of Cr_2O_3 is formed, but they also do not internally oxidize (Fig. 5.22b).

The presence of an external scale is obvious, and its formation has a direct consequence in that it wastes the alloy away. The presence of internal oxidation is not at all obvious and has more subtle consequences which arise principally because the surface layers of the alloy remaining are depleted of the element which has oxidized internally. These changes in composition persist, moreover, when the surface scale has been removed. Surface-sensitive properties, such as fatigue resistance, toughness, and color, may be affected. Bulk properties, such as electrical properties, may also be affected in proportion to the cross-sectional area that has been affected.

Alloys which are proof against internal oxidation could, in principle, be devised by confining alloying to more noble metals. But this is not always practicable. The next possibility is to raise the alloy content to a level above the oxidation transition value, as was done for the alloys illustrated in Fig. 5.22(b). If this has the consequence that some other property required of the alloy is not maintained at an acceptable level, the possibility remaining is to substitute an alloying element which can be used satisfactorily at a content above its transition level.

Catastrophic Oxidation

Reduction of the kinetics of oxidation to acceptable levels is contingent on a protective layer of a solid oxidation product being formed on the metal surface. There are metals and alloys on which protective layers do not form and for which unusually high oxidation rates then ensue.

First, the oxide in some systems evaporates as soon as it forms, in which event oxidation proceeds as fast as oxygen is made available. The oxide of tungsten, for example, evaporates at temperatures above 1000 °C (1830 °F), and that of molybdenum evaporates at even lower temperatures. Paradoxically, both of these metals have unusually high melting points and so have the potential to be used at unusually high temperatures. But they can be so used only if they are maintained in an inert atmosphere. The tungsten wire used in domestic incandescent lamps runs at temperatures above 1500 °C (2730 °F) and can be used successfully for this purpose only when it is contained in a glass bulb which has been evacuated of air and backfilled with an inert gas (argon). The wire disintegrates with almost explosive violence if air happens to get into the bulb.

Secondly, the oxide may melt at a temperature below that of the metal. This occurs in a number of metals (e.g., iron), and the oxide melting point then sets a limit on the temperature at which reasonable oxidation resistance can be expected. The effect may become more serious when the oxide melts at an unexpectedly low temperature, which can occur with some oxides when they become contaminated either from elements in the alloy itself or from those in the oxidizing atmosphere. Atmospheres containing vanadium or molybdenum may, for example, contaminate iron oxides in this way, a phenomenon which can assume practical importance because fuel oils from some sources contain unusually large amounts of vanadium. Oils of this nature have to be avoided as fuels for steam boilers.

Oxidation is always an exothermic reaction — that is, heat is generated by the reaction. An oxidizing piece of metal thus heats up if the rate of heat production during oxidation exceeds the rate at which heat is lost from the oxidizing region. A chain reaction is then set in motion. The temperature continues to increase, and the oxidation may build up to a catastrophic rate. This rarely occurs with industrial metals, but one case of interest is encountered when steel is ground. The shower of sparks that typically is then seen is a consequence of catastrophic oxidation of the small chips of steel that are removed during grinding. The temperature of the chips increases due to oxidation heating until the oxide melting point is reached. Catastrophic oxidation then ensues, and the oxide particles glow brightly and are seen as sparks.

Controlling Oxidation

Oxidation cannot be prevented in common metals when they are exposed to air or to an oxidizing atmosphere, but the rate of oxidation usually can be kept within acceptable limits. This control of oxidation rate relies on a protective layer of a solid being formed over the metal surface by the products of oxidation themselves. Much is required of this layer if it is to confer good oxidation resistance.

Oxidation usually becomes severe in practice only when metals are heated to comparatively high temperatures. The oxidation which then occurs during fabrication may be no more than a technological and economic nuisance, and can usually be kept within reasonable bounds by good process control. There are occasions, however, when little or no oxidation is acceptable during fabrication, and positive control measures then become necessary even though they may be costly. The most direct method is to remove the oxidizing environment, and this can be done in a number of ways. The normal atmosphere can be flushed out of the heated enclosure and replaced by a specially prepared atmosphere which is neutral to the metal at the temperatures concerned (this is called a *controlled atmosphere*; see p. 52). Another possibility is to pump the air out of the enclosure and maintain a certain level of vacuum during the heating cycle (*vacuum heat treatment*; p. 53). A third is to immerse the metal in a heated neutral liquid, often a molten salt.

Oxidation control has to take a different approach, however, when the metal is required to operate for long periods in an oxidizing atmosphere. Special alloys into which extra oxidation resistance has been built then are required, and most of these depend on the addition of chromium as the key alloying element. Nickel-chromium and nickel-cobalt-chromium alloys are commonly used in more arduous service conditions, and iron-chromium and iron-nickel-chromium alloys in less arduous conditions. The principles underlying the effectiveness of chromium in nickel-base alloys have been discussed above. The scales of Cr_2O_3 that are formed on these alloys have a good balance of oxidation-inhibiting characteristics. In particular, they are highly adherent. Chromium also rarely adversely effects other important properties. However, disadvantages of chromium are its high cost and its limited availability. Silicon and aluminum would in principle be better choices for conferring oxidation resistance, because they are both effective and readily available. But their protective scales are brittle and so tend to spall more readily, and as alloying elements they often adversely affect important mechanical properties of the alloy. Nevertheless, both are used as alloying elements in this role, either alone or as supplements, in appropriate circumstances.

The problems become more severe when the metal is required to carry high stresses in service, because the alloy then has to be developed to have, above all, adequate mechanical properties at the operating temperature. The alloys used in the hot-end components of gas turbines, particularly the turbine blades which are subjected to high centrifugal stresses in service, are cases in point. These alloys are in effect basic heat-resistant nickel-chromium alloys which have been modified considerably to improve their creep resistance. The chromium content has been juggled around with other alloying elements as the alloys have been improved, the objective being to balance creep strength, oxidation resistance, cost, and a number of other factors. Oxidation resistance has tended to lose out as the strengths of these alloys have been improved and as their operating temperatures have been increased. It has become a life-limiting factor. The last line of defense, as for many corrosion-control problems, is to change the composition of the surface layers. This is the only region where good oxidation resistance really is required, and hopefully the surface changes can be made without affecting the bulk mechanical properties of a component too seriously. Gas-turbine blades are routinely treated in this way. In the simplest process, aluminum is evaporated in a vacuum onto the finished surfaces and is then diffused into the surfaces by a heat treatment. A layer of a compound of the NiAl type is produced and develops a scale layer of the highly protective Al_2O_3 during oxidation. The blades have to be recoated at intervals as the surface oxidizes away, but this is cost-effective because the blades are very expensive and many are required in an engine.

Internal oxidation occurs simultaneously with oxidation scaling in many industrial alloys. This phenomenon can be controlled only by increasing the content of the element that is oxidizing internally to above that at which the transition to external scaling occurs. Measures of this nature commonly are not possible because the composition would be deoptimized from other, more essential points of view. Perhaps the important thing is that the possible existence of this phenomenon, and its consequences, should be recognized.

FURTHER READING

M. G. Fontana and N. D. Greene, *Corrosion Engineering*, McGraw-Hill, New York, 1967.
H. L. Logan, *The Stress Corrosion of Metals*, Wiley, New York, 1966.
P. Kofstad, *High Temperature Oxidation of Metals*, Wiley, New York, 1966.
Metals Handbook, 9th Ed., Vol 13, *Corrosion*, ASM International, Metals Park, OH, 1987.

· 6 ·

Metallography: Observing and Characterizing the Internal Structures of Metals

SUMMARY

Physical metallurgy is based on the concept that many of the important properties of a metal can be correlated causally with its structure. To be useful, this concept requires that techniques be available to detect and investigate the relevant structures — structures ranging from those that can be seen by the unaided eye to the details of the atomic arrangements which make up the crystals of the metal. A wide range of these metallographic techniques is now available for such purposes, each filling a niche in the required spectrum of needs. They include microscopic methods using optical microscopes, transmission electron microscopes, and scanning electron microscopes; diffraction methods employing x-rays or electrons; and techniques by which the local chemical compositions of the smallest features being investigated can be determined. These investigational techniques are powerful but destroy the specimen, and there is a need for others with which at least some information can be obtained by nondestructive methods. This need arises particularly in routine examination of components in production. However, at present only discontinuities which are comparatively large can be detected by nondestructive methods. Discontinuities which outcrop at the surface can be outlined in ferromagnetic materials by applying a powder of magnetic particles while the component is held in a magnetic field (magnetic-particle inspection). They can be detected in all metals by allowing a liquid to penetrate into the discontinuity and subsequently allowing it to seep out (liquid-penetrant inspection). Internal discontinuities are detected by radiographic methods or by beaming ultrasonic waves into the component and detecting waves that are reflected from the discontinuities (ultrasonic inspection).

The observation and characterization of the internal structures of metals are the bases on which physical metallurgy is built. These, together with the development of an understanding of the origins of structures, constitute the branch of metallurgical science and technology known as *metallography*. We shall in the present chapter outline in a broad way the investigational techniques used for this purpose, giving indications of their fields of use and their limitations.

Structures have to be characterized on a wide range of scales, which requires a wide range of investigational techniques. In one group of techniques it is required that a specimen be cut out of the material being investigated, and so these techniques are destructive. Sometimes, however, it is not acceptable for a useful or valuable component to be destroyed, and thus there is a need also for nondestructive methods of investigation by which as much information can be obtained as possible. Several such methods have been developed with this end in mind, and the more important of these will also be outlined in this chapter. Many of the nondestructive methods can also be used for routine inspection of components.

OPTICAL MICROSCOPY

The Problem of Seeing Inside a Metal

It is by no means obvious how the structure inside an opaque and apparently impervious object such as a piece of metal might be discerned. Very rarely can anything of significance be seen by inspecting the natural surface of a metal. The early microscopists — including Hooke,* who was perhaps the most perceptive of them all — tried this with little success. Hooke, for example, examined the edges of shaving razors and saw no more than that they were not as sharp as had been imagined. Others tried another obvious strategem of examining the surfaces produced when a metal was broken open by fracturing. We have seen (in Chapter 3) that much useful information can be obtained about the fracturing process in this way, but little can be obtained about the structure of the metal itself.

It would also have been of no use to have attempted to imitate the early biological microscopists who had established techniques for examining thin slices of specimens. No matter how thin, a slice of a metal is completely opaque to light and so cannot be examined in a transmitted-light microscope of the type used by biologists. For metals, the only recourse is to examine by reflected light the surface of a section cut from a specimen. But nothing can be seen directly on such a surface, no matter how well it has been machined, except perhaps major discontinuities such as cavities if they happen to be present. Success comes only when the machined surface is treated with a chemical re-

---

*This is the same Hooke who is commemorated in Hooke's Law. He was a man of many parts. He was, for example, a leading member of the team led by Christopher Wren that was responsible for rebuilding many of London's churches, including the masterpiece St. Paul's Cathedral, after the Great Fire of 1666.

agent which dissolves the metal; this is called *etching*. The different structural features of the metal may then be attacked differently by the etchant and so become distinguishable.

Aldis von Widmanstätten was the first to take this seminal step in 1813 when, as Curator of Meteorites at a museum in Vienna, he developed the technique to investigate the structures of metallic meteorites. An example of the arresting types of structures that he saw is given in Fig. 6.1. This general type of regular geometric arrangement of plates (seen as needles or bands in a two-dimensional section) of one constituent in another is still known as a Widmanstätten structure (see also Fig. 6.3).

Macroscopic Methods of Investigating Structures

Investigational techniques which trace back to that pioneered by Widmanstätten are characterized by sectioned surfaces which are prepared by normal machine-shop practices and which are comparatively heavily etched. They are capable only of delineating structural features which can be observed by the unaided human eye or at best using modest magnification (say up to 10×). They

The section surface has been prepared to a good workshop-standard finish and then etched with a strong acid. The etching has revealed a pattern which delineates the macrostructural arrangement of the various constituents of the material. This particular geometrical type of structural pattern is known as a Widmanstätten structure, after Aldis von Widmanstätten, Curator of Meteorites at a museum in Vienna, who first observed it in a meteorite in 1813. This meteorite is a single crystal (p. 240), as evidenced by the fact that the Widmanstätten plates extend completely across the section. They would change direction if they encountered a grain boundary. Optical macrograph; 1½×.

Fig. 6.1. Section of a meteorite found in central Australia. (Source: M. Hatherly)

are called *macroetching* as techniques of macroscopic examination. Nevertheless, they fulfill a useful role at one end of the spectrum of investigational techniques that are available to metallurgists. Many significant large-scale structures that could not be investigated in any other way can be characterized, examples of which are given throughout this book (e.g., Fig. 2.22, 2.26, 8.9, and 8.10).

Microscopic Methods of Investigating Structures

Two refinements in this general technique are necessary, however, before it can be used to examine metals successfully at higher magnifications in an optical microscope. The first is that the surface must be prepared to a much better finish — in fact, to a flat, mirrorlike finish. Moreover, the microstructure of the material must not be altered in the process. This is more difficult to achieve than might be thought, and special laboratory techniques have to be used for the purpose. Secondly, the etching procedure also has to be more carefully controlled. The surface must not be attacked too severely; otherwise, its necessary reflectivity will be lost. The etchant must also be sufficiently selective to differentiate fine structural features clearly in a predictable way. A wide range of etching procedures have been developed for this purpose, but experience and skill are required to select the most appropriate one for a particular application and to use it successfully. Most of these etching procedures involve chemical reactions between a solution and the polished surface. Some, however, involve physical phenomena, such as the deposition of films on the prepared surface, but they still are usually described as being etching techniques.

The required refinements in preparation procedures were first achieved by Henry Clifton Sorby, who can fairly be said to be the founding father of metallography. He carried out his first successful experiments in 1863, which was some 50 years after Widmanstätten discovered how to carry out macroetching and which says much about the difficulty of making the transition from macroscopic to microscopic examinations. Moreover, it was not until about 1900 that his techniques began to be widely applied, which says much about the conservatism and technological state of the metals industry of the time. When they finally were applied, they turned out to be the breakthrough that established physical metallurgy as a technology based on sound science. Metallurgists for the first time began to appreciate the importance of structure as well as chemical composition in determining the properties of metals and alloys.

Actually, a third refinement is needed for effective application of this technique — a refinement in the optical microscope. An optical microscope consists basically of three units. The first is an *objective lens*, the purpose of which is to carry out the primary magnification of the object. It must, however, do more than this. It must produce a sharp image in which closely spaced points in the object can be distinguished as being separate, or *resolved*, as it is called, in the image. The second unit is an *eyepiece*, which further magnifies the primary image. It produces a final image which can be either viewed directly by eye or projected onto a photographic film for permanent recording. The third unit is an *illuminating system* consisting of a light source and several lenses. Its role, which in many ways is just as important as that of the objective lens or the eyepiece, is to ensure that the maximum possible resolution is achieved by the objective lens and that the field of view in the object is uniformly illuminated and illuminated brightly enough to be seen or photographed.

The specimens that had been examined up until Sorby's time were of types that were reasonably transparent to light. With such specimens the illuminating system can be placed on the far side of the specimen from the objective lens and light transmitted through the specimen into the objective lens. This is the type of transmission microscope that is still used by, for example, biologists. A transmission arrangement is not practicable for metals because, as we have already noted, a section of a metal, no matter how thin, is completely opaque to light. The most obvious way of modifying the illuminating system to overcome this problem is to place it above and directed toward the highly reflecting specimen surface. The idea is to use the specimen surface to reflect the light back into the objective lens. Sorby used this technique but soon found that it has serious limitations. For a start, it is difficult to do, but worse still are its fundamental shortcomings which seriously degrade the performance of the microscope. One of these shortcomings results from the fact that the specimen surface is not all that flat on a microscopic scale, particularly after it has been roughened by etching. So some areas might be oriented to reflect light into the objective lens, but many others usually are not so favorably oriented and reflect the light away from the objective lens. The latter areas consequently appear black in the image and cannot be examined effectively. The higher the magnification, the more serious these difficulties become.

To examine effectively a polished metal surface, the illuminating light has to be incident normal to the surface. This can be achieved only by directing the light through the back of the objective lens onto the specimen surface, from which it is reflected back into the objective lens. The type of arrangement necessary in a *reflecting microscope* is sketched in Fig. 6.2. Those portions of the specimen surface that have not been much attacked during etching then reflect most of the incident light into the objective lens and appear bright in the image. Those that have been roughened or covered by a light-absorbing film during etching do not reflect quite so much light back into the objective lens and so appear as shades of gray in the image, or even black in extreme cases (Fig. 6.3). A few metals reflect white light in colors, (e.g., copper and gold), in which event color contrast also is introduced into the image. This illuminating method is known as *vertical bright-field illumination*. There are other methods of obtaining image contrast which depend on other phenomena of optical reflectivity, but the bulk of the optical microscopy of metals is carried out using bright-field illumination.

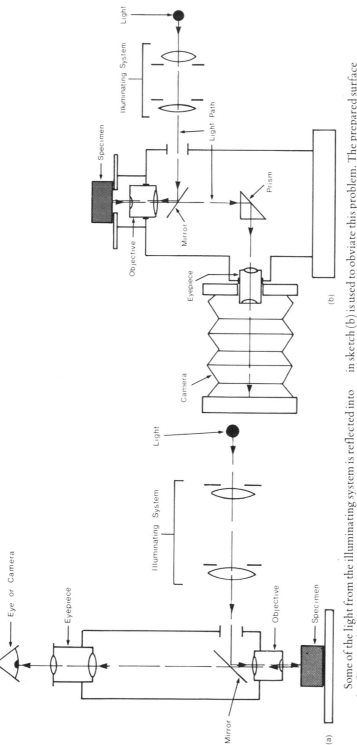

Some of the light from the illuminating system is reflected into the objective lens by a mirror which is only partly reflecting. The objective lens first acts as the condenser lens of a transmission microscope and focuses the light onto the specimen surface. It is then reflected back into the objective lens, which this time acts as a magnifying lens. Some of the light emerging from the objective lens now has to pass through the partly reflecting mirror, which is why it has to be only partly reflecting, and then into the eyepiece.

Sketch (a) illustrates a simple upright microscope, the main disadvantage of which is that the alignment of the specimen surface has to be adjusted to ensure that it is normal to the optical axis of the microscope. Commonly, the inverted arrangement illustrated in sketch (b) is used to obviate this problem. The prepared surface of the specimen can now be placed directly onto the specimen stage with the assurance that it will be normal to the optical axis. In sketch (b), a prism is shown as directing the light into a projection eyepiece and then into a photographic camera. Other prisms can be used to direct the light into an alternative eyepiece for visual examination. Mechanisms not shown are also necessary to translate the specimen relative to the optical axis so that different areas on the surface can be viewed. Others are necessary to adjust the objective lens to a position of sharp focus. These features are common to microscopes of all types, including electron microscopes.

Fig. 6.2. Schematic diagram of the arrangement of the components of an optical microscope used to examine metals by reflected light.

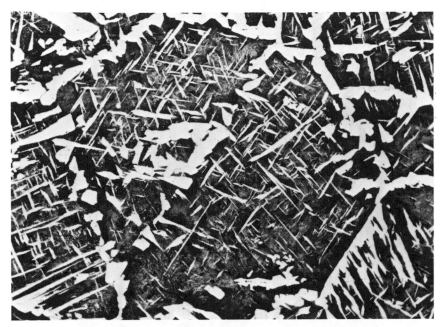

The section surface has been much more carefully prepared than the one in Fig. 6.1, using special laboratory techniques, and has been etched more lightly in very dilute acid. The white areas are bright because they have been little attacked during etching and so reflect almost all of the incident light back into the microscope. The dark areas appear dark, on the other hand, because they have been roughened during etching and scatter much of the incident light out of the collection range of the microscope objective. The metal shown here is a steel. The white areas are regions of comparatively pure iron (known as ferrite). The dark areas can be resolved at higher magnifications, as shown in Fig. 6.4, into approximately parallel plates of ferrite and iron carbide (cementite). This structure is known as pearlite. Note that some of the ferrite is arranged as plates in a geometrical pattern similar to that in Fig. 6.1. This too is known as a Widmanstätten structure. Photomicrograph; 100×.

Fig. 6.3. A representative optical photomicrograph of a metal.

As we have already noted, a microscope has to do more than magnify. It also must resolve fine detail, and the resolution of any imaging device is limited by the wavelength of the radiation used to form the image. The design of the objective lens used primarily determines how closely this limit is approached, and the lenses intended for use at higher magnifications are designed to approach the limit most closely. However, in the ultimate, the resolution is limited by the shortest wavelength of light that can be used, and this in turn is limited to the shortest wavelength that the human eye can see. The wavelength at which the eye works best is about 0.5 μm (20 μin.), and this limits the resolution of an optical microscrope to about 0.2 μm (8 μin.). The conse-

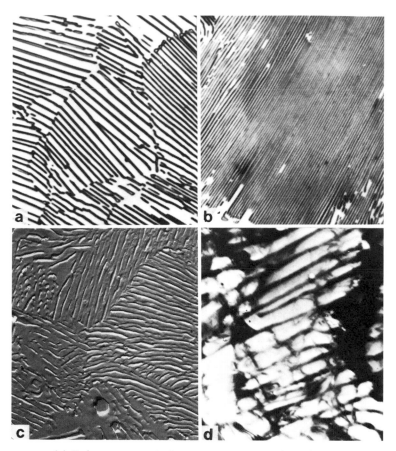

(a) Only a comparatively coarse structure such as that illustrated here can be resolved easily by optical microscopy, even when the best optical components are used. Optical photomicrograph; 1000×.

(b) The structure shown here is about the finest that can be resolved by optical microscopy. The spacing of the pearlite plates is 0.2 to 0.3 μm (8 to 12 μin.). Optical photomicrograph; 2000×.

(c) Finer detail can be seen by electron microscopy when used to examine replicas of specimens prepared as for optical microscopy. The plate spacing of the pearlite illustrated here is about 0.5 μm (20 μin.), a spacing which could only just be resolved by optical microscopy. Replica electron micrograph; 5000×.

(d) Much finer detail can be resolved by an electron microscope when it is used to examine thin foils by transmission. The spacing of the plates in the pearlite illustrated here is about 0.1 μm (4 μin.), and the resolution potential of the technique has by no means been fully utilized. The cementite plates are seen as dark lines which effectively are shadows of the plates in the foil. Transmission electron micrograph; 28,000×.

Fig. 6.4. Photomicrographs illustrating that there is a limit to the fineness of detail that can be resolved by optical microscopy. The subjects are all pearlites in steel, a lamellar mixture of two constituents, but with decreasing spacing of the two constituents.

quences of this limitation are illustrated in Fig. 6.4. Nevertheless, an optical microscope can improve the performance of the human eye by a factor of as much as 500 times, and magnifications of up to 2000 times may be used effectively. This improvement is a very worthwhile one indeed, allowing much information to be obtained about the structures of metals, but a wealth of information remains undisclosed.

ELECTRON MICROSCOPY

To improve matters, we have to seek some way of using in a microscope an electromagnetic radiation whose wavelength is much shorter than that of visible light. Beams of electrons turn out to be the most useful contenders, because they have some of the characteristics of an electromagnetic radiation. They can be focused by electromagnetic lenses and have an effective wavelength about 50,000 times shorter than that of light. So far, however, design problems associated with the use of such a short-wavelength radiation in a microscope have not permitted the full benefit of the short wavelength to be realized. Currently, the resolution obtained in the best electron microscopes is about 10,000 times better than that of the best optical microscopes. This makes them about five million times better than the unaided human eye. Magnifications of over five million can be used effectively, although most usage is confined to magnifications in the hundreds of thousands.

Electron microscopes were first developed to operate in a transmission mode in which the electrons are transmitted through the specimen in the same way as in a transmission optical microscope. They are composed, as sketched in Fig. 6.5, of the same basic units as those of an optical microscope. An electron gun generates a beamed source of electrons which is focused by an electromagnetic condenser lens onto the specimen. A system of electromagnetic magnifying lenses, comprising an objective lens and a projection lens, forms an image from the electrons that are transmitted through the specimen. This image is projected either onto a fluorescent screen, where it can be viewed by eye, or onto a photographic film, on which it can be recorded permanently.

Replica Electron Microscopy (REM)

An electron microscope is a much more complex instrument than an optical microscope and has its own inherent limitations. A major one from our point of view centers again around methods of preparing specimens of metal which can in fact be examined. Metals are transparent to electrons, but only in very thin sections. In the beginning, specimens which were thin enough to be adequately transparent could not be prepared. The strategem adopted was to make a replica of a surface which had been prepared as for optical microscopy. The rep-

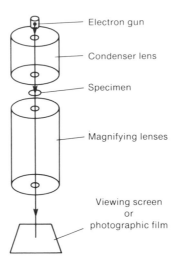

The layout is similar in principle to that for an optical microscope. A beam of electrons is generated and then focused by an electromagnetic condenser lens onto the specimen, which must be thin enough to be reasonably transparent to electrons. A selected field of the specimen is then magnified by electromagnetic objective and projector lenses, the image being projected onto either a fluorescent screen (for direct viewing) or a photographic film (for recording). All of these units have to be located in a container which can be pumped out to a high vacuum; the electrons would be absorbed if many gas molecules were present in the microscope column. The column also has to be much longer than for an optical microscope, and so it usually is convenient to use the inverted arrangement sketched here.

Fig. 6.5. A highly stylized diagram illustrating the principles of a transmission electron microscope.

lica is made as a thin film of a material, such as plastic or carbon, which has no detectable structure of its own and which is highly transparent to electrons.

The contrast in the image that is formed from such a replica is related to topographical features of the surface that has been replicated. These features introduce variations in the thickness of the replica film, and these in turn cause variations in the absorption of the electrons which pass through the replica in the microscope. This technique is, in a sense, an extension of optical microscopy, hopefully with improved resolution. But the full potential of electron microscopy has not been realized. Resolution is restricted by factors such as the resolution achieved by the specimen-preparation processes and the faithfulness with which the topography of the surface is reproduced in the replica. A significant improvement in resolution compared with that of the optical microscope is achieved nevertheless, and this improvement can help in elucidating structures that cannot quite be resolved by optical microscopy (Fig. 6.4c). However, the introduction of replica electron micrscopy hardly resulted in a revolution in metallography.

Transmission Electron Microscopy (TEM)

A revolution did occur, however, when techniques were finally developed to produce thin slices, often called *foils*, from bulk metal specimens — foils which were thin enough to be examined directly in transmission in a standard electron microscope. The mechanisms by which contrast is produced in the microscope image are then quite different from those responsible for image contrast in optical or replica electron microscopy. They are complicated and not as obvious as they might seem to be. However, for our purposes we can describe two of the more important mechanisms as follows:

1. Constituents in the matrix of the foil absorb electrons more or less strongly than the matrix. "Shadows" of the structural constituents are then seen in the image (Fig. 6.6a).
2. The transmission of the electrons is deflected by irregularities in the arrangement of the atoms in the crystals of the constituents. Image contrast is then related to features of the crystal structure (Fig. 6.6b and c).

The first mechanism of image formation in effect allows microstructural constituents to be observed in the same general way as for optical microscopy but with greatly improved resolution. It has allowed the exploration of structures the existence of which could not even have been suspected from optical microscopy; the precipitates so clearly seen in Fig. 6.6(a), for example, are much too small to be resolved by an optical microscope. The second mechanism has enabled exploration of some of the details of the arrangements of atoms in the crystals of which metals are composed, details which have a determining influence on many properties of metals (see Chapters 7 and 9). For example, lines along which the crystal defects known as dislocations are located can be seen (Fig. 6.6b). The most advanced electron microscopes can even obtain information about the arrangement of individual planes of atoms in a crystal (Fig. 6.6c).

Transmission electron microscopy of metals is, however, not an easy technique to use experimentally, and the images are not always easy to interpret properly. Moreover, it is by no means always possible to apply this technique because it is not always possible to prepare a satisfactory thin foil. Even when this is possible, there may be doubts about whether the structure of the metal has changed during the preparation of the foil. Moreover, the volume of material that is examined in any field of view is minute — perhaps a thousandth part of the point of a pin. Consequently, it is sometimes a matter for conjecture as to how representative are the observations that are made on such a small sample. Nevertheless, transmission electron microscopy undoubtedly has been a most valuable addition to the armory of the metallurgist and has made an enormous contribution to our understanding of the structures of metals. It is not a panacea, as some thought it to be in its first boom of enthusiasm, but it should be regarded as an important tool of the metallographer.

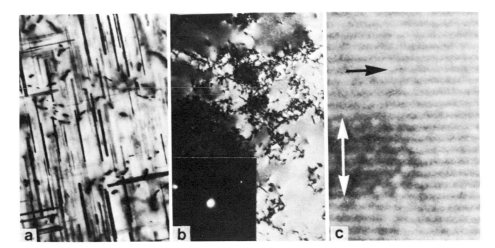

(a) Small precipitates in an aluminum alloy are seen here effectively as shadows because they have absorbed electrons more strongly than has the matrix. These particles are much too small to be detected by optical microscopy, let alone for their shapes to be so clearly revealed. Magnification, 10,000×. (Source: I. J. Polmear)

(b) Dislocations (p. 236) in plastically deformed copper, each line in the image indicating the location of a dislocation line running through the thickness of the specimen foil. Magnification, 15,000×. The inset is an electron diffraction pattern obtained from the same area of the foil. Certain characteristics of the structure in the area of the crystal seen in the micrograph can be deduced from the positions and shapes of the spots which comprise the diffraction pattern.

(c) The lines of which one is indicated by the black arrow are images developed from individual crystal planes, seen on edge, in an aluminum alloy. These planes are known to be spaced only 233 pm (2.33 Å) apart. A portion of a precipitated particle (located within the bounds of the white arrows) is present in the field of view, and traces of crystal planes can be distinguished in this particle also; they can be seen to be parallel to and continuous with those in the matrix crystal, from which observation conclusions can be drawn about the likely mechanism of formation of the phase and its effects on properties. This photomicrograph is an example of electron microscopy of metals at its zenith. Magnification, 1,000,000×. (Source: I. J. Polmear)

Fig. 6.6. Examples of the types of images that are obtained when metals are examined by transmission electron microscopy.

Scanning Electron Microscopy (SEM)

It would obviously be desirable to be able to examine directly the surface of a bulk metal specimen in an electron microscope. The difficult techniques of specimen preparation involved in transmission electron microscopy would be avoided and natural surfaces could be examined. Such a procedure would require that the electron microscope be modified, as was necessary for the optical microscope, so that an image could be formed from electrons that were, in a sense, reflected off the specimen surface.

This concept has been realized in the *scanning electron microscope*, the principles of which are outlined in Fig. 6.7. An electron beam is generated in an evacuated chamber and accelerated toward and focused on a very small spot on the specimen surface. This spot is made, by electronic means, to scan an area of the specimen surface in a raster pattern, as in a television system. Electrons and a number of other types of radiations are produced in each volume of specimen material as it is irradiated, and some fraction of these radiations emerge into space above the specimen surface. The intensity of any selected one of the emergent radiations can be measured by means of a suitable detector (Fig. 6.7), and the signal developed by the detector can then be used to modulate a cathode-ray tube which is scanned in synchronism with the scanning of the

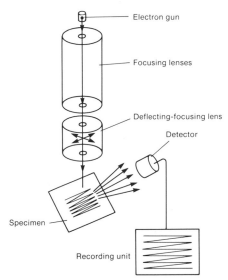

An accelerated beam of electrons is focused by electromagnetic lenses on a very small spot on the specimen surface, the beam being deflected by an electronic system to paint the spot over an area of the specimen surface in a raster pattern of the type that is used in television systems. Radiations of various types are emitted from each spot on the surface as it is scanned, and the intensity of one or more of these radiations can be measured by a detector appropriate to each. The output of the detector is used to modulate a display unit (which again is similar to that used in a television system) or to operate a recording unit. The whole unit, with the exception of the display/recording unit, is enclosed in a chamber which is pumped out to a high vacuum. The specimen can be translated and tilted so that different regions can be examined.

Fig. 6.7. A highly stylized diagram illustrating the principles of a scanning electron microscope.

specimen surface. An image which can either be viewed on the screen or recorded photographically is produced.

The final image is similar to that obtained by a television system, but the image contrast is related to the way in which the specimen surface reflects the incident electrons; more specifically, it is related to the type of "reflected" radiation that is "seen" by the detector in use at the time. Sensors are available to detect any one of the possible reflected radiations. Each will "see" a different characteristic of the surface, but all will see a characteristic which is quite different from that seen by other methods of microscopy. Thus a weapon is added to the metallographer's armory which can produce new information.

Three radiations are used routinely to produce images in metallography, namely: *backscattered electrons*, which are incident electrons that have been scattered by elastic collisions within the specimen crystal; *secondary electrons*, which are new electrons formed by interactions between the incident electrons and orbiting electrons of atoms of the specimen material; and *x-rays*, those of greatest interest being produced by atoms of the specimen material which have been excited during collisions with incident electrons. Leaving the x-rays aside for the moment, the intensity of both backscattered and secondary electrons is determined by characteristics of the specimen surface which give rise to what are known as *topographic contrast* and *atomic-number contrast*, respectively.

Topographic contrast is the effect that produces the most pronounced contrast and arises because the intensity of both backscattered and secondary electrons decreases markedly as the inclination of the surface to which they are incident increases. The effect is noticeably stronger with secondary than with backscattered electrons. It induces contrast which is related to variation in the inclination of the surface across the field of view — that is, to the topography of the specimen surface. A strong stereoscopic effect is obtained (Fig. 6.8), the contrast being analogous to an optical image in which the light comes from the direction of the detector and the observer is looking down the electron beam.

Atomic-Number Contrast. The intensity of both backscattered and secondary electrons increases with the *atomic number** of the atom responsible for their presence at the detector. The resultant image contrast consequently is related to surface variations among the types of atoms present — that is, to the chemical composition of the spot irradiated. Atoms of higher atomic number tend to produce brighter images (Fig. 6.9). This enables phases, which invariably have different compositions, to be differentiated and permits useful qualitative information to be obtained about their compositions. However, the effect is normally swamped by topographic contrast and usually is noticeable only in favorable situations such as those that can be achieved in sections prepared for

---

*The individuality of any type of atom is defined by, and can be identified by, its atomic number, which is the number of protons which are present in its nucleus. This number progresses from 1 for hydrogen, the lightest element, to 92 for uranium, the heaviest of the commonly available metals, and beyond for unstable radioactive elements.

(a) Natural outer surface of an oxide layer on nickel. A strong stereoscopic impression is obtained of systems of facets on the surface. Magnification, 100×.

(b) A section of an aluminum-silicon alloy from which the aluminum phase has been removed from a surface layer by preferential etching. An optical micrograph of this material is shown in Fig. 8.14b (p. 273). The three-dimensional morphology of the silicon phase can be seen here, which incidentally is only possible because of the large depth of field of the microscope (it would not be possible to deduce this morphology from Fig. 8.14b). Magnification, 1000×.

Both of these micrographs are examples of new avenues of investigation opened up by this branch of microscopy. Both images were produced by secondary electrons.

Fig. 6.8. Examples of images produced in a scanning electron microscope by topographic contrast. The large depth of field of this type of microscope is also illustrated.

optical microscopy. These sections are often flat enough to suppress topographic contrast considerably. It may not be eliminated, because small differences in level usually develop between grains and phases during preparation of the surface, but it may be suppressed sufficiently for the contribution of atomic-number contrast to the image contrast to become detectable (Fig. 6.9). The most marked effects are then obtained with backscattered electrons, because their intensity varies more markedly with atomic number than does the intensity of secondary electrons.

The electronic and recording systems of a scanning electron microscope are arranged so that a magnified image of the specimen surface is obtained, as is to be expected of a microscope. The magnification typically is variable over a wide range, from actual size up to 10,000× or even 100,000×. The resolution in the image is determined, however, not by the effective wavelength of the electron beam, as for transmission electron microscopy, but, at the limit, by the diameter of the scanning spot. The closeness with which this limit is ap-

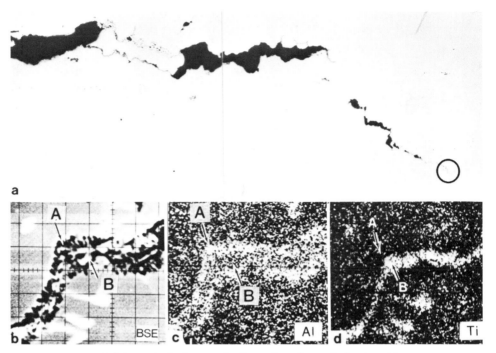

(a) An optical micrograph of a section of a crack developed in service in a jet-engine turbine blade manufactured from a nickel-base superalloy. The crack can be seen to be filled with foreign matter at its tip (circled). Scanning electron microscopy was used to identify the filling material.

(b) A scanning electron micrograph of the circled area in (a) obtained when using a modulating detector sensitive to backscattered electrons. Image contrast here is related to the variation in atomic number of the elements present. It is apparent that two constituents are present in the filling material, the atomic number of the major element in one (arrowed B) being substantially higher than for the other (arrowed A). The atoms of higher atomic number have generated more backscattered electrons and so produce a brighter image in the micrograph.

(c) and (d) The electron microscope has been operated here as a microanalyzer in an area-scan mode, the detector used being sensitive to the element aluminum in (c) and to titanium in (d); of the two, titanium has the higher atomic number. The dot plans indicate that constituent A is aluminum rich, and B titanium rich, in agreement with the predictions made from (b). They are probably aluminum oxide and titanium oxide, but the presence of oxygen could not be confirmed because oxygen cannot be detected by an energy-dispersive analyzer. This could have been confirmed, on the other hand, by a microanalyzer using a fluorescence analyzer (Fig. 6.10B).

Magnifications: Optical micrograph, 100×; scanning electron micrographs, 2000×.

Fig. 6.9. Examples of images produced in a scanning electron microscope by atomic-number contrast supplemented by energy-dispersive microanalysis.

proached is determined by the quality of the electronic system. Resolution consequently varies with the microscope design but never reaches the level possible in transmission electron microscopy. Scanning electron microscopes fit in the gap between optical and transmission electron microscopes. They are perhaps 10 to 100 times better than optical microscopes and 5000 to 50,000 times better than the human eye.

But the system has one distinguishing characteristic which is of great significance: it has a very large *depth of field** — as much as several millimetres at low magnifications and even a few micrometres at the highest magnifications. This is at least 100 times better than the depth of field of an optical microscope and allows rough surfaces which could not possibly be examined by other methods of microscopy to be investigated (for example, natural surfaces and fracture surfaces). This also opens up whole new fields of investigational possibilities (see Fig. 6.8b) which are, in fact, responsible for much of the wide use of scanning electron microscopy in contemporary metallurgy. You will notice many examples throughout this book.

MICROANALYSIS

We mentioned earlier that x-rays are included among the radiations that are emitted when an accelerated beam of electrons impinges on a metal. These x-rays have two characteristics which are uniquely related to the atoms that were present at the point irradiated. More specifically, they are related to the *atomic number* of the atom and so provide the equivalent of fingerprints on which positive identifications can be based. First, certain energy levels are emitted particularly intensely, the levels concerned being characteristic of the emitting atom; examples are given in Fig. 6.10. Second, x-rays of certain wavelengths are emitted particularly strongly (*x-ray fluorescence*), and the spectrum of these wavelengths is also characteristic of the emitting atom. The wavelength spectrum can be determined experimentally and from this the atom(s) present can be identified. For both methods, the concentration of the atoms concerned can be estimated from the relative intensities of the characteristic radiations.

These phenomena provide methods for carrying out a chemical analysis of a chosen volume of metal if it can be irradiated with an electron beam, as can be done in the type of systems used in electron microscopes. Such an analyzer based on the determination of the x-ray energy spectrum is called an *energy-dispersive analyzer* (commonly abbreviated to *EDAX*) and one based on the wavelength spectrum is called an x-ray fluorescent analyzer. Since we can irradiate very small volumes of material in an apparatus such as a transmission or a scanning electron microscope, we have available methods of carrying out

---

*Depth of field is the vertical distance (depth) between points in the object field of view which can be seen in adequately sharp focus in the image. It is often erroneously called depth of focus.

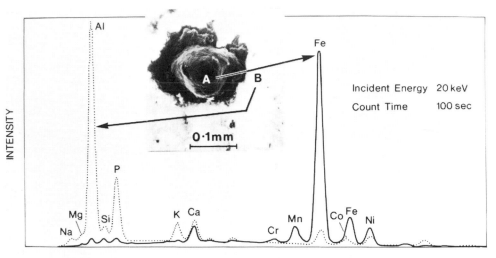

X-RAY ENERGY SPECTRUM

The inset is a scanning electron micrograph (magnification, 1500×) of a foreign inclusion (A) lodged in the surface of a jet-engine turbine blade made from a nickel-base superalloy (B). The energy spectrum obtained from the two areas (main figure) indicate that A is iron rich and almost certainly a fragment of steel. The surface at B is aluminum rich because an aluminide coating has been formed on the blade for oxidation protection (p. 181).

Fig. 6.10. An example of constituent identification by spot microanalysis using a scanning electron microscope fitted with an energy-dispersive analyzer system.

analyses (*microanalyses*) on a very small volume of material — a volume of material, moreover, which can be identified and correlated with structural features observed in the microscope (Fig. 6.10). This may be called *point analysis.* As we shall soon point out, the analysis using the x-ray fluorescent technique can be made to be fully quantitative (Fig. 6.10B).

Each of the analytical techniques has its advantages and disadvantages. The EDAX system can be fitted to a standard scanning electron microscope (or even a transmission electron microscope) and operated as a normal adjunct to it. Thus any feature observed microscopically can be investigated (Fig. 6.9 and 6.10). However, only elements with an atomic number greater than about nine can be detected with normal equipment. This means that the elements sodium, magnesium, aluminum, and silicon among the lightest elements can be investigated but not oxygen, nitrogen, carbon, and boron. A more serious limitation often is that the results cannot be made better than semiquantitative. An x-ray fluorescent system, on the other hand, can detect oxygen (Fig. 6.10B), nitrogen, and carbon and is generally a better performer. It can be made to be reliably quantitative. The x-ray fluorescence apparatus is, however, more

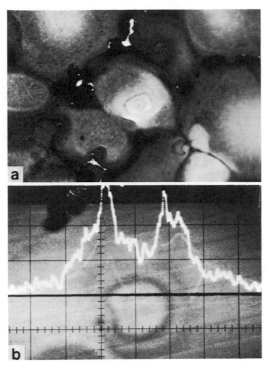

(a) An optical micrograph of a section of a cast magnesium-zinc-zirconium alloy etched to reveal concentric segregation shells in the magnesium-rich matrix grains. Magnification, 500×.

(b) An image of a similar area in the same section formed in an electron-probe microanalyzer. Superimposed on the image of a segregated grain is a line scan indicating the variation in zirconium content across that grain, the scan being produced by a fluorescence analyzer. The segregation shell is rich in zirconium.

Fig. 6.10A. An example of a microanalyzer used in a line-scan mode to explore microsegregation in a constituent.

complex and usually has to be operated in a dedicated instrument which may be called an *electron-probe microanalyzer* (EPMA). Due to fundamental reasons, it has comparatively limited resolution, more like that of an optical microscope than a scanning electron microscope.

In either case, a microanalyzer can be operated in several modes. First, in an *area-scan mode*, in which the field of view is scanned as during the normal operation of a scanning microscope and a modulating x-ray detector is set to be sensitive to a particular element. The resultant image is composed of discrete dots, the concentration of the dots giving a rough indication of the local concentration of the selected element (Fig. 6.9 and 6.10B). The image, which is often called a *dot plan* of the element, can be compared with a standard micrograph, be it obtained by an optical microscope (Fig. 6.10B) or a scanning elec-

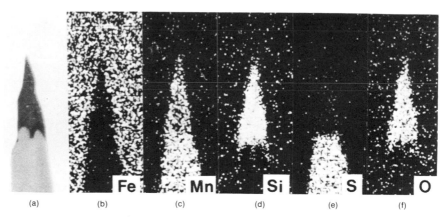

An optical micrograph of an inclusion in steel is shown in (a),
two constituents being discernible in the inclusion. Images of the
same area produced in a microanalyzer in the area-scan mode are
shown in (b) to (f), the element for which each scan was made
being indicated. These scans suggest qualitatively that the light
area of the inclusion is manganese sulfide and that the dark area is
a manganese silicate. Quantitative point analysis showed that the
compositions (wt %) actually were:

Element	Light area	Dark area
Iron	2.0	2.5
Manganese	61.8	49.7
Silicon	Nil	16.9
Sulfur	36.2	2.6
Oxygen	Nil	28.3

These compositions correspond closely to those of MnS and
$MnO \cdot SiO_2$, respectively. Note that analysis for oxygen has been
possible. Magnification of all micrographs, 1500X.

**Fig. 6.10B. An example of the use of a dedicated
electron probe microanalyzer in area-scan and point-
analysis modes, x-ray fluorescent methods of analysis being employed.**

tron microscope (Fig. 6.9). Dot plans can be prepared for a range of elements,
from which a good idea often can be obtained of the identity of a constituent
(Fig. 6.9 and 6.10B).

In a second mode of operation, called the *line-scan mode*, the scanning beam is
made to traverse across a chosen line in a field of view, the output being presented as a line graph depicting semiquantitatively the variation in the concentration of a chosen element along the scan line (Fig. 6.10A). The location of the
scan line can be identified on a micrograph and thus be correlated with features
of the microstructure. Moreover, the results can be made to be reasonably
quantitative with EPMA-type apparatus (Fig. 8.5, p. 258).

The development of microanalyzers has been of considerable importance in
metallography as an adjunct to microscopy. The identities of the phases and
constituents observed in any of the microscopes described earlier in this chap-

ter have to be inferred from indirect evidence, experience, or perhaps even educated guesses, procedures which are not always completely reliable to say the least. The information obtained by microanalysis is positive information on which identifications can be based with much more assurance. Moreover, compositional deviations of a phase from the norm, or compositional variations with microanalysis, compositional deviations of a phase from the norm, or compositional variations within a phase, can be explored in a way which previously had not been possible.

X-RAY AND ELECTRON DIFFRACTION

The techniques which we have described so far establish the presence of phases and constituents and give information about their shapes, sizes, distributions, and compositions. Information about how the atoms are arranged in these constituents (their crystal structures; see p. 222) can be obtained by the techniques of *x-ray diffraction* and *electron diffraction.*

These techniques are based on the fact that electromagnetic radiations which can penetrate a crystal can in effect be reflected by the planes of atoms in the crystal (p. 223). This occurs when, but only when, certain conditions are met. One of these necessary conditions is that a certain relationship exist among the wavelength of the radiation, the angle at which it is incident to the crystal plane, and the distance between the planes. The circumstances under which reflection does occur can be established experimentally under conditions where enough is known about these variables to allow a number of characteristics of the reflecting crystal to be established, such as the type of crystal (the crystal structure; see p. 224), the distance at which the atoms are spaced in the crystal (the lattice constant; see p. 223), and the orientation of the crystal with respect to an external axis of the specimen.

Either x-rays or electrons of various wavelengths, the values of which are well established, can be used as investigating radiations. X-ray diffraction techniques require the use of special equipment, and the more refined applications may require special specimens such as powders or fine wires. Electron diffraction techniques are almost exclusively applied these days to foils that are being examined in transmission in electron microscopes. Information about the crystal structure of any field being viewed can thereby be obtained, as indicated by the example given in Fig. 6.6(b).

NONDESTRUCTIVE INSPECTION METHODS

All of the methods of investigating structures that have been discussed so far require that the specimen or component being examined be destroyed. This is acceptable in many laboratory investigations but clearly is not acceptable for finished useful components. What then can be discovered about structures, particularly unwanted structures or flaws, without destroying a component?

Major flaws, such as wide cracks or holes, which outcrop at the surface may be discernible by simple visual inspection, particularly when the surface is finely finished. The visual acuity of an experienced inspector can be quite remarkable, but even so there is still a great need to enhance the visibility of surface flaws. We shall first discuss several commonly used methods of doing this. But what we also need to be able to do is, in effect, to see inside a component to determine whether or not it is sound. Techniques are available for this purpose, and we shall discuss the more important of them, but they are limited in range and in scope. In general, only comparatively major features which involve physical discontinuities (e.g., cavities or cracks) can be detected.

Nondestructive investigation of microstructure is a much more difficult proposition. Small areas on the surfaces of components sometimes can be polished and etched and then examined by normal methods of microscopy without doing an unacceptable amount of damage, but this is difficult and costly. Moreover, there is no guarantee that the structures of internal regions will be adequately represented. Techniques are available which enable indirect assessments to be made of some internal microstructural features of magnetic materials, but only by comparison with a standard. We shall not discuss these techniques.

METHODS OF ENHANCING THE VISIBILITY OF DISCONTINUITIES OPEN TO THE SURFACE

Liquid-Penetrant Inspection

The general principle used in these methods, illustrated diagrammatically in Fig. 6.11, is to coat the surface of the component with a liquid (a *penetrant*) which will be drawn into any discontinuity which outcrops at the surface. The excess liquid is cleaned off the surface without removing much from within the discontinuity, and the penetrant remaining in the discontinuity is then allowed to seep out to the surface. This process is assisted by coating the surface with an absorbent powder (a *developer*). The presence of penetrant on the surface then identifies the presence of a flaw. The penetrant may contain a colored dye to make its presence more easily visible. Alternatively, it may contain a fluorescent material which makes its presence easy to distinguish by visual examination in ultraviolet light.

Although these principles are simple enough, the practice is more complicated if optimum reliability and sensitivity are to be achieved. Specially compounded penetrants, cleaners, and developers are required and are best obtained in proprietary systems. These systems are available in forms ranging from simple packs suitable for investigating individual components in the field

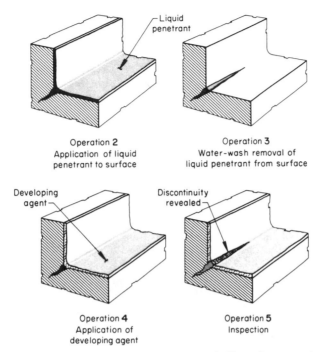

Operation **2**
Application of liquid
penetrant to surface

Operation **3**
Water-wash removal of
liquid penetrant from surface

Operation **4**
Application of
developing agent

Operation **5**
Inspection

The penetrant in this case is water-washable. A first operation, that of degreasing the surface to be inspected, is not illustrated. From *Metals Handbook*, 8th Ed., Vol 11, *Nondestructive Testing and Quality Control*, American Society for Metals, Metals Park, OH, 1976.

Fig. 6.11. Diagrammatic sketches illustrating the steps involved in a liquid-penetrant inspection system.

to large production installations. The advantages of a satisfactorily designed system are that it is relatively simple to use and that it can be applied to any alloy and to products of any shape. Quite fine discontinuities can be detected, but only those that are open to the surface and that are not covered by a coating, such as a layer of paint. Also, the technique is not easily applicable to components with rough surfaces.

Magnetic-Particle Inspection

This is a method for locating surface and slightly subsurface flaws in ferromagnetic materials, including most of the alloys of iron, nickel, and cobalt. The material is magnetized, and any discontinuity which lies in a direction generally transverse to the magnetic field generates a leakage field at and above the surface of the material (Fig. 6.12a). Finely divided particles of a ferromagnetic material applied to the surface during or after magnetization gather at, and are held by, the leakage field (Fig. 6.12b). They thereby form an easily observable outline, or *indication*, as it is called, of a flaw (Fig. 6.12c).

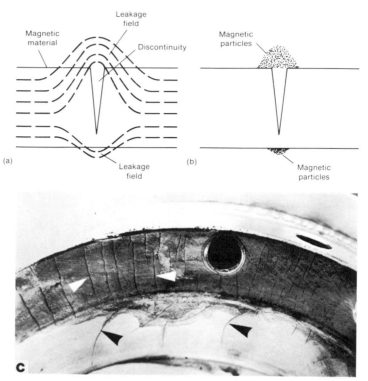

(a) The application of a suitable magnetic field induces a leakage field at a discontinuity which is open to the surface, and a weaker leakage field at one which is slightly subsurface. (b) Particles of a ferromagnetic powder accumulate in the regions of the leakage fields when they are applied to the surface. The particles remain in place when the magnetizing field is removed. (c) An example of the type of indications developed by a magnetic-particle inspection method. The indications reveal the presence of many fine cracks which developed in the cheeks of a crankshaft in service. Photograph; 1×.

Fig. 6.12. Sketches and a photograph illustrating the principles of magnetic-particle inspection.

The crux of this procedure is the generation of a magnetic field of adequate strength and in an appropriate direction in the areas under test. This may be done in one of several ways. The material may be placed in the field of either a permanent magnet or, more commonly, an electromagnet. Alternatively, a large electrical current may be passed through either the test material itself or a conductor located adjacent to the test area. Direct or alternating fields may be used, each having their own advantages. A few examples of the many magnetizing methods used in practice are illustrated in the sketches in Fig. 6.13 and 6.14. It is important to know the direction in which the magnetic field is induced in each case, which is indicated in Fig. 6.13 and 6.14, because the best results are obtained only with discontinuities that are oriented in directions

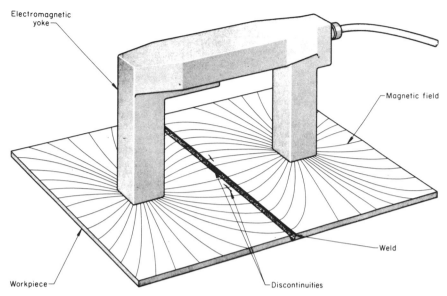

The electromagnet illustrated here is a small hand-held device with which a localized region of a test piece can be magnetized. The lines on the surface of the test piece indicate the directions of the magnetic field established in the test piece. Only discontinuities which are nearly perpendicular to these lines are likely to be detected. Small test pieces can be magnetized completely by placing them between the poles of a large fixed electromagnet. From *Metals Handbook*, 8th Ed., Vol 11, *Nondestructive Testing and Quality Control*, American Society for Metals, Metals Park, OH, 1976.

Fig. 6.13. Diagrammatic illustration of a method of magnetizing a test piece by placing it in the field developed by an electromagnet.

perpendicular or nearly perpendicular to the direction of the field. Several magnetizing methods which induce fields in different directions consequently may be necessary in some instances to ensure that discontinuities of all possible orientations are detected. A magnetic field of substantial strength is required, and much practical experience has been built up to establish the requirements under specific sets of circumstances.

The magnetic characteristics, size, and shape of the particles applied to delineate the discontinuity all are important in determining sensitivity. Color is significant in that it affects the visibility of the indications, depending on the color of the background material. Proprietary formulations generally are used. Some are dry powders which are blown at low velocity onto the inspection surface; these types are available in a range of colors and are most useful on rough surfaces and for detecting subsurface flaws. Others are used as suspensions in a fluid, either water or a light petroleum distillate, which is run over the surface being inspected. The suspensions are more suitable for detecting fine discontinuities. They commonly are black, which is suitable for examining

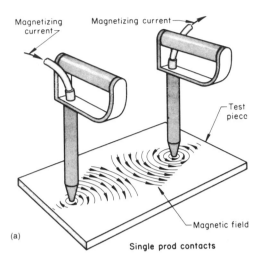

Single prod contacts

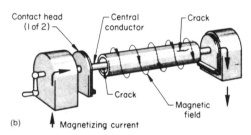

In (a), the current is passed through the test piece itself via two hand-held probes. In (b), the current is passed through a conductor adjacent to the test piece. In both cases, the arrowed lines indicate the directions of the induced magnetic fields. From *Metals Handbook*, 8th Ed., Vol 11, *Nondestructive Testing and Quality Control*, American Society for Metals, Metals Park, OH, 1976.

Fig. 6.14. Diagrammatic illustration of the principles of two methods by which a test piece can be magnetized by passing a current through a conductor.

clean, bright surfaces (Fig. 6.12c) but possibly not so when the surface is covered with, say, a black oxide scale. Red-colored suspensions are available which are then more suitable.

The magnetic-particle method is a particularly sensitive means of locating small and shallow cracks — but, remember, only in ferromagnetic materials. It is widely used for inspecting components for cracklike flaws produced during manufacture (e.g., heat treatment cracks) or developed in service (e.g., fatigue cracks). However, many variations in procedure affect the sensitivity of the process, and inspection procedures have to be established with care. Skill and experience are also necessary for interpreting the indications that are developed; distinguishing true from false indications due to magnetic anomalies arising from extraneous factors such as changes in section; identifying the type of abnormality detected; and assessing whether the abnormality is surface or subsurface.

Varying degrees of magnetization are left in the specimen material after the magnetizing field has been removed, depending on the magnetic property of the material known as its retentivity. The components may have to be demagnetized after inspection because the residual magnetism might impair its subsequent performance, such as by attracting foreign particles.

NONDESTRUCTIVE METHODS OF DETECTING INTERNAL STRUCTURES

Radiographic Inspection

Industrial radiography is a nondestructive testing method in which the object being inspected is penetrated by a radiation from a source of small area. The intensity of the radiation that emerges from the object varies if the radiation is absorbed differentially during its transit through the object (Fig. 6.15a). A shadow image of the features responsible for the variation in absorption can then be obtained if the variations in intensity of the emergent radiation can be

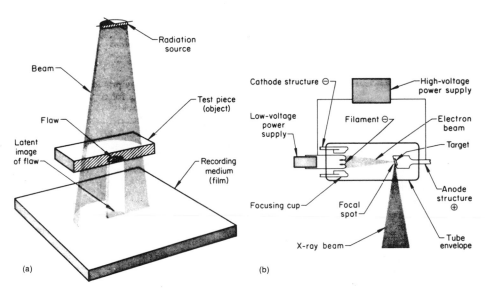

In (a), the detection of an internal flaw in a plate of uniform thickness is illustrated. From *Metals Handbook*, 8th Ed., Vol 11, *Nondestructive Testing and Quality Control*, American Society for Metals, Metals Park, OH, 1976.

Fig. 6.15. Diagrams of the basic elements of (a) a radiographic inspection system and (b) the principal components of an x-ray generating unit.

recorded. A radiographic inspection system thus consists of three units — namely, a source of penetrating radiation, an object, and a recording system. The radiation source and the recording system have to be designed so as to obtain the maximum amount of information about the interior of the object.

Either x-rays or gamma (γ) rays are typically used as the penetrating radiation. X-rays are generated by a unit the principles of which are illustrated in Fig. 6.15(b), the minimum wavelength and the intensity being the two important characteristics of the radiation. The minimum wavelength is determined by the voltage applied between the cathode and the anode, which may range from tens to thousands of kilovolts (higher voltages produce shorter wavelengths). The intensity of the radiation is determined by this voltage and by the current supplied to the electron-generating filament. Both accelerating voltage and filament current are made to be adjustable in commercial x-ray units, but within limits which are set by design compromises. For example, most of the energy of the beam of electrons incident on the target anode is converted to heat, and the target might melt if the target area is too small, no matter how well the anode structure is cooled. Hence, one major design compromise centers around the dissipation of heat.

Radioactive materials are the sources of gamma rays, which actually are physically identical to x-rays but with wavelengths at the short end of the x-ray spectrum. The wavelength is a characteristic of the radioactive element used in the source. The intensity of the radiation is related to the size and radiation history of the particular sample of material in the source and varies as the radioactivity of the source decays. Radioactive sources are small, cheap, and readily transportable compared with x-ray installations but have technical limitations in use which will soon emerge.

The recording device most commonly used is a film emulsion similar in principle to that used in light photography but designed specifically for this application. After being exposed and developed, the film is viewed directly as a negative. Thus relatively poorly absorbing features in the object, such as cavities, appear as relatively dark areas in the film (Fig. 8.18, p. 278). Conversely, relatively highly absorbing features in the object appear relatively light (Fig. 6.16). Fluoroscopic screens, sometimes assisted by electronic image intensification, can be used to view the image, but this is a less common practice. The image is seen in real time on fluoroscopic screens but with reduced resolution; moreover, a permanent record is not obtained. With either recording method, only a two-dimensional representation of a three-dimensional object is obtained, although there are special techniques for obtaining indications of the locations of flaws in the third dimension.

A primary requirement of a radiographic system is that the radiation be able to penetrate the object well enough for the image to be recorded in an acceptable time. The difficulty of achieving this depends, firstly, on the thickness and absorption characteristics of the metal of the object. The absorption

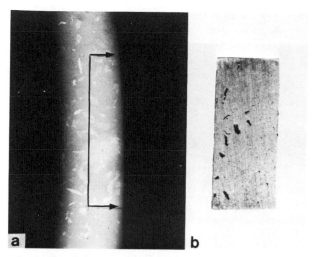

A section of the same region of the casting, confirming the presence of the inclusions, is shown at right. The inclusions are revealed as light areas in the radiograph because iron absorbs x-rays more strongly than does aluminum.

Fig. 6.16. Reproduction of a radiograph of an aluminum alloy casting which contains inclusions of iron oxide.

characteristics depend on a number of complicated factors but increase in a general way with the density of the metal. Thus higher accelerating voltages are required in x-ray units used for dense metals such as steel than in those used for less dense metals such as aluminum, and for thick components than for thin components (Table 6.1). X-ray units designed to operate at higher accelerating voltages become increasingly costly, so manufacturers supply a range of sets capable of operating at increasing accelerating voltages from which the optimum can be chosen for a particular need. An indication of the range available is given by the listing in Table 6.1. Most gamma-ray sources are highly penetrating, being equivalent to x-ray units operating at high accelerating voltages.

The second requirement of a radiographic system is that maximum sensitivity be achieved — that is, that it be possible to detect differences between regions which have the smallest possible difference in absorption. In practice, the limit is an absorption difference of about 2%. This corresponds to a flaw thickness of 2% when the flaw is filled with a gas, which is the usual case, but even this sensitivity is obtained only when all operating parameters are optimized. In particular, the exposure of the recording film has to be optimized within quite narrow limits; a consequence of this, incidentally, is that only a limited range of object thicknesses can be examined effectively in the one exposure. It may also be necessary to use a radiation of the shortest possible wavelength.

This limited sensitivity of radiography means in practice that the technique

Table 6.1. Useful Penetrating Thicknesses in Steel(a)
for Some X-Ray and Gamma-Ray Sources

Source	Maximum thickness	
	mm	in.
X-ray tube with		
maximum accelerating potential of:		
150 kv	20	0.79
250 kv	60	2.36
1000 kv	80	3.14
Gamma-ray source with		
radioactive element:		
Iridium 192	75	2.95
Cobalt 60	230	9.06
Radium	125	4.92

(a) As further examples, the penetrable thicknesses for aluminum alloys and copper alloys would be, respectively, 5 to 10 times greater and 10 to 50% smaller than those given here.

finds its main application in the inspection of castings and welds, which are prone to internal cavities of various types of the necessary thickness (see Chapter 8, p. 265). Only occasional applications are found with wrought products. In the fields of castings and weldments, however, radiography is a most important and widely used method of inspecting components and structures. This is particularly so when the performance of a structure has to be ensured by keeping internal flaws within certain limits.

Even here, however, there is a limitation in that planar flaws, such as cracks, cannot be detected with certainty. The gap spacings of most cracks are much less than 2% of the section thickness, so that there is a reasonable probability of detecting a crack only if its plane happens to be aligned strictly parallel with the beam of radiation. Cracks perpendicular to the inspection surface and located immediately below the radiation source may be detected, but not if they are located even a short distance away (Fig. 6.17a). Cracks and similar flaws aligned at angles to the inspection surface or, even worse, parallel with the surface certainly will not be detected.

The third requirement of a radiographic inspection system is that the shadows viewed in the recording device must indicate the sizes, shapes, and distributions of the flaws in the object as faithfully as possible. This is necessary to enable diagnosis of the nature and the significance of the flaws. It is also necessary to ensure that adjacent flaws are separated, or resolved, as clearly as possible in the image.

The shadow of any flaw is necessarily enlarged due to the geometry of the shadowing system (Fig. 6.17b). This enlargement can be reduced by locating the recording film as closely as possible to the back of the object, but there are

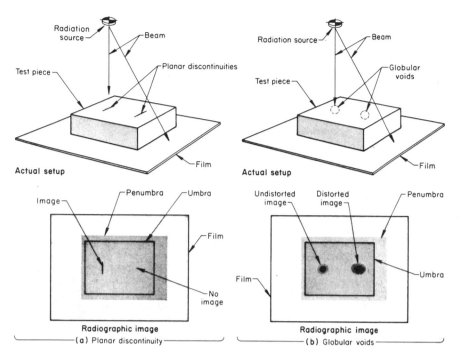

These diagrams illustrate that planar discontinuities are detected only when they are aligned strictly parallel with the direction of the beam of radiation. Otherwise, they do not appear to be thick enough to the radiation. In the case illustrated, they would have to be aligned perpendicular to the plate surface and be located immediately beneath the radiation source. Globular voids are detected at all positions but their images are geometrically distorted when they are not located immediately beneath the radiation source. These sketches also illustrate the formation of penumbras when the radiation source has finite size.

Fig. 6.17. Diagrams illustrating the effect of the direction of the beam of penetrating radiation on the appearance in a radiograph of (a) planar discontinuities and (b) globular voids.

physical limits to this. A long source-to-object distance is also desirable but can be achieved only at the expense of increased exposure time (an inverse square law applies, of course, to the decrease in intensity of the incident radiation with distance). Another problem arises because the edges of the image shadows are blurred due to a partial shadow, or penumbra, formed because the radiation source always has a finite size (Fig. 6.17b). The formation of a penumbra impairs resolution and so places a premium on the use of sources of minimum effective area. The source area of an x-ray unit is determined by the area on which the electron beam is focused on the anode target, and this is limited by the needs of heat dissipation. X-ray tubes which are required to operate at higher accelerating voltages thus tend to have larger source sizes, from which it follows that it is desirable to use a unit that is no more powerful than is really needed for penetration. The size of a source of gamma rays is generally larger than that of an x-ray unit, being determined by the volume of radioactive material that is needed to give a source of the required strength. Stronger sources tend to have larger source sizes.

Radiography basically is useful because of its ability to detect internal flaws (Fig. 6.16 and 8.18). Much information can be obtained about the nature and origin of the flaws, but cracks and similar planar defects cannot be detected with certainty. A permanent record can be produced for future reference. However, both capital and operating costs are high, and the operation is potentially hazardous. Great care in the design of an inspection operation and in the training of operating staff is essential to ensure that the health of personnel is not jeopardized. Special training of operating staff is also required to obtain optimum results, and even more so to interpret the results soundly.

As you will no doubt have realized by now, industrial radiography is based on exactly the same principles as medical diagnostic radiography. There are, however, considerable differences in the operating parameters of the radiation sources and recording devices because of the differences in the test objects.

Ultrasonic Inspection

The basic technique of ultrasonic inspection uses a pulse-echo system in which a pulse of mechanical (elastic) waves whose frequency is well beyond the audible range (*ultrasonic*) is introduced into one surface of the object from a suitable generator, usually a transducer. The transducer search unit is held in contact with the surface of the test object (lower left in Fig. 6.18), and the ultrasonic waves travel down through the metal, with some attendant loss of energy (attenuation), until they are reflected at (echo from) an interface. In a simple plate of the type sketched in Fig. 6.18, but without a flaw, the reflection occurs at the back surface of the plate. The reflected wave then travels back to, and is detected by, a receiver-transducer. This may be the same unit that was used to send out the ultrasonic pulse in the first place, as indicated in Fig. 6.18, or it may be an independent receiver. In any event, the time taken for the pulse to travel back and forth is measured by an electronic system and the result displayed on a cathode-ray oscilloscope (upper right in Fig. 6.18). This travel time is related to the thickness of the object and the velocity of sound in the metal concerned. Thus the technique can, by calibration, be used to measure thickness even when access is available only to one surface, which in itself can be useful.

If, however, a flaw is present internally in the plate, as sketched in Fig. 6.18, some or perhaps even all of the ultrasonic energy is reflected by the flaw, producing a second intermediate peak in the oscilloscope trace. The proportion of the energy reflected depends, firstly, on the area of the flaw compared with

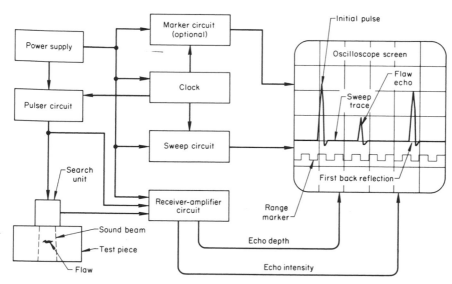

A pulse of ultrasonic radiation is generated by a transducer in the search unit, from which it is transmitted into the test piece. It is then reflected by the back surface of the test piece and also by a flaw if a suitable one happens to be present. The reflected radiations are detected when they reach the search-unit transducer. The relative intensities of the initial, flaw-echo, and reflection pulses are recorded on an oscilloscope as peaks separated by a distance related to the times at which the three events occurred. The boxes represent the various electronic units required to achieve these functions.

Fig. 6.18. Diagrammatic sketch of the basic elements of a pulse-echo system of ultrasonic inspection, using an A-scan display mode.

that of the search beam and, secondly, on the reflecting characteristics of the flaw. The interface must be oriented such that the ultrasonic pulse is reflected into, and not away from, the area of the receiving unit. This means, in the arrangement sketched in Fig. 6.18, for example, that a substantial portion of the interface has to be aligned nearly normal to the direction of propagation of the pulse. Thus the position of the second peak is related to the depth of the flaw, and its height is related to the plan area and to the reflecting characteristics of the flaw. Quantitative estimates of both can be made by comparison with reference standard test pieces. Note that a second flaw beneath, and smaller than, the first reflecting flaw cannot be detected.

The thickness of a flaw has no effect on its reflecting characteristics. Thus a crack or a lamination is as easy to detect as any other type of flaw, provided that it is aligned appropriately. As we have just seen, this in effect means that it must be aligned nearly parallel to the surface being inspected with the type of arrangement illustrated in Fig. 6.18. This is the reverse of the situation for radiographic inspection. However, other but more complicated types of probing arrangements, which need not concern us here, are available that are capable of detecting cracks and flaws of other orientations. Techniques are also available for searching regions remote from the transmitting and receiving units.

There are also methods of displaying the echo data other than the one just described, which is called an *A-scan* display. One system, called a *B-scan* display, indicates the positions of reflecting interfaces along a section through the object (Fig. 6.19a). Another indicates the plan position and outline of reflecting flaws over an area of the test object, and is called a *C-scan* display (Fig. 6.19b).

One of the experimental difficulties with ultrasonic techniques is that of ensuring that the acoustic energy generated by the search transducer actually gets into the test object. Even a very thin air gap between search unit and test-piece surface would severely retard the transmission of the ultrasound, so some coupling medium has to be interposed. Materials such as water, oils, greases, pastes, and rubbers are used to suit various needs. It is also possible to immerse the test object in a tank of liquid which wets its surface and to suspend the search unit above the test surface. The search unit can then be scanned easily over the test surface, which is convenient when large areas have to be investigated. Even at best, however, acoustic energy is lost at the transducer/test piece interface, and more is lost during transmission through the object material due to a range of physical phenomena. The magnitude of the total attenuation determines the depth to which inspections can usefully be carried out. This depth varies with the material, an indication of the range being given in Table 6.2. Note that the useful inspection depth is much greater than for radiography.

Ultrasonic inspection methods are useful for detecting internal macroscopic flaws with a comparatively high penetrating power. These methods also have reasonably high sensitivity for detecting small flaws, including planar flaws, and can establish accurately the positions and depths of the flaws. Only one surface needs to be accessible, and the operation is not hazardous. Extensive knowledge is required, however, to develop reliable inspection procedures, and careful attention by experienced operators is needed to implement these procedures satisfactorily and to interpret the results. A permanent record usually is not obtained, which increases the need for reliable operators. Rough or thin parts are difficult to inspect, as are the very surface layers of a workpiece.

The ultrasonic technique had its forebears in the sonar methods used to detect submerged submarines, although elastic waves of sonic rather than ultrasonic frequencies are used in this application. Analogous techniques are also used extensively in medical diagnostic imaging, as most modern mothers know.

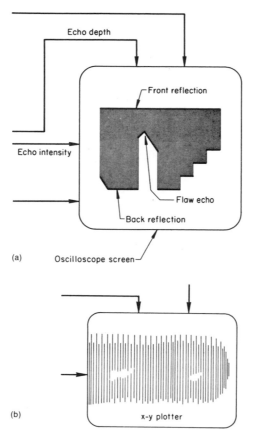

(a) Line-scan mode. The search probe scans along a chosen line across the test piece, and the major reflection signals are displayed in synchronism to delineate the front and back surfaces of the test piece and the surface of an internal flaw. The display is analogous to a section of the test piece.

(b) Area-scan mode. The search probe scans one surface of the test piece in a raster pattern, and the display records whether or not an echo was received from the back surface. A blank area in the display thus indicates a plan view of an internal flaw.

Fig. 6.19. Diagrammatic illustrations of the displays obtained in an ultrasonic inspection apparatus operating in line-scan (B-scan) and area-scan (C-scan) display modes, respectively.

Acoustic-Emission Inspection

A high-frequency stress wave is generated by the rapid release of strain energy when certain events occur within the body of a solid. The emission can be picked up by attaching a suitable sensor almost anywhere on the surface of the body concerned, and the characteristics of the emission can be analyzed by a

**Table 6.2. Useful Depths of Ultrasonic Inspection for
Various Metals**

Useful inspection depth, mm	Metals
1000 to 10,000	**Cast:** aluminum, magnesium **Wrought:** steel, aluminum, magnesium, nickel, titanium
100 to 1000	**Cast:** steel, high-strength cast iron, aluminum alloys, magnesium alloys **Wrought:** copper, lead, zinc
0 to 100	**Cast:** low-strength cast iron, copper, zinc

recording system. From this record, it may be possible to determine the position and nature of the initiating event. This is the principle of an inspection technique known as *acoustic emission*, or sometimes known by other terms such as *stress-wave emission* or *microseismism*. The last term is an indication that this technique has similarities to those used to locate and characterize earthquakes.

The main source of such acoustic waves in metals is, as for the earth, the generation and propagation of a crack. A large amount of elastic energy is released during this type of event (p. 86). The technique may thus be used to locate and monitor a growing crack. It may be used for the continuous surveillance of a structure (such as a pressure vessel in service or a weld during deposition; see Fig. 6.20), or to establish the presence and location of a static crack by applying a stress to the component or structure. In both cases, the stress must be large enough to cause some propagation of the crack but not large enough to cause catastrophic failure. Observations of this type must be made while a stress is being applied, because only active flaws can be detected. But then flaws can be detected which are at least an order of magnitude smaller than those detectable by any other method of nondestructive inspection. Moreover, the orientation of the crack is not significant (cf. radiographic and ultrasonic methods).

A number of microstructural changes which are much less dramatic than the growth of a crack may also be triggered in metals by the application of a stress. These too may generate elastic waves. Twinning, the movement of dislocations, sudden reorientation of grain boundaries, and martensitic phase transformations (all features the nature of which will emerge in our later discussions) are examples. The study of these phenomena by acoustic emission is also possible but more difficult because the emission is continuous and of low intensity compared with the burst of high-intensity emission characteristic of crack growth.

Useful though it is in some circumstances, a major difficulty with the acoustic-emission method is that of distinguishing signals of interest from those generated by foreign sources. Particular difficulties arise with noises generated

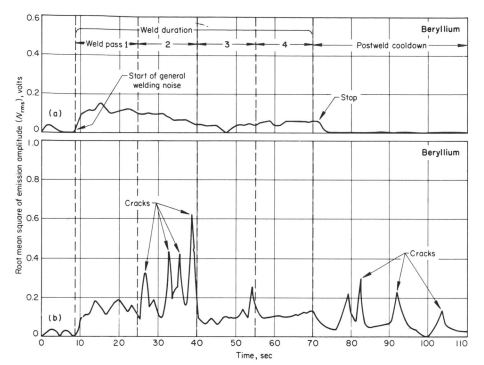

(a) The variation in the intensity of the emissions during the deposition of a sound weld. The emissions in this case were generated by events occurring during the welding process itself.

(b) Sharp extra peaks in the emission detected during the deposition of this weld correlated with the formation of cracks in the weld metal. This establishes that the cracks formed during or soon after the deposition of the weld metal and not at low temperature during subsequent cooling.

Fig. 6.20. An example of the use of acoustic emission to monitor the development of cracks during the welding of beryllium.

by the mechanical movement of various joints within a structure. There is no positive foolproof way of overcoming this difficulty, and so skill and care are necessary in interpretation. The mechanical noises may even be so continuous as to render the technique ineffective. Likewise, the technique may be rendered ineffective by noises generated by events which are of no interest, the cracking of a surface scale on the test object being an example.

THE METALLOGRAPHER'S ARMORY

Physical metallurgists now have at their disposal a wide range of tools to use in their attempts to decipher the structures of metals and alloys. The range starts with those which reveal only gross structures visible to the human eye, and some of them even permit internal structural features of this nature to be studied nondestructively. The range ends with those which give information about the arrangement of atoms and small groups of atoms in a crystal. Each investigational technique has its special niche that matches its advantages and disadvantages and in which it can produce information that cannot be obtained by any other. In this respect, the techniques compliment one another, and the application of a number of them is necessary more often than not. All require specially trained staff for optimum application, some more so than others.

Nevertheless, the structures with which a metallurgist has to be concerned are so subtle and complex that even this powerful armory does not always suffice to resolve the problems which arise. New weapons are constantly being added to the armory, and each addition seems to be more complex and esoteric than the last. A saving grace is that many significant structures can be recognized by a comparatively simple technique once the nature of the structure has been elucidated. We shall use the simplest possible technique for our illustrations, but you should always remember that the information summarized usually is the result of many careful and laborious investigations which have brought to bear a range of sophisticated techniques.

Thus, after a late start, great progress has been made in metallography in the 120 years since Sorby carried out his pioneering investigations. But by no means have all problems been solved. Many important subtleties and complexities of the structures of metals remain unexplained.

FURTHER READING

C. S. Smith, *A History of Metallography*, University of Chicago Press, Chicago, 1960.

Metals Handbook, 9th Ed., Vol 9, *Metallography and Microstructures*, American Society for Metals, Metals Park, OH, 1986.

Metals Handbook, 9th Ed., Vol 10, *Materials Characterization*, American Society for Metals, Metals Park, OH, 1986.

Metals Handbook, 8th Ed., Vol 11, *Nondestructive Testing and Quality Control*, American Society for Metals, Metals Park, OH, 1976.

· 7 ·

Crystals, Grains, and Phases: Perfect and Imperfect

SUMMARY

One of the basic characteristics of metals is that they are crystalline, which means in effect that the atoms of which a metal is composed are arranged in a regular three-dimensional array called a crystal lattice. The array is a comparatively simple and symmetrical one in a metal, and the atoms are bonded together in a characteristic way by a cloud of electrons drifting among the atoms instead of by direct atom-to-atom forces as in chemical compounds. Many of the distinguishing properties of metals depend on these factors. Atoms of a different kind added as alloying elements in the first instance fit themselves into the crystal lattice of the parent metal, usually by substituting for the parent atoms on a one-for-one basis. In a few instances, however, they fit in the spaces between the parent atoms. There is often a limit, however, to the number of foreign atoms that can be accommodated in either of these ways. Further additions then cause a new phase to form, a phase which has a different crystal structure. The new phase is first formed in addition to that based on the parent metal but eventually replaces it. A sequence of phases may form in this way.

Atoms in real crystals are never arrayed with absolute perfection, defects of a number of different types being common. The first important type is one in which atoms are absent from some of the expected lattice sites (vacancies). These defects have an important influence on the ease with which atoms can move, or diffuse, through the crystal. A second type of defect results from a disorder in the arrangement of the planes of atoms, of which there are a number of possibilities. All are called dislocations. Their existence is responsible for many of the important mechanical characteristics of metals, including strength, ductility, and toughness.

Industrial metals actually are composed of many small crystals, or grains, which are variously oriented. The grains contact one another intimately on an atomic scale along a surface called a grain boundary. Grain boundaries have special structures.

PERFECT CRYSTALS OF PURE METALS

The Atomic Structures of Crystals

We defined in Chapter 1 a package of properties that characterize metals, among these being that they are crystalline. Crystalline does not mean that they are brilliantly transparent to light, as in "crystal" glass: they aren't. It also does not mean that they grow with a regular geometric external shape as do many minerals, particularly those treasured by collectors: occasionally they do (Fig. 7.1), but in general they don't, and this doesn't make them any less crystal-

Gold has a cubic crystal structure, and these pieces have a generally cubic shape. Each is a single crystal. The development of external indications of crystallinity of this nature is, however, extremely rare in metals. Magnification, 6×.

This photograph is reproduced with the permission of the Gold Museum, Ballarat Historical Park Association, in which museum the crystals are displayed. Ballarat was one of the richest alluvial gold fields in history, and its wealth contributed greatly to the development of the Australian colony in the middle of the 19th century.

Fig. 7.1. Two pieces of gold found in an alluvial gold field, exhibiting well-developed geometrically shaped faces, slightly worn, which give an external indication of the crystallinity of the metal.

line. To a solid-state physicist, the only necessary characteristic of the crystalline state is that the atoms of which the material is composed be arranged in a pattern which is repetitive in three dimensions over long distances relative to the size of the atoms.

It is convenient to illustrate this by representing the atoms as hard spheres. Atoms are, of course, not like this at all, but the sphere can be taken to represent the volume over which the influence of an atom extends. The radius of the sphere then represents the closest distance to which another atom can ap-

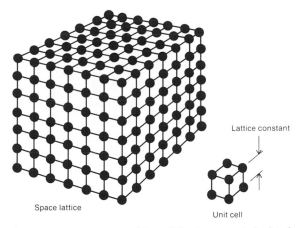

Lattice constant

Space lattice

Unit cell

> The atoms are represented by solid spheres, and the binding forces between them by lines. The lattice in this instance has atoms located at the corners of cubes which are repeated indefinitely in three dimensions in space.

Fig. 7.2. Diagrammatic illustration of a simple crystal lattice.

proach, a distance which is on the order of 10^{-10} m* but which varies for each type of atom. The arrangement of atoms in a crystal can then be represented by the manner in which these spheres are packed together (Fig. 7.2). This too is a simplification. For a start, the atoms are not stationary at the temperatures with which metallurgists deal, but are vibrating about a mean position. The amplitude of vibration increases with temperature, but is on the general order of one tenth of the atom diameter. This does not matter, however, as far as the representation of the atoms in a crystal structure is concerned.

More strictly also, it is the ions of metallic atoms that are represented, an ion being the positively charged residual of the atom after one or more of its outer electrons have become detached. A metal can be thought of as an assemblage of positive ions immersed in a cloud of these detached electrons. The electrons of the cloud are relatively free in the sense that they are not bound to any particular ion, but move rapidly through the metal in such a way that their density throughout is always approximately uniform. Metal crystals are held together by electrostatic attraction forces between the free electrons (negatively charged) and the ions (positively charged); this is called *metallic bonding*. Moreover, the free electrons confer electrical conductivity on metals. They can drift in the direction of an applied electrical potential, and this drift constitutes an electrical current. By contrast, the predominant binding forces in nonmetallic substances act directly from one atom to another, from a positive ion to a negative ion (*ionic bonding*). Electrons are not then free to move easily

---

*Crystallographers call this distance an angstrom unit (Å) although this is not strictly an SI unit. It is a unit of convenient size for them. See Appendix 2, p. 477.

through the crystal to carry an electric current, and this is why most nonmetallic crystals are poor conductors of electricity.

The simplest arrangement of atoms that could be imagined is illustrated in Fig. 7.2. The atoms here are sketched as being located at the corners of cubes which repeat themselves indefinitely in three dimensions in an arrangement which is called a *space lattice*. Each cube is called a *unit cell*, and the edge length of this cell is referred to as the *lattice constant* of the crystal. The *orientation* of the crystal can be defined in terms of the orientation in space of either one of the faces of the cell or one of its edges.

The simple *basic cubic* crystal structure represented in Fig. 7.2 is not actually the most desirable one because the atoms are not as closely packed together as they could be. This is illustrated in two dimensions by the sketches in Fig. 7.3(a) and (b). If you carry out a close-packing experiment of this nature with spheres in three dimensions, you will find that you have a choice between two possible close-packing positions in the third row of spheres. One choice results in the stacking pattern being repeated in the fourth layer and gives rise to a pattern of cubes in which an atom is located at each corner of a unit cell and one at the center of each face of the cube cell (Fig. 7.4a). Each cell then contains four atoms, but the face of the cube is not the close-packed plane. This close-packed plane is a pyramidal face of the cube (Fig. 7.4b), of which there are eight in each cell. This is the *face-centered cubic* (fcc) structure and is the most highly symmetri-

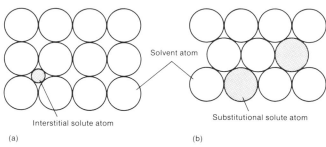

(a) The basic cubic arrangement sketched in Fig. 7.2. (b) A more closely packed arrangement. These sketches also illustrate two ways in which foreign atoms might be accommodated in a crystal lattice. In (a), the foreign solute atom (black) fits in a space between the host solvent atoms; this is representative of an interstitial solid solution. In (b), the foreign solute atoms (shaded) substitute for the host solvent atoms; this is representative of a substitutional solid solution. In reality, foreign interstitial atoms are always larger than the spaces available rather than fitting neatly as indicated in this sketch. This causes considerable distortion of the surrounding host lattice. Also, the size of a foreign substitutional atom is always different from that of the host atom it replaces, and this also causes distortion of the surrounding host lattice.

Fig. 7.3. Two possible arrangements, sketched in two dimensions, in which spheres might be stacked together.

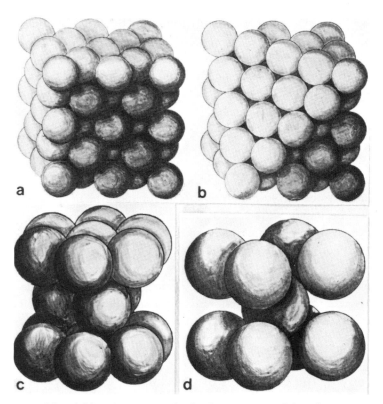

(a) and (b) A face-centered cubic lattice. Some of the spheres have been removed in (b) to expose the plane of close packing. (c) A close-packed hexagonal lattice. The basal plane is the close-packed plane here. (d) A body-centered cubic lattice. There is no close-packed plane in this lattice. From W. Hume-Rothery, *The Structure of Metals and Alloys*, Institute of Metals, London, 1945.

Fig. 7.4. Sketches of ball models of some of the crystal lattice types commonly found in metals.

cal crystal structure found in nature. Many common metals have fcc structures — for example, copper, aluminum, nickel, lead, silver, and gold. Note that these are all very ductile metals, and we shall see later that this is a direct consequence of their crystal structures.

Adoption of the second choice of close packing on the third plane gives a structure of hexagonal symmetry in which the stacking pattern repeats every third layer. The close-packed plane this time is the basal plane for each unit cell (Fig. 7.4c). Each cell again contains four atoms, but this time there is only one close-packed plane in each. This structure is known as a *close-packed hexagonal* (cph) structure. The common metals which have cph crystal structures include zinc, magnesium, and titanium. These metals are not as ductile as fcc metals, which also is a consequence of their crystal structures.

The third important crystal structure for metals is the *body-centered cubic* (bcc) structure, which is illustrated in Fig. 7.4(d). It is a more open structure (it is not close packed), and is the structure exhibited by iron, chromium, molybdenum, and tungsten. It is less symmetrical than either the fcc or the cph structure. It contains only two atoms in each cell.

There are many possible arrangements of atoms in crystal structures other than those just described. Seven crystal systems are found in nature, and most of them have several space-lattice variants. For example, the cubic system has the three variants that we have already described — basic, face centered, and body centered. Each metal adopts a particular type of space lattice which has a lattice constant unique to itself. These lattice constants are well established and are listed in standard handbooks such as those listed at the end of this chapter. Only a minority of less-common metals adopt space lattices other than the simpler fcc, bcc, and cph structures. Examples of the minority are tin, antimony, uranium, and plutonium.

The mass density to be expected of a metal clearly can be calculated once the dimensions of its unit cell, the number of atoms in each cell, and the weight of the atoms are known, and this value of density corresponds exactly, or at least nearly exactly, with the density measured in bulk specimens. Very precise measurements may reveal some very small discrepancies which are attributable to some of the lattice imperfections we shall discuss later in this chapter.

Anisotropy of the Properties of Crystals

You will notice that the arrangement of the atoms in a crystal varies with direction in the crystal lattice, the more so the less symmetrical the crystal lattice. It is to be expected, therefore, that some properties of the crystal would also be assymetrical (*anisotropic*). Physical properties in fact do not vary with direction in cubic metals but do vary considerably with less symmetrical crystals. The coefficient of thermal expansion, for example, varies by factors of two to three in hexagonal metals, such as zinc, and the specific electrical resistance varies by 20 to 30%. Mechanical properties, both elastic and plastic, are anisotropic even in cubic metals. Young's modulus (p. 64) may vary by factors from one-half to four. The plastic properties vary even more. These matters will be discussed in Chapter 9.

Allotropy

Some materials can exhibit different crystal structures at different temperatures. Each variety then is stable over a particular range of temperatures, and the transformation from one variety to another occurs at a definite characteristic temperature. This phenomenon is known as *allotropy* in the metals and alloys with which we shall mostly be concerned, although *polymorphism* is a more general term. One important feature of an allotropic change is that it must be associated with a change in density, and hence in volume. This is to be expected because the volume of, and number of atoms in, the unit cells of the two allo-tropes are sure to be different.

The most important examples of allotropic changes in metals occur in iron. Pure iron has a body-centered cubic structure at temperatures from absolute zero (−273 °C, or −459 °F) up to 912 °C; this form is called *alpha iron* or *ferrite*. Iron adopts a face-centered cubic structure (called *gamma iron* or *austenite*), however, at temperatures between 912 and 1394 °C (1674 and 2541 °F), above which temperature it reverts to a body-centered cubic form (called *delta iron*). These *transformation temperatures* change when alloying elements are added. The many heat treatments to which iron alloys respond, and which so profoundly affect their structures and properties, are possible only because iron undergoes these allotropic changes. The transformation from the alpha to the gamma allotrope is of particular importance and will be discussed in Chapter 11 (p. 282). Titanium is another metal which undergoes an allotropic change. It has a close-packed hexagonal structure below 882.5 °C (1620.5 °F) and a body-centered cubic structure above this temperature, and this allotropic transformation enables many structural changes to be induced by varying the thermal cycles to which titanium alloys are subjected.

Not all allotropic transformations are a blessing, however. Uranium, for example, has a number of allotropes, the allotropic transformations concerned being associated with large volume changes. This causes serious distortion of uranium components if they are subjected to repeated heating cycles through certain temperature ranges. Major problems consequently arise when uranium is used in the metallic form in atomic reactors, and thus its use in this form had to be abandoned. Tin can undergo a transformation at 13.2 °C (55.8 °F), and the transformation to the low-temperature form called *gray tin* causes the metal to disintegrate. This transformation is, however, more a curiosity than an event of practical importance.

PERFECT CRYSTALS OF IMPURE METALS AND OF ALLOYS

A metal of 99.9999% purity (an impurity content of one part per million) is regarded as being exceptionally pure, but even a metal of this purity contains one foreign atom in a block of crystal cells (such as the one sketched in Fig. 7.2) which consists of 100 atoms per edge. A metal of typical commercial purity would contain 100 foreign atoms or more in a block of this size. Deliberately alloyed metals might contain up to a 50–50 mixture of atoms. These remarks refer to metals in which only one type of foreign atom is present, but in principle they apply also to the more normal case where a mixture of a number of types of foreign atoms is present. What we now need to consider is where these foreign atoms fit into the crystal lattice of the host metal.

Substitutional Solid Solutions

There are several possibilities. The first is that a foreign (solute) atom simply replaces the host (solvent) atom on a one-for-one basis (Fig. 7.3b). The result can be called a *solid solution*, using as an analogy the liquid solutions with which we are all familiar. More specifically, it is called a *substitutional solid solution*. Usually, the substituted atoms are randomly located, and the solid solution is said to be *disordered*. In certain cases, however, the foreign atoms may be located at regularly spaced positions in the host lattice, in which event the solid solution is called an *ordered* solution.

The solute atom is never the same size as the solvent atom, so they cannot substitute for one another exactly. The host atoms surrounding each foreign atom must be either forced apart a little or collapsed together a little. In either event, strains are introduced locally in the host lattice. Moreover, the average spacing of the atoms (the lattice constant) for the whole crystal must change, the magnitude and direction of the change depending on the relative sizes of the two atoms. The lattice constant of the host lattice continues to change as more and more foreign atoms are introduced, usually in a way which is directly proportional to the number introduced. Once the relationship is known, the alloy content of a crystal can be estimated by determining its lattice constant by, say, x-ray diffraction methods (p. 203).

Interstitial Solid Solutions

A second possible way of introducing foreign atoms is to fit them into the spaces between the host atoms, forming what is called an *interstitial solid solution* (Fig. 7.3a). Clearly, only atoms which are much smaller than the host atoms can be fitted in this way, the only likely contenders actually being hydrogen, boron, carbon, nitrogen, and oxygen. Moreover, interstitial solutions are most likely in crystals which have comparatively large spaces available between the atoms — that is, in crystals which are not close packed. Nevertheless, the presence of foreign atoms in interstitial positions greatly distorts and expands the host lattice locally because the size of the foreign atom is always much larger than the space in the host-atom lattice which it has to occupy. The expansion may be anisotropic, in which event the crystal system may be changed to a less symmetrical one.

The most important examples of interstitial solutions are those of carbon and nitrogen in alpha and gamma iron.

Solubility Limits

We have just noted that the incorporation of foreign atoms inevitably strains a host lattice. You might well ask how much of this a host lattice will put up with. The answer depends on a number of factors, the most important of which are the similarity between the crystal structures of the two elements and the relative sizes of their atoms.

Complete solid solubility (that is, uninhibited substitution of one type of atom by another) occurs only when the two metals have the same crystal structure and very similar atomic radii. For example, copper and nickel are completely soluble in one another. They both have a face-centered cubic structure and their atomic radii differ by only about 2%. On the other hand, copper and aluminum also have the same crystal structure but their atomic radii differ by about 10%; it is found that only about 15 atoms of aluminum can be substituted for every 100 atoms of copper in a solid solution. To go to the extreme, copper and lead also have the same crystal structure but their atomic radii differ by nearly 50%; lead is virtually insoluble in copper. As an example of the influence of crystal structure, the atomic radii of copper and zinc differ by only about 4%, which is favorable, but the two metals have different crystal structures. Even though both these crystal structures are close packed, only about three atoms of zinc can be substituted for every ten atoms of copper. These concepts were developed into a set of guidelines by Hume Rothery of Oxford University, England. His guidelines actually include several factors in addition to the structural and size factors that we have just discussed, and are useful for at least giving a feel for the likely results of attempts to alloy one metal with another.

The number of foreign atoms that can be held in interstitial solution depends on the size of the interstitial hole that is available for it. The holes in the alpha iron lattice are very small, the diameter of the hole that carbon atoms prefer to occupy being only about a quarter of the diameter of a carbon atom. Alpha iron consequently can happily dissolve only very small amounts of carbon* — about one carbon atom for every ten million iron atoms at room temperature and a maximum of one for every thousand at higher temperatures. However, the spaces available in gamma iron are much larger, approaching the size of the carbon atom. Up to ten carbon atoms per hundred iron atoms can then be held in solution, a factor which also is of great significance in the heat treatment of steels.

All of our considerations up to this point apply to equilibrium conditions. Larger numbers of foreign atoms can often be obtained in a supersaturated solution, with a corresponding increase in the strain on the host lattice. A most important example that we shall be discussing later (p. 401) is the supersaturated solution of carbon in alpha iron that can be obtained, giving rise to the constituent known as *martensite*. A supersaturated solution is, however, metastable and will rid itself of the excess solute atoms if given a chance. This would involve the rearrangement, or diffusion, of the excess atoms in the host crystal lattice, and diffusion is a process which normally does not occur at a reasonable

---

*The differences in crystal structure and atomic radii between iron and carbon are so great that the alternative of substitutional solid solutions is out of the question. It's interstitial solution or nothing.

rate at room temperature. Some heating consequently is usually required to achieve the rearrangement. For example, alpha iron supersaturated with carbon in the manner that we have just described undergoes a complicated sequence of changes when it is heated, in the treatment known as *tempering* (p. 408). Many of the more important commercial heat treatments of metals are based on the sequence of obtaining a supersaturated solid solution and then controlling the manner in which the excess solute atoms are precipitated out of solution. We shall discuss some of them in Chapter 11.

Intermediate Phases

There comes a stage sooner or later in most alloy systems where the lattice of the host metal cannot accommodate any more foreign atoms. Our next step will be to inquire what happens if more are in fact added. The general answer is that the additional atoms have to be accommodated in a new phase which has a different crystal structure — one that is more able to cope with them.

There are many ways in which this can be achieved. The simplest is one in which crystals of the foreign metal itself constitute the new phase. More commonly, however, the new phase is a solid solution of the foreign metal. The two *terminal solid solutions* (i.e., solutions the crystal structures of which are based, respectively, on the two solvent metals) commonly then separate out simultaneously from the molten alloy as a mixture called a eutectic.* A eutectic is an intimate mixture of two phases on a microscopic scale in which one phase appears to be embedded in a matrix of the other. They are discussed in more detail on p. 270. The succession of structures then found in a simple system when, say, metal B (P as Cu_3P in Fig. 7.5) is added to metal A (Cu in Fig. 7.5) in increasing amounts, as illustrated in Fig. 7.5, are:

1. Phase based on A (Fig. 7.5b).
2. Phase based on A plus a eutectic mixture of phases A and B (Fig. 7.5c).
3. Eutectic mixture of phases A and B (Fig. 7.5d).
4. Phase B plus a eutectic mixture of phases A and B (Fig. 7.5e).
5. Phase based on B.

A second common occurrence is for one or more intermediate phases to form, the crystal structures of these phases being different from those of the solvent metals. Often they are *intermediate solid solutions* of the two metals, but, as we shall soon see, other types of phases are also possible. Each intermediate phase typically is stable over a particular range of compositions, and alloys which have compositions between these ranges consist of appropriate mixtures of the two bounding intermediate phases. The mixture may form a eutectic, as described above, or the phases may be arranged in a less regularly organized

---

*The various components of a structure are called constituents. Thus a constituent may be a phase or, as is a eutectic, a characteristic mixture of phases.

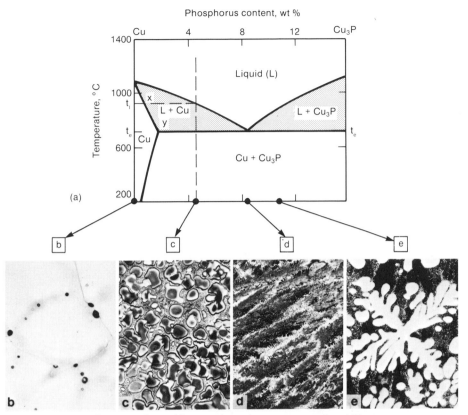

The example used here is the copper – copper phosphide system, and all the alloys are cast. (a) The relevant portion of the phase diagram. (b) Terminal copper-rich solid solution. (c) Primary dendrites of terminal copper-rich solid solution among areas of eutectic. (d) Eutectic structure. This structure is shown resolved at higher magnifications in Fig. 8.13(a) and (b), p. 272. (e) Dendrites of copper phosphide surrounded by eutectic. Optical micrographs; 100× (shown here at 79%).

Fig. 7.5. An illustration of a typical series of structures encountered in a eutectic alloy system.

way resulting from a *peritectic* reaction between solid and liquid (p. 274). The succession of structures found when metal B is added to metal A in increasing amounts and when a eutectic is not involved are (as illustrated in Fig. 7.6, in which metal A is Cu and metal B is Zn) typically as follows:

1. Terminal solid solution or phase based on A (Fig. 7.6b).
2. Mixture of terminal solid solution or phase based on A and intermediate phase (Fig. 7.6c).
3. Intermediate phase (Fig. 7.6d).
4. Mixture of two intermediate phases (Fig. 7.6e).
5. Second intermediate phase etc., until:
6. Terminal solid solution or phase based on B.

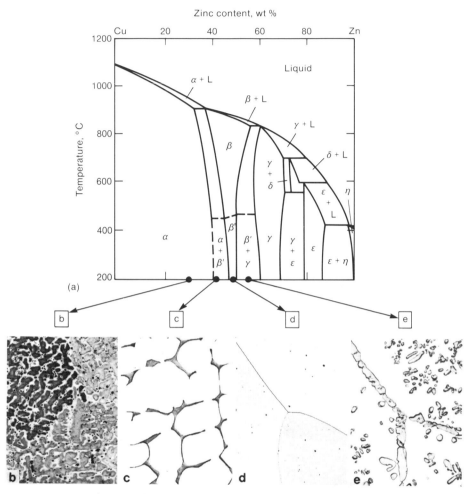

The example used here is the copper-zinc system. All the alloys are cast. (a) The relevant portion of the phase diagram. This diagram is simplified in that transformations in the solid state at low temperatures are ignored, which is a reasonable thing to do in practice. (b) Copper-zinc terminal solid solution, called the alpha (α) phase. The etching contrast within individual grains results from segregation developed in the dendrites during solidification (see p. 255). (c) Mixture of alpha phase (light) and an intermediate solid solution called the beta (β) phase (dark). (d) Beta phase. (e) Mixture of beta phase (light) and a second intermediate phase called the gamma (γ) phase. Optical micrographs; 100× (shown here at 79%).

Fig. 7.6. An illustration of a typical series of structures encountered in an alloy system in which a succession of solid-solution phases are formed by peritectic reactions.

The intermediate phases are of a variety of crystal types, and are frequently identified by lower-case letters of the Greek alphabet in succession (alpha, beta, gamma, etc., as in Fig. 7.6) or by a chemical formula (as for Cu_3P in Fig. 7.5). A range of increasingly complex crystal structures and types of atomic bonding occur, varying from those characteristic of a metal to those characteristic of a chemical compound. Physical and mechanical properties vary accordingly. In many systems, the first one or two phases to form beyond the terminal solid solution (the β and γ phases illustrated in Fig. 7.6, for example) have compositions which vary over a range about one that can be expressed in terms of a fairly simple chemical formula (for example, $CuZn$ for the β phase and Cu_5Zn_8 for the γ phase in the system illustrated in Fig. 7.6). These phases have crystal structures more complex than that of a terminal solid solution but still have metallic-type atomic bonding and metallic-type properties. They are comparatively hard, but the less complex of them (such as the β phase above) are still reasonably ductile. The more complex phases (such as the γ phase above) may, however, fairly be described as being brittle. Phases of this general type are known as *electron compounds*.

In other cases, particularly phases in middle compositional ranges, the phase may have a composition which even more closely fits a chemical formula, has an even more complex crystal structure, and exhibits ionic bonding. Such phases are frequently identified by chemical formulas, are characteristically hard and brittle, and are poor conductors or semiconductors of electricity. An example is the Cu_3P phase referred to in Fig. 7.5. This type of phase is called a *valence compound*.

The range of compositions and temperatures* over which the various intermediate phases and mixtures of phases that we have just been describing are stable can be plotted out as areas on a *phase diagram* (also known as a *constitutional diagram* or an *equilibrium diagram*) which has composition and temperature as its two ordinates (Fig. 7.5a and 7.6a). Each area in the diagram is labeled with the name of the phase or phases that are stable over the compositions and temperatures defined by the boundaries of the area. The diagram also includes information on the regions over which solid and liquid are stable in the presence of one another — that is, on the melting ranges of the alloys (p. 255). Phase diagrams are the foundation on which is based our understanding of the influence of composition and of thermal treatments on the phases and constituents present in alloys. This important topic is discussed in other publications in more depth than is possible here, although we shall discuss specific features of phase diagrams at various points throughout. We need to note here, however, that the diagrams attempt to describe the state reached when true equilibrium has been attained, a state which is rarely achieved in practice. This has a number of consequences, one of which is that the intermediate phases may be found over

---

*The discussion here has concentrated on the effects of composition on the phases present in an alloy. The effects of temperature will be discussed in Chapter 11 (p. 345).

somewhat different ranges of composition from those indicated by the diagram. These could be due to failure to reach equilibrium during solidification (Chapter 8), during subsequent cooling to room temperature, or during cycles of heating and cooling (Chapter 11). Moreover, phase diagrams of the type illustrated in Fig. 7.5 and 7.6 describe alloys containing only two elements, whereas many industrial alloys contain three or four or more significant alloying elements. The additional elements not only may affect the phase relationships of the original binary system but also may introduce entirely new phases and relationships. These effects become increasingly difficult to represent in a diagram, and are too difficult for us to discuss here.

IMPERFECT CRYSTALS

The crystal lattices of real metals do not repeat themselves perfectly over indefinite distances. Occasional mistakes are made in the positioning of the atoms, mistakes which confer some of the most important and useful characteristics on the crystals.

Point Defects: Vacancies

The simplest type of irregularity is one where an atom is missing from an expected site (Fig. 7.7). This constitutes a point defect which is called a *vacancy*. The concentration of vacancies at room temperature in metals of high melting point is very small: perhaps one in 10^{20} lattice sites. The concentration increases with temperature to become as high as one in 10^4 sites at temperatures close to the melting point, but these concentrations are not usually retained for any length of time when the material is quenched to a lower temperature.

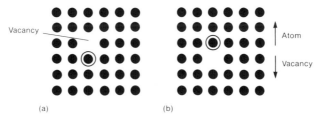

(a)   (b)

The atom circled in (a) can move comparatively easily to the other side of the vacancy, as in (b). The net result is that the vacancy has moved in one direction and the atom has been transported in the opposite direction. The movement of atoms of the same kind, as sketched here, is called *self-diffusion*.

Fig. 7.7. Diagrammatic sketch of a vacant lattice site in a crystal lattice, illustrating how the presence of vacancies can expedite the diffusion of atoms.

Vacancies are also created when a metal is plastically deformed, and these vacancies are mostly retained at room temperature.

The importance of vacancies is that they allow atoms of both like and dissimilar kinds to move, or diffuse, through the crystal lattice. It is very difficult to see how an atom could move through the lattice if all of the sites were occupied, but not so when vacancies are present. One of the atoms bounding a vacancy, such as the one circled in Fig. 7.7, could easily slip into the vacancy irrespective of whether it was an atom of the same type or a foreign atom in substitutional solution. In this event, the vacancy moves in one direction and atoms move effectively in the opposite direction. Some energy input would be required to initiate the event, the higher the temperature and so the greater the amplitude of vibration of the atoms, the smaller the amount of energy likely to be required. Also, the number of vacancies available increases with temperature. Less energy is also likely to be required to transpose atoms that are smaller than the host atoms than to do the reverse.

Even so, the over-all situation would remain unchanged if successive movements occurred in random directions. Net transport of atoms can occur only when the movements are weighted statistically in one direction. An impetus is required for this, and one possibility is a gradient in the concentration of the atoms that are to diffuse. The impetus then is the need to minimize energy by achieving an equilibrium uniform distribution. Many of the reactions that occur in solids and which we shall discuss later can occur only if diffusion occurs, particularly diffusion driven by a concentration gradient. The critical thing then is the rate at which the diffusion occurs. This will determine whether the reaction can actually complete itself in the time that is available in practical heat treatment schedules. Note first, however, that these reactions would never occur if diffusion did not occur and that, in turn, diffusion would not occur if vacancies were not present in crystals. Without their lattice vacancies, metals would lose many of their attractions as engineering materials.

The diffusion relationships can be quantified. Considering here only the simplest situations, diffusion rate can be defined in the following way:

$$\frac{dn}{dt} = -D \frac{dc}{dx}$$

where n is the number of atoms diffusing through a unit area in time t, in a gradient of a change in concentration c, over a distance x. D is a constant known as the *diffusion constant* and also as the *diffusivity*. This relationship is called *Fick's first law* and describes only a steady state. More complex relationships have to be developed to describe a generalized dynamic state, but they need not concern us here. D varies with temperature in the following way:

$$D = D_0 \exp[-Q/kT]$$

where T is temperature in kelvins, k is the ubiquitous Boltzman's constant, and Q is a measure of the energy required to activate diffusion. Q is called the *activation energy for diffusion* and is related to the energy required for an atom of the required type to move into a vacancy and the probability that the movements will occur in the required direction. Q thus varies with the type and concentration of the diffusing atom.

Physical chemists will recognize the above relationship as an adaptation of the Arrhenius relationship. However, the activation energy is much, much higher for reactions in the solid state than for those to which chemists are used to applying this relationship — namely, reactions in liquids or gases. In fact, it is so high that the diffusivity in industrial metals at room temperature for all practical purposes is zero, with only a few exceptions. Hence most metals are highly stable at ambient temperatures. Diffusion commonly starts to become significant only at temperatures approaching the melting point. The usual Arhenius relationship holds that the diffusion rate approximately doubles for each 10 °C (18 °F) rise in temperature.

Line Defects: Dislocations

Now let us consider what happens if a part of a plane of atoms, rather than a single atom, is removed from its expected site. The atoms have to rearrange themselves at the edge of the missing portion of the plane (or at the edge of the extra half plane if you prefer to look at it that way), producing a defect which has an atomic arrangement something like that sketched in Fig. 7.8(a). The defect extends as a line defect normal to the plane of the paper in the region labeled ⊥ in the sketch. This general type of defect is called a *dislocation*. The particular type sketched in Fig. 7.8(a) is called an *edge dislocation*, a defect which usually is present in profusion in real metal crystals. The presence of these defects plays a central role in the plastic deformation of metals, phenomena which we shall discuss in Chapter 9. It can also be imagined from Fig. 7.8(a) that they provide a channel down which diffusion of atoms might occur, and so their presence is another factor that increases diffusion rate.

A second type of line defect is also known to be present in real crystals. This is the *screw dislocation*, which can be imagined to be formed by making a partial cut in a perfect crystal, shearing the two sides of the cut relative to one another instead of inserting the extra plane of atoms between the two, and then welding the cut faces together again. The end result is that a particular plane of atoms is arranged as a helical ramp, the dislocation line being the axis of the ramp (Fig. 7.8b).

It is possible for a dislocation to lie on any plane in a crystal, but a number of factors tend to make it lie on a definite crystallographic plane. This is usually the plane of densest atomic packing. For simple geometric reasons, a dislocation cannot terminate inside a crystal but must end on a free surface, on another

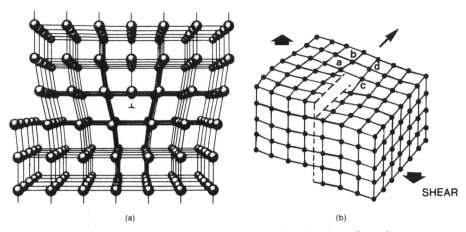

(a) (b)

The edge dislocation (a) can be imagined to have been formed
by removing part of a plane of atoms and rejoining the lattice.
Alternatively, it can be imagined to have been formed by partially
splitting the crystal and inserting an additional plane of atoms in
the split. For a screw dislocation (b), the split can be imagined to
have been sheared by the distance of a plane of atoms and the split
then rejoined. Source for (a): A. G. Guy and J. J. Hren, *Elements of
Physical Metallurgy*, Addison-Wesley, Reading, MA, 1971.

**Fig. 7.8. Diagrammatic illustration of (a) an edge
dislocation and (b) a screw dislocation.**

dislocation, or on itself. The result is that dislocation lines mostly are arranged
in networks on the closest-packed plane, as sketched in Fig. 7.9. The line segments of this network are curved and are composed of the mixture of edge and
screw dislocations (as indicated by the inset in Fig. 7.9) that is necessary to
achieve the particular degree of curvature.

The existence of both of these types of dislocations was originally proposed
as pure hypotheses, the edge type by Taylor* to explain the plastic properties of
crystals (see Chapter 9) and the screw type by Frank* to explain their growth
characteristics. Their existence has subsequently been confined by a large body
of experimental evidence, including the direct observation of the arrangement
of atomic planes sketched in Fig. 7.8(a). The distribution of dislocation lines
throughout a crystal can also be observed directly by transmission electron
microscopy; examples are given in Fig. 6.6(b), p. 194, and Fig. 9.7, p. 297.

The study of dislocations, their structures, and their properties has almost
become a self-sufficient science in itself, an intellectual exercise in three-
dimensional chess as it has somewhat unkindly been called. The development of
these concepts has, however, been of the greatest significance to the develop-

---

*Both were distinguished English academics. Sir Geoffrey I. Taylor was Professor of Mathe-
matics at Cambridge University, and F. C. Frank was Professor of Physics at Bristol University.
Both had a general interest in the solid state, and Sir Geoffrey had the widest interest in matters
mathematical.

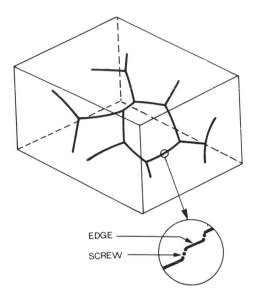

Fig. 7.9. Sketch illustrating the arrangement of dislocation lines as a network in three dimensions. The inset shows how a curved dislocation line would be made up of segments of edge and screw dislocations.

ment of physical metallurgy as an applied science. It has enabled an excellent understanding to be developed of why metals behave as they do in a number of important ways, particularly during deformation. On the other hand, it is doubtful whether dislocation theory has resulted directly in any improvements in the behavior of metals in an engineering sense, but it has enabled sound explanations to be developed of the many improvements that have been made. The hope and expectation are that it will provide a firmer basis on which to build future improvements.

Planar Defects: Stacking Faults

We have already noted that there are alternative possibilities regarding the manner in which the atoms can be stacked on every third close-packed plane (p. 224), the two possibilities resulting in quite different crystal structures. The wonder perhaps is that the atoms of a particular metal continue to choose the alternative which results in their own particular crystal structure being developed. Well, occasionally they don't. A plane on which an error in stacking is made is called a *stacking-fault plane*. Stacking faults obviously can occur only in close-packed structures and must lie on the close-packed planes.

Stacking faults can be produced during the growth of a crystal. They can also be introduced into the crystals of some metals when a particular type of dislocation moves through the lattice, which may occur during plastic deformation (p. 290). Metals in which this can occur easily are said to have a low stacking-fault energy; the addition of an alloying element to a metal often reduces its stacking-fault energy significantly. Additions of zinc, aluminum, and silicon to copper, for example, have this effect.

TWINS

A crystal is said to be twinned if it is composed of portions that are joined together in a definite mutual orientation. The orientation relationship is either that of a mirror image about the *twinning plane* or one that can be derived by rotating one portion about a *twinning axis* relative to the other. Rotational twins of the latter type usually also have reflection symmetry. A twin boundary (cf. grain boundary, below) is flat on an atomic scale and is located on a crystal plane that is characteristic of the metal and its crystal structure.

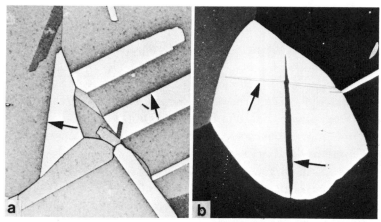

(a) Growth twins produced in a Cu-30Zn alloy during recrystallization. The arrows indicate typical twin boundaries. The slightly curved lines are grain boundaries. (b) Deformation twins (arrows) produced in zinc during plastic deformation. Optical micrographs; 100×.

Fig. 7.10. Illustrations of typical twins in polycrystalline metals.

Twins may be produced during either crystal growth (Fig. 7.10a) or plastic deformation (Fig. 7.10b). The formation of extensive deformation twins is common in metals which do not have face-centered cubic structures (see p. 224); their formation is also important in the deformation of some metals which have fcc structures, but then the twinning occurs on a scale that is on the order of the spacing of the crystal planes (p. 223). Growth twins are common only in metals with fcc structures.

GRAINS AND POLYCRYSTALS

Single Crystals

Very large *single crystals* of many metals can be grown, "single crystal" being the term used for a specimen which is occupied completely by one crystal of the type that we have just been discussing. The meteorite illustrated in Fig. 6.1 (p. 185) is a single crystal. Engineering components (specifically, turbine blades for jet engines) have been manufactured successfully as single crystals, and single crystals of some semimetals (silicon in particular) are routinely produced in large quantities in the semiconductor industry. However, with these exceptions, single crystals are of the nature of laboratory curiosities. They are used extensively in this context to study the basic properties of crystals, and the variation of these properties with crystal orientation. Such studies then are not complicated by the influence of adjoining grains and the boundaries between them that we shall soon be discussing.

Polycrystals

The metals and alloys used in industry are composed of many crystals which abut against one another intimately. Such a material is said to be *polycrystalline*. Each crystal is called a *grain*, and each grain contains defects of the same types as those found in single crystals. The size of the grains (the *grain size**) varies over a wide range, from at least 10 μm to 10 mm (390 μin. to 0.39 in.). There may be tens of millions of grains in each cubic centimeter of a metal, with each grain containing many millions of crystal cells.

The grains are, in the general case, *randomly oriented* crystallographically — that is, their crystal lattices are aligned randomly. In this event, anisotropy in the properties of the individual crystals is averaged out so that the properties of the bulk material are homogeneous. In some materials, however, the grains have a *preferred orientation* or, as it is sometimes called, a *crystal texture*. This means that a certain crystal direction or plane of a proportion of the grains tends to be aligned in a direction parallel, or nearly parallel, with some external reference in the bulk specimen (e.g., the rolling direction or the rolling plane of a sheet). The properties of a textured material are anisotropic in a manner which reflects the anisotropy of the properties of the basic crystal and which to an extent depends on the strictness of the preferred orientation.

---

*The grain size is usually defined as the diameter of an equivalent sphere which has the same volume as the grain.

Grain Boundaries

The individual grains in a polycrystalline material are not physically separated from one another, as are the grains in a lump of sugar. They are joined together intimately on an atomic scale at surfaces called *grain boundaries*. The structure of grain boundaries has been the subject of much study, but the matter still has not been fully resolved because the structure turns out to be complex and variable. In particular, it depends on the magnitude of the difference in orientation between the adjoining grains.

Grain boundaries have to be regions in which the atoms are less well organized than in the adjoining crystals; the atoms cannot all lie on both crystal lattices, particularly when there is a large difference in orientation between adjoining grains. On the other hand, the degree of departure from crystal order cannot be too great, because polycrystalline materials do not pull apart at the grain boundaries unless some special effect is operating. Direct observations indicate three key things about grain boundaries:

1. They behave like surfaces. They have free energies which, however, are considerably smaller than those of free surfaces (see p. 87).
2. They are very narrow, probably only two to three atoms wide, and so the adjustments to the two adjoining lattices have to be made over this small distance.
3. There are regions of good fit in a typical boundary where atoms could be placed on, or close to, the sites of both adjoining crystal lattices; these regions may also house dislocations captured from within the grains. There are also periodically regions of bad fit which can be analyzed in terms of special dislocations which are only possible in the boundary itself. The proportions of the two types of structures depend on the difference in orientation between the adjoining grains.

The locations of grain boundaries can be revealed readily enough by optical microscopy. They are seen as dark lines after treatment by many etchants (Fig. 7.11), or they may be apparent as abrupt changes in the color shading of the grains in other cases (Fig. 7.10). However, optical microscopy gives an erroneous impression of the width of the boundaries. They are seen so readily because of secondary effects. For example, the adjoining grains may be etched to different levels, and it is the step developed at the boundary that is seen; or a groove may be etched at the boundary, the groove being much wider than the boundary itself.

Grain boundaries are certainly regions containing large concentrations of crystal defects, and consequently they have some special properties. Solute atoms may segregate along the boundaries because they fit in more easily than in the more perfect crystal. They are preferred sites for the precipitation of a second phase during solid-state transformations (see Chapter 11). The site of a

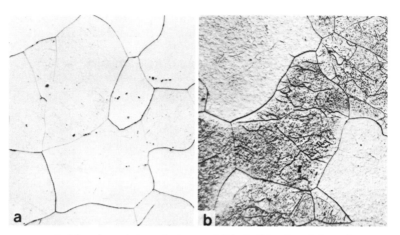

In (a), only grain boundaries are visible. In (b), subgrain boundaries have been made visible within these grain boundaries by
etching the specimen more deeply. Optical micrographs: (a) 100×;
(b) 250×.

**Fig. 7.11. Illustrations of grain and subgrain bound-
aries in high-purity iron.**

pre-existing boundary may as a result be indicated in a metallographic examination by the location of the precipitated phase (see, for example, Fig. 6.3, p.
189, and Fig. 7.6e). In reverse, the solution of a phase during heating tends to
occur first in those particles which are located at the boundaries. Thus, melting
characteristically also commences at the grain boundaries, and the boundaries
are the first regions to weaken when a metal is heated close to its melting point.
Grain boundaries also provide an even easier route for diffusion than vacancies
(p. 234), the activation energy being about halved. This does not contribute
much, however, to general diffusion because the boundary section area is comparatively small. But it does mean that diffusion fronts characteristically penetrate preferentially along the boundaries. The boundaries may also, under special circumstances, be attacked preferentially during corrosion (pp. 149 and
163).

Grain boundaries normally are comparatively strong mechanically in the
sense that plastic deformation is inhibited in the regions of the boundaries (see
p. 296). Moreover, metals do not fracture along the boundaries except under
special circumstances. Examples of these special circumstances include the
presence of a film of a brittle phase or segregate at the boundary (p. 333),
deformation by creep (p. 118), and certain cases of the simultaneous application
of stress and corrosion (p. 163).

Grain Shape and Size

The grains in a polycrystalline material grow from individual nuclei, and can grow in this way either in a liquid (p. 250) or in a solid (p. 308). If the conditions allow them to grow equally well in all directions, the grains finish up with an approximately *equiaxed* shape. They are also of equal size if they are nucleated simultaneously. Normally, however, the grains are nucleated over a period of time, in which event the grain size also varies over a range.

The grain boundary has surface energy so that the grains must tend to assume a shape which has minimum surface area. The solid with the smallest surface area per unit volume is a sphere, but the grains in a polycrystalline material cannot assume spherical shape because they must fit together intimately, and a stack of spheres would not do so (Fig. 7.3). The simplest shape which best meets both criteria of minimum area and close fit is sketched in Fig. 7.12(a). This is a 14-sided figure which is simply an octahedron the points of which have been truncated by a cube. This shape is known formally by the imposing name *tetrakaidecahedron*, but its useful characteristic is simply that objects (grains) of this shape can be stacked to fill completely and uniformly a three-dimensional space, as indicated in Fig. 7.12(b) and (c). This is the ideal grain shape under equilibrium conditions.

A three-dimensional view of the grain shape of a metal can be obtained under any of the special circumstances where an *intercrystalline* or *intergranular fracture* can be produced. It is then seen that the grain shape is indeed approximately that sketched in Fig. 7.12, as indicated in Fig. 7.13. In a section, the

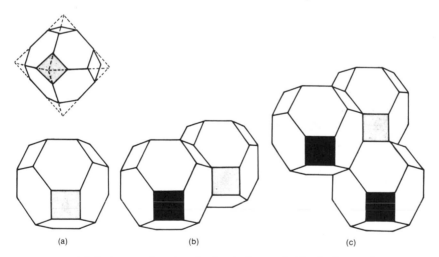

(a)   (b)   (c)

(a) At bottom is a sketch of a single tetrakaidecahedron and above it a sketch showing that this apparently complex 14-sided figure is simply a tetrahedron the points of which have been truncated by the faces of a cube. (b) Two tetrakaidecahedra stacked together. (c) Three stacked together.

Fig. 7.12. Sketches illustrating how tetrakaidecahedra can be stacked together to fill space completely.

An impression can be gained from this macrograph of the three-dimensional shapes of the grains. The shapes are similar in principle to those illustrated in Fig. 7.12, although they are much less regular. Scanning electron macrograph; 5×.

Fig. 7.13. The boundary surfaces of grains in a copper-zinc alloy exposed by fracture.

grains are seen as polygons with varying numbers of sides, the sides generally being approximately straight or slightly curved (see Fig. 7.10 and 7.11). The equivalent diameters of the polygons seen in sections vary over quite a range even when the grains are of the same size in three dimensions. This can be understood by visualizing random sections through grains of the shape sketched in Fig. 7.12. Statistical treatments, however, enable the true grain diameter to be determined from measurements made on sections. The grain boundaries meet at triple points in section. The included angles at the triple points vary over a range in random sections, but statistically the included angle is 120° at equilibrium.

The movement of a grain boundary would require that atoms move through a crystal lattice — in effect, from one adjoining crystal to another by self-diffusion. Thus the grain boundaries in most industrial metals are static at room temperature, where the diffusion rate is vanishingly small. However, they can and do migrate constantly at higher temperatures at which self-diffusion becomes reasonably rapid. The tendency then is for the grains to grow to reduce the boundary surface energy to a minimum. In practice, they tend to grow rapidly to a size that is characteristic of the particular alloy and the particular temperature, but thereafter they usually grow slowly. During the growth process, smaller-than-average grains are absorbed to enlarge the larger-than-average grains (the rich get richer and the poor get poorer, of course). Once coarsened, a structure does not revert to a finer grain size on cooling and reheating to a lower temperature unless a new nucleation-and-growth cycle is initiated. This may occur during recrystallization after cold working (p. 308) or during a phase transformation (p. 358).

Although grains are typically equiaxed in the manner that we have just discussed, there are exceptions. For example, a migrating grain boundary may be held up locally by a small particle of a second phase, causing a local perturbation in the boundary. Layers of such particles of either microscopic or submicroscopic size cause the migration of the boundaries to be slowed down generally in a direction normal to the layers, with the result that the grains end up with a shape that is more like a pancake than a sphere. Grains may also grow, particularly during solidification (p. 262), with an elongated columnar shape. The shapes of the grains are also distorted during plastic deformation (p. 297).

With few exceptions (one of which is resistance to creep, as we shall see later), small grains are desirable. Reducing the grain size, for example, desirably increases both yield strength and toughness (p. 98). Another example is that the large surface area associated with a fine grain size can spread out an impurity which otherwise might embrittle the grain boundaries, so reducing the embrittling effect. Small grain size has a similar mitigating influence on phenomena in which the grain boundaries are corroded preferentially (p. 149).

Subgrains

Grains of the type that we have just discussed may be subdivided further into a number of subgrains (Fig. 7.11b) which differ in orientation by only a few degrees. The resulting subgrain boundaries can consist of simple arrays of normal dislocations by which the changes in orientation can be accommodated (Fig. 7.14).

Boundaries Between Phases

The boundary between two phases can be simpler than those between grains because simple relationships between the crystal structures of the two phases are possible at the interface. A range of boundary structures of varying complexity can result.

In their theoretically simplest forms, the crystal structures and the lattice constants of the two phases may be the same, in which event the structure of the phase interface can be represented diagrammatically as in the sketch in Fig. 7.15(a). An equally simple interface is possible if the crystal structures are the same, the lattice constants are different, but the arrangements of the atoms in the two crystals can be matched on some common plane, as is sketched in Fig. 7.15(b). Either arrangement produces a *strain-free coherent interface*. "Coherent" is used to indicate that there is continuity between the two crystal lattices, and

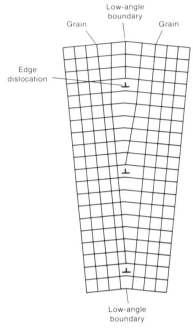

Small orientation changes involving only tilts can be accommodated by an array of edge dislocations. Twists in orientation require screw dislocations to be present as well.

Fig. 7.14. Sketch illustrating the structure of a simple low-angle tilt boundary between two grains or subgrains.

"strain-free" to indicate that the atoms adjoining the interface are not disturbed from their normal positions.

As the next possible step in complexity, the crystal lattices may be of the same type, the lattice constants may be different, but the atoms of the two crystals can be made to nearly, but not quite, match on a common plane. A coherent interface can then form, but the atoms in the two crystals in regions adjoining the interface have to be displaced a little from their normal positions to make a match (Fig. 7.15c). This means that these regions of the adjoining crystals are strained elastically (p. 63), in compression in one crystal and in balancing tension in the other depending on the original relative lattice spacings. An example can be seen in Fig. 6.6(c) (p. 194). The crystal planes imaged in this photograph can be seen to be continuous from matrix to precipitate, but the planes are expanded a little in the matrix adjacent to the end of the precipitate. The region containing these *coherency strains* may be considerably larger than the included phase when the phase particle is small (Fig. 7.15d; see also Fig. 11.11a and b). There is a limit, however, to the amount of mismatch that can be accommodated in this way. A *semicoherent interface* then has to form in which there are regions of good fit and periodical regions of dislocations to accommodate the misfit (Fig. 7.5e). The structure is then rather similar to that of a

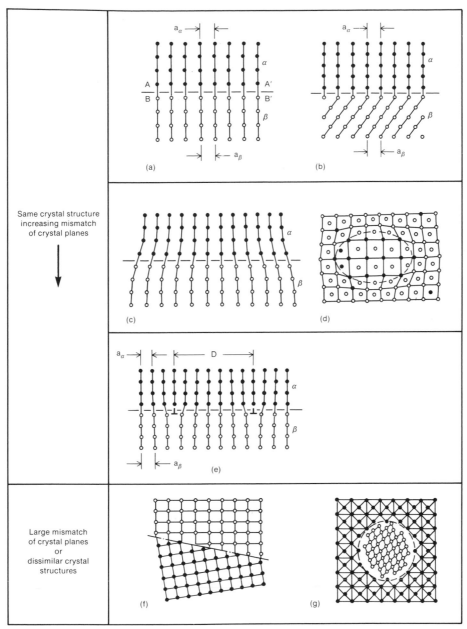

(a) and (b) Interfacial planes of the two crystals can be matched exactly on some plane of each, forming a strain-free coherent interface. (c) and (d) Interfacial planes of the two crystals can be matched with some displacements in each, contraction in one and expansion in the other. A coherent interface is still formed but with coherency strains in a surrounding region. The region containing coherency strains may be large compared with a small precipitate particle (d). (e) Interfacial planes can be matched if an array of dislocations is present at the interface, producing a semi-coherent interface. (f) and (g) Interfacial planes cannot be matched in any of the above ways and an incoherent interface having a complex structure develops.

Fig. 7.15. Sketches illustrating possible structures of interfacial boundaries between two phases.

low-angle grain boundary. A coherent interface also tends to lose full coherency in this way as a precipitate grows and the interfacial strains increase.

As a final step, a fully *incoherent interface* has to form when the two phases have different crystal structures and/or little or no atomic matching is possible across an interface (Fig. 7.15f and g). The interface is then analogous to a high-angle grain boundary.

Coherent interfaces have a relatively low interfacial energy, and the more perfect and strain-free the interfacial match, the lower this energy. Precipitates try to grow in such a way that the interface that is developed has minimal energy, and hence the above order is also the one in which phase interfaces tend to develop.

FURTHER READING

ASM Metals Reference Book, American Society for Metals, Metals Park, OH, 1983.

C. S. Barrett and T. S. Massalski, *Structure of Metals, Crystallographic Methods, Principles, and Data*, 3rd Ed., Pergamon, New York, 1980.

· 8 ·

Solidification and Casting

SUMMARY

The first step in the production of a usable piece of metal typically involves the solidification of a molten liquid. This is a more complicated process than might be imagined. Nuclei of the solid first have to be created in the melt, and this is a difficult process. It is, in fact, the rate-determining stage of the solidification sequence and has a major influence on the grain structure of the resultant solid mass. The nuclei grow, eventually to consume all of the melt. The microscopic details of the growth of the new grains are determined by the crystal structure of the metal, the anisotropic characteristics of which cause the crystals to grow in a three-dimensional branching, or dendritic, pattern. Variations in composition usually develop during growth of impure or alloyed metals with the result that fully solidified grains are segregated on a fine scale (microsegregation). The macroscopic shape of the grains is determined partly by these same factors and partly by the direction of, and the rate of, heat flow during solidification. Segregation on a macroscopic scale may also develop. Metals characteristically shrink during solidification, and the distribution of this volume shrinkage is determined by factors which determine the manner in which the solidification front moves through the liquid. The shrinkage may cause a large open cavity, or pipe, to develop in the last region to solidify. It may also cause small cavities to form throughout the mass if volumes of liquid are cut off from the main volume of the melt. Some gases (hydrogen in particular) to which molten metals are commonly exposed, dissolve in the molten metal but are not soluble in the solid which forms from it. The excess gas has to be evolved during solidification, and some may be trapped among the solidifying grains. Gas cavities of various shapes are thereby formed. A number of different phases form during the solidification of some alloys. They may form either in succession or simultaneously. In the latter event, the phases often form in characteristic geometrical arrangements. Many of these structural features developed during solidification persist to a greater or lesser extent in wrought products.

THE PHYSICAL PROCESSES OF SOLIDIFICATION

The Solid-Liquid Transformation

Solidification is a process by which a solid grows at the expense of a liquid with which it is in contact. There is a temperature for a pure metal above which the liquid is the stable form of the metal and below which the solid is the stable form, and this phase-transformation temperature is defined as the *melting point*. This definition is such that the term "melting" is used irrespective of whether the intent is to melt the solid by exceeding the melting point slightly or to solidify the liquid by reducing the temperature to slightly below the melting point.

There is not a great difference between the arrangements of the atoms in a liquid metal and in the solid crystal that is formed from it. There are at any instant in a liquid small volumes in which the atoms are part of a cluster which has substantial short-range order, exactly as they would be in a solid crystal (p. 222). These crystal-like clusters contain perhaps a few hundred atoms, are oriented randomly, and are separated by regions in which the atoms are more disorganized. The clusters form and disperse constantly and quickly, and atoms change neighbors frequently, aspects which differ from the solid state. The amplitude of vibration of all atoms is also greater than for the solid state. The solid thus has the lower energy of the two, but not by a great amount. Consequently, although there is a driving force toward solidification when a molten metal is cooled to below its melting point, the force is not a very strong one.

Two physical phenomena which are important to the process of solidification follow from this. They are:

1. Heat, called the *latent heat*, is evolved as a result of the difference in energy between the solid and liquid states.
2. The volume decreases by an amount between 2 and 6%.* This results from the closer packing of the atoms in the crystalline than in the molten state.

Solidification of Individual Crystals

Let us consider first the solidification of a pure metal cooled slowly at a steady rate. Ideally, the metal commences to solidify as soon as the melting point is reached. Heat is then evolved which compensates for that being lost to the cooling system. Solid metal forms at just the rate necessary to balance this heat loss. The temperature consequently remains constant until solidification is complete, at which time cooling resumes (curve A in Fig. 8.1). This delay in the

---

*There are a few exceptions to this — notably antimony and bismuth, which expand slightly.

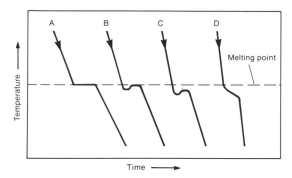

A uniform rate of heat extraction is assumed in all cases, but the rate is assumed to increase progressively from curve A to curve D. Details are given in the text.

Fig. 8.1. Diagrammatic sketches of temperature-time cooling curves for a pure metal, illustrating various types of thermal arrests during solidification.

fall in temperature in spite of continuing extraction of heat from the system is called a *thermal arrest*.

A sharp thermal arrest at the melting point, of the type implied by curve A in Fig. 8.1, rarely occurs in practice. This is because stable nuclei have to form in the melt before the grains can start to grow and liberate heat, and these nuclei do not form easily. In fact, they rarely form in practice until the melt has cooled to well below the melting point; it is then said to be *supercooled*. With low cooling rates, the heat that is evolved once the crystals start to grow raises the temperature back to the melting point, and thereafter the solidification proceeds at constant temperature, as described previously (curve B in Fig. 8.1). With higher cooling rates, however, the melting point may not quite be regained, in which event solidification completes itself at a temperature somewhat below the true melting point (curve C in Fig. 8.1). With still higher cooling rates, the temperature may not rise at all, the thermal arrest then taking the form of a reduction in the rate of fall of temperature (curve D in Fig. 8.1).

Nucleation is such a difficult process that undercooling by hundreds of degrees Celcius may be necessary to achieve true nucleation without any external assistance (*homogeneous nucleation*). This degree of undercooling can be achieved, but generally only in rapidly cooled regions such as at chilled surfaces or surfaces exposed to the atmosphere. Fortunately, however, extraneous particles which serve as nucleants (*heterogeneous nucleation*) usually are floating about in a melt. These particles seem mostly to be pieces broken from crystals that have started to grow elsewhere, such as at a rapidly cooled surface where large amounts of undercooling and hence some degree of true homogeneous nucleation have been able to take place. In any event, higher cooling rates, which result in larger amounts of undercooling, typically increase the number of nucleants that become active per volume of melt.

In addition, completely foreign particles, usually particles of nonmetallic substances, sometimes act as nucleants. These particles may have been present accidentally or added deliberately, and a number of effective nucleants for the latter purpose have been discovered. For example, additions of small amounts of titanium have been found to increase greatly the nucleation rates in aluminum and aluminum alloys, apparently because small particles of titanium carbide or titanium nitride form which actually are the nucleants. Nucleating additives of this nature have been discovered for a number of alloy systems, but all by empirical methods because it is not yet clear what chemical, crystallographic, or physical characteristics are required to make a successful nucleant.

Metallic crystals characteristically grow from a melt in an anisotropic manner as long, thin spikes which have a specific crystallographic direction as their axis. This direction is the one of maximum growth rate for the crystal and is, for example, a cube axis in cubic crystals. The primary spikes lengthen and thicken and also bud secondary spikes at roughly regular spacings; these in turn bud tertiary and quaternary spikes, etc. The branch spacing decreases for each successive order, but all orders grow in the family of preferred crystallographic directions. The arms bud more frequently, and hence the arm spacing is smaller, the faster the cooling rate. The net result is that a crystal grows with the morphology sketched in Fig. 8.2, and illustrated as a real example in the inset, until the entire space is filled with solid. This structure is seen as a branched tree-like pattern in a two-dimensional section (Fig. 8.3a), as a consequence of which it is known as a *dendrite*.

Each grain so formed grows, on average, equally in all directions when it is surrounded by a uniform temperature gradient until its growth is stopped when it butts up against adjoining growing grains. This produces grains which are *equiaxed.* The mean diameter of the grains depends on the number of nuclei that become active per unit volume of melt; the larger this number, the smaller the space that is available for each grain and so the smaller the grain size. Hence, for a given alloy, higher cooling rates during solidification result in smaller grain sizes. So do additions of nucleants. For a given cooling rate, smaller grain sizes are obtained in alloys in which nucleation occurs more readily. The grains grow with a somewhat different shape when the temperature gradient is strongly directional, as it sometimes is in practice. This is because grain growth then is more rapid in the direction of maximum heat extraction (that is, in the direction of the maximum temperature gradient) than in perpendicular directions. The grains consequently assume an approximately cylindrical, or *columnar*, shape. In any event, the external shape of the grains is determined by the accident of how they make contact with adjoining grains (or external surfaces), modified by the drive to achieve minimum grain-boundary area (p. 243), and is not determined by the crystallography of the growth process. The result is called an *allotriomorphic* grain shape.

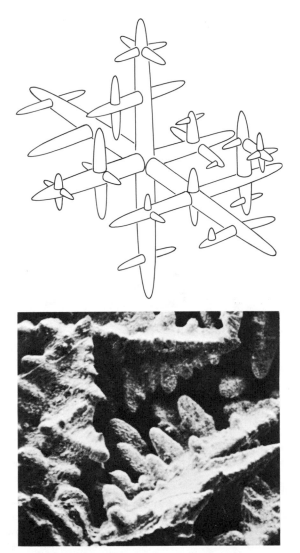

The sketch is an idealized illustration of the manner in which the primary arms of a dendrite grow from a nucleus in preferred crystal directions. The arms thicken and at the same time bud successive generations of arms until all the available space has been filled. The photo is a scanning electron micrograph (magnification, 10×) of partly grown dendrites of iron. These dendrites have grown freely into a cavity in the casting.

Fig. 8.2. Illustrations of the dendritic nature of the growth of a solid metal crystal from a molten liquid.

Nevertheless, the microstructural consequences of the dendritic mode of growth can be discerned when the growth of the dendrites is interrupted before they contact one another. One example of this is given in the inset in Fig. 8.2, where the dendrites have grown into free space. Another example is found when, for reasons given later, the growth of the primary phase stops a part of

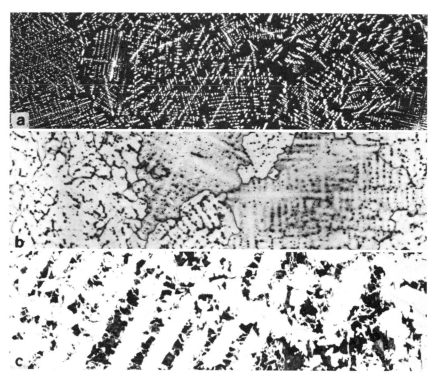

Dendrite growth patterns may be discernible when: (a) a second phase or constituent forms around the dendrites when they
are only partly grown; (b) variations in composition developed in
the dendrites result in differential attack of each crystal during
etching in a manner which outlines the dendrite arms; or (c) variations in composition developed during the growth of the dendrites
affect transformations that occur during subsequent cooling of the
solid. The metal shown in (a) is an aluminum-silicon alloy. The
dendritic shapes (light) are crystals of aluminum formed early in
the solidification sequence. The interdendritic regions (dark) are
occupied by an Al-Si eutectic, the formation of which followed
the growth of the dendrites and the structure of which is shown at
a higher magnification in Fig. 8.14(b). Note the random orientation of the dendrites. Magnification, 10×. The metal shown in (b)
is a cast Cu-30Zn alloy in which etching has developed variations
in color contrast reflecting the segregation of zinc in the dendrites. Magnification, 100×. The material shown in (c) is a low-
carbon steel which has solidified as segregated dendrites of austenite. The austenite transformed during cooling to ferrite (light)
and pearlite (dark). The ferrite has formed at the cores of the
dendrites and the pearlite in the regions between them, so outlining a dendritic shape. Magnification, 100×.

**Fig. 8.3. Optical micrographs illustrating situa-
tions in which dendrite growth patterns often can be
discerned in cast metals by variations in microstructure even though the physical outlines of the dendrite arms no longer exist.**

the way through solidification, the remaining liquid solidifying as a different phase (p. 230). The outlines of the dendrites of the primary phase can then be seen in sections of the solidified material (Fig. 8.3a). Similarly, dendrites can be seen partly in outline when shrinkage cavities are formed (see Fig. 8.12b). Other microstructural consequences of the dendritic mode of growth arise because of the segregation of alloying elements that can develop during the growth process (Fig. 8.3b and e), phenomena that will be discussed in the immediately following section.

Although it is generally true that metal crystals grow from the melt in a dendritic form, there are some exceptions (e.g., bismuth). There are also exceptions with the phases known as *valence compounds* (p. 233) which form in alloy systems. Sometimes valency compounds do grow in a dendritic form; the Cu_3P compound illustrated in Fig. 7.5(e) is an example. Sometimes, however, these compounds grow with a geometrically regular external shape, the bounding surfaces of which are simple crystallographic planes (an example is shown in Fig. 8.16). A crystal of a phase which grows so as to develop a crystallographic shape in this way is said to be *idiomorphic*.

Solidification of Solid Solutions: Microsegregation

The grains of a single-phase alloy solidify in much the same way as those of a pure metal, with the following modifications. Most alloys of this nature solidify over a range of temperatures rather than at a specific melting point, a fact which is indicated in phase diagrams in the manner sketched in Fig. 8.4. This

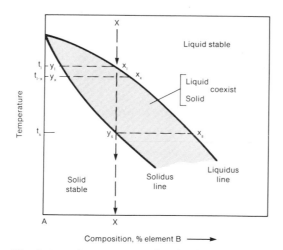

Fig. 8.4. A diagrammatic sketch of a portion of a phase diagram describing the solidification of an alloy of two metals which are soluble in one another in both the liquid and solid states.

diagram tells us that a molten alloy of composition X begins to solidify when cooled to a temperature t_l (known as the *liquidus temperature*) and that solidification is not complete until temperature t_s (the *solidus temperature*) is reached. The loci of these temperatures for alloys of varying composition are plotted in the phase diagram as the *liquidus line* and the *solidus line*, respectively. Nucleation during this type of solidification process occurs after some undercooling, as for pure metals, and dendrites then grow from these nuclei, liberating latent heat. In this case, however, because not all the liquid can solidify at the liquidus temperature, there cannot be a full thermal arrest but merely a reduction in cooling rate as the temperature falls from the liquidus to the solidus.

Figure 8.4 also tells us that a liquid of a particular composition can be in equilibrium only with a solid of another particular composition. For example, liquid of composition x_l in Fig. 8.4 is in equilibrium with solid of composition y_l at temperature t_l when it starts to solidify. The solid that starts to grow when solidification commences thus has the composition y_l, which consequently is the composition of the cores of the dendrites that begin to grow from the melt. When the temperature falls by an increment x to x_{l-x}, however, the equilibrium solid has the composition y_x, and this is the composition of the solid that is now deposited on the rims of the growing arms of a dendrite. If equilibrium is to be maintained, the alloy content of the already-formed core of the dendrite now has to be increased to y_x by virtue of element B diffusing through the dendrite arm to the liquid and so impoverish the core. The liquid remaining thereby should increase its content of element B to x_x. This process should continue as the temperature falls, and the dendrites should grow until the last solid to be deposited has the composition $y_s = X$, deposited from liquid of composition x_s. The entire dendrite would then have this composition X.

However, the diffusion process required to adjust continuously the composition of the solidified dendrite usually is too slow to keep up with this solidification sequence. Consequently, there may not be sufficient time for the solid to adjust its composition to that predicted for the equilibrium state. The extreme case would be one where no diffusion occurred at all. The cores of the arms of the completely solidified dendrite would then have the composition y_l, and the rims would have the composition x_s. In fact, the rims would be even richer in element B than x_s because undercooling would occur and solidification would not be complete until a temperature below t_s had been reached. The composition of the solidified grain would then vary from y to at least x_s, changing progressively across and along the dendrite arms. In practice, something between the two extremes is found, depending on the cooling rate, the characteristics of the alloy system, and the diffusivity (p. 235) of the atoms involved. The variation in composition is likely to be greatest when the cooling rate is highest and the liquidus and solidus lines of the alloy system are mostly widely separated. The over-all result is that the dendrites of the solidified alloy are segregated on a microscopic scale; this effect is called *dendritic microsegregation*, or, more simply, *dendritic segregation*, or, sometimes and more colloquially, *coring*.

The presence of microsegregation in a dendritic pattern can frequently be revealed by optical microscopy because many etchants attack regions of different compositions selectively, so outlining dendritically shaped contours in grains in which segregation has occurred (Fig. 8.3b). Sometimes the presence of segregated dendrites may be apparent because transformations which occur in the solid state during subsequent cooling (Chapter 11) are affected by the variations in composition. This causes, for example, the ferrite phase that forms in cast low-carbon steels to be located mostly in the central regions of the dendrite arms and the pearlite constituent to be confined to the interarm regions (Fig. 8.3c). Once the positions of the dendrites have been identified in this way, the presence and magnitude of the segregation can be established quantitatively by x-ray microanalysis. As indicated in Fig. 8.5, the magnitude of the segregation can be quite substantial.

We shall see later that phases other than the primary dendrites form late in the solidification sequence in many alloys. These phases have to be located in spaces between the primary dendrites (Fig. 8.6). Likewise, any small particles of material that are not soluble in the melt tend to be swept in front of the solidifying dendrites and so finally end up trapped in chains at the edges of the arms. Some particles of this nature are visible in Fig. 8.3(b). These insoluble particles result mostly from bits of dirt and refractory, refining slag and reaction products, and melting fluxes which inevitably get into a commercial melt of metal. They are described by the generic term *nonmetallic inclusion*. Their final distribution is not primarily the result of microsegregation during solidification but has been influenced in a similar manner by the dendritic mode of growth.

The converse to all of the above occurs during heating of a solid. Melting commences in the grain-boundary regions and the melt progressively consumes the cores of the grains as the liquidus temperature is approached.

Dendrite segregation can be smoothed out by annealing the cast material at high temperatures for long times in what is known as a *homogenizing annealing* treatment (Fig. 8.7). Annealing at temperatures as close as possible to the solidus temperature is desirable to achieve maximum diffusion rates. Even so, the effectiveness of a homogenizing anneal depends on the diffusivities of the elements involved, which vary considerably. Thus the segregation of copper in aluminum can be completely eliminated (Fig. 8.7) whereas that of, say, manganese in iron (Fig. 8.5) can at best only be reduced in severity. Certainly, only elements which are soluble in the matrix can be homogenized. Elements which form insoluble constituents, such as those in the material illustrated in Fig. 8.6, cannot be homogenized. The probability that a degree of homogenization will be achieved is greatest when the distance over which the diffusion has to occur is smallest — that is, when the dendritic arm spacing is smallest. Ingot-casting

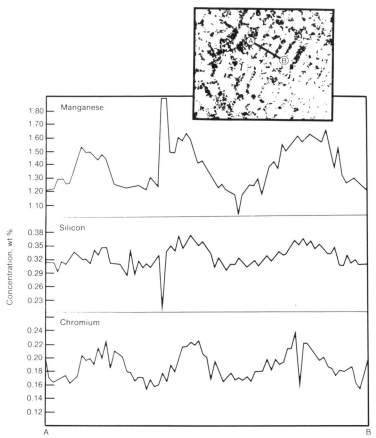

The material is a cast low-carbon steel containing, on average, 1.13% Mn, 0.32% Si, and 0.12% Cr. All of these alloying elements form solid solutions in iron when present in these amounts. The inset is an optical micrograph (magnification, 100×), the white regions being the cores of dendrite arms (cf. Fig. 8.3c). The curves show the variations of the alloying elements along the line AB marked on the micrograph through several dendrite arms. The analysis was carried out by electron probe microanalysis methods, using an x-ray fluorescence analyzer system (p. 199). The concentrations of all alloying elements are much higher, by a factor of up to two, in the centers of the arms than in the regions between them.

Fig. 8.5. An illustration of the magnitude of the segregation that can develop across the arms of dendrites during their growth.

techniques have been devised to achieve this end, mainly by ensuring high cooling rates. Continuous casting (p. 25) is an example.

Nevertheless, a significant degree of microsegregation typically persists in an ingot when it is worked into a semifinished product — more so in some alloys, and in some batches of a given alloy, than in others. This microsegregation is not eliminated during the subsequent working of the ingot; it is merely

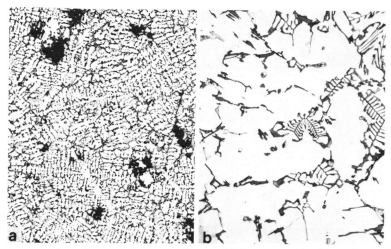

It can be seen at low magnification (a) that primary dendrites of aluminum (white) have solidified first and that some other constituents have formed later in spaces left between the dendrite arms. It is apparent at higher magnification (b) that several types of constituents are present in the interdendritic regions. They are compounds of aluminum, copper, silicon, and iron. Optical micrographs: (a) 20×; (b) 100×.

Fig. 8.6. Microstructure of a cast aluminum alloy containing 5% Si, 2.5% Cu, and some iron and manganese.

redistributed. The shape of a segregated region in the ingot is in fact changed in the same way as the shape of the ingot as a whole is changed. Thus, in a plate, the arms of dendrites are reshaped into "pancakes" greatly elongated in the direction of rolling, with the interarm regions forming interleaving pancakes of a different composition. In a bar, the two are shaped into interleaving rods. The pancakes or rods are not, of course, sharply distinguished but shade into one another as did the segregation in the parent dendrites. An indication of the persistence and the change in morphology of segregated dendrites is given in Fig. 8.8(a). Ferritic arms such as those illustrated in Fig. 8.3(c), which were present in the ingot from which this plate was rolled, are now seen in section as elongated bands. Compositional variations of the type indicated in Fig. 8.5 persist to a considerable degree in these bands.

Phases or nonmetallic inclusions that were located between the dendrites in the ingot are also relocated during working of the ingot and are reshaped if they are plastic under the circumstances. They then tend to be grouped in elongated stringers of pancake-shape particles if they were plastic or as stringers of isolated irregularly shaped particles if they were not (Fig. 8.8b). The consequences of this structural banding can be of considerable significance because directionality is introduced into the properties of the product, a phenomenon to which we have referred on several occasions. The bands can constitute planes of weakness when a metal is stressed normal to the bands, and

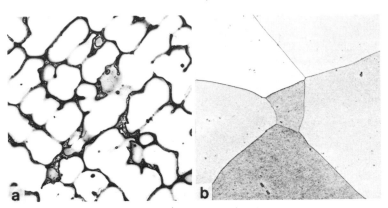

The variation in etch shading in (a) indicates that extensive dendritic segregation was present in this Al-5Cu alloy when it first solidified. There is no similar shading in (b), which is of the same material after annealing for 16 h at 525 °C (975 °F) and water quenching. The copper is now homogeneously distributed in the aluminum. Note that annealing for a long time at a temperature near the melting point was required to achieve the homogenization. The difficulty in achieving homogenization varies with the alloy system, depending on the diffusivity of the segregated elements. The interdendritic constituent (arrow) visible in (a) is the compound $CuAl_2$. This compound was taken into solution during the homogenization anneal and was retained there by subsequent rapid cooling. The homogenization anneal thus doubled as a solution treatment (see p. 370). Optical micrographs; 100×.

Fig. 8.7. Micrographs illustrating the homgeniza-tion of dendritic segregation by annealing at high temperature.

this can be responsible for reduced fracture toughness of steel and aluminum alloy plate in the transverse direction (p. 98). Objectives of the production of high-quality plate, particularly plate which might be stressed in the transverse direction, have to include reduction of the size and elongation of the shape of the phases and inclusions and their uniform distribution. This can be done only by close attention to ingot-making practice, principally to the control of the composition and cleanness of the melt and to the adoption of solidification procedures which reduce the dendrite-arm spacing.

THE GRAIN STRUCTURES OF INGOTS AND CASTINGS

We saw in Chapter 2 that typically a metal may be solidified (cast) into a shape whose length is great in comparison with its lateral dimensions and which has a simple cross section. Such a casting is intended to be shaped subsequently by plastic deformation, and is known as an *ingot*. An ingot may be cast statically by

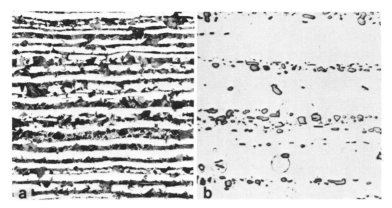

(a) Longitudinal section of a steel plate hot rolled from an ingot which had a microstructure similar to that shown in Fig. 8.3(c). The dendrite arms (white in Fig. 8.3c) are now elongated into a pancake shape, shown as white bands here in section. This is because the microsegregation that was present in the dendrite arms has persisted and has influenced the locations at which the ferrite and pearlite formed during cooling of the plate after hot rolling.

(b) Longitudinal section of a plate of an aluminum alloy rolled from an ingot which would have contained constituents like those in Fig. 8.6 but in smaller volume fractions. The constituents are now grouped in stringers.

Optical micrographs; 100×.

Fig. 8.8. Illustrations showing that the segregated alloying elements and constituents in a cast ingot are not removed during subsequent working but are reshaped into bands elongated in the direction of the main working strain.

pouring the molten metal into a simple open-ended mold (p. 24) or it may be cast continuously (p. 25). Alternatively, a metal may be solidified into a more or less complicated shape which, perhaps with some modification by machining, is intended to be its final useful shape. This product is known as a *casting* (p. 14). There is also a third possibility — one which occurs, for example, during fusion welding (p. 45) — in which a volume of metal is solidified *in situ* as part of a composite structure. The cast volume then can usually be regarded as an ingot; often it is a continuously cast ingot.

Ingots

Let us consider first the simplest case of an ingot cast in an open-ended metal mold (see Fig. 8.11a). The cooling rate adjacent to the ingot wall is high at the beginning of the solidification sequence, and the temperature gradient is at a maximum in a direction normal to the ingot wall. The high cooling rate of the portions of the melt that contact the ingot wall usually results in homogeneous nucleation, in which event many small, randomly oriented grains are produced in a *chill zone* (the left-hand region in Fig. 8.9). The formation of homogeneous

Surface   Center

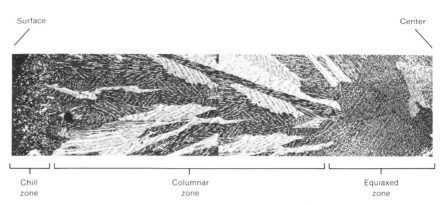

Chill   Columnar   Equiaxed
zone   zone   zone

This is an etched transverse section of a chill-cast cylindrical
bar of a tin bronze, with only half of the cross section being
shown. Note the succession from small equiaxed grains in the surface layer, to a layer of long columnar grains, and then to a central
region of large equiaxed grains. The dendrite arms can be discerned in each columnar grain and in each central equiaxed grain.
Note that one of the arm systems in each columnar grain is aligned
approximately normal to the surface of the casting. Those of the
central equiaxed grains, on the other hand, are randomly oriented. Optical micrograph; 15×.

**Fig. 8.9. Illustration of the classic succession of
crystal structures developed across the section of a
casting.**

nuclei soon stops, however, because the cooling rate falls, but some of the chill
grains that are located at the liquid/solid interface at that time continue to
grow along the direction of the temperature gradient. These are the grains that
happen to be oriented with a dendrite arm (i.e., in one of the dendrites' preferred growth directions) aligned parallel with the temperature gradient. The
survivors of this competitive growth process elongate, at the same time thickening a little, to produce a *columnar zone* of roughly cylindrical or columnar
grains (the central region in Fig. 8.9). The columnar grains have, as a result of
this mode of growth, a similar or *preferred* crystal orientation along their axis
but a random orientation about this axis. This is known as a *fiber texture*.

Eventually, a new generation of heterogeneously nucleated dendrites
forms in the melt ahead of the growing columnar grains. These grains grow to
produce a core of randomly oriented equiaxed grains (the *equiaxed zone*) such as
those in the right-hand region in Fig. 8.9. These equiaxed grains eventually
impinge against the growing columnar grains and stop their further growth.
The size of the equiaxed grains depends on the number of heterogeneous nuclei
that become active per unit volume of melt. Thus the grain size of the equiaxed
zone is reduced by higher cooling rates or the addition of nucleants (*grain refin-
ers*) to the melt. Likewise, any process which causes pieces of partly solidified
dendrites to be broken off and redistributed to places in the melt where they

can act as nucleants also refines the grain size. Vigorous shaking or stirring of the ingot may have this result.

The classic arrangement of these three zones of grains in a statically cast ingot is illustrated in Fig. 8.9 and 8.10(a). However, the three zones are not necessarily always present. Any one of them may occupy the full volume of the ingot, and any one of them may be absent. It all depends on a complex interplay among the cooling rate, the availability of nuclei, the directionality of the temperature gradient, and the growth characteristics of the metal concerned. For example, chill grains and columnar grains are most likely to predominate in ingots cast in water-cooled molds. Columnar grains are most likely to predominate in metals, such as zinc, which have strongly anisotropic growth characteristics. Columnar grains are also most likely to predominate when the nucleation required to start the growth of the equiaxed zone is difficult; aluminum and aluminum alloys are examples (Fig. 8.10a), as also are welds (Fig. 8.10b).

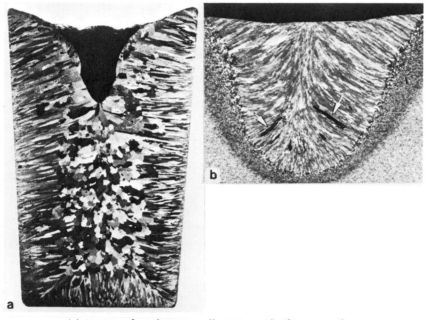

(a) Section of an aluminum alloy ingot. The fine-grained surface region of this ingot is too thin to be clearly resolved, but the region of columnar grains and the central region of equiaxed grains can be seen clearly. A pipe cavity is also evident at the open end of the ingot. (b) Transverse section of a weld deposit in steel. The solidified metal in this deposit is composed entirely of columnar grains, the long axes of which are aligned normal to the weld metal/parent metal interface. Several wormhole gas cavities (arrows) are present between the columnar grains. Optical macrographs: (a) ½×; (b) 1×.

Fig. 8.10. Examples of representative grain structures in an ingot and a weld deposit.

The direction of alignment of the long axes of the columnar grains depends only on the direction of the temperature gradient. The columnar grains in the ingot illustrated in Fig. 8.10(a) have grown equally from the base and sidewalls of the mold, and hence intersect along a 45° plane at the bottom corner of the ingot. Columnar grains in an ingot that has solidified directionally from the base only, as sketched in Fig. 8.11(g) to (i), extend vertically from the bottom to the top of the ingot. Those in an ingot, such as a continuously cast ingot, which has solidified from the sidewalls only, extend only from the side to the center of

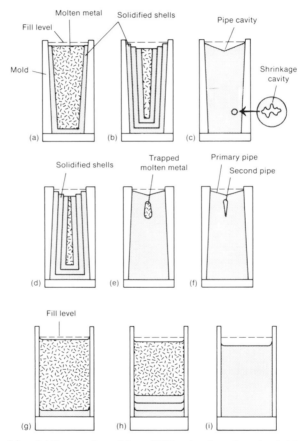

(a) to (c) Progression of the solidification front in a simple big-end-up ingot, illustrating the formation of a primary pipe cavity. (d) to (f) Progression of the solidification front in a simple small-end-up ingot, illustrating how the surface of a primary pipe cavity might bridge over while some liquid metal remains beneath the bridge. An internal, or secondary, pipe cavity then forms. (g) to (i) Progression of the solidification front in an ingot which is solidified directionally from the base up. Little or no pipe cavity forms.

Fig. 8.11. Sketches illustrating the progression of solidification fronts in ingots, and the development of pipe and shrinkage cavities.

the ingot. In many cases, the columnar grains in continuously cast ingots are slightly curved because the direction of the temperature gradient changes as the solidification front approaches the center of the ingot. The weld metal shown in Fig. 8.10(b) is an example.

Castings

The range of grain structures in die castings can be expected in principle to be similar to that in ingots. Chill and columnar grains are particularly likely to predominate in the many die castings that are made in thin sections. In sand castings, on the other hand, the cooling rate is lower and the temperature gradients less steep than in ingots. The formation of chill and columnar grains is not as likely, except in specially susceptible metals, and the equiaxed crystals are likely to be comparatively large unless grain-refining additions are made.

In Situ Castings

A special feature of metal cast *in situ*, such as in fusion welds, is that the unmelted parent metal provides the nuclei for crystal growth during solidification. Moreover, a comparatively small volume of material is molten at any instant, and this material is in intimate contact with comparatively cold metal. This results in fast and highly directional cooling. The growth of columnar grains during solidification is encouraged by all of these factors, and columnar grains commonly occupy the full solidified zone in a weld (Fig. 8.10b). Exceptions are found only when large volumes of metal are melted at any one time and when the weld puddle progresses slowly. Equiaxed structures may then be produced.

SHRINKAGE DURING SOLIDIFICATION

Pipe Cavities

We noted earlier that a molten metal shrinks a significant amount when it solidifies. The consequences of this in practice are most easily seen by again considering first the solidification of an ingot cast in a simple open-ended tapered mold; ingot molds have to be tapered to enable easy separation of the ingot and mold after solidification. A mold tapered in the manner sketched in Fig. 8.11(a) is known as a big-end-up mold. Let us assume that solidification occurs progressively in shells from the mold surfaces. The shrinkage of each shell then lowers its own level and that of the remaining liquid, and hence progressively that of each new shell as it solidifies (cf. Fig. 8.11a and b). A conical-shape depression known as a *pipe cavity* is thereby formed in the top of the ingot as sketched in Fig. 8.11(c), an actual example being shown in Fig. 8.10(a). The volume of the pipe cavity is equal to the total volume contraction of solidification, minus the small volume of the shrinkage porosity developed throughout the ingot (see below). Consequently, a pipe cavity is much larger in aluminum (solidification shrinkage, 6%) than in, say, iron (solidification shrinkage, 2.2%). Note also that contraction of the melt during cooling to the melting point adds to the volume of the pipe cavity. Hence high pouring temperatures result in larger pipe cavities.

The situation that we have just described is the simplest possible one. It supposes that the growing dendrites have uninhibited access to liquid at all times and that the solidification front advances smoothly and progressively. Several complicating factors arise in practice. The first is on a macroscopic scale and affects the shape of the pipe cavity. The second is on a microscopic scale and causes some of the shrinkage to be distributed throughout the ingot.

The macroscopic effect is related to the manner in which the solidification front advances through the melt. One extreme case occurs when the surface of the *primary pipe* solidifies over before solidification of the entire melt is complete, leaving an enclosed volume of liquid metal beneath the primary pipe and within the ingot (Fig. 8.11d and e). A *secondary pipe* then forms below the primary pipe when this pocket of liquid finally solidifies (Fig. 8.11f). The likelihood of the formation of a secondary pipe is increased by high radiation from the free surface of the ingot and by use of ingot molds which taper inward toward the pipe region instead of outward (small-end-up ingots, as sketched in Fig. 8.11d to f).

Both of these factors increase the probability that the primary pipe will close over before solidification of the melt is complete. The likelihood of the formation of a secondary pipe is decreased, on the other hand, by factors which delay solidification in the pipe region. This can be achieved, for example, by placing a collar of insulating material at the mouth of the ingot (Fig. 2.8c, p. 24). It is then said that a *hot top* has been used, and that a *hot-topped ingot* has been produced.

At the other extreme, an ingot in which the solidification front is made to be approximately planar and to advance in one direction only toward the open end of the ingot usually will not develop a pipe; at worst, the pipe formed will be a very open one (Fig. 8.11g to i). The solidification shrinkage is then taken up by a length contraction of the ingot. This situation can be approximated in practice by, say, water cooling the base of the ingot mold and insulating its sides, as implied in the sketch in Fig. 8.11(g). The same result is achieved in continuously cast ingots (p. 26) in which, in effect, the pipe cavity is continuously kept filled with molten metal. A pipe is then produced only when the casting line stops.

The formation of pipes is important in practice because the piped length of an ingot has to be cropped off before the ingot is worked. The pipe, if left, would close up during working but would not weld because its surfaces are oxidized. The result would be an undesirable central discontinuity running through a considerable length of the final product. It is consequently the shape of the pipe that is important as much as its volume, because its depth determines the length of the ingot that has to be removed, and wasted, by cropping. Secondary pipes are particularly deleterious. They are deep, and some may be left inadvertently if only the primary pipe is cut off. We have already seen that the depth of a pipe can be reduced by several methods, but the viability of these techniques depends in practice on the relative cost of applying the particular technique and on the savings achieved by reducing the amount of material lost in cropping. The use of open-ended ingots and of hot topping are economical in steel production, for example, only for the more costly grades of steel. The elimination of piping is, however, a considerable economic and quality advantage of continuous casting.

Shrinkage Cavities

The effect which occurs on a microscopic scale and which modifies our first simplified approach to solidification shrinkage occurs when a small volume of liquid is cut off from the main body of the melt. A small cavity is then left when this entrapped liquid solidifies and shrinks, the cavity forming as an interstice between dendrites (Fig. 8.11c). The resultant cavities are called *interdendritic shrinkage cavities* or *microporosity*. They are most likely to develop in metals with a short solidification range (small distance between the liquidus and solidus lines in Fig. 8.4) because solidification occurs over a shorter period of time and so the supply of liquid to some regions is more likely to be cut off during solidification.

Cavities of this nature in an ingot usually weld up during subsequent working, unless they happen to break out to a surface and become oxidized. They may cause difficulties, however, if aligned so as to constitute a plane of weakness. This may happen, for example, along a plane of intersection of two systems of columnar grains, which might be present at the bottom corner of an ingot (Fig. 8.10a). The ingot may be made with a rounded bottom (Fig. 2.8c) to avoid this problem.

Castings

The principles that we have just been discussing for ingots apply equally well to the solidification of castings, although the complexity of shape of a casting may introduce complications.

First, complexities of shape work against a smooth progression of the solidification front from a remote point to that at which the metal entered the mold. Volumes of melt, large and small, consequently are likely to become isolated from the main body of the melt. A macroscopic internal pipe cavity must form

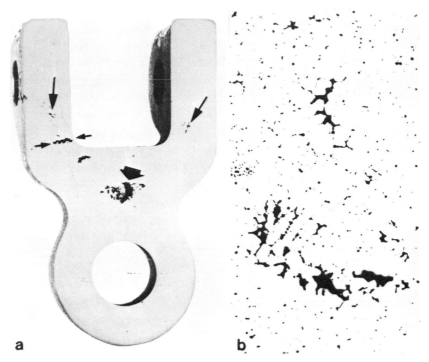

The section in (a) shows an internal pipe cavity (large arrow) and patches of internal shrinkage cavities (intermediate-size arrows). The casting also contains a shrinkage crack or tear (small arrows). The shrinkage cavities are shown in more detail in (b), the interdendritic morphology of the cavities being apparent. (a) Optical macrograph; ½×. (b) Optical micrograph; 100×.

Fig. 8.12. Sections of a tin bronze casting containing shrinkage cavities and shrinkage tears.

wherever a large volume of melt becomes isolated in this way (Fig. 8.12a), and areas of microporosity are also likely (Fig. 8.12a and b), particularly if a volume of melt becomes isolated late in the solidification sequence. Large areas of shrinkage porosity significantly impair the strength of a casting locally because the cavities are interconnected and so in effect reduce considerably the section area. They may also seriously impair the pressure or leak tightness of a casting because they may provide an interconnected path that extends completely through the wall thickness of the casting. They may not do this in a casting initially but could well do so after a machining operation which breaks into the flawed material.

A major skill required of a foundry technologist, therefore, is that of designing a mold system which ensures that the solidification front progresses through the casting without significant volumes of melt becoming isolated from the main body of liquid metal. This is achieved by a variety of techniques which include the provision of feeder heads, risers, and perhaps metal chill inserts at appropriate positions throughout the mold (Fig. 2.4, p. 16). These appurtenances may be needed to the extent that they add considerably to the complexity and volume of the mold system. A considerable amount of extraneous material then has to be removed from the casting after it has been removed from the mold. All of this adds to the cost of producing sound castings. As likely as not, a foundry technologist will not get the mold system completely right on the first attempt, in which event the cause of the deficiency has to be diagnosed and the casting technique modified accordingly. A number of attempts may be necessary in difficult cases. It follows that quality castings usually are not inexpensive castings.

The general shrinkage that occurs during solidification results in a casting being smaller than the mold cavity into which it was cast. The amount concerned is significant, varying from about 2 to 6% depending on the alloy, and has to be allowed for in making the mold pattern. Another possible consequence arises with complex castings, the contraction of which may be inhibited by the mold both during and after solidification. An example would be a casting which consists of a length of uniform section terminated by two shoulders. In this instance, the contraction of the uniform section would be restricted when the shoulders bore against the mold. The casting might then crack (develop a *shrinkage tear*) at a point of stress concentration adjacent to the shoulders (Fig. 8.12a). This possibility is compounded by the fact that the strength of a metal is at its lowest at temperatures close to its melting point, particularly if films of liquid remain between the already solidified grains for some time (p. 113). This situation is likely when there is a long solidification range or when a small volume fraction of a low-melting-point eutectic has to solidify late in the sequence.

A final consequence of solidification shrinkage worthy of mention is that it limits the sharpness of detail that can be reproduced in a casting because the metal shrinks away from sharp corners in the mold during solidification. This adds to a limit imposed by the surface tension of the liquid metal, which prevents very fine detail from being filled completely in the first place. An example of a situation where this is important is found in the casting of type for printing. Up to 30% antimony is added to lead-tin alloy *type metals* to reduce the solidification shrinkage, antimony being one of the few metals that expand during solidification. These types of metal alloys are, however, the victims of technological change in the printing industry. The printing processes for which they are required are not much used any more.

SEGREGATION ON A MACROSCOPIC SCALE

We have already seen that the composition of a liquid alloy (and that of the solid which forms from it) usually changes during the course of solidification; this change occurs because the melt is cut off from the solid which has formed during the earlier stages of solidification. We also saw earlier that this effect causes segregation to develop on a microscopic scale because the core of a dendrite loses contact with the melt. The effect can in addition cause segregation on a macroscopic scale (*macrosegregation*) because the composition of the melt adjacent to the solidification front changes in a similar way when solidification occurs in shells in the manner sketched in Fig. 8.11. Variations in composition may then develop on a gross scale from surface to center and from top to bottom of an ingot that solidifies in the manner sketched in Fig. 8.11(a) to (e). Variations in composition of this nature do not occur along the length of a continuously cast billet, but variations from surface to center are also possible to a lesser extent in continuously cast ingots, although they are likely to be much less marked because melt is added continuously and so the solidifying pool is stirred vigorously. These are further advantages of the continuous casting route.

Any macroscopic variations in composition in an ingot persist in products made from that ingot; the distance over which the segregation has occurred is much too large for the variations in composition to be evened out by any practical homogenizing anneal. Compositional variations are therefore possible throughout any wrought material produced from a statically cast ingot. Moreover, the composition at any point in the material may be slightly but significantly different from the average composition of the melt from which it came. Consequently, a chemical analysis of a sample removed from a melt does not necessarily fully characterize all portions of the products made from that melt. One of the objectives of quality control in metal production is to keep these variations within limits that are acceptable for the intended end use.

SOLIDIFICATION OF MULTIPHASE ALLOYS

Many of the alloys used in practice contain two or more phases after solidification. These phases may solidify concurrently in a so-called *eutectic* system (p. 230), or they may solidify in sequence (p. 231), in which event a *peritectic reaction* between solid and liquid phases usually is involved. These two basic phase reactions were mentioned briefly in Chapter 7, but the first reference under "Further Reading" at the end of this chapter should be consulted if a more complete understanding of these topics is needed.

Eutectics

Let us look briefly at the solidification sequence that occurs in a eutectic system, using as our example the copper – copper phosphide system* illustrated in Fig. 7.5 (p. 231). Consider first the alloy containing 4.6% phosphorus. A molten

*This system has a valence compound (Cu_3P) as one of its boundary phases, but this does not affect the principles being illustrated.

alloy of this composition, when cooled, begins to solidify at the temperature t_ϱ indicated in Fig. 7.5(a). Dendrites of copper-rich solid solution nucleate and grow in the manner described earlier in this chapter until the eutectic temperature (t_e) is reached. There is then a thermal arrest similar to that for a pure metal (p. 251) while all of the remaining liquid solidifies as the eutectic mixture of two phases — in the case being used as an illustration, as a mixture of copper-rich solid solution and the copper phosphide phase (Cu_3P). Consequently, all of the spaces between the primary dendrites of copper-rich solid solution are occupied by eutectic in Fig. 7.5(c). Note that the eutectic solidifies (and melts on heating) at the lowest temperature of any liquid in that particular region of the phase system.

An alloy of eutectic composition (8.4% phosphorus) does not begin to solidify until the eutectic temperature is reached, when there is a thermal arrest while all of the melt solidifies as eutectic (Fig. 7.5d; cf. a pure metal). An alloy containing more phosphorus than the eutectic composition commences to solidify as dendrites of Cu_3P, but the same eutectic as before forms when the temperature falls to the eutectic temperature (Fig. 7.5e).

Many eutectics grow as a colony from a single nucleus, the two phases then growing and branching in such a way as usually to produce a fairly regular geometric pattern. One of the phases is continuous throughout a colony and the second phase appears to be embedded in it. The pattern adopted by the two phases is generally a characteristic of the alloy system, although a bewildering array of patterns is encountered in different alloy systems. The simplest is a regular lamellar arrangement (Fig. 8.13a). Although this is commonly described as being the classic arrangement in eutectics, it is not very common in practice. In fact, regular lamellar eutectics form only under unusual conditions of slow directional solidification. Distorted lamellae, often arranged as colonies, are more common (Fig. 8.13b), but patterns of degenerate plates, of rods (Fig. 8.13c), of script shapes (Fig. 8.13d), of spirals, and even of trigons are also found. Nevertheless, a characteristic of all of these eutectics is that the two phases grow edgewise into the liquid with an approximately planar solid/liquid interface. It is difficult to explain this type of solidification pattern and why the growth should occur in a particular pattern in a particular alloy system. Cooling rate during solidification often affects the morphology of a eutectic and always affects the spacing of the constituents, higher cooling rates giving rise to smaller spacings.

There are some eutectics, moreover, in which the embedded phase is discontinuous and randomly arranged (Fig. 8.14). This means that each particle of the discontinuous phase has to be nucleated separately and that the nucleation has to occur repeatedly, phenomena which are particularly difficult to explain. An interesting example of a discontinuous eutectic is found in the aluminum-silicon system, alloys from this system being used commonly in foundry practice. The interesting thing about this eutectic is that its morphology can be

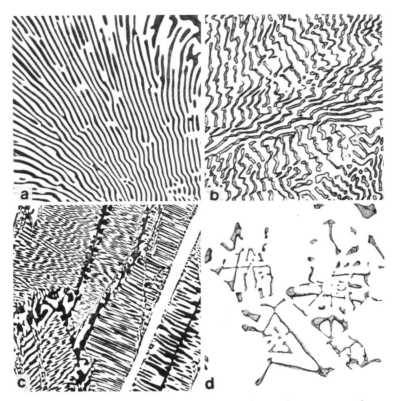

(a) Eutectic between copper (light) and the valence compound CuP_3, exhibiting a reasonably regular lamellar morphology. Optical micrograph; $1000\times$. (b) Same eutectic shown in (a), but exhibiting the more irregular morphology that is more typical of "lamellar" eutectics; the lamellae are considerably distorted. Optical micrograph; $500\times$.

(c) Rod eutectic between cementite (white) and austenite (appearing dark here) in a chilled cast iron. The austenite grew as rods in a continuous matrix of cementite, the rods being seen here as bands when sectioned longitudinally, as at lower right, and as ellipses when sectioned transversely, as at upper left. Optical micrograph; $500\times$.

(d) Eutectic between aluminum (light matrix) and the valence compound Mg_2Si, in which the Mg_2Si has grown in the aluminum matrix with a branched script morphology. Optical micrograph; $250\times$.

Fig. 8.13. Examples of the range of morphologies found in continuous eutectics.

changed from discontinuous plates (Fig. 8.14a) to continuous branching rods, which are seen as circular particles in sections (Fig. 8.14b; see also Fig. 6.8b, p. 197). This *modification* of the eutectic, as it is called, is achieved by adding small amounts of sodium or strontium to the melt. This phenomenon has been known for some time, but no explanation of it is yet available.

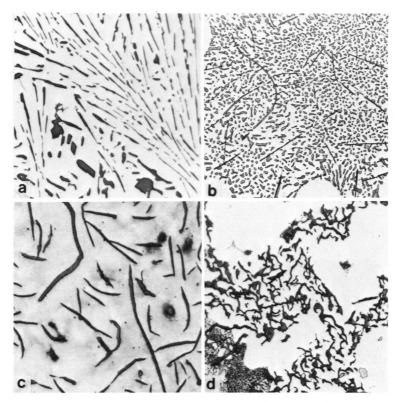

The silicon (gray) in the eutectic between aluminum and silicon illustrated in (a) is present as randomly arranged, disconnected plates. The morphology of this eutectic can be changed dramatically, however, by adding small amounts of elements such as sodium and strontium. The silicon is then present as branched rods (see Fig. 6.8b, p. 197) seen in section in (b) as circles. The long, thin laths visible in both (a) and (b) are a phase resulting from the presence of iron as an impurity.

Eutectics between graphite (gray) and austenite in gray cast iron are illustrated in (c) and (d). The graphite is present as discontinuous, randomly oriented flakes. As illustrated here, the size of the flakes can vary markedly, depending on the composition of the iron and the cooling conditions, and this has a considerable influence on the mechanical properties of the iron.

Optical micrographs: (a and b) 250×; (c) 100×; (d) 500×.

Fig. 8.14. Examples of the morphologies found in discontinuous eutectics.

We have so far discussed only binary eutectics (eutectics involving two phases). Eutectics can involve three or more phases. In this event the third phase is always nucleated discontinuously, but these are complications which we shall not pursue.

The morphology of the constituents of a eutectic can have a considerable influence on the mechanical properties of an alloy. An extreme example is the aluminum–silicon eutectic just mentioned, the mechanical properties of the modified and unmodified versions being discussed on p. 338. Gray cast irons are an even more important example. The flakes of graphite in gray cast irons are formed during solidification of a eutectic, and the shape and size of the flakes can be changed considerably in a number of ways from very course to very fine (cf. Fig. 8.14c and d), which has a critical influence on the strength and toughness of cast irons. Cast irons with a graphite structure of the type illustrated in Fig. 8.14(c) have low strength because the large graphite flakes provide an easy fracture path through the material, but those with a structure like that in Fig. 8.14(d) have much greater strength. These phenomena are important in practice.

Peritectics

It is a common feature of alloy systems for alloys over certain ranges of composition to undergo a *peritectic reaction* during solidification. A typical peritectic portion of a phase diagram is sketched in Fig. 8.15 and a real example is given in Fig. 7.6 (p. 232), this system having a number of peritectics in succession. The various possibilities that then arise during solidification are:

1. Alloys of X containing up to a% of Y solidify as solid solution A in the manner described earlier (p. 255).
2. Alloys with B contents between a% and b% solidify as in possibility 1 above until temperature t_p is reached. There is then a thermal arrest while the melt reacts with solid solution A to form solid solution B until all of the liquid has been consumed. This is the peritectic reaction. The

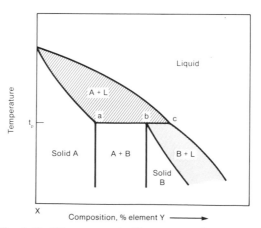

Fig. 8.15. **Diagrammatic illustration of portion of a phase diagram describing the solidification of alloys involving a peritectic reaction. Details are given in the text.**

solid produced is then a mixture of phase A of composition a% Y and phase B of composition b% Y, in the proportion required to give the correct mean composition (see Fig. 7.6c, for example). Thus alloys of compositions close to a% Y contain only a little B phase, and alloys of compositions close to b% Y are composed almost entirely of B phase.

3. In alloys with Y contents between b% and c%, the already-solidified A phase transforms completely to B phase (Fig. 7.6d, p. 232). The reaction just completes itself at composition b% Y, but some liquid remains with larger contents of Y. In the latter event, cooling recommences and the remaining liquid solidifies as B phase in the normal manner for a solid solution.

4. Alloys with Y contents exceeding c% Y commence to solidify as solid solution B and continue to solidify as B phase in the normal way for a solid solution.

This sequence assumes that equilibrium is maintained throughout. The reaction by which B phase is formed from A phase requires diffusion to occur in the solid state. Atoms of Y have to diffuse into the solid A phase to change its composition from a% Y to b% Y. This is just as difficult to achieve in practice as the diffusion required to even out the segregation in the dendrites that grow in simple solid solutions (p. 256). Consequently, there is a tendency for the peritectic reaction to cease once the A phase has been well covered by a rim of B phase. That is all that the liquid needs for things to seem to be right for solidification to proceed to produce more B phase, is leaving some A phase remaining. An example is given in Fig. 8.16. The result is that alloys solidified under practical conditions tend to contain B phase at a lower alloy content than the phase diagram would predict, and that richer alloys tend to contain more B phase than expected. Added to this is the segregation effect discussed on p. 256, which also encourages the early presence of liquids of a composition that could produce B phase.

Note also that nuclei do not have to form in a liquid before the crystals of B phase can grow when B phase is already present as a result of the peritectic reaction. The new material simply deposits on the old. Consequently, the second phase forms as massive areas of a single phase that fit in between the dendrites of the primary phase, as in Fig. 7.6(c).

GASES IN METALS

We have so far considered only the interactions between solids and liquids. Under most practical melting conditions, metals come in contact with a range of gases some of which are absorbed into solution in the melt. Those with which we shall now be concerned are gases for which the solubility in solid metal is much less than it is in liquid metal. Any gas held in solution in the melt

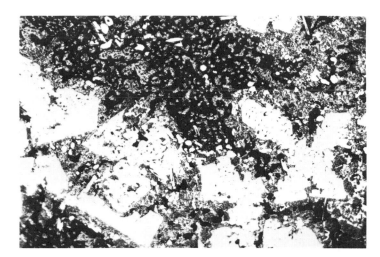

The material is a lead-base alloy containing 40% tin and 12% antimony. Indications can be seen of the original outlines of the cubes of a tin-antimony compound which had formed directly from the melt early in the solidification sequence. Much of the rims of these cubes subsequently reacted with the melt, but cooling continued before the whole cubes were consumed by the peritectic reaction. Optical micrograph; 250×.

Fig. 8.16. An example of a microstructure resulting from an incomplete peritectic reaction.

which is in excess of the solubility limit in the solid must then be evolved during solidification. The gas which is of special concern here is hydrogen, because it dissolves in a number of common metals (e.g., copper, nickel, aluminum, magnesium, and iron). Oxygen, which dissolves in silver and some other precious metals, is also sometimes of concern.

The possible magnitude of the effect of a gas can be judged by using iron as an example. Iron can dissolve up to 8 cm^3 of hydrogen per 100 g of metal at atmospheric pressure when molten but only about 6 cm^3 per 100 g when solid. The remaining 2 cm^3 or so, which is about 15% of the volume of 100 g of iron, has to be evolved when the iron solidifies if the melt is saturated with hydrogen. Melts are rarely saturated, which reduces the magnitude of the problem, but can contain enough gas in solution to raise the question of what happens to this large volume of excess gas during solidification. Some of the gas may escape by diffusion to an external free surface, a process which is encouraged in directionally solidified and continuously cast ingots. It may also escape to internal surfaces, such as those associated with shrinkage cavities. In this event, the shrinkage cavities become enlarged, perhaps significantly so. Most often, however, the excess gas forms bubbles within the melt.

These bubbles have to be nucleated, as for any other phase, and the means by which the nuclei are provided is again not clear. It is known, however, that nucleation occurs at the solid/liquid interface and that it occurs easily. The subsequent growth behavior of a bubble depends on a number of factors. The bubble may float to a free surface and escape from the melt, thus causing no trouble. It may, however, be trapped among the crystals that are growing in the surrounding melt, in which event it ends up as a cavity in the solidified mass. The solid/liquid interface will be advancing while such a trapped bubble is growing, and so the final shape of the cavity that is formed depends on the relative rates of the advance of this front and of the growth of the bubble. Three possibilities are indicated schematically in Fig. 8.17. The cavities that form in a casting in this way are known as *gas cavities* or *gas porosity*. The particular type of cavity sketched in Fig. 8.17(b) frequently is called a *wormhole*. Examples of approximately spherical gas cavities are shown in Fig. 8.18, and wormholes in a weld deposit are shown in Fig. 8.10(b).*

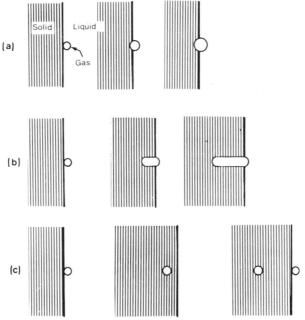

(a) A small number of large cavities form when the growth of the bubbles is relatively fast. (b) Elongated cavities, known colloquially as wormholes, form when the bubbles and solidification front have comparable growth rates. (c) A large number of small cavities form when the bubble growth is slow compared with the advance of the solidification front.

Fig. 8.17. Sketches illustrating how the relative rates of growth of gas bubbles and the advance of the solidification front affect the shape and size of gas cavities formed in metals during solidification.

*Sometimes cavities of this shape can be observed in blocks of ice. The cavities in ice are formed by the same mechanisms, because air is soluble in water but less soluble in ice.

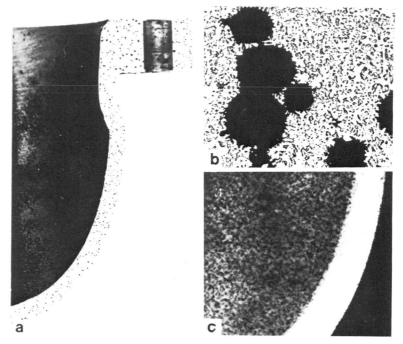

The distribution of the gas cavities (dark spots) in the aluminum alloy casting shown in section in (a) and in a radiograph in (c) is uniform because the cavities were nucleated uniformly throughout the melt. The detailed view of the cavities shown in (b) indicates that they are approximately spherical in shape. This is in contrast to the shrinkage cavities illustrated in Fig. 8.12(b). (a) Optical macrograph; 1×. (b) Optical micrograph; 25×. (c) Reproduction of radiograph; 1×.

Fig. 8.18. Examples of approximately spherical gas cavities in an aluminum alloy casting.

Hydrogen is all too readily available in many metal-melting environments because water is readily reduced to hydrogen by the metals in which hydrogen is soluble. The first precaution that is desirable to reduce the probability of the formation of gas cavities as a result of the absorption of hydrogen in the melt, therefore, is to ensure that all equipment and feed material are dried as thoroughly as possible. Some alloying elements help by inhibiting the absorption of hydrogen. For example, the presence of comparatively small amounts of oxygen greatly reduces the solubility of hydrogen in molten copper. Much of the copper used in electrical conductors is deliberately made to contain small amounts of oxygen, when it is known as *tough pitch copper*. There is some loss of conductivity, but fabrication is made much easier because gas cavities do not form in the ingots from which the conductor wires are drawn. Flux coverings over the melt may also reduce mechanically the ingress of hydrogen.

However, it frequently is not possible to be sure that some hydrogen has not been absorbed in a melt. The gas then needs to be removed from the melt immediately before it is cast. The dissolved gas can usually be made to diffuse out of a melt if the melt is held for a period in a vacuum or an inert atmosphere. Even large melts of steel are now commonly degassed in this way. The same result can sometimes be more easily achieved by bubbling an inert gas through the melt. Thus hydrogen can be removed from copper using nitrogen as the inert gas. It can be removed from aluminum by using chlorine or a mixture of chlorine and nitrogen; adding solid but volatile chloride salts to the melt has the same effect. The reason why chlorine is used with aluminum is that it is needed to flux away the film of aluminum oxide which soon coats the inner surfaces of the bubbles of flushing gas and which would prevent the diffusion of gas from the melt into the bubbles. The addition of zinc to copper in the alloys known as brasses keeps the melt free from hydrogen because zinc vapor is evolved continuously and has a scavenging action. The precautions that can be taken to prevent the development of excessive amounts of gas porosity in cast metals are thus numerous, and the most appropriate one to use varies with the alloy being cast. Foundry operators ignore these problems at the risk of producing flawed castings like the one illustrated in Fig. 8.18.

Chemical reactions between the constituents of a melt are another source of gas evolution when a gas is among the products of the reaction. This does happen in some alloys, an example being the reaction between carbon and iron oxide in steel. Both are likely to be present in a melt of steel, and one of the reaction products is the gas, carbon monoxide. The evolution of this gas can be used to advantage in ingot casting when properly controlled* but otherwise can result in very undesirable amounts of gas cavities. Control of the degree of oxidation of the melt consequently is a critical factor in steel casting technology.

Gas porosity is undesirable in castings because it reduces strength, but essentially only to the extent that the section area of the casting wall is reduced. It is generally less deleterious in this respect than shrinkage porosity (p. 267) because gas cavities, unlike shrinkage cavities, are not interconnected and so reduce the section area less. For the same reason, gas porosity is not likely to cause a casting to leak. The undesirability of gas cavities in ingots, on the other hand, depends almost entirely on their location. Cavities which are located well inside an ingot are of little consequence because they mostly weld up during subsequent working of the ingot and so disappear. Cavities located at or

---

*The concentration of carbon and oxygen can be arranged so that the reaction does not start until a thick shell of solid metal has formed around the ingot, as in Fig. 8.11(a). The gas cavities (called *blowholes* in this context) that then form compensate partly for shrinkage solidification and hence reduce the size of the pipe cavity. The surfaces of these buried cavities are not oxidized and so weld up during subsequent working of the ingot. Low-carbon steels of particularly good surface quality and with other characteristics that are adequate for many purposes can be produced economically in this way. They are called *rimmed steels.*

close to the surface can, however, be serious because they open during subsequent working, their surfaces oxidize, and they do not weld up. A rough surface results that perhaps is associated with surface laminations.

RAPIDLY SOLIDIFIED METALS

From the point of view of obtaining an improved microstructure, there are advantages in cooling a metal as rapidly as possible during solidification. The grain size is reduced, and so is the dendritic arm spacing, from which it follows that segregation is less severe. Moreover, the formation of undesirable intermediate phases may be reduced or even suppressed. Casting in a cooled metal mold (chill casting) goes some way along this path, but only a short way, because there are physical limits on the rate at which heat can be extracted from thick sections.

A major step can be taken, however, if an alloy is solidified in very thin sections, 20 to 30 μm (0.8 to 1.2 mils) thick or less. This can be done, for example, by atomizing a thin stream of liquid in a blast of inert gas, producing globular particles, or by forcing a stream against the rim of a rapidly rotating metal disk, producing a ribbon. Not only is a more desirable microstructure produced, but it may also be possible to produce alloys of more desirable compositions which could not be made by conventional methods. The problem is, however, to fabricate these thin particles or strips into the desired useful shape without destroying their special microstructural characteristics. This is at present difficult and expensive but can be done and shows promise of becoming economically competitive. For example, the fabrication route gives hope for developing aluminum alloys which are competitive with titanium alloys at higher temperatures than at present. Improved nickel-base high-temperature alloys are also a possibility.

The application of these principles of rapid solidification can be taken one considerable step farther with some special alloys that are not crystalline at all but are amorphous in the sense that the long-range ordering of atoms that is characteristic of the crystalline state (p. 222) is not present. The atoms have only short-range order similar to that of liquids (p. 250) and the transparent solids that we know as glasses. Metallic alloys in this state are known as *glassy metals* because of the structural analogy, although they are certainly not transparent.

Glassy metals have a number of unusual properties. Some, for example, have magnetic properties which suggest that they have potential for application in magnetic devices such as core laminations for electrical transformers, recording heads for tape recorders, and magnetic shields. Many have outstandingly good corrosion and wear resistance. They all behave completely elastically under stress right up to fracture, having a very high yield strength but a low Young's modulus. All of these properties appear to be related to the atomic and electronic structure of the glassy state. The usefulness of these glassy metals is again limited in many applications by the difficulties in obtaining them in a usable shape. It is also limited to date by the range of alloys that can be obtained in the glassy state. All of them are rather unusual alloys, including, for example, compounds of iron and boron, alloys of zirconium and nickel, and alloys of calcium and magnesium.

A LITTLE ABOUT PHASE DIAGRAMS

Without tackling the subject directly, we have by now discussed most of the basic phenomena which govern the relationships between liquid and solid metal phases, at least so far as alloys of two metals are concerned. We have discussed the solidification of pure metals (p. 250), alloys which solidify as solid solutions (p. 255), alloys which solidify as eutectics (p. 270), and alloys the solidification of which involves peritectic reactions. We have also used diagrams to define the ranges of temperatures and compositions over which the various phase changes occur. These are the diagrams that are called *phase diagrams*, or *constitutional diagrams*, or *equilibrium diagrams*.

The various phenomena have been discussed in isolation, and there are alloy systems in which just one of them is involved. But more frequently several are involved, and the phase diagram builds up to one which at first glance appears to be complex. You should not be put off by this, however, because the complexities are more apparent than real. Even the most complex phase diagram can be broken up into a number of simple units, each one of which we have considered here. For example, the phase diagram for the copper-zinc system shown in Fig. 7.6 (p. 232), with its many lines, comprises only five units of peritectic reactions in succession with solid-solution units in between. The portion of the copper-phosphorus system shown in Fig. 7.5 comprises a solid-solution unit and a eutectic unit. They can be comprehended easily enough if they are looked at one unit at a time in this way. These remarks apply, of course, only to the uppermost portions of phase diagrams which describe the relationships between liquids and solids. The solid phases may become unstable at lower temperatures, decomposing into one or more phases with different compositions and different crystal structures. These phenomena will be discussed in Chapter 11.

Although phase diagrams are invariably determined experimentally, a good understanding has been developed of the theoretical principles which underlie them. But these principles need not concern us. Moreover, it should always be kept in mind that a typical diagram attempts, as nearly as is practicable, to describe equilibrium relationships. This is a state that is rarely reached in practice, particularly at low temperatures. Indeed, many of the techniques used in physical metallurgy to manipulate the structures of metals are based on arranging things so that this state is not reached and so that specific nonequilibrium structures are obtained.

FURTHER READING

B. Chalmers, *Principles of Solidification*, Wiley, New York, 1964.

Metals Handbook, 8th Ed., Vol 5, *Forging and Casting*, American Society for Metals, Metals Park, OH, 1970.

P. R. Beeley, *Foundry Technology*, Wiley, New York, 1972.

The Making, Shaping and Treating of Steel, 10th Ed., edited by W. T. Langford, N. L. Samways, R. F. Craven, and H. E. McGannon, Association of Iron and Steel Engineers, Pittsburgh, 1985.

Binary Alloy Phase Diagrams, edited by T. B. Massalski, J. L. Murray, L. H. Bennet, and H. Baker, American Society for Metals, Metals Park, OH, 1986.

· 9 ·

Deformation and Annealing

SUMMARY

Metals subjected to increasing stress deform first in a reversible, or elastic, manner (Chapter 4). The atoms of the metal are pulled apart from one another or forced closer together depending on the direction of the stress, and so the resistance to elastic deformation is determined by the manner in which the bonding forces between the atoms change with atom spacing. Thus elastic characteristics are an immutable property of the metal concerned. Fracture strength would be very high if deformation occurred only in this way, and it would not be possible to shape a metal. However, another mode of deformation — a plastic mode during which a permanent change in shape occurs — soon intervenes. Plastic deformation occurs by blocks of crystal shearing, or slipping, past one another due to the cooperative movement on particular crystal planes of the crystal defects known as dislocations. Typically, it is comparatively easy for dislocations to move in metals but difficult for them to do so in most other types of crystalline solids. In any event, the stress required to cause a permanent change in shape, or plastic yielding, at any stage is the stress that is required to move a number of dislocations. Features which impede the movement of dislocations, and grain boundaries are a prime example, thus increase this stress. Hence yield strength increases with a decrease in grain size. Dislocation movement is also impeded when dislocations have to move on intersecting crystal planes, as they have to do at higher strains. Thus, yield strength increases with deformation strain, which is the basic cause of the phenomenon called strain hardening. Strain hardening is particularly rapid in polycrystalline materials because complex slip processes become necessary to keep the grains in contact with one another as they are strained. Many other structural features can affect dislocation movement. Thus yield strength and strain hardening, unlike elastic properties, are open to metallurgical manipulation. However, full explanations of these phenomena are difficult to arrive at, particularly after larger strains when the situation becomes very complex and deformation occurs inhomogeneously. The ease with which dislocations can move in metals is, nevertheless, undoubtedly the essential reason why metals have those important characteristics of toughness and ductility, even if at the same time theoretical strength is not attained.

The effects of deformation can be removed by heating for a period of time at a sufficiently high temperature. New strain-free grains nucleate and grow to consume the old strained grains. The size of the new grains depends principally on the amount of prior deformation and the temperature and time of heating, but there is an opportunity to reduce the grain size. An important side effect, however, is that the recrystallized grains may not be randomly oriented, the type and degree of any preferred orientation depending on the alloy and the deformation and annealing schedules. Strain hardening and recrystallization occur virtually simultaneously when the deformation is carried out at a high enough temperature, in which event there is no net hardening. This is the objective during hot working operations.

Deformation at low strain rates and high temperature, such as during creep, also occurs mainly by fine slip, but in addition the grains slide past one another on their boundaries. It is this sliding that causes fractures to initiate along the boundaries. Deformation by cyclical stresses, such as during fatigue, occurs by complex mechanisms involving the concentration of slip in bands. It is known that the fatigue cracks initiate and grow along these bands, but their cause is obscure.

We introduced in Chapter 3 a number of concepts characterizing the mechanical behavior of metals, concepts which included elastic deformation, plastic deformation, strain hardening, and recrystallization. We shall now discuss these phenomena in a little more detail, and develop some understanding of the mechanisms by which they occur and of their significance in practice.

ELASTIC DEFORMATION

We saw in Chapter 3 (p. 63) that any piece of metal changes its shape a little when it is subjected to a stress, even quite a small one. The metal extends a little under tensile stresses, shortens a little under compressive stresses, and distorts a little under shear stresses. Moreover, at least up to a certain point, it returns to its original shape when the stress is removed. This is called elastic behavior. We shall now show that elastic behavior is a direct manifestation of the forces that hold the atoms together in a metal crystal.

We described in Chapter 7 the manner in which the atoms of a crystal are arranged in a regular pattern in three dimensions, and how they are bound together by the forces of the free electrons that drift among them. These bonding forces are directed along lines between the atoms, such as the lines drawn in the sketches in Fig. 7.2 (p. 223) and 9.1(a). As an aid to understanding, imagine that small tension springs are located along these lines. The magnitude of the bonding forces (the stiffness of the imaginary springs) can be calculated by atomic physicists and they have found that it varies with the distance of separation between the atoms in the general manner sketched in Fig. 9.1(b). The spacing of the atoms changes when a crystal is strained elastically, as indicated in Fig. 9.1(a), and it is easy to establish experimentally not only that this is so but also that the change in spacing corresponds exactly with the elastic strain measured in a bulk specimen.

Let us take, as the simplest example, a crystal which has a basic cubic lattice

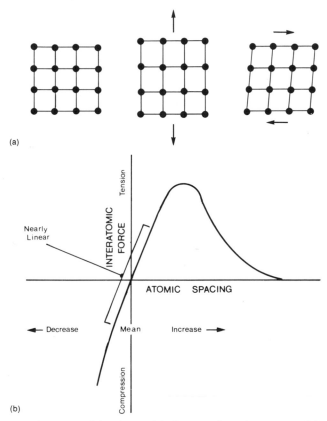

(a)

(b)

The series of sketches in (a) illustrates how the spacing of the atoms in a lattice is changed by tensile strains (center) and shear strains (right). The diagram in (b) illustrates the manner in which the bonding force between neighboring atoms changes with variations in the spacing of the atoms from that at equilibrium.

Fig. 9.1. Diagrammatic sketches illustrating the manner in which elastic deformation is controlled by the bonding forces between the atoms in a crystal lattice.

and which is stressed in tension along one of the cube directions (Fig. 9.1a). The atoms separate along the cube direction when a small stress is applied, and this increases the bonding forces between the atoms in the manner described in Fig. 9.1(b). Consequently, the stress has to be increased progressively to separate the atoms progressively. On this argument, the slope of the nearly linear portion of the curve in Fig. 9.1(b) should have the same value as Young's modulus measured on bulk specimens, and it does. Moreover, the atoms should be pulled back to their equilibrium portions by the atomic bonding forces when the external force is removed, and they are. Explanations of all elastic phenomena can be based on these concepts. For example, the variations in elastic behavior with crystal direction in single crystals (p. 249) are due to the directionality of the atomic bonding forces, and elastic properties are structure-insensitive (p. 64) because they are determined only by the properties of the atoms themselves.

The form of the curve in Fig. 9.1(b) also indicates that there is a theoretical limit on the strain that can be obtained by deformation in an elastic mode. In compression, the stress that has to be applied to achieve an additional decrease in the spacing of the atoms begins to increase very rapidly beyond a certain point. To use the hard-ball analogy of atoms, it becomes next to impossible to force the atoms any closer together once a certain spacing has been reached. In tension, on the other hand, a stage is reached, and this is after only 5 to 10% strain, where the force required to separate the atoms further begins to decrease. Any further increase in stress would cause the atoms to separate to infinity — that is, would cause the crystal to break apart. The forces required to do this can be calculated approximately on quite simple arguments, and they turn out to be very large — at least ten times, and more often a hundred times, the fracture stresses that are observed in practice. Moreover, the elastic strains of 5 to 10% that would be needed to cause fracture in this way cannot be achieved in bulk material in practice because the phenomenon of plastic deformation intervenes. This usually occurs at strains of only 1 to 2%. The obvious implication is that the intervention of plastic deformation is the factor which causes metals to be weaker than theoretical considerations would predict.

Remember, however, that it is this very ability of metals to deform plastically that makes them tough, and that toughness is more important in structural materials than straight-out strength (p. 84). We might be able to make metals much stronger, to approach more closely the theoretical strength, by inhibiting the onset of plastic deformation, but a point would soon be reached where they would become useless as structural materials. The ability of metals to deform plastically is also the characteristic that makes it possible to change their shape relatively easily so as to form them into useful components. It would in many applications be of no avail to have a superstrong metal which could not be shaped into a useful component.

PLASTIC DEFORMATION

One possible means by which the external shape of a crystal could be changed substantially would be for blocks of the crystal to move bodily past one another. The manner in which the shape of a deck of cards can be distorted by sliding the cards over one another is a rough analogy. This general type of shearing does occur in metal crystals, as can be established by viewing a polished external surface of a crystal after it has been plastically deformed. Systems of parallel lines that usually are reasonably straight can then be seen on the surface, the number of lines increasing with increasing plastic strain (Fig. 9.2a and b). Each of these lines is the site of a step in the surface at which one block of a crystal has moved bodily past its neighboring block (Fig. 9.2c). The simplest of these block-shearing processes is known as *slip*.

The basic characteristics of slip processes are best understood by studying first the behavior of single crystals (p. 249) of either pure metals or simple solid solutions. The additional complexities that arise due to the many grains and other phases that are present in engineering alloys can then be based on this foundation.

Plastic Deformation of Single Crystals

The most important characteristics of the basic slip process are, in summary, as follows:

1. Slip always occurs on a specific crystal plane and in a specific direction on that plane, the *slip system* being a characteristic primarily of the crystal system to which the metal belongs. Slip does not occur, however, on all possible crystal planes of the preferred type, but only on a selected few which are widely spaced by atomic standards.

2. In close-packed crystals (p. 224), the *slip plane* is always one of the close-packed planes and the *slip direction* one of the close-packed directions on that plane. In other crystal systems, the slip plane and slip direction usually are the ones which most nearly approach being close packed. A consequence is that there are many slip systems (12 in all) available in face-centered cubic crystals, which have four close-packed planes and three close-packed directions in each of these planes. Close-packed hexagonal crystals, on the other hand, have only one close-packed plane which has three close-packed directions (p. 225), and so this type of crystal has only three slip systems. Three nearly close-packed planes may serve as slip planes in body-centered cubic metals, depending on the conditions, with a close-packed direction serving as the slip direction in all three. These crystals are consequently an intermediate case, and typically have five slip systems.

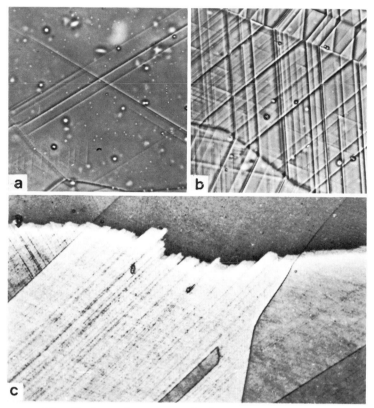

The side faces of test pieces of polycrystalline Cu-30Zn brass were polished, and the test pieces were then compressed to plastic strains of 0.03% (a) and 0.85% (b). The systems of lines discernible in these two micrographs are steps in the surface formed where blocks of crystal have sheared past neighboring blocks. These lines are known as slip lines. Note that the number of slip lines increases markedly with an increase in strain. The fact that the lines in (a) and (b) are steps can be confirmed by sectioning the surfaces, as shown in (c). Note that the height of each step is substantial on an atomic scale. This section has been etched in a reagent which reveals planes on which dislocation movement has occurred by developing rows of etch pits along the traces of the planes. The slip plane steps developed at points where the planes on which the dislocations moved emerged at the surface. Optical micrographs: (a) and (b) 250×; (c) 500× (horizontal) and 5000× (vertical).

Fig. 9.2. Micrographs illustrating plastic deformation that has occurred by slabs of crystal sliding past one another.

3. Macroscopically observable shear starts to occur on a slip system when the external stress, resolved as a shear stress (p. 79) on the slip plane and in the slip direction, exceeds a certain value. This value is called the *critical shear stress*. Of the available slip systems, shear occurs first on the one for which the resolved shear stress is highest. A consequence is that the yield stress is highly dependent on crystal orientation.

4. The value of the critical shear stress is, at least to a first approximation, a characteristic of the metal. However, the values are somewhat similar for all metals with close-packed crystal structures and somewhat similar for all metals with non-close-packed structures, the latter group having the lower values of the two. The critical shear stress is always very small compared with the yield stress of a polycrystal of the same material.

5. Only a limited amount of slip occurs on any one system, which then stops operating, often for good. Further strain is achieved by slip starting on another plane (cf. Fig. 9.2a and b). The new slip occurs first on a series of parallel planes but eventually on an intersecting plane which contains the next most favorably oriented slip system.

6. The orientation of the crystal changes progressively with increasing strain, the most favorably oriented slip plane and direction tending to rotate to become parallel to the tensile-stress axis. This is a simple geometric consequence of the shearing of a block that is restrained, an effect which is illustrated schematically in Fig. 9.3.

7. The critical shear stress increases with increasing plastic strain, and the rate of this *strain hardening* increases markedly when slip starts on an intersecting system. *Multiple slip* of this nature commences early in the deformation process in face-centered cubic metals. These crystals have many suitably oriented slip systems, and one or more of them soon rotate into an orientation for which the resolved shear stress is close to that for the primary system. It occurs later in crystals which have few available slip systems, and so strain hardening is less rapid than for face-centered cubic crystals.

The elucidation of these principles enabled much of the mechanics of deformation to be explained, and this constituted a considerable advance. However, it did not enable an insight to be developed into the mechanisms by which the slip-shearing process occurred. The problem in the days when crystals were thought of as being perfect was that it was difficult to see how complete planes of atoms could be forced to slide past one another at small shear stresses. Remember that the atoms have to be regarded as being fairly rigid spheres and that slip typically occurs on a close-packed plane, such as that illustrated in section in Fig. 7.3(b) (p. 224). The atoms would, in a perfect-crystal model, have to ride over one another for slip to occur, and this can be imagined to be a difficult process at best. The shear stresses required to achieve this bumpy ride can in fact be calculated, and turn out to be at least a hundred times, and sometimes a thousand times, the value of the shear stress actually required to cause slip. An additional problem is that there seems to be no particular reason why slip should not occur on all atomic planes with roughly equal probability, as for a pack of cards, instead of being confined to a few comparatively widely spaced planes. Slip on any plane might also be expected to run away catastrophically once it has started.

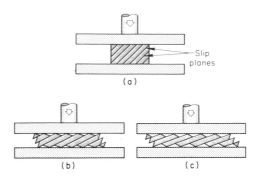

A single crystal is imagined to be compressed while being constrained by two parallel platens, slip occurring on a single system of planes. The slip planes have to rotate, tending to become normal to the compression direction, when the crystal is compressed in one direction and elongates in a perpendicular direction. Source: *Metals Handbook*, 8th Ed., Vol 8, *Metallography, Structures and Phase Diagrams*, American Society for Metals, Metals Park, OH, 1973.

Fig. 9.3. Schematic diagram illustrating how the orientation of the slip planes in a crystal has to change when the crystal is deformed plastically and when there is external constraint.

This dilemma was resolved by the brilliant suggestion made by G. I. Taylor that real crystals contain the line defects known as dislocations, which were discussed in Chapter 7 (see p. 236). The brilliance of Taylor's suggestion lay in the realization that minor cooperative adjustments of the atoms surrounding a dislocation would be necessary to cause the position of the dislocation to move, and that these adjustments should be comparatively easy to achieve. This is illustrated diagrammatically for an edge dislocation in Fig. 9.4. The movement of atom A_1 to position A_2 and of atom B_1 to position B_2, etc., would cause the dislocation in this sketch to move from the base of plane A to the base of plane B — that is, one atom spacing to the right. A component of shear stress in the direction of the arrows is required to do this. But note that this movement can occur easily only in the plane in which the dislocation line lies and only in a direction at right angles to the dislocation line. The latter direction is the one in which the slip shear occurs. Note also that the dislocation line cannot easily be moved onto an adjoining slip plane. Atoms would have to be added to or subtracted from the base of the half-plane of atoms A, a point to which we shall return.

Consider now a screw dislocation, such as the one sketched in Fig. 7.8(b) (p. 237). An extra half-plane of atoms is not present in this case, and so a different type of rearrangement of atoms is needed to cause a dislocation line to move. Imagine that the atoms labeled a and b in Fig. 7.8(b) move up a little, that those labeled c and d move down a little, and that appropriate coordinated readjustments are made to adjoining atoms. The dislocation line would thereby move in the direction of the smaller arrow. A component of shear stress in the direction

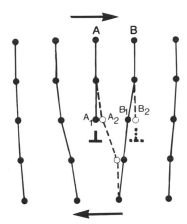

Movement of the atoms at positions A_1 and B_1 to positions A_2 and B_2 would cause the dislocation ($\perp$) to move from the base of plane A to the base of plane B ($\perp$).

Fig. 9.4. Sketch illustrating diagrammatically that comparatively small cooperative movements of the atoms around a dislocation cause the dislocation effectively to move in the crystal lattice.

of the large arrows would be required to achieve this. Note that this time the dislocation line moves in a direction perpendicular to the dislocation line and the shear stress, but that the slip shear occurs in a direction parallel to the dislocation line. To summarize, the line of an edge dislocation has to be perpendicular to a component of shear stress for it to be moved, but the line of a screw dislocation has to be parallel to the shear stress. In both cases, however, the resultant slip shear occurs in a direction parallel to the shear stress. Also, note that a screw dislocation can move with comparative ease in any direction perpendicular to the dislocation line. In Fig. 7.8(b), for example, imagine the atoms readjusting their vertical positions cooperatively in some direction other than a-b. Thus, screw dislocations can move easily onto adjoining crystal planes — which edge dislocations cannot do.

The dislocations in real crystals typically are composed of alternate segments of edge and screw dislocations, with the dislocation lines at right angles (p. 238). A mixture of this type can move through a crystal under the influence of the same shear stress because the differences just mentioned complement one another and enable the two components to move in perpendicular directions but still produce a slip shear in the same direction. Thus a mixed dislocation line can sweep through a crystal until it reaches free surfaces, producing an external step of one atom spacing (Fig. 9.5). The dislocation itself is eliminated. Dislocations which have the characteristics that enable them to move easily in this way are called *glide dislocations*.

These simple considerations of the characteristics of dislocations allow qualitative explanations of many of the basic characteristics of deformation by

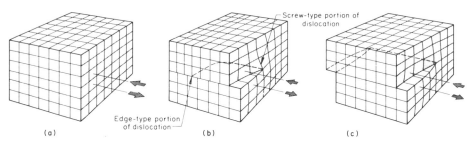

The dislocation line sketched in (a) can move, as in (b), under the influence of a shear stress acting in the direction of the large arrows. The edge components can move parallel with this direction and the screw components in a perpendicular direction. The dislocation line eventually emerges at the free surfaces of the crystal to produce a shear step of one atom spacing (c). Source: *Metals Handbook*, 8th Ed., Vol 8, *Metallography, Structures and Phase Diagrams*, American Society for Metals, Metals Park, OH, 1973.

Fig. 9.5. Schematic representation of the manner in which a crystal can be sheared by a dislocation which consists of a mixture of edge and screw dislocations.

slip, such as: the reason why the slip phenomenon occurs at all (only small stresses are required to move suitable dislocations); the reason why it occurs only on certain types of crystal planes (glide dislocations have to lie on these planes); and the reason why it occurs only on a few of the many available crystal planes (it occurs only on those particular slip-type planes on which glide dislocation lines actually lie). But deeper considerations of the properties of dislocations are required to explain other basic characteristics of slip, particularly to explain them quantitatively. We need to be able to predict the magnitude of the stress required to initiate slip, to explain why this stress increases with strain, and to explain the extent to which it so increases — that is, to explain strain hardening.

However, before dealing with these matters we need to note that the movement of the dislocations that were originally present in a previously undeformed crystal is never enough to produce detectable slip. Taking one example, the movement of about a million dislocations would have been required to produce each of the slip steps visible in Fig. 9.2. Yet the crystals would have contained only about this number of dislocation lines threading through each square centimeter of section before deformation started. Many many slip shears of similar magnitude occurred in this volume of material, and so clearly many many extra dislocations had to be generated during the slip process itself to produce even this small strain. The ingenuity of our colleagues the solid-state physicists did not fail us here either, and they proposed mechanisms by which a single dislocation might multiply and generate large numbers of new dislocations. These proposals by no means turned out to be figments of their imaginations, but have been well supported by experimental evidence. In fact, it soon becomes apparent that plastic deformation in the engineering sense cannot be detected until the sources generate dislocations at a higher-than-critical rate.

We have already seen that shear stress, which is actually called the *lattice friction stress*, has to be applied to a dislocation in order to achieve the atomic rearrangements that are necessary to get it moving and keep it moving. As might be expected, the force of this stress depends firstly on the extent to which the atoms around the dislocations have to be readjusted to cause movement of the dislocation line. This in turn depends on the distances by which the atoms around the dislocation are displaced from their nominal positions. As is almost intuitively obvious, the force required to move the atoms to the new positions is small if the atoms are already displaced a long way from their nominal positions, and this is generally the case for metals. The force required is, by the same token, smaller in some metal crystals than in others, and in some planes and directions in a given crystal than in others, because the initial displacements were larger. Hence the existence of planes and directions of easy glide. On the other hand, very large shear forces can be expected to be needed to move a dislocation when the atoms around it are located close to their nominal positions. This is the situation for many nonmetals. Dislocations are present in these materials, but it is well nigh impossible to move them. Consequently they do not yield easily under an externally applied force, which is why they are described as being hard and as being brittle.

We have just seen, however, that merely getting the dislocation to move is not enough to cause yielding in an engineering sense. The engineering yield strength actually corresponds to the stress required to get dislocation sources generating dislocations at the critical rate required for detectable slip. This has to be taken into account when predicting yield strength from dislocation theory. A further complication is that the atoms in a crystal lattice are not frozen at precise positions, as the model developed to this point indicates, but are vibrating about a mean position, the amplitude of vibration increasing with temperature. This increases the probability that atoms will at some instances of time be located at positions better suited to easy dislocation movement. The yield strength is thereby reduced. Movement of dislocations on additional slip systems may even become possible. All of this applies to nonmetallic materials as well; many of them do indeed become ductile at sufficiently high temperatures.

Strain hardening is the other important phenomenon that has to be explained. This is simple enough to do in principle but is difficult to do quantitatively and in detail. Many factors are involved. One factor is that the force fields around dislocations interact when the dislocations pass one another on closely adjoining crystal planes, so impeding relative movement. More important, however, are the effects that occur when dislocation lines moving on different slip planes intersect one another. Remember that strain hardening becomes rapid only when slip starts to occur on two or more intersecting planes. In general, a step or jog forms in one dislocation line when it passes through another one, and the formation of the jog in itself requires energy. Moreover, the jog does not have the characteristics required for it to move easily in the same direction as the dislocation and so has to be dragged along. This requires a larger stress. Other even more seriously complicating structures can be introduced into the dislocation during an intersection, including some which prevent further movement altogether. The immovable ones are called *sessile dislocations*. Segments of sessile dislocations may lock together a complex tangled forest of dislocations as more and more move through the crystal as the strain increases. Larger and larger stresses are then required to maintain the dislocation's movement that is necessary to produce larger strains.

There are, however, mechanisms by which succeeding dislocations can bypass an obstacle, such as a tangle of other dislocations, which would alleviate these problems. Screw dislocations can actually do this with ease because they can move easily perpendicular to their length and so skip around an obstacle. Edge dislocations can do so only with difficulty by special mechanisms. One of these is called *cross slip*. The part of the dislocation line which encounters the obstacle moves laterally on one of its alternative slip systems until it is clear of the obstacle (Fig. 9.6). The entire line can then move forward again. It is the connecting slip that is called cross slip. The dislocation line now contains a rectangular corrugation and may soon develop many such corrugations as it moves. For reasons which we need not go into, cross slip occurs in metals which have face-centered cubic crystal structures, and it then occurs most easily in

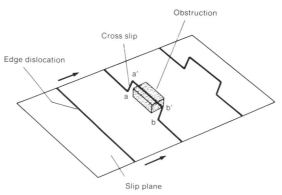

Segment ab of the dislocation line cross slips on an alternative slip plane (aa', bb') to a'b' when the line encounters the obstruction. The dislocation line is then able to continue to move past the obstruction, but thereafter contains a rectangular corrugation.

Fig. 9.6. Sketch illustrating the manner in which an edge dislocation may move past an obstruction by cross slip.

those metals in which stacking faults can form most readily and which consequently are said to have a low stacking-fault energy. These are the metals which strain harden the least rapidly.

Edge dislocations can also move onto an adjoining slip plane by a process known as *climb*. Suppose that the row of atoms at the base of plane A of the edge dislocation sketched in Fig. 9.4 were removed. The dislocation would then move up, or climb, one atom spacing. It would move down one atom spacing if a row were added. This movement is in a direction perpendicular to that which would ordinarily occur. The realization of the process, however, requires that energy be provided to move the atoms, and this usually has to be in the form of thermal energy. The requirement is, in effect, one of self-diffusion (p. 235), which in most metals occurs only at relatively high temperatures. Thus dislocation climb is not usually important during deformation at room temperature but can be important at high temperatures in processes such as creep (p. 116).

All in all, there is little wonder that at higher strains, crystals seek methods of deformation other than simple slip. Most start to do this at engineering strains of about 10%. Deformation then concentrates progressively in localized microscopic bands of material which can be regarded as being plastic instabilities. Several different types of these strain inhomogeneities are known to be generated, the particular one which operates depending on the metal and the magnitude of the strain. In general, the inhomogeneities at intermediate strains are related to crystal orientation and are probably a complex manifestation of the slip process. At high strains, however, the alignment of the strain inhomogeneities is dictated for the most part by the direction of the principal applied stress, and shear mechanisms other than slip probably are involved. Slabs of crystal several micrometres wide shear past one another, the shear being confined to a band, a few tenths of a micrometre wide, that is aligned on the plane of principal shear stress. This band is called a *shear band*. The nature and properties of these strain inhomogeneities are only beginning to be understood.

In summary, we can say that we now have a reasonably good understanding of the mechanisms by which elastic and plastic deformations occur in single crystals at small strains. These are the strains that are involved when the stresses that are applied to engineering structures are considered. We are, however, only beginning to develop an understanding of the more complex phenomena that occur during deformation at the larger strains involved in operations such as those performed in shaping and forming of metals.

Plastic Deformation of Polycrystals

The metals used in industry are composed of an aggregate of small crystals, called grains, which are everywhere in intimate contact with one another and which in the general case are randomly oriented (p. 240). The atomic structure of each grain is the same as that of a single crystal, defects and all. The basic mechanism by which plastic deformation occurs in each grain consequently also is the same in principle, but quantitatively there are significant differences. In particular, the stresses required to cause the various deformation mechanisms to occur are much higher in polycrystals than in single crystals. This is because neighboring grains have a considerable influence on the deformation of one another.

In the simplest analysis, the yield strength of a polycrystal of random orientation might be expected to be the average of the yield strengths of single crystals of all possible orientations. But polycrystals are much much stronger than that. Moreover, the yield strength (σ) varies with the grain size, increasing with decreasing grain size in a manner described by the *Hall-Petch relationship*:

$$\sigma = \sigma_0 + kd^{-\frac{1}{2}}$$

where σ_0 is the average yield strength of a large single crystal, k is a constant characteristic of the material, and d is the average diameter of the grains. There is much doubt about the validity of the details of this relationship (e.g., whether k is really a constant, whether the power function of d is always $-\frac{1}{2}$, and even the physical significance of d), but it is nevertheless widely used as a general indicator of the effect of grain size on yield strength. Moreover, the first term is so small in comparison with the second that it can usually be ignored. The effect is a significant one: the yield strengths of ferritic steels can be doubled over the range of grain sizes that can be achieved in practice.

Grains have this influence because a dislocation cannot simply move through a grain boundary when it encounters one and then continue on its merry way unimpeded into an adjoining grain which has a different orientation. The first dislocation that tries to do so is stopped by the grain boundary, which you will remember has a relatively disorganized structure (p. 241), and the dislocations that follow *pile up* behind it (Fig. 9.7). Soon a logjam develops and the resulting back pressure causes all dislocation movement on that particular plane to cease — that is, slip on that plane stops. Consequently, the dislocation sources that are by now operating within the grains have to become more active than ever if slip is to continue, and this requires a larger stress. It follows that a higher stress is required for yielding to be detected in a bulk specimen. The smaller the grain size, the smaller the distance that a dislocation can move before a boundary is encountered, and so the more marked the effect, as described by the Hall-Petch relationship. Things get a lot more complicated at higher strains, but this simple explanation is adequate at strains up to the yield point.

In addition, the various grains of a polycrystal have to remain in contact during deformation and, for a reason that we shall soon discuss, this forces slip to occur on intersecting slip systems at much smaller strains than for single

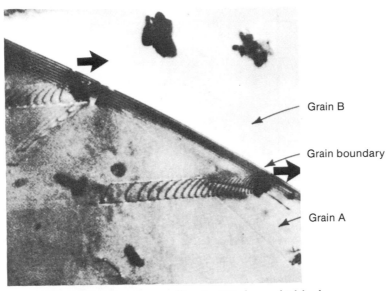

The specimen is a polycrystalline Cu-30Zn brass which has been plastically deformed only a small amount. Dislocations had been moving in grain A in the direction of the large arrows but were stopped by and piled up against a grain boundary. They have not been able to enter grain B although their stress field might cause new dislocations to be initiated in this grain. Transmission electron micrograph; 200,000×. Source: G. Dupouy and F. Perrier, *J. de Microscopie*, 1962, *1*, 167.

Fig. 9.7. An illustration of a situation where movement of dislocations in a polycrystal has been impeded by the presence of a grain boundary.

crystals. Multiple slip actually starts to occur almost as soon as yielding can be detected (Fig. 9.2). Locking of dislocations at intersections consequently occurs at smaller strains and with greater frequency than for single crystals. This makes it more difficult for slip to occur and increases the rate of strain hardening.

Each grain in a polycrystalline aggregate is strained in nearly the same manner and to nearly the same extent as the workpiece as a whole (Fig. 9.8), yet the grains remain fully in contact with one another during deformation. A little thought will indicate that this is no mean feat considering the complicated shape of the grains in three dimensions (p. 242). Simple geometrical analysis of the situation suggests that it is a feat which could be achieved by slip processes only if slip occurred on at least five systems simultaneously. This is likely enough in face-centered cubic crystals, which have 12 slip systems available (p. 287), but it is the reason why slip is forced to occur on a number of intersecting systems even at very small strains with the consequences that we have already mentioned. Metals which have this type of crystal structure are indeed the ones which are easiest to deform and which can deform plastically right up to frac-

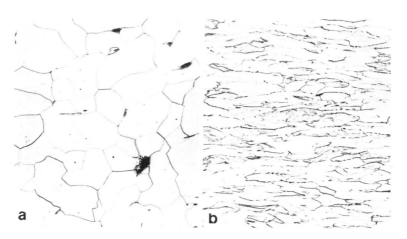

These micrographs show longitudinal sections of a low-carbon steel. The grains in the material are equiaxed before rolling (a) but elongated after cold rolling (b). The sheet has been reduced in thickness by about 50%, and so has each grain. Optical micrographs; 250×.

Fig. 9.8. An illustration of how the shapes of the individual grains in a polycrystalline metal are changed during plastic deformation in the same manner as the specimen in bulk, and yet the grains remain in intimate contact.

ture (e.g., aluminum and copper). The requirements for multiple slip cannot be quite so easily met, however, in metals with body-centered cubic crystal structures (e.g., iron), because they generally have only five slip systems available. Consequently, metals of this type are less easy to deform. These requirements are impossible to meet in metals with hexagonal crystal structures, which have only three slip systems available. The assistance of another deformation mechanism is then needed, the main one being the formation of comparatively large mechanical twins (p. 239). The twins provide regions which have different orientations from that of the parent crystal and in which additional slip systems can operate. Nevertheless, this is not a particularly good solution to the problem, and metals with these structures (e.g., zinc) are considerably more difficult to deform than metals with either face-centered cubic or body-centered cubic crystal structures. They are also prone to fracture by cleavage as a result (p. 70).

Actually, mechanisms other than simple slip assist in achieving the complicated grain-shape change in all metals, including those which have many slip systems available. Neighboring grains butt and jostle against one another, and some planes of the crystals literally bend and buckle (Fig. 9.9), producing regions which differ significantly in orientation from the main crystal and in which differently oriented slip systems can operate. This occurs in practice at quite small strains, and the more so the more difficult it is for multiple slip to

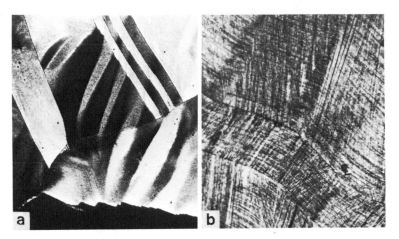

Individual grains have to deform in a complex way to maintain contact during deformations. It is difficult for them to do this by simple slip processes alone, but one way in which slip can be helped is for the slip planes to buckle as the grains jostle against one another. Regions which differ slightly in orientation from the main grain are produced. The micrographs shown here are of sections of a Cu-30Zn brass which was deformed 15% in compression. The section in (a) has been etched to distinguish the regions at the edges of which kinking of the slip planes has occurred. The section in (b) has been etched in a reagent which reveals traces of the planes on which slip has occurred. The nearly horizontal set of these planes has kinked in a coordinated manner at the edges of bands much as those shown in (a). Optical micrographs: (a) 100×; (b) 500×.

Fig. 9.9. Examples of buckling and kinking of slip planes within grains which have occurred during plastic deformation of a polycrystalline metal.

occur. In addition, extra slip systems which would not ordinarily be active often are forced to operate in regions immediately adjacent to grain boundaries by the pressure of piled-up dislocations in the adjoining grain.

However, at large strains, if they can be achieved, the factors that we have just been discussing come to be of diminishing importance. The orientation of each grain changes during deformation in such a way that all grains tend to have the same orientation, as we shall soon discuss. Consequently, the importance of differences between adjoining grains, and hence of grain size, diminishes. Moreover, deformation progressively tends to involve the formation of strain inhomogeneities (p. 295) which are controlled primarily by the stress system and which are not influenced by the individuality of grains. Shear bands of the same type as those that develop in single crystals can now extend through many grains (Fig. 9.10a) without paying any heed to grain boundaries; they may even shear the boundaries by large amounts (Fig. 9.10b). They then dominate the deformation process.

We noted earlier that single crystals undergo changes in orientation during

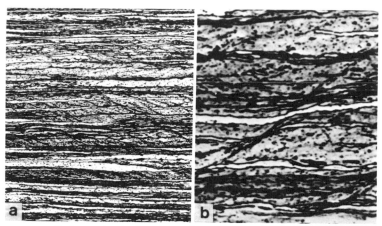

Deformation in this sheet of copper, which has been reduced 70% by cold rolling, has occurred inhomogeneously in bands which extend through a number of grains. The orientation of the bands is related to the geometry of the macroscopic strain system. The shear bands can be seen as diagonal etch markings in (a). The intersections of bands with grain boundaries are shown in (b). Note the marked shear displacements of the boundaries. Optical micrographs: (a) 250×; (b) 1000×.

Fig. 9.10. Examples of shear bands formed in a poly-crystalline metal at a large strain.

deformation, tending toward an orientation determined by the slip system that operates irrespective of their original orientations. So do the individual grains of a polycrystal. They develop a preferred orientation, known in this instance as a *deformation texture*. This means that the grains, instead of being randomly oriented, as they ideally would be before deformation, tend after deformation to have similar orientations with respect to external reference axes of the workpiece. The grains do not end up with precisely the same orientation, but with orientations scattered around some characteristic mean. The mean defines the type of the preferred orientation; the scatter defines its degree. The effect is marked only after comparatively large strains.

The texture developed after modest strains is that to be expected from simple predictions based on the behavior of single crystals. In sheet, for example, the crystal plane that is the characteristic slip plane of the crystal tends to become parallel to the sheet surface, and the characteristic slip direction parallel to the rolling direction, as indicated diagrammatically in Fig. 9.11(b). In wire, on the other hand, the slip direction tends to become parallel to the longitudinal axis, but the slip plane remains randomly oriented (Fig. 9.11a) because of the circular symmetry of the drawing process. This arrangement is called a *fiber texture*. In some metals, however, different textures develop at higher strains, textures that are determined by the type of strain inhomoge-

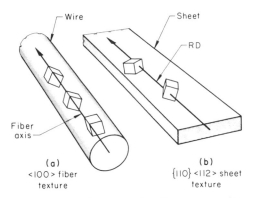

(a)
<100> fiber
texture

(b)
{110} <112> sheet
texture

The cubes indicate the manner in which the crystal lattices of the different grains might be oriented. Source: *Metals Handbook*, 8th Ed., Vol 8, *Metallography, Structures and Phase Diagrams*, American Society for Metals, Metals Park, OH, 1973.

Fig. 9.11. Sketches illustrating the types of preferred orientations that are developed in wire (a) and sheet (b) during cold working.

neity that now dominates the deformation process. These are matters of importance because a consequence is that the mechanical properties vary with direction in a manner which is related to the type of texture and its degree. This may be an advantage in a product, such as a wire, which is to be stressed only in one direction if the texture is such that the best properties are obtained in the stressing direction. On the other hand, it may be a disadvantage in, say, a sheet which has to be formed by bending around a sharp radius. It may not be possible to bend the sheet safely to nearly as sharp a radius in some directions as in others because of a variation in tensile ductility.

Plastic Deformation of Multiphase Alloys

At comparatively small strains, particles of a second phase in an alloy behave roughly in the same way as the matrix grains when they are the same order of size, if allowance is made for any difference in intrinsic strength (p. 233). However, the proportionate effect on strengthening may be quite different when the second phase is very small, a matter which will be discussed in Chapter 10 (p. 340). Particles of a ductile second phase tend to elongate in roughly the same way as the matrix after deformation to large strains. Particles of a phase which would be quite brittle in bulk may also do so if the straining conditions are such that the particles are restrained in hydrostatic compression by a surrounding ductile matrix. Otherwise, particularly when the particles are small and isolated, they may be fractured where they are intersected by a shear band propagating in the matrix (cf. Fig. 9.12a and b). A particle may thus be broken up into

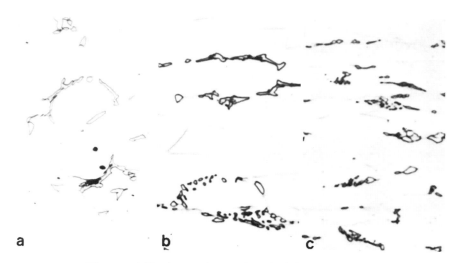

The material is a low-carbon steel in which the main microconstituent is ferrite; films of cementite are also present among the ferrite grains (a). The cementite films began to break up when the material was reduced 40% by cold rolling (b), each film having fractured where it was intersected by a shear band developed in the ferrite. The films were broken up further into small, nearly spherical particles after 70% reduction (c), the particles became aligned in stringers in the rolling direction. Optical micrographs; 1000×.

Fig. 9.12. Examples of how, under some circumstances, the shear bands developed during large plastic strains can break up particles of a second phase.

a number of fragments. These fragments become separated and the matrix squeezes in between them and welds together. The end result is that a single particle is redistributed as a stringer of small, roughly spherical particles (Fig. 9.12c).

Residual Stresses Induced by Plastic Deformation

We have seen that plastic deformation is not homogeneous on a microscopic scale. Frequently, it is not homogeneous on a macroscopic scale either. For example, the surface layers of a cold rolled sheet typically are plastically deformed more than the central regions during rolling because they are more directly influenced by the rolls (Fig. 9.13b). The yield strength of the surface layers consequently is greater than that of the core and so can prevent the core from recovering elastically (p. 63) as much as it would otherwise do (Fig. 9.13c).

In other words, the core is held permanently elastically extended, and this

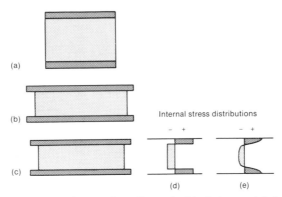

This example depicts the rolling of a block into a slab by a procedure which deforms the surface layers more than the core. For simplicity, consider only an outer rim, shaded dark in (a), which extends more than the core during cold rolling (b). Both rim and core begin to recover elastically after deformation (c), but full recovery of the core is prevented by the now-stronger rim. The recovered core consequently is longer than it would have been had it been able to recover fully, and a tensile stress is necessary to maintain this extension (d). Remember that tensile stresses by convention are regarded as being negative. The residual stresses in the rims, by the same type of argument, are compressive, balancing the tensile stresses in the core (d). In a more realistic case, the distribution of the residual stresses would be like that sketched in (e).

Fig. 9.13. Sketches illustrating how residual stresses develop during plastic deformation when the deformation is macroscopically inhomogeneous.

means that it is in the same condition as if it were subjected to a tensile elastic stress. An elastic stress maintained permanently within a body in this way is called a *residual stress.** The residual stress at the surface in our example is tensile. By the reverse argument, residual compressive stresses are developed in the core, and, to maintain equilibrium, the integrated tensile and compressive stresses have to balance (Fig. 9.13d), giving a final stress distribution something like that sketched in Fig. 9.13(e). Each type of deformation process induces its own characteristic pattern of residual stresses.

Residual stresses add to externally applied stresses and, depending on the relative signs of the two, may cause the actual stress to be higher than it might be thought to be. This can be important in service, such as during fatigue (p. 104) or stress corrosion (p. 158). The residual stresses may also cause a body to distort if the balance of the stresses is disturbed. For example, the sheet illustrated in Fig. 9.13 would bend if material were removed from one surface but not from the other.

---

*Such stresses are sometimes known as *internal stresses*, but this is not a preferred term.

Practical Consequences of Plastic Deformation

The hardening that can be achieved by cold working constitutes one of the basic methods used in practice to strengthen metals. Tensile strength can usually be about doubled, and yield strength can be increased by a factor of three to five, as a maximum. Intermediate increases can be obtained by controlled adjustments to the amount of cold work. The actual magnitudes of the increases vary with base metal and alloy, an example for one alloy system being given in Fig. 9.14. The cold worked material is said to be in a particular *temper*,* and each branch of the metal industry has developed a convention of its own to designate degrees of temper. The system indicated in Fig. 9.14 is that used by the copper industry. In general, the grades are based on the proportion of the increase in tensile strength that has been obtained relative to the maximum that can be

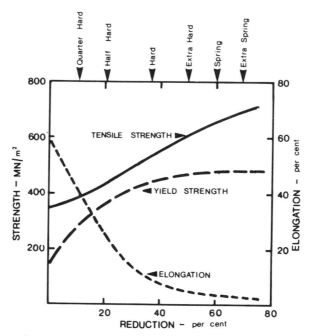

The terms used in the copper industry to designate the degree of work hardening (the work-hardening temper) are indicated at the top of the graph.

Fig. 9.14. Variations in yield strength, tensile strength, and tensile elongation to fracture of a Cu-30Zn brass with reduction by cold rolling.

---

*"Temper" is another one of those terms that are used in metallurgy in a number of ways. The literal meaning of the word is "to modify properties". It is used in metallurgy to signify a degree of strain hardening (as above), a particular type of working process (p. 330), or a particular heat treatment to soften steel (p. 408). All do modify properties, but so do many other treatments which are not called tempering. We have to live with these problems of terminology.

obtained. The penalty for the increase in strength is a reduction in tensile ductility, as is also indicated in Fig. 9.14.

Industry takes advantage of this method of strengthening in semifabricated products, such as sheet, tube, and wire. Many alloys are available in a range of temper grades as well as the annealed grades that we shall soon discuss. Although they cannot be further formed as much as the annealed material, advantage can be taken of the higher strengths of the temper grades — particularly the higher yield strengths — in the final product. Copper alloy wires used in springs are examples. Copper alloys have reasonably high values of Young's modulus, which make them good candidates for spring materials, but low yield strengths in the annealed condition, which means that springs made from annealed wire will deform permanently after only small spring deflections (p. 67). Springs made from cold drawn wire, on the other hand, do not have this deficiency as markedly and are widely and effectively used in this role.

The second way in which practical advantage is taken of strain hardening is to strengthen a product while it is being formed. Local areas can be strengthened in this way. You will find around you many products which have been so strengthened, one well-known example of which is the now-common aluminum beverage can. The bodies of these cans are produced by a cold drawing process, and it is the work hardening that occurs in the body walls during drawing that strengthens the material sufficiently to enable the walls to be made so thin. This in turn permits less aluminum to be used, a feature which has contributed to the economic success of this type of can. These cans are, incidentally, produced at an eye-boggling speed and are a triumph of production and metallurgical engineering.

WORKABILITY AND FORMABILITY

As we noted in Chapter 3 (p. 73), rather vague terms are used to describe the ability of a metal to be deformed plastically without cracking or fracturing. "Ductility" is the general term used. The terms *workability* and *formability* are also used in the specific contexts of working and shaping; the term *malleability* is used as well, but is restricted to shaping by hammering. Ductility in tension is reduced during cold working, and so it is to be expected that a reduction in workability is the price that has to be paid for the benefits of work hardening.

It is not possible fully to predict formability from reasonably fundamental material properties of the type discussed in Chapter 3. The forming processes that are used in industry are many and varied (Chapter 2), and the stress systems and a host of other factors are different in each process. Many of these factors, as well as material characteristics, influence the severity with which a metal may be worked without developing fractures. For example, most metals can be worked extensively if they are subjected solely to compressive stresses, although the forces required to do so may vary considerably depending on the yield strength and the rate at which the material work hardens. On the other hand, it is possible to work many metals only to a limited extent when they are subjected to tensile stresses, or to components of tensile stresses, because fractures may develop after comparatively small strains. Some regions of a workpiece may be subjected to tensile strains even though most regions are being deformed in compression, and it is these regions that are vulnerable and that determine formability in the particular forming process. Special laboratory tests have been devised to assess the formability of a metal in various specific types of forming operations through approximate simulation of the operations. For example, simple cup drawing tests are used to assess formability in deep drawing operations. The development of sophisticated analyses of the mechanics of forming operations and of the laboratory tests used to simulate them has greatly improved the reliability of interpretation of these tests. But many forming operations are so complicated that ultimately the formability of a particular alloy or a particular batch of that alloy has to be established by production trials.

The reduction in tensile ductility that accompanies strain hardening often sets a limit on the amount of forming that can be achieved in any one operation. If further forming is required, the ductility has to be restored by an annealing treatment of the type that we shall discuss later in this chapter. This costs money, and so one of the skills in metal-forming production planning is to optimize the sequence of working and annealing operations. Working has to be continued for as long as possible without risk of fracture, and annealing treatments must be effective enough to restore optimum ductility. It is these annealing treatments that we shall now discuss.

ANNEALING OF DEFORMED METALS: RECOVERY AND RECRYSTALLIZATION

The crystals of a deformed metal are in a state which clearly is far removed from equilibrium, so it might be expected that they would revert to an equilibrium lattice if given a chance. To get this chance, however, they usually have to be heated to a temperature at which the atoms acquire reasonably good mobility, because only then can the atoms whose positions were disturbed during deformation move back to more stable positions in a finite time. Several distinct steps can be recognized in the changes that then occur, namely:

- *Recovery,* which embraces all changes that do not involve the formation of new grains or the movement of grain boundaries.
- *Recrystallization,* in which a population of new grains is nucleated, the new grains growing at the expense of the deformed grains until the latter have all been consumed. This strictly should be called *primary* recrystallization to distinguish it from *secondary* recrystallization.

- *Secondary recrystallization,* in which a few of the grains formed during primary recrystallization take off and consume the first population of recrystallized grains. The new grains then grow to a very large size. This phenomenon need not concern us here.

The temperatures at which these various phenomena occur differ among different metals and alloys, depending primarily on melting point but also on several other factors. Ignoring these latter factors for a moment, the temperature in kelvins (K) at which recrystallization occurs is usually somewhat below half the melting point of the metal, also expressed in kelvins. That is, recrystallization occurs in all metals at approximately the same *homologous temperature* of approximately $0.4\,T_m$ (T_m denotes melting point in kelvins). Recovery occurs at a temperature lower than that at which recrystallization takes place. Secondary recrystallization occurs at temperatures considerably above the primary recrystallization temperature.

Recovery

There are really two separate effects that can occur in the recovery temperature range. The first is the relief of internal stresses. This is able to occur because the yield strength of the material is lowered at the higher temperatures and so the local variations in strain that were responsible for the residual stresses (p. 303) can equalize themselves. The second effect is the true recovery process in which the deformed crystals retain their identities but in which the density and distribution of the crystal defects that were introduced into the crystals during deformation are changed. Thus there are no changes in grain shape that can be discerned but many changes on an atomic scale that can be demonstrated by transmission electron microscopy or by other techniques that are sensitive to atomic structure.

The vacancies created during deformation are always annihilated during the recovery phase, but the likelihood of changes in the deformation-created dislocations depends on the stacking-fault energy (p. 239) of the metal. Little change occurs in metals of low stacking-fault energy (such as copper alloys), but a considerable change is the norm for metals of high stacking-fault energy, such as pure copper and iron. In these metals, the dislocations rearrange themselves to form *cells,** the cores of which contain few dislocations and the boundaries of which are composed of narrow bands of tangled dislocations (an example is given in Fig. 6.6b, p. 194). There is, as a consequence, a small random misorientation of 1 to 2° between the cells. The cells may develop into what are described as subgrains (p. 245) if the boundaries become narrow. This process is called *polygonization.*

---

*There is a difference in degree rather than kind between these cells and the subgrains described on p. 245. The boundaries are narrower and the orientation differences across the boundaries are smaller for subgrains than for cells.

The relaxation process may occur during or immediately after deformation at room temperature in some metals, such as aluminum and copper. It occurs more rapidly and becomes more complete the closer the temperature approaches that at which recrystallization takes place. Relaxation is also accelerated considerably if a large enough stress is applied while the material is at the recovery temperature, the accelerated process being known as *dynamic recovery*.

Structural recovery of this nature is associated with recovery, to a greater or lesser extent, of mechanical properties. The practical result is that metals of high stacking-fault energy usually soften somewhat when heated to higher temperatures in the recovery range, but that the properties of those of low stacking-fault energy remain much the same. A complication sometimes arises because other phenomena which cause slight hardening can occur at the same time. This occurs, for example, in high-zinc brasses, such as cartridge brasses, and in low-carbon steels (Fig. 9.17). In the former case, the hardening is probably due to a rearrangement of the zinc atoms in the crystal lattice. In the latter case it is due to precipitation hardening (p. 367).

Recrystallization

New, strain-free grains appear at various positions within the deformed grains when a deformed metal is heated above a certain temperature (Fig. 9.15b), and these grains grow with time until they consume completely the deformed grains (Fig. 9.15c). This is the process called *recrystallization,* the thermal treatment that produces the recrystallization being called a *recrystallization annealing* heat treatment. Be warned again that the term "annealing" is applied to several quite different thermal treatments, the common feature being that they cause softening. The use of a qualifying adjective, such as "recrystallization", consequently is desirable whenever "annealing" is used, but this practice unfortunately is not always followed because it is too inconvenient. The type of annealing treatment being referred to then must be worked out from the context.

Recrystallization involves two separate processes — namely, the *nucleation* of grains and the *growth* of these grains, as is illustrated diagrammatically in Fig. 9.15. The nature of the nuclei from which the new grains grow has long been debated, but it is now recognized that each nucleus was present in the deformed structure. They are regions which have a structure, which perhaps was modified during heating through the recovery range, that is particularly favorable for growth into surrounding regions. The possible sites of these nucleating regions are numerous, the most important ones being the shear bands that form during deformation at high strains. Points of intersection of these bands are particularly likely sites. Other important sites include grain boundaries and boundaries with other phases, both of these being regions at which deformation strains concentrate. Many nucleation sites are thus available after even modest strains.

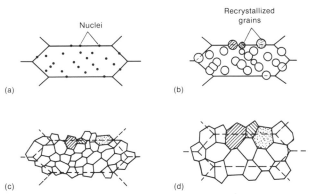

The locations of recrystallization nuclei in a cold worked grain are indicated in (a), shown here as being located at grain boundaries and along deformation shear bands. The nuclei have started to grow in (b), producing approximately spherical, strain-free (recrystallized) grains in a range of orientations. The new grains have grown in (c) until their boundaries have met and all of the strained material has been consumed. Some grains continue to grow, absorbing others, as indicated in (d). Four grains which survived to this stage are shaded. At stage (c), the grain size is smaller than for the original material because the cold working has been severe enough to activate many recrystallization nuclei in each original grain. Whether the grain size will also be smaller finally depends on how much grain growth is allowed to occur in stage (d).

Fig. 9.15. Sketches illustrating diagrammatically the stages of recrystallization during the annealing of a cold worked metal.

Nevertheless, nucleation is a difficult process — more difficult than grain growth. Grain growth consequently becomes the dominant process once nucleation is well in hand. The boundaries of the newly nucleated grains sweep through the deformed grains (Fig. 9.15b), being driven by the general tendency for the disorder of the system to be reduced, until eventually they encounter the boundaries of other new grains that have originated from other nuclei. An equiaxed grain structure is typically established when the grains have impinged fully on one another (Fig. 9.15c), but this is not the final structure unless the heating time is very short. Grain boundaries migrate continuously at the temperatures at which recrystallization occurs, and this enables the grains to cannibalize one another. The surviving grains gradually increase in mean size (Fig. 9.15d), the driving force this time being the tendency for the surface energy of the grain boundaries to be reduced to the minimum. This means in effect that the surface area of the boundaries is reduced to the minimum, which is achieved by the grains increasing their diameter.

The boundaries of either newly established nuclei or well-established equiaxed grains can move only at a finite rate because, in effect, they move by a process of self-diffusion (p. 235). This rate is higher at higher temperatures but also is a characteristic of the metal. In general, boundaries move more slowly in solid solutions than in pure metals, and they can be further delayed by structural features such as stable inclusions or precipitates. In any event, it follows that the development of a fully recrystallized structure takes time, the higher the temperature the shorter the time. The time is also shorter in some metals than in others. If, then, we follow the consequence of heating a deformed metal at increasing temperature for a constant time, we find that at first only a portion of the deformed grains are replaced by recrystallized grains (cf. Fig. 9.16a and b). The volume fraction of recrystallized grains then increases with anneal-

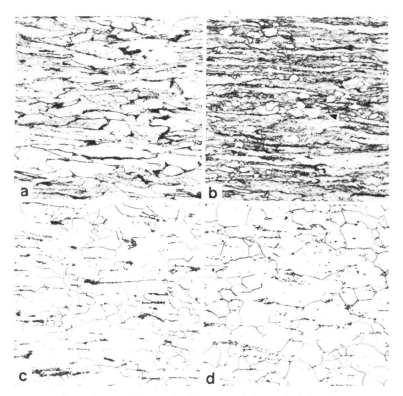

Shown here are longitudinal sections of sheet which have been reduced 70% by cold rolling and then heated for 30 min at temperatures of: (a) 500 °C (930 °F); (b) 525 °C (975 °F); (c) 550 °C (1020 °F); and (d) 700 °C (1290 °F). The material in (a) has substantially the same structure as that of the as-rolled sheet (cf. Fig. 9.8b). Small recrystallized grains have begun to appear in (b) in material which has been heated to a temperature only slightly higher than for (a). Complete recrystallization has occurred in (c). Grain growth has taken place in (d). Variations in the hardness and grain size of this sheet with annealing temperature are shown in Fig. 9.17. Photomicrographs; 500×.

Fig. 9.16. Examples of the stages of recrystallization during annealing of a cold rolled low-carbon steel sheet.

Table 9.1. Approximate Recrystallization Temperatures of Some Commercially Pure Metals

Metal	Recrystallization temperature °C	Recrystallization temperature K	Melting point, K
Lead	<0	<273	600
Zinc	20	293	692
Aluminum	150	323	933
Gold	200	473	1336
Copper	200	473	1356
Iron	450	723	1809
Nickel	600	873	1726
Tungsten	1200	1473	3683

ing temperature until a fully recrystallized structure is obtained, as in Fig. 9.16(c). Thereafter, the grain size of the recrystallized structure increases (cf. Fig. 9.16d), a point to which we shall return.

The lowest temperature at which complete recrystallization occurs is defined as the *recrystallization temperature*. This temperature is, as we have already noted, a characteristic of a metal or alloy in a general way, being about half its melting point in kelvins (Table 9.1). It is, however, affected in a secondary way by a number of other factors. First of all, the prior strain has to exceed a certain value before recrystallization will occur at all. The value of the critical strain varies a little with the alloy and the annealing conditions but typically ranges from 2 to 4% engineering strain. Thereafter, the recrystallization temperature decreases with increasing prior strain and with increasing annealing time. It is increased by the presence of alloying elements and impurities, more so in some cases than in others.

The mechanical properties of cold worked material recover during recrystallization in direct proportion to the amount of recrystallization that has occurred. The strength and hardness thus decrease sharply once recrystallization has commenced (Fig. 9.17), the ductility increasing at the same time, to the values that pertained before cold working began. The formability of the material is restored. There are, however, several important permanent changes in structure that result from any deformation-recrystallization cycle. The first concerns the size of the recrystallized grains, and the second their orientations.

It is generally desirable that as fine a grain size as possible be achieved in an annealed product. First, reducing the grain size increases the yield strength of the material and it does this while also improving toughness (see p. 98), which is a most desirable state of affairs. Secondly, grain size can have a significant effect on the surface finish that is obtained in a forming operation in which the material is stretched. Consider, as an example, a sheet formed into a cup by stretching. The grains exposed at the surface of the sheet deform more in their central regions than in restrained regions adjacent to their grain boundaries, and

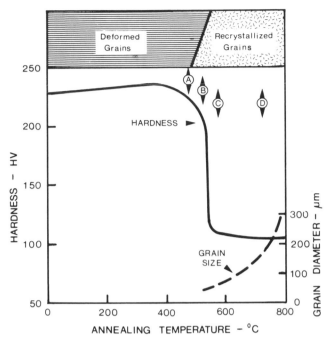

The sheet was heated at temperature for 30 min. The microstructures of sheets annealed at the temperatures identified by the
double-ended arrows are shown in Fig. 9.16.

**Fig. 9.17. Variations with annealing temperature of
the hardness and grain size of a cold rolled low-
carbon steel sheet.**

so a bump develops in the center of each grain. These bumps always develop, but
are not too noticeable when the grain size is small (Fig. 9.18a and b), because the
bumps are close together and not very high. They may become obvious, however, when the grain size is large, because the height of the bumps is greater and
the bumps are more widely separated (Fig. 9.18c and d). This effect tends to spoil
the appearance of an object and, for obvious reasons, is known as *orange peel*. A
prolonged and expensive buffing treatment may then be required to smooth out
the surface if the product has to have a bright surface finish.

The size of the grains that are produced during recrystallization depends on
a balance between the rate at which nuclei form (fewer nuclei means that each
grain has to grow larger to fill a given volume) and the rate at which they grow
(high growth rates enable the grains formed immediately after recrystallization to grow to a larger size in the time available). Only a few nuclei become
active when the prior deformation has only just exceeded the critical strain for
recrystallization, and the resultant grain size is large. Sometimes very large
grains may grow — a phenomenon which is known as *critical grain growth*. Progressively more nuclei become active with increasing strain, and soon large
numbers become active. So, other things being equal, the recrystallized grains

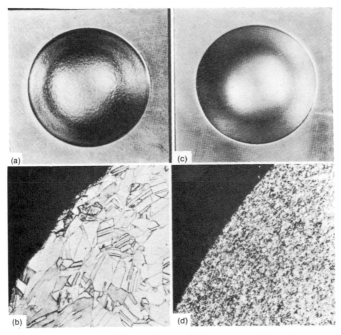

The surface roughening known as "orange peel" can be seen on the hemispherical cup shown in (a), which was drawn from sheet having a coarse grain size, as shown in (b). The similar cup illustrated in (c), on the other hand, appears to the eye to have a smooth surface. The material used for this cup had a much finer grain size, as illustrated in (d). The photographs in (a) and (c) are of the surfaces of the cups shown at actual size. Sections of the walls of these cups, etched to reveal the grain structure, are shown in optical micrographs in (b) and (d) (magnification, 75×). The surface roughness visible in (a) can be seen in (b) to be due to the development of a bump in the center of each surface grain. Similar bumps have developed in the fine-grain material but are too small to be discerned. Source: *Metals Handbook*, 8th Ed., Vol 7, *Atlas of Microstructures of Industrial Alloys*, American Society for Metals, Metals Park, OH, 1972.

Fig. 9.18. Illustrations of the development of an orange peel surface on Cu-30Zn sheet during forming.

rapidly decrease in size. The recrystallized grain size thus is determined firstly by the last deformation and annealing cycle, rather than by the prior grain size. The effect of prior strain on the recrystallized grain size is illustrated qualitatively, but dramatically, in Fig. 9.19, and is shown more quantitatively by the curve on the back panel of the diagram in Fig. 9.20. A result of the type illustrated in Fig. 9.19 is not altogether a laboratory curiosity, but is indicative of what may occur in practice in formed components in which strain gradients are established varying from zero to some high value. The critical strain will occur somewhere, and a large grain size will develop in this region during annealing. Both forming properties and mechanical properties are then impaired locally.

A sheet of high-purity aluminum was penetrated by a bullet and then annealed and etched to reveal the grain structure. Recrystallization has occurred in a circular region surrounding the bullet hole, this being the region that was plastically deformed during penetration. The recrystallized grains are smallest close to the hole, where the plastic strains were largest. The grain size increases with distance from the hole, with decreasing strain, and is particularly large at the periphery of the recrystallized region, where the strains were just sufficient to promote recrystallization. Macrograph; actual size.

Fig. 9.19. An illustration of the effect of deformation strain on the grain size developed in aluminum sheet during annealing. Critical grain growth during annealing is also illustrated.

A factor that counterbalances the effect of increasing deformation on the recrystallized grain size is the tendency for grains to grow with increasing time, more particularly at higher temperatures. This factor results in larger grains finally being produced for a particular value of prior strain. The interplay between the parameters of prior deformation, annealing temperature, and grain size is complex, but can be illustrated by a single perspective diagram known as a *recrystallization diagram* (Fig. 9.20). Such a diagram illustrates the general trends, but at best is semiapproximate because it would vary a little with different batches of an alloy. More seriously, it applies to only one heating time and to a material with a particular initial grain size. Prior grain size also has an effect, secondary to those mentioned earlier, because grain boundaries are preferred sites for nucleation during recrystallization and there is relatively more grain boundary in fine-grain materials. Thus smaller recrystallized grains are, other things being equal, produced from finer-grain stock. A desirable consequence of this last effect is that the grain size is reduced progressively by a sequence of similar working and annealing treatments. Producers of sheet plan their sequences of operations to take advantage of this progressive reduction in grain size.

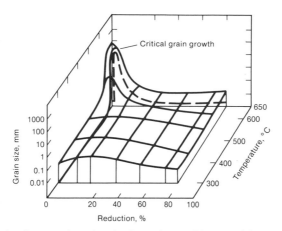

This diagram describes the dependence of the size of the recrystallized grains produced during annealing on, first, the amount of deformation to which a sheet has been subjected and, secondly, the annealing temperature. The annealing time is constant. The heavy broken curve on the back panel illustrates the variation of grain size with deformation strain for a constant annealing temperature and time. The peak of this curve defines the regime of critical grain growth.

Fig. 9.20. An annealing diagram for pure aluminum.

Another technique which producers of sheet use to the same end is, however, not quite so obvious. Strip is passed continuously at high speed through a furnace held at a temperature well above the normal annealing temperature. It is then cooled rapidly on exit. If the temperature and speed are right, the material is held in the recrystallization range for only a very short time (a matter of seconds); new recrystallized grains form but do not have time to grow, and a very fine grain size is obtained. Continuous annealing processes of this type are used extensively in the production of thin low-carbon steel strip, but special equipment, and large production runs to justify its use, are required.

It is also possible in some instances to build features into an alloy which greatly reduce the rate of grain growth during annealing. A grain size finer than usual can then be obtained. Things can even be arranged so that the growth rate is reduced more in one direction (say across a sheet) than in others. Grains with a pancake shape, which can have advantages in some instances and disadvantages in others, are then produced. These effects are usually achieved by dispersing very small particles of a stable phase throughout the grains. The grain boundaries have difficulty moving past the particles and so their rate of migration is reduced. This is sometimes done deliberately in steels by adding aluminum, which produces a dispersion of small particles of aluminum nitride that restrain the growth of the austenite grains during austenitizing heat treatments (p. 396). The same result is achieved in aluminum alloys, but rather more inadvertently this time, by insoluble particles resulting from the presence of certain impurities.

We need also to mention that parallel-sided annealing twins (Fig. 7.10a, p. 239) form during growth of the recrystallized grains of certain metals (copper-group metals and their solid solutions, lead, and austenitic steels). The tendency for twins to form is determined by the relative surface energies of the twin boundaries and grain boundaries. The mechanism of their formation is well understood, but is a little complicated.

Now we will take up the second permanent structural change caused by a deformation-recrystallization cycle. This concerns the orientations of the grains in the material. Assuming a starting material in which the grains are randomly oriented, cold working causes the grains to change their orientations in such a way that they tend to take up similar orientations; the resulting texture is called a *deformation texture* (p. 300). The new grains formed when such a material is recrystallized also have a preferred orientation, which now is called the *recrystallization texture*. This recrystallization texture may be identical to that of the parent deformation texture, but more often it is quite different. The recrystallization texture develops because the orientation of a recrystallized grain is determined by the orientation of the nucleus from which it grew, and not by the orientation of its parent grain. In highly strained material, the nuclei tend to have the same orientation — one that is characteristic of the mechanism of deformation at the strain and in the material concerned. These are the factors that determine the type of recrystallization texture that is developed.

The presence of a recrystallization texture and the extent to which it is present (the scatter about the ideal texture) are of great importance in sheet that is to be pressed or deep drawn. The strength and ductility of a sheet varies with direction when a texture is present, and as a result the sheet stretches differently in different directions. The consequences are most easily seen in the simple case of a deep-drawn cylindrical cup. Waves, known as *ears* (Fig. 9.21), develop around the mouth of the cup when a recrystallization texture is present in the feed-stock sheet, because the sheet stretches more in some directions than in others. The depth of the ears is related to the severity of the preferred orientation of the recrystallization texture — that is, they are sharper and deeper when there is less scatter about the mean orientation. Their number and locations are related to the type of recrystallization texture — that is, to the directions in which the ductility is at a maximum.

The development of ears may have to be accepted in deep-drawing practice, because some degree of recrystallization texture often cannot be avoided in the feed stock. This can be coped with if it is limited in extent. The mouth of a pressing has to be trimmed anyway, and a reasonable allowance can be made to remove some of the earing. Trouble obviously arises, however, if too large an allowance has been made or if the depth of the ears exceeds this allowance. In severe cases, the sidewall of the pressing may even split. The control of recrystallization texture consequently is central to the production of sheet for pressings. Indeed, the viability of much of the large deep-drawing industry, not the

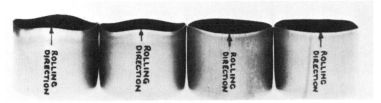

Simple cylindrical cups drawn from low-carbon steel sheet are shown here. The waves at the mouths of the cups, called ears, were formed because the sheet was stretched more in some directions than in others. This occurred because the grains in the sheet had a preferred orientation. The cups were drawn from four different types of steel, the two at the left being commercial steels and the two at the right being experimental steels. Ears have developed along the rolling direction and the perpendicular direction in the commercial steels, but at positions 45° to the rolling direction in the experimental steels. The depths of the ears also differ. Both differences are due to differences in the types and degrees of the preferred orientations in the two types of sheet.

Fig. 9.21. Illustration of the development of ears during deep drawing.

smallest segment of which is the automobile body industry, relies on such control. This is a difficult metallurgical specialty, but one which has achieved much success.

There are occasions, however, when a recrystallization texture is advantageous, and there are even situations in which it is produced deliberately. The prime example perhaps is the sheet used for the laminations of the magnetic cores of electrical devices, such as transformers and motors. These laminations are stamped from sheets of a special iron which has a low carbon content and which contains up to 4.5% silicon. High-performance grades of sheet are made to have a highly developed recrystallization texture of a specific type which is known to result in low energy losses when the sheet is placed in an alternating magnetic field. Developments of this nature are responsible for the considerable reduction in the size of electrical transformers and motors which has occurred progressively over recent decades and which, among other things, has made the many domestic appliances of the modern world so much more effective.

DEFORMATION AT HIGH TEMPERATURES AND HIGH STRAIN RATES: HOT WORKING

Plastic deformation at the temperatures and strain rates that are used in practical hot working operations involves the same basic mechanisms as those which were discussed earlier in this chapter. There are, however, marked differences in degree. One is that the yield strength of the material being worked is much reduced (pp. 67 and 113) and hence that smaller forces are required to achieve a given strain. The other important difference is that recovery and/or recrystallization occur simultaneously with the deformation. Thus strain hardening is at best transient, which further reduces the forces required to achieve a given strain. Massive equipment which can apply very large forces is still used in hot working (p. 30), but it can achieve much larger strains than those produced during cold working. A second consequence of simultaneous recovery or recrystallization is that the structure of the material after cooling to room temperature is similar to that obtained after cold working and annealing.

Recovery or recrystallization does not, however, occur instantaneously during hot working. The grains of the workpiece first distort, as they do during cold working, and then either recover or recrystallize (or some combination of the two) after a short period. Recovery is the dominant phenomenon in some metals (e.g., aluminum) in which the original grains are replaced by networks of subgrains. The size and perfection of the subgrains depend on the strain, strain rate, and temperature. The orientation differences between the subgrains can be large. However, recrystallization is the dominant phenomenon in most metals (e.g., copper alloys and steels) when working is carried out at temperatures above 0.6 to 0.7 T_m (T_m stands for melting point in kelvins). Alloying elements or microstructural features which slow down the movement of grain boundaries thus impede recrystallization and so effectively strengthen the material a little during hot working. As an extreme example, alloys which have been designed to be strong at high temperatures obviously must be expected to be difficult to hot work.

The final grain size of a metal which has been hot worked under conditions where *dynamic recrystallization* does occur depends primarily on the temperature at which the working was finished. As might be expected from our previous discussions, the grain size is smaller the lower the finishing temperature. Moreover, the grain size produced at a given finishing temperature normally is smaller than that which would result from a comparable static recrystallization anneal, because the new grains have less time in which to grow. The repeated deformation and recrystallization that occur while the material continues to be worked as the temperature falls, which happen during most practical hot working operations, also help to produce fine grain sizes, particularly if working is continued down to the lowest possible temperature. Such precise control of hot rolling schedules is one of the techniques used to produce the very-fine-grain steels which have improved fracture toughness (p. 95).

Although it usually is possible to achieve large strains during hot working, metals do not necessarily have infinite hot ductility and workability under these circumstances. This means that it may be safe to attempt only a limited strain during each deformation pass. In fact, the hot ductility of metals varies considerably. Concentrated solid solutions, for example, are considerably less ductile than pure metals over the hot working temperature range. The reasons for this are not understood, but it does have consequences in determining the manner in which different metals have to be handled in practice. Other phenomena which limit hot ductility may also intrude, one important one being hot shortness (p. 115).

DEFORMATION AT HIGH TEMPERATURES AND LOW STRAIN RATES: CREEP

We saw in Chapter 4 (p. 116) that the term "creep" is used to describe a phenomenon in which a material continues to strain with time under a constant stress. The phenomenon becomes significant at temperatures somewhat above the usual recrystallization temperatures, and three characteristic regions of creep strain can be distinguished (see Fig. 4.8, p. 117). The first is a stage of primary creep in which the creep rate decreases with time; this can be ascribed to work hardening of the type discussed earlier in this chapter. The next is a stage of secondary creep in which the creep rate is constant; this is a stage during which strain hardening is balanced by thermal recovery, both of which we have already discussed. The final stage, tertiary creep, is one of rapidly increasing creep rate during which the processes of fracture are developing.

Slip process of the normal type occurs during the primary and secondary stages of creep, but the slip characteristically occurs on a finer scale than at room temperature and on an increasingly fine scale as the temperature increases. Moreover, a different deformation phenomenon, referred to as *grain-boundary sliding*, occurs in addition to the slip process during secondary creep. The grain boundaries literally slide over one another with the result that they can wriggle past each other without changing their shape as the workpiece is elongated. A shear stress along the grain boundary is required for sliding to occur. Thus boundaries aligned at 90° to the tensile axis do not slide, and those aligned at 45° slide at the maximum rate. Nevertheless, slip is still the dominant deformation process. A distinguishing feature of the slip mechanism is, however, that dislocation climb (p. 295) occurs to a significant extent. The result is that dislocations do not lock up so easily because they can climb around obstacles and continue to move on adjoining planes. Recovery with the formation of subgrains also occurs in most metals, and even recrystallization occurs in a few.

Although grain-boundary sliding makes only a small contribution to creep elongation (about 10% of the total), it is crucial to the two characteristic mechanisms by which creep fracture occurs: development of chains of cavities, and development of planar cracks, at the grain boundaries (p. 118). The accelerated creep rate under constant load that occurs during tertiary creep results from a reduction in the load-bearing area, principally by the formation of these internal discontinuities. The growth of either one or a mixture of the two eventually leads to fracture. The cavities develop when minute steps in the grain boundaries, which may be only about 10 nm (100 Å) high, are separated by grain-boundary sliding (Fig. 9.22). These steps are known to be present naturally in boundaries and also to be formed where slip which has been initiated within the grains impinges on the boundaries. Second-phase particles at the boundaries are also potent initiating sites, and so their presence greatly reduces creep ductility. Once formed, the cavities can grow either by further boundary sliding or by the accretion of vacancies. Intergranular cracks, on the other hand, form when grain-boundary sliding is held up by a barrier of some sort, so building up large shear stress along the nonsliding boundaries. One way in which this may occur is illustrated in Fig. 9.22, which depicts a situation where two boundaries at a triple point are sliding but the third one is not. Other situations which are similar in principle can be imagined. Note that cavities form along boundaries which shear during creep, and this is why they are most likely to occur on those at 45° to the tensile axis. Cracks, on the other hand, form on boundaries that do not slide, and hence are most likely on those that are normal to the tensile stress axis.

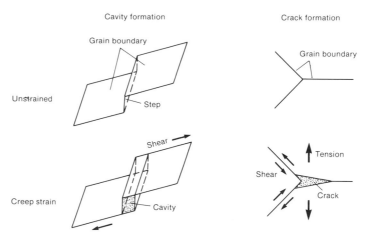

The formation of a cavity by the sliding of a grain boundary which contains a step is illustrated at the left. At the right, three grain boundaries are sketched as meeting at a point. Two of the boundaries are oriented so that they can slide, but the third is not. This third boundary has to separate in order to allow sliding on the other two.

Fig. 9.22. Diagrammatic sketches illustrating two mechanisms by which intergranular fractures develop during creep deformation. Both occur because grains slide past one another.

DEFORMATION BY CYCLICAL STRESSES: FATIGUE

In the early stages of cycling at constant stress, slip occurs in a manner which is similar to that for unidirectional deformation. After a comparatively small number of cycles, however, slip begins to concentrate inhomogeneously into characteristic broad bands. Then, typically, thin ribbons of material are extruded out of the surface, each extrusion being accompanied by a contiguous intrusion, as sketched in Fig. 9.23. Complicated cross-slip processes, or slip on alternate systems, seems to be necessary for the formation of these bands, but the reason why they form under these particular stressing conditions is not known.

The actual fatigue cracks are initiated at the intrusions (Fig. 9.23), and the mechanism for this is not known either. The cracks then advance discontinuously, one step for each stress cycle, the mode of fracture for each step depend-

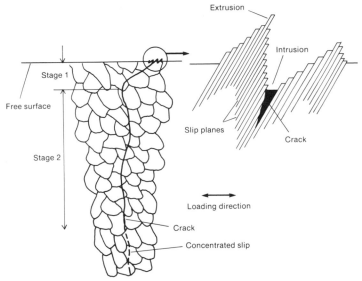

The crack is shown to initiate at an intrusion that is part of an intrusion-extrusion complex sketched in more detail in the inset at upper right. The complex develops along a crystal plane where slip has concentrated. The crack propagates first in a direction determined by shear components of the applied stress (Stage 1), but soon deviates to one determined principally by the applied tensile stresses (Stage 2). On a microscopic scale, however, the crack propagates along a particular crystal plane in each grain — again along planes on which slip has concentrated. This highly localized concentration of slip is a phenomenon that is a unique characteristic of fatigue. The microdeviations in crack path can be seen in Fig. 4.2. (p. 107).

Fig. 9.23. Sketches illustrating diagrammatically the initiation and growth of fatigue cracks.

ing on the stress. Fracture at high stresses occurs by dimple formation and growth as for a normal ductile fracture (p. 70). At lower stresses, however, fracture seems to be induced by the same complicated slip processes that are responsible for the formation of the concentrated slip bands. In any event, the cracks propagate along an active slip system in each grain. On a more macroscopic scale, they propagate at first in a general direction influenced by shear stresses (stage 1 in Fig. 9.23) but soon in a general direction normal to the applied tensile stress (stage 2 in Fig. 9.23). Hence the characteristic slightly zig-zag paths of fatigue cracks (see Fig. 4.2, p. 107).

FURTHER READING

R. W. K. Honeycombe, *The Plastic Deformation of Crystals*, 2nd Ed., Edward Arnold, London, 1984.

S. L. Semiatin and J. J. Jonas, *Formability and Workability of Metals: Plastic Instability and Flow Localization*, American Society for Metals, Metals Park, OH, 1985.

R. W. Cahn, "Recovery and Recrystallization", in *Physical Metallurgy*, edited by R. W. Cahn and P. Haasen, Elsevier Science, 1983, p. 1596.

M. Hatherly, "Deformation at High Strains", in *Strength of Metals and Alloys (ICSMA 6)*, edited by R. C. Gifkins, Pergamon, Oxford, 1983, p. 1181.

· 10 ·

Methods of Strengthening Metals While Keeping Them Tough

SUMMARY

One of the basic methods of strengthening metals is to add an alloying element, and an alloying element may either dissolve in the base metal, partly or completely, or introduce new phases into the structure. Elements which dissolve usually do so substitutionally. In this event, they have small but useful strengthening effects, usually without strongly affecting ductility or toughness. The strengthening obtained after strain hardening tends to be somewhat more marked. This solid-solution strengthening occurs because the crystal lattice is distorted locally by foreign atoms and these distortions impede the flow of dislocations. In a few cases, the alloying element dissolves interstitially and then may have an especially strong effect in delaying the onset of yielding. This is because the alloying element tends to segregate at the dislocations that are present initially and to lock them in place. A comparatively high stress is then required either to drag them loose or to generate fresh dislocations. Alloying elements generally have more marked effects on strengthening when they introduce additional phases into the structure, the magnitude of the effect depending on the strength of the new phase, the volume fraction present, and the way in which it is distributed. The first two factors are determined by composition, but the third often is determined by the thermal history of the material. Large, widely spaced particles of a stronger phase cause only the modest strengthening to be expected of a mixture. However, reducing the spacing of the particles increases the strengthening effect, eventually markedly so. This is because the mean free path of possible dislocation movement is reduced by the interphase boundaries blocking such movement in the same manner as it is blocked by grain boundaries. Phase particles can be too small to actually stop dislocations but still impede their movement because the dislocations have to bow and lengthen as they wriggle through the forest of phase particles. The impedance, and hence the strengthening, increases as the particle spacing decreases. But dislocations cannot move even in this way when the particles are too closely spaced. They then have to shear each particle encountered as they continue on their way. The magnitudes of either the bowing or the shearing effect is small per particle but can be large in total when large numbers of particles are present. This is a situation which is arranged in strong alloys. Closely spaced particles of a brittle phase, however, tend to have an adverse effect on toughness because the particles can crack in a cooperative way during straining, thus providing an easy fracture path; hence the need to balance strengthening and toughness. Moreover, the strengthening structures are not necessarily stable under complex straining conditions, such as during fatigue or creep. Mechanisms by which dislocations can easily bypass phase particles also operate at higher temperatures, as during creep. Extra difficulties consequently arise in the development of fatigue- or creep-resistant alloys. Nevertheless, metallurgists now have available some sound guidelines on which to base the development of strong alloys.

We have already discussed two basic methods by which the strength of a metal may be increased. The first is by plastic deformation at a relatively low temperature (p. 304) when strengthening is achieved at the expense of loss in ductility. This technique was employed by the earliest metalsmiths, who hardened the edges of tools and weapons by hammering. It has been widely used ever since in more controlled ways. The second method is by reducing the grain size (p. 296), which improves ductility and also toughness. The effectiveness of this method of strengthening is a much more recent discovery. We have discussed one method of controlling grain size (cold work followed by recrystallization annealing), and others will emerge in our later discussions.

We shall now discuss several other basic methods of strengthening, all of which depend on the addition of alloying elements to the metal — in particular, the addition of elements which cause new phases to be introduced into the microstructure. The important point that will now be demonstrated is that the types, sizes, shapes, and distributions of these phases are of crucial importance as well as their properties. We need to remind ourselves here that it is not acceptable merely to increase the strength of a metal regardless of the effect on other properties. Frequently there is a need to increase strength without affecting ductility or, sometimes more importantly, toughness (p. 83) to an unacceptable extent.

SOLID-SOLUTION HARDENING

Hardening by addition of alloying elements which dissolve in the base metal was also one of man's earliest metallurgical discoveries. In the early Bronze Age, copper was alloyed with arsenic, which may have been accidental rather than intentional. It was very effective, although it certainly did not improve the health of our metallurgical forebears who had to melt these alloys. Later, copper was alloyed with tin, which certainly was intentional. It was about as effective as arsenic, and was much safer for all concerned. All of these alloys essentially were substitutional solid solutions (p. 228), although additional phases were introduced into some. Then iron became the dominant industrial metal, and its effectiveness in more demanding applications also was based on alloying. The important alloying element in this case was carbon, which can form interstitial solid solutions (p. 228) in iron.

Substitutional Solid Solutions

We saw in Chapter 7 (p. 227) that it is often possible to replace the atoms in the crystal lattice of a solvent metal with foreign solute atoms, forming a so-called substitutional solid solution. However, we also saw that this must distort the crystal lattice whenever the solute and solvent atoms differ in size, as they always do to a greater or lesser extent. These distortions add extra impediments to the movement of dislocations through the crystal (they increase the lattice friction stress; see p. 293), a concept which is simple enough in principle but which actually is complex in detail. In any event, the resultant effect on yield strength is illustrated in Fig. 10.1 for an alloy system in which a continu-

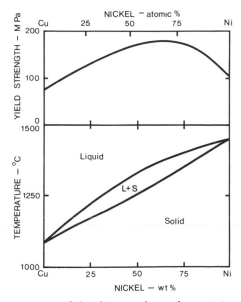

The upper part of this diagram shows the variation of yield strength with alloy content in the copper-nickel system. These two metals dissolve in one another in all proportions, as indicated by the phase diagram below.

Fig. 10.1. Diagram illustrating the order of strengthening that can be obtained in substitutional solid solutions.

ous series of solid solutions is formed. The maximum increase in yield strength occurs with about a 50:50 mixture of the two atoms, at which the disturbance of the crystal lattice might be expected to be at a maximum. The yield strength is then about doubled. Ductility is not greatly affected in this particular system.

In most systems, however, a solvent metal can dissolve only a limited amount of a solute element, and the opportunity for solid-solution hardening then has a limit. The maximum solid-solution strengthening that can be obtained in the alpha phase of copper-zinc alloys, for example, is limited to that obtained at zinc contents of about 30% and is comparatively small (Fig. 10.2). A compensating advantage, however, is that ductility is not greatly affected and is even improved a little in copper-zinc brasses (Fig. 10.2). A further advantage is that the rate of strain hardening is greater in solid solutions than in the pure metal, so that more marked improvements in strength are obtained in cold worked than in annealed alloys (Fig. 10.3). Advantage can be taken of this in

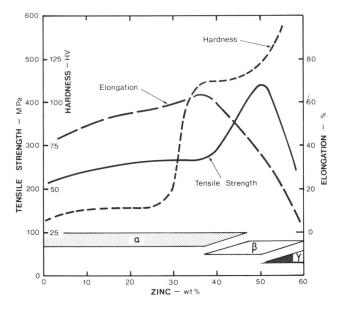

The copper-zinc alloy system, for which the phase diagram and representative microstructures are presented in Fig. 7.6 (p. 232), is used in this illustration. The phases present in the various alloys are indicated by the bands below the curves. Solid-solution hardening increases the strength of the alpha phase, but only slightly. It also increases ductility. The introduction of beta phase increases strength more markedly, but reduces ductility. The introduction of the brittle gamma phase, however, reduces both strength and ductility, particularly when it is present as films around the grain boundaries of the beta phase.

Fig. 10.2. Diagram comparing the hardening obtained in a substitutional solid solution and with that obtained by introducing a second phase.

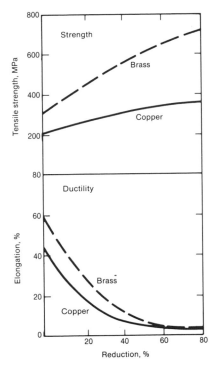

Cold working has different relative effects on the strengths and ductilities of pure metals and their solid solutions. In the comparison illustrated here, strength increases more rapidly with cold work in the solid solution (brass) than in the pure metal (copper). Ductility, on the other hand, decreases more rapidly in the solid solution.

Fig. 10.3. Comparison of the effects of cold working on the strengths and ductilities of pure copper and an alpha brass containing 30% Zn.

parts which are pressed or drawn after cold working, but the advantage is not obtained without cost. The comparative improvement in ductility due to alloying is gradually lost after cold working (Fig. 10.3).

The magnitudes of these changes in strength and ductility vary with the alloy system, the greater the difference in size between the solvent and solute atoms the greater the magnitudes of the effects, atom for atom of alloying element. To take one example, tin is more effective than zinc in strengthening the alpha solid solution of copper, and aluminum is more effective still. But other factors also have to be taken into account when assessing the best alloying element to use. Thus more zinc can be dissolved in copper than either tin or aluminum and so similar total hardening can be achieved. Zinc is also more effective in improving all-around formability and is cheaper: hence the wide use of copper-zinc brasses in formed and deep-drawn articles. The competitive advantage of tin bronzes, on the other hand, is that they have better corrosion resistance. They also strain harden more rapidly, which makes them more difficult to form but makes them useful in, for example, springs when used in the cold worked condition. This is a good example of the complex balance of characteristics which determines the niche that an alloy can best occupy.

Even though small, the effects of solid-solution hardening are frequently still worthwhile. An alloying element may be added primarily to effect strengthening by some other mechanism, but the additional contribution that it makes by strengthening the matrix solid solution is still usually worthy of consideration and optimization.

Interstitial Solid Solutions

The solute atoms in an interstitial solid solution are able to fit in between the solvent atoms instead of substituting for them, but they always do this rather uncomfortably (p. 228). Interstitial solid solutions are rare, the most important cases being the solutions of carbon and nitrogen in the alpha and gamma phases (p. 379) of iron.

Let us consider first the alpha phase (ferrite) present in low-carbon steels, such as those used in automotive sheet. As can be imagined from the sketches in Fig. 7.8 (p. 237), there is more room for interstitial atoms to fit at the cores dislocations than at general sites throughout the crystal lattice, and in fact this is just where they do try to fit. Carbon and nitrogen atoms diffuse fairly rapidly through the iron lattice at room temperature, and quite rapidly at temperatures of 100 °C (212 °F) or more, and so these interstitial atoms are constantly wandering through the crystal lattice. They eventually reach dislocations, where they tend to stay put because they fit there more comfortably and because there is not much incentive for them to move on. Moreover, they diffuse more rapidly along the dislocation line than they do through the body of the crystal. Thus the cores of the dislocations become lined with carbon and nitrogen atoms. Only about 0.000001 wt % of carbon would be necessary to line all the dislocations in an average piece of annealed iron in this way. The interstitial atoms also concentrate in the regions surrounding the dislocations, where there is still a little more room for them than in the general lattice.

This phenomenon is responsible for the anomalous yield-point behavior described in connection with Fig. 3.2 (p. 63). Dislocations to which interstitial atoms are segregated can move only if they either drag all the segregated atoms along with them or tear themselves away from the segregated atmospheres. Both courses of action are difficult, and so the dislocations tend to become locked in place. This locking inhibits the onset of plastic deformation, and an unusually high stress has to be reached before plastic deformation can commence: hence the extension of curve AB to x in Fig. 3.2. However, many fresh dislocations are soon generated at the higher stress, and perhaps some of the old dislocations are pulled away from their locking atmospheres of carbon atoms.

The movement of these fresh and freed dislocations allows a sudden increase in strain, and typically also a drop in stress (xy in Fig. 3.2), to occur. The dislocations resume normal behavior once this phase has been exhausted, and a normal stress-strain relationship is then resumed (yC in Fig. 3.2).

This phenomenon can sometimes be beneficial in low-carbon structural steels because it increases yield strength, tensile strength, and fatigue strength. However, it has severely adverse effects in steel sheet and strip used for drawing and pressing. The sudden yielding in material in which the dislocations are locked inevitably starts at some local region where the stress happens to be a little higher than usual (Fig. 10.4a). Further yielding then has to concentrate in this region, causing a strain front to propagate in a band across the sheet in a direction determined by the resolved shear stress (Fig. 10.4b). The thickness of the sheet is reduced slightly in the band, enough for the band to be seen by eye. Such a visible band is known as a *stretcher strain* or a *Lüders line*. Additional Lüders lines form in succession as the original bands are exhausted (Fig. 10.4c to e) and patterns such as those sketched in Fig. 10.4 and illustrated in Fig. 10.5 are developed. Eventually, when the whole region has passed through yield (when y in Fig. 3.2 has been reached), the bands amalgamate and are then no longer visible (Fig. 10.4f). So stretcher strains are seen only in sheet which has been subjected to small strains within a certain range, but many pressings have some regions where this is the case. The result is very disfiguring. Although their presence

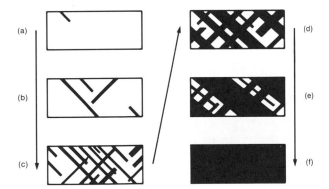

These sketches are of a unit area stretched horizontally by progressively increasing amounts. In (a), the material has been strained only just beyond the yield point (just beyond x in Fig. 3.2, p. 63), and a single stretcher strain has formed at a point where the stress was slightly higher than average. This line grows and new lines initiate and grow (b to e), gradually filling in the area as the over-all strain is increased. No localized strain lines are visible (f) when all of the area has been strained beyond the yield-point perturbation (beyond y in Fig. 3.2).

Fig. 10.4. Sketches illustrating the formation of stretcher strains (also called Lüders lines) during stretching of steel sheet which exhibits a yield point.

The region shown has been strained to about the equivalent of sketch (c) in Fig. 10.4. From *Metals Handbook*, 8th Ed., Vol 7, *Atlas of Microstructures*, American Society for Metals, Metals Park, OH, 1972.

Fig. 10.5. Stretcher strains (also called Lüders lines) on the surface of steel sheet which has been stretched just beyond the yield point of the steel.

would not affect its service performance, you would not be pleased if you noticed a pattern of markings like those shown in Fig. 10.5 extending across the fender of your automobile or the door of your refrigerator. Great efforts consequently are expended by the industries which produce and press steel sheet to ensure that markings of this type do not develop.

Commercial low-carbon steels inevitably contain enough carbon and nitrogen for the effect to be noticeable, for which only about 0.002% C or 0.01% N in solution in ferrite is required in practice. So we have to live with the fact that deleterious elements are always present and to seek ways of circumventing their influence. One solution is to cold roll the sheet to a stage where the critical strain range has just been exceeded. Nothing untoward can then occur during pressing. This is called *temper rolling*. The catch is that carbon and nitrogen atoms diffuse to the sites of the newly created dislocations within a few days to a few weeks, and relock them. The yield-point phenomenon then returns. This effect is called *strain aging* (cf. quench aging, p. 380). For success with temper rolling, therefore, the schedules for rolling and forming operations have to be coordinated, and this is not always easy to do, particularly if the temper rolling mill and the press shop are not located close to one another.

The second solution is to reduce the amounts of carbon and nitrogen in solution in the ferrite, where they do the damage, as distinct from the total amounts that are present. This can be done in principle for carbon by cooling the material very slowly after the recrystallization annealing treatment used after final cold rolling. Most of the carbon present is then precipitated out of solution as particles of iron carbide (cementite), and very little is left in solution (p. 379). This type of treatment is not, however, always completely effective. While carbon and nitrogen can both cause strain aging, the higher solubility of nitrogen in ferrite means that it causes the greater problem. Steps are taken in steelmaking to keep nitrogen levels down, but the easiest and surest way of minimizing the effect of nitrogen is to add small concentrations of elements, such as titanium and vanadium, which have greater affinities for nitrogen than does iron. These elements remove the nitrogen as innocuous nitrides so that very little is left to dissolve in the ferrite. This is the preferred way of making high-quality drawing sheet and strip.

It is possible in some systems to obtain interstitial solid solutions which have much higher concentrations of solute than those which we have just been discussing, the important example again being found in steels. We shall see in Chapter 11 that the high-temperature allotrope of iron known as austenite (p. 384) can dissolve much more carbon in interstitial solid solution than the low-temperature allotrope, ferrite, that we have just been discussing. Austenite can dissolve up to 2% C, and this carbon can be retained in solution if the austenite is cooled to room temperature quickly enough in a heat treatment procedure known as *quench hardening* (p. 388). The face-centered cubic crystal structure of austenite reverts to a body-centered cubic structure characteristic of ferrite, but to a very distorted form of this structure (p. 404). The distortion is caused by the extra carbon atoms that have been forced to stay in the interstices between the iron atoms, and so it increases with carbon content. The distorted phase is called *martensite*. Although several factors other than solution hardening are involved (p. 406), it is difficult for dislocations to move in martensites, the more so the greater the distortion. Martensites consequently are strong, at least in compression. This certainly means that they can be described as being hard (p. 80). Whether it also means that they are strong in tension is determined by their toughness (p. 72). The hardness of martensites increases rapidly with carbon content (Fig. 10.6), but their toughness is reduced rapidly at the same time. The result is that it may not be possible to use martensites directly in structural applications involving tensile stresses, certainly not without modifying their structures by a further heat treatment known as *tempering* (p. 408).

STRENGTHENING WITH ADDITIONAL PHASES

A metal is often able to take only a limited amount of an alloying element into solid solution, additions beyond this point causing a new phase to be introduced into the microstructure (p. 228). The new phase characteristically has a crystal structure more complex than those of the pure metal and its terminal solid solutions. It is typically stronger but less ductile, because dislocation movement

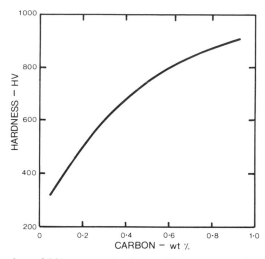

The three-fold increase in hardness with increasing carbon content is directly or indirectly due to the carbon being present in interstitial solid solution in the iron. This effect is much greater than that which can be obtained in any substitutional solid solution.

Fig. 10.6. Variation in hardness with carbon content for quench-hardened steels.

is more difficult in complex crystal structures. The introduction of such a new phase consequently can be expected to increase strength more markedly than does solid-solution hardening. The important point, however, is that the magnitude of the strengthening depends very much on the way in which the phase is distributed in the microstructure.

Strengthening With Large Particles of a Second Phase

Consider first the situation where the new phase is present as particles which are comparable in size to the matrix grains, approximately equiaxed in shape, and randomly distributed. An example would be the brass illustrated in Fig. 7.6(c)(p. 232), which contains particles of beta phase randomly distributed in a matrix of alpha phase. The new phase then acts rather as do normal but stronger-than-average grains. Strength is increased more rapidly than if the alloying element had continued to enter a terminal solution (Fig. 10.2), but more or less only directly in proportion to the amount of new phase. A limit on the increase is set by the properties of the new phase. Ductility is reduced at the same time (Fig. 10.2) because the new phase is less ductile than the terminal solid solution. Toughness will be even more adversely affected if the new phase is distinctly brittle, as it can be. The brittle particles tend to crack easily when the material is strained, and the toughness is then determined largely by the ease with which these cracks can join up to form a continuous fracture path (p. 88). This effect also may eventually reduce tensile strength, because fracture occurs prematurely soon after yielding has commenced (p. 72).

The deterioration in toughness at the structural level which we are currently discussing depends on the extent to which the brittle phase itself provides a continuous fracture path. The worst situation arises when the new phase is present as continuous, or nearly continuous, films around the grain boundaries of the matrix; the alloys illustrated in Fig. 7.6(e) (p. 232) and Fig. 10.7(a) are examples. Unfortunately from this point of view, grain boundaries are just the places where new phases prefer to form. Sometimes such structures cannot be avoided, as is the case for the high-zinc brasses (Fig. 7.6e), in which both tensile strength and ductility fall drastically with the introduction of gamma phase into the structure (Fig. 10.2). An alloy of this type cannot be used in practice. Sometimes, however, a susceptible alloy can be salvaged by a heat treatment which converts the morphology of the brittle phase to a more acceptable one of randomly distributed particles (cf. Fig. 10.7a and b). The alloy may then be usable in some applications, but only if care is taken to ensure that the more desirable structure is maintained.

Large plates constitute another undesirable type of morphology for brittle phases. An aluminum-silicon alloy containing 12% Si is potentially a good

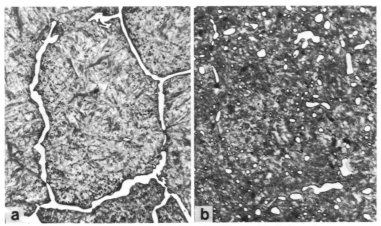

This material is a tool steel used in files and is designed to contain particles of free cementite in a quench-hardened matrix. The particles of free cementite improve wear resistance. This material is too brittle to use if the cementite is present in the form shown in (a), because a continuous crack path develops easily along the films of cementite. Toughness is improved to an acceptable level when the cementite is present as nearly spherical particles, as in (b). This tool steel is used only in the latter condition. Optical micrographs; 1000×.

Fig. 10.7. An example of the deleterious effects of continuous films of a brittle phase on toughness.

foundry alloy because it is easy to cast and has good corrosion resistance, but the silicon naturally is present as large plates of the type illustrated in Fig. 8.14(a) (p. 273), and as a result the tensile strength and ductility of the alloy are rather poor. Both properties are improved significantly, however, when the structure is modified so that the silicon is present as branched rods (Fig. 8.14b and 6.8b); the alloy then finds wide application. The trick usually is achieved by adding small amounts of sodium or strontium to the melt — a fairly spectacular operation but one which can be coped with.

Dispersion Hardening With Strong Particles of a Second Phase

The individual particles of a new phase introduced by alloying may be smaller and more complicated in shape than those which we have just discussed. Particles of this nature can have considerable strengthening effects because they tend to interfere with the movement of dislocations in the matrix phase more frequently than do larger, and hence more widely dispersed, particles. In this sense, there is an analogy with strengthening resulting from a reduction in the grain size of a single-phase metal or alloy (p. 296), but the influence on toughness can be different. Toughness can this time be impaired, to a degree that is dependent on the shape and dispersion of the strengthening phase as well as on its properties.

We can take as an illustrative example the addition of 15 vol % of iron carbide (cementite) to alpha iron (ferrite). The size and morphology of the cementite particles can be varied by various heat treatment procedures that will be discussed in Chapter 11. Ferrite, the structure of which is illustrated in Fig. 10.8(a), is comparatively weak, having a tensile strength of about 450 MPa (65 ksi) when containing the alloying elements normally present in commercial steels. Addition of the cementite particles, which are comparatively large and evenly distributed, as illustrated in Fig. 10.8(b), increases the tensile strength by a factor of about 1.4. This is the order of increase to be expected from the immediately preceding discussion. However, reducing the size, and hence the spacing, of the particles progressively increases the strengthening effect to a factor of a little over 2.0 for the finest particle sizes that can be achieved in practice (as illustrated in Fig. 10.8c and d). Consider now the effect of adding the same volume fraction of cementite as parallel plates in the structure known as lamellar pearlite. Widely spaced plates, such as those illustrated in Fig. 10.9(a), have a strengthening factor of about 1.4, which is roughly the same as that for spherical particles of about the same spacing. The strengthening factor again increases progressively as the spacing of the plates decreases. However, strengthening factors as high as 2.5 can be achieved in this case because it is possible to obtain more finely spaced structures, such as those illustrated in Fig. 10.9(b) and (c), by straightforward heat treatments. Strengthening factors of

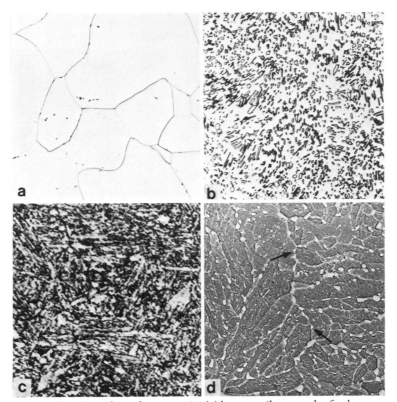

Ferrite without the cementite (a) has a tensile strength of only about 450 MPa (65 ksi). Addition of cementite as large spherical particles of the type shown in (b) increases the tensile strength of the ferrite by a factor of about 1.4, to approximately 630 MPa (90 ksi). The same amount of cementite when present as smaller spheres (c) has a strengthening factor of about 2.0, increasing the tensile strength to about 900 MPa (130 ksi). The cementite particles can barely be discerned in this optical micrograph, which can be compared to (b). An electron micrograph of the material in (c), in which the cementite particles (arrows) are now resolved, is shown in (d). (a) to (c) Optical micrographs: (a) 100×; (b) and (c) 1000×. (d) Replica electron micrograph; 10,000×.

Fig. 10.8. A qualitative illustration of the strengthening of ferrite by addition of 15 vol % cementite distributed as spheroidal particles.

5 or more can in fact be achieved in special circumstances because it is possible to reduce the lamellar spacing even further by cold drawing a fine pearlite (Fig. 10.9d).

Yield strength is increased in the same general way as tensile strength by the presence of cementite in ferrite, and it can be established for the lamellar structures that this increase is inversely proportional to the square root of the spacing of the cementite plates. This is the same type of relationship as that which governs the effect of grain size (p. 296), which is the reason for conclud-

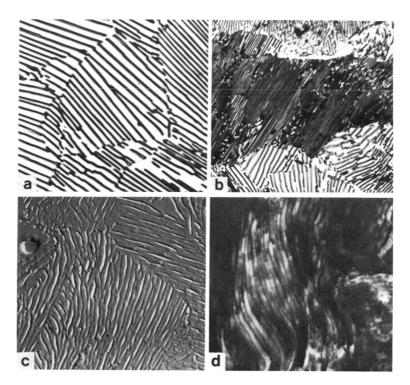

A large plate spacing of about 1 μm (40 μin.), such as that shown in (a), has a strengthening factor of about 1.4, increasing tensile strength from about 450 MPa (65 ksi) to about 630 MPa (90 ksi). This is the same nominal factor as that of the structure in Fig. 10.8(b). A finer spacing (b) of less than 0.5 μm (20 μin.) has a strengthening factor of about 1.7, increasing tensile strength to about 750 MPa (110 ksi). A still finer spacing (c) of about 0.2 μm (8 μin.) has a strengthening factor of about 2.5, increasing tensile strength to about 1100 MPa (160 ksi). This structure is about the finest that can be obtained normally by heat treatment. The structure shown in (d) is the finest lamellar structure that can be obtained in practice. It has a plate spacing of less than 0.1 μm (4 μin.) and a strengthening factor of over 5.0, increasing tensile strength to as high as 2500 MPa (360 ksi). Such a structure can be obtained only by cold drawing of a bar which has a structure similar to that in (c). This process is known as *patenting*, and the resulting products can have tensile strengths of up to nearly the theoretical strength of iron, which is the highest strength that can be obtained in a metal. Such structures can, however, be obtained only in wire, which is used in applications such as wire rope and cable, springs, and the strings of musical instruments.

(a) and (b) Optical micrographs; 1000×. (c) Replica electron micrograph; 5000×. (d) Transmission electron micrograph; 200,000×. Micrograph (d) is from *Metals Handbook*, 9th Ed., Vol 9, *Metallography and Microstructures*, American Society for Metals, Metals Park, OH, 1985.

Fig. 10.9 A qualitative illustration of the strengthening of ferrite by addition of 15 vol % cementite distributed as parallel plates.

ing that the same type of strengthening mechanism is operating — namely, blocking of dislocation movement by boundaries. The blocking boundaries in the present instance are those between the ferrite, in which the yielding has to occur, and the cementite, which is comparatively rigid, and hence the mean free path for dislocation movement is determined by the thickness of the ferrite plates in the pearlite, which is approximately a constant fraction of the lamellar spacing. The same thing applies to other morphologies of distribution of the second phase, including the spheroidal one. The spacing of the blocking particles on the operative slip planes of the matrix is thus the essential factor that determines strengthening. Particle size and shape have secondary influences insofar as they affect spacing.

Why then bother about the morphology of the strengthening phase? Why not, for example, settle for structures of lamellar pearlite in steels when they have such excellent strengthening effects? The reason is that phase morphology has a considerable influence on toughness if the phase is at all brittle, and many strengthening phases are brittle. Cementite certainly is. Consider lamellar pearlite again. Shearing forces fracture the brittle cementite plates easily in tension, and, perhaps more importantly, the presence of a crack in one plate causes a crack to be initiated in an adjoining plate. Thus a crack spreads from plate to plate in a chain reaction (Fig. 10.10). This occurs more easily the more closely the cementite plates are spaced. The propagation of cracks in this manner is a process which absorbs little energy that causes poor toughness. The detrimental effects of the cementite are not nearly so severe, however, when the cementite is present as spherical particles. The particles themselves are stronger mechanically, and so it is more difficult to initiate cracks in them. Moreover, longer and more convoluted crack paths have to develop in the ferrite between cracked cementite particles before a crack can propagate to fracture. Spheroidal structures consequently are much tougher than lamellar structures at comparable strength levels. Other structural arrangements of a brittle phase can, on the same grounds, be expected to have their own characteristic effects on toughness in a manner which is not necessarily related directly to their effects on strength.

Thus the strengthening potential of lamellar structures has to be applied with discretion when toughness is an important criterion. As an example, we saw in an earlier discussion on the development of high-strength weldable steels (p. 94) that the earliest attempts to do this were abortive because toughness was impaired too severely. This occurred because the strengthening methods adopted increased the volume fraction of pearlite in the steels and also reduced the lamellar spacing of the pearlite. The best combinations of strength and toughness are obtained in steels which contain no pearlite but in which strengthening is obtained by some other mechanism, such as dispersion hardening by small spheres of the type illustrated in Fig. 10.8(d) and (e). Nevertheless, there are applications where the toughness of lamellar structures is acceptable, and a useful range of steels is based on strengthening by this means.

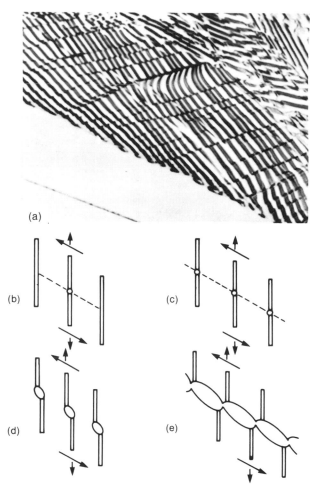

(a) A section through a test piece which has been strained in tension in a vertical direction. Crack systems have extended from plate to plate along shear planes. Optical micrograph; 1500×. (b) to (e) Diagrams illustrating a suggested mechanism for the shear cracking shown in (a). In (b), a crack initiates in a single cementite plate. In (c), plastic shear in the interleaving ferrite initiates cracks in adjoining cementite plates. In (d) and (e), fracture occurs in the interleaving ferrite by void formation and coalescence.

Fig. 10.10. An illustration of how cracks can propagate easily through a pearlite colony in steel.

There are many other important instances where practical use is made of this strengthening method. Many aluminum casting alloys, for example, are compounded to contain insoluble constituents which are responsible for much of the strengthening but which are themselves rather brittle. We have already mentioned the effect of the morphology of one of these — namely, the silicon phase in aluminum-silicon alloys containing large amounts of silicon. The example in Fig. 10.11 illustrates how the size and spacing of this phase can have a

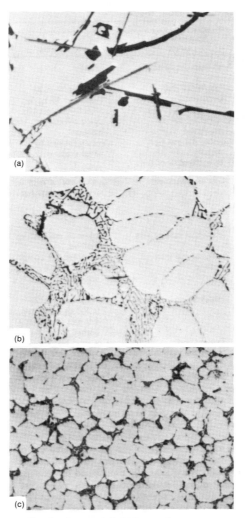

An Al – 5Si alloy has been solidified at cooling rates increasing from (a) to (c). The main phase present is silicon, and the size and spacing of the particles of this phase decrease with increasing cooling rate. As shown below, tensile and yield strengths increase as the structure is refined, but the tensile ductility remains substantially constant. Tensile strength, MPa: (a) 130; (b) 160; (c) 230. Yield strength, MPa: (a) 55; (b) 60; (c) 110. Elongation, %: (a) 8; (b) 10; (c) 9. Source: *Metals Handbook Desk Edition*, American Society for Metals, Metals Park, OH, 1985.

Fig. 10.11. An illustration of the effect on mechanical properties of the size and spacing of a comparatively large brittle constituent in an alloy.

marked effect on mechanical properties even when there is no major change in morphology. The strength of the alloy increases from structure (a) to (c) in Fig. 10.11 because the spacing of the phase particles decreases. Moreover, there is not a commensurate decrease in ductility, because the particles decrease in size.

The change in microstructure in this instance can be achieved by changes in solidification conditions which are at the direct control of the foundryman. This is the type of control that has to be exercised to produce premium-quality castings (see p. 17).

Dispersion Hardening With Weak Particles of a Second Phase

The particles of the second phase may be able to retard the movement of dislocations in the matrix phase but may be too small or too weak to block them completely. Particles which have this characteristic are much too small to be seen by optical microscopy, but usually are large enough to be detected by transmission electron microscopy (the particles illustrated in Fig. 6.6a and c, p. 194, are representative).

Larger particles of the type now under consideration can be bypassed by a moving dislocation, but to do so the dislocation has to bow out between the particles and then re-form by a mechanism such as that sketched in Fig. 10.12. A loop of dislocation is then created and left around each particle by each dislocation that moves past it. An additional force is required to move dislocations in this tortuous way, which means that the yield strength is increased. The magnitude of the effect increases with decreasing spacing of the particles down to a limiting spacing of about 50 atom layers. Dislocations cannot bypass particles

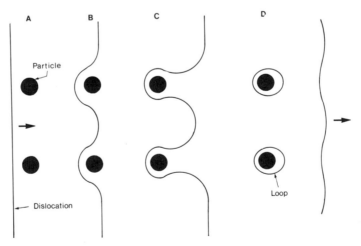

These diagrams represent various stages in the progress of the dislocation line past two adjacent particles. Note that a loop of dislocation is left around each particle, so that new lengths of dislocation are created

Fig. 10.12. Schematic diagrams indicating how a dislocation line, moving from left to right, has to bow and re-form if it is to move through a forest of small particles of a strong second phase.

which are more closely spaced than this because they cannot bend sharply enough to perform the convolutions sketched in Fig. 10.12. At best, however, the effect of this phenomenon on yield strength is small. Its effect on the rate of strain hardening, on the other hand, is large because new dislocation lengths are generated and become tangled with one another. Tensile strength consequently is increased more than yield strength (p. 74), which is another trap for those who pay more attention to tensile strength than to yield strength (p. 72).

Particles have to be very small in order to be so closely spaced that they cannot be bypassed by the bowing mechanism, and particles of this nature typically can be sheared by an impinging dislocation (Fig. 10.13). Once this has been done, the dislocation can continue on its way unaffected. Nevertheless, an additional force is required to shear the particles, and energy is required to create the new interface surface that is formed between the particles and the matrix (Fig. 10.13). Consequently, yield stress is increased. This effect is small for any individual particle, but it adds up to a considerable one when a very large number of particles are present, which can be the case (p. 371). The rate of strain hardening is not additionally affected in this case because new dislocation lengths are not created and a dislocation can continue to move unaffected once it has sheared a particle. Consequently, tensile strength is increased in proportion to the increase in yield stress produced by this hardening mechanism.

There is a final possibility — namely, that the particles are too closely spaced to permit the progression of dislocations by the bowing mechanism and are too strong to be sheared. The dislocations then have to move bodily over or under the particles to bypass them, which requires cross slip (p. 294) or even movement on slip planes other than those of easy glide. Both yield strength and rate of work hardening are then increased markedly. This is the most effective strengthening structure of all, but is only rarely obtained in practice. Perhaps it is a strengthening mechanism on which future alloy design will be based.

There are also situations that happen to be important in practice in which the formation of a discrete particle of a precipitate phase is preceded by a clustering of groups of atoms in the crystal lattice of the parent solid solution (p. 372). The same principles that apply to the strengthening influence of discrete particles apply also to these clusters. In practice, cluster zones typically are too closely spaced to permit bowing of dislocations and too weak to resist shearing by the dislocations. Strengthening consequently occurs by the shearing mechanism illustrated in Fig. 10.13. The clusters in suitable alloys tend to be very numerous, in which event they become potent sources of strengthening.

The influence of these small particles and clusters on toughness is more subtle than that for the larger particles discussed earlier. The dominating factor here is the ease with which cracks can be propagated through the matrix of the alloy rather than by the initiation of internal cracks in brittle particles. The energy absorbed during the propagation of a crack is determined by the amount

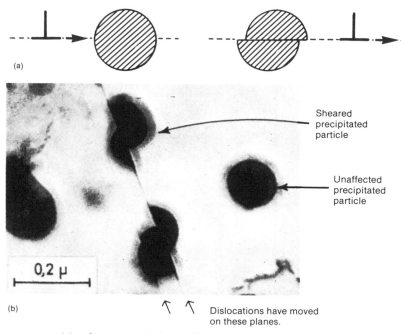

(a)

Sheared precipitated particle

Unaffected precipitated particle

0,2 μ

(b)

Dislocations have moved on these planes.

(a) A diagrammatic sketch illustrating how a spherical precipitate may be sheared by an edge dislocation moving from left to right. Note that additional interfacial area is created between the two phases and that the dislocation proceeds unaffected once it has sheared the particle.

(b) An actual example in a nickel-base superalloy. The traces of two planes on which dislocations have moved and sheared precipitated particles are visible in the center of the field. Transmission electron micrograph; 100,000×. The micrograph shown in (b) is from H. Gleiter and E. Hornbogen, *Acta Met.* 1965, *13*, 577; also in C. R. Brooks, *Heat Treatment, Structure and Properties of Nonferrous Alloys*, American Society for Metals, Metals Park, OH, 1982.

Fig. 10.13. Particles which are too closely spaced to allow passage of dislocations by the bowing mechanism illustrated in Fig. 10.11 have to be sheared by a dislocation which encounters them if the dislocation is to continue moving.

of plastic deformation that occurs in a small region surrounding the tip of the crack. The very aim of the strengthening mechanisms that we have just been discussing is to inhibit plastic deformation. It follows that all of them are likely to impair toughness. So it is easy to see in general terms that a compromise between strength and toughness may have to be reached. But analysis of these phenomena in a more specific way is a much more difficult matter. It is particularly difficult to predict means by which strength can be increased with minimum detriment to toughness. In the ultimate, the optimum combination for a particular alloy has to be determined by experiment. What metallurgists are seeking are guidelines by which to work that will allow this expensive experimentation to be reduced to the minimum.

SOME OTHER LIMITATIONS OF STRENGTHENING MECHANISMS

In this chapter we have discussed one factor (toughness) which can impose a limit on the advantage that can be taken in practice of certain strengthening mechanisms. There are other factors which can be equally restricting in particular circumstances. There are, for example, occasions where increasing static strength does not improve fatigue resistance commensurately. In this event, the strengthening may have little advantage in components which are susceptible to failure by fatigue (this is the case with some high-strength aluminum alloys). There are also occasions where the stronger alloys become highly susceptible to failure under static loading in the presence of a corrosive medium. This phenomenon of stress corrosion (p. 158) is also a common cause of failure in service, and sensitivity to stress corrosion can seriously inhibit the usefulness of a particular alloy.

A laboratory program which has the objective of developing stronger alloys consequently has to include a testing program designed to reveal deficiencies of these types. This is expensive. Moreover, laboratory tests are not always completely reliable indicators of behavior in actual service. A crippling deficiency might be revealed only after service in the real world, an event which unfortunately does occur. The real engineering world is so complex that progress often includes an element of trial and error. The lot of those who attempt to develop high-strength alloys is not an easy one.

FURTHER READING

Alloy and Microstructural Design, edited by J. K. Tien and G. S. Ansell, Academic Press, New York, 1976.

R. W. K. Honeycombe, *Steels, Microstructure and Properties*, Edward Arnold, London, 1981.

C. R. Brooks, *Heat Treatment, Structure and Properties of Nonferrous Alloys*, American Society for Metals, Metals Park, OH, 1982.

· 11 ·

Changing the Structures
of Metals by Heat Treatment

SUMMARY

Most of the heat treatment procedures used in metallurgical practice are based on the central tenet of physical metallurgy that a causal relationship exists between structure and properties. The objective of a heat treatment then is to produce a structure which is known to be associated with the properties desired in the intended application, having first chosen an alloy which is of a suitable composition. As a generalization, the alloy is heated to a temperature at which a selected phase change occurs and then cooled in a controlled manner through the range of temperatures in which the reverse phase changes should occur. This primary cycle may in some instances be supplemented by further heating cycles to modify the structure obtained. These principles can be illustrated in copper alloys containing about 40% zinc in which the proportions and distributions of two phases, alpha and beta, can be varied over a considerable range. Heating the alloy to a high temperature transforms it completely to beta phase. Both the proportion and the morphology of the alpha phase that then forms during cooling can be varied by varying the cooling rate. These effects can be understood in terms of the transformation of beta phase at constant temperature in the range over which it is unstable. A highly supersaturated solution of beta phase can be obtained at room temperature by cooling at a high enough rate to suppress completely the formation of alpha phase. Heating the supersaturated solution at low temperature then causes alpha phase to precipitate and thus to reattain equilibrium. These particles have their own special morphology. In some alloy systems, moreover, extremely fine precursors to the equilibrium precipitates can be obtained in this way, and they can have strong hardening effects when the precipitation conditions are chosen correctly. This precipitation hardening, or age hardening, constitutes one of the basic practical heat treatment procedures. High-strength aluminum alloys are, for example, based on it. Another important group of practical heat treatment procedures are based on eutectoid phase systems. In steels, for example, the distribution of the two equilibrium phases, ferrite and cementite, can be varied by first form-ing the high-temperature-stable phase austenite by heating to a tempera-ture at which austenite is stable. The equilibrium phases then re-form dur-ing comparatively slow cooling, but in a form and distribution which vary with the cooling rate. Cooling at a high enough rate suppresses this trans-formation altogether, and a metastable phase (martensite) which has spe-cial characteristics is formed. This phase tends to be hard and strong, but brittle. Toughness can be restored by decomposing the martensite into a fine distribution of the equilibrium phases by heating at a comparatively low temperature. All of these phenomena can be understood in terms of the transformation of austenite at constant temperatures in the range in which it is unstable. A third generic type of heat treatment involves heat-ing at a temperature at which a phase transformation does not occur but at which one phase may change its shape under the influence of interphase surface tension forces. The phase particles then tend to adopt a spherical shape. All treatments in which a new phase is formed by heating to a high temperature establish a new grain size and provide an opportunity to re-fine the grain size.

You have by now been thoroughly indoctrinated with the concept that many of the important properties of a metal are governed by its microstructure — that is, by the phases that are present and the way in which these phases are ar-ranged. Properties can be changed for good or bad by changes in microstruc-ture, but the objective in physical metallurgy must be, of course, to produce the most desirable types of microstructures with deliberate intent. The question that remains is how to do this. One of the basic ways is to subject an alloy to a specific and characteristic thermal cycle in which the metal is heated to a chosen temperature and then cooled back to room temperature in a particular way. The heat treatment of metals in this way actually has been carried out for a long time, but formerly only as an art. Metalsmiths, for example, have known for over 2000 years how to harden the edges of steel tools, but only in much more recent times have metallurgists as scientists and technologists come to understand what was actually being done to the steel, and to know how to obtain optimum results in a controlled way.

Not all alloys are candidates for heat treatments of the types which we shall discuss. Pure metals and many terminal solid solutions are examples of metals that are not. They are amenable to heat treatments that are intended to homog-enize chemical composition (p. 257), or to recrystallize grains after straining (p. 306), but none of these heat treatments changes or rearranges the phases that are present. Candidate metals have to be alloys which can contain two or more phases, the relative stabilities of which change with temperature. These are features which can be recognized in the *phase diagram* (alternatively called the *constitutional* or *equilibrium diagram*) of the particular alloy system which, in general terms, describes graphically the effects of composition and temperature on the stability of various phases in the system. We have discussed earlier (p. 281) the manner in which diagrams of this type can be used to describe graphically various interactions between solid and liquid metals. We now have to discuss how they can be used to describe interactions which occur within a solid mass of metal. Anyone who has studied physical chemistry has learned to accept that reactions between gases and liquids, and between either of these and a solid, are part of the natural order of things, but it is more difficult to accept that similar reactions occur within solids. Nevertheless, they do, and it is the fruits of these reactions that we are about to discover.

SOME BASIC CHARACTERISTICS OF PHASE TRANSFORMATIONS IN SOLIDS

Phase Diagrams

An understanding of the theory and use of phase diagrams is one of the foundations on which the science of physical metallurgy is based. But for our purposes a simple understanding of how they may be used will suffice. Phase diagrams are available for all conceivable binary combinations of metals and for combinations between metals and some appropriate nonmetals. Diagrams are also available for some multiple-element systems or portions thereof. Many of these diagrams appear to be complex, but, as with the liquid-solid reactions discussed earlier (p. 281), they usually can be broken down into a number of easily understandable units. Indeed, specific heat treatment procedures characteristically involve only one or two such units. So the approach that we shall adopt will be to deal with some of the more important of these units individually.

Let us look at the information contained in one of the simplest types of units, as represented by that portion of the copper-zinc phase diagram which describes the relationships between the alpha and beta phases of the system (Fig. 11.1). This diagram simply states that at, say, temperature t_1 the amount of zinc that can be dissolved in copper as a terminal solid solution (p. 230; it is called *alpha* phase) is limited to a%. Alloys containing more than a% zinc consist of a mixture of alpha phase and an intermediate solid solution known as *beta* phase. The alpha phase of this mixture always contains a% zinc and the beta phase always contains b% zinc, the volume fraction of alpha phase decreasing from 100% to 0% as the zinc content increases from a% to b%. The actual volume fraction is determined by the requirement to maintain the particular average composition. Alloys containing more than b% zinc consist at this temperature solely of beta phase, the zinc content of this phase increasing with the zinc content of the alloy up to a limit at which another phase, the *gamma* phase, is introduced into the structure. We shall not be concerned with the gamma phase. At temperature t_2, on the other hand, the limits for the stability

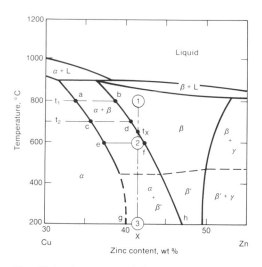

Fig. 11.1. A portion of the copper-zinc phase diagram. Details are discussed in the text.

of the alpha and beta phases are c% and d% zinc, respectively, and so on. Loci of the two sets of points, such as aceg and bdfh, can be drawn to define the composition/temperature boundaries of the fields in which the various phases are present in the structure. The boundaries are called *phase boundaries*, and the fields which they enclose are labeled to identify the phases that are stable within them. The intersections of the phase boundaries with the room-temperature ordinate indicate the phases to be expected in an alloy as it is normally used — if it has attained an equilibrium structure, which is a big if.

A phase diagram sets out, in fact, to describe *only* the states of true equilibrium, or at least those as close to true equilibrium as it is humanly possible to achieve. Normally, in practice, only conditions that are a fair way removed from this ideal can be achieved, particularly at lower temperatures. Consequently, we have to learn to interpret the diagrams in such a way as to derive indications of the structures likely under more realistic conditions. Moreover, the user generally is interested only in the structure that is to be expected at room temperature, and so you might well ask why we should bother at all about the structures that are possible at higher temperatures. The answer is that the high-temperature portions of a diagram indicate the structural changes that have to occur during any excursion of heating and cooling through the region concerned. This gives vital clues as to the possible consequences of any heat treatment or temperature cycle, be it intentional or accidental, which in turn gives clues as to what the room-temperature structure is really likely to be. What a phase diagram does *not* do is to give any indication of the ease with which equilibrium structures are likely to be attained — that is, the kinetics of the reactions that are involved. Neither does it give any indication of the shapes and distributions of the phases that are likely to be present. These two critical factors have to be considered separately.

Transformation at Constant Temperature

Let us now return to Fig. 11.1 and use it as an example of how a phase diagram might be used as a basis for the study of phase transformations. Let us study an alloy of the composition X indicated on this figure and assume for a start that equilibrium is maintained at all times. Let us assume that the alloy is at 800 °C (1470 °F), where, as the phase diagram indicates, it consists entirely of beta phase (point 1 in Fig. 11.1). The structure would then be similar to that illustrated in Fig. 11.2(a) if we could examine it at temperature. We could do this actually, but here we have retained the structure by rapid cooling to room temperature and examined it there. You will see later that this is a valid technique. It is certainly a much more convenient one experimentally.

The phase diagram indicates that when this alloy has been cooled at 600 °C (1110 °F) (point 2 in Fig. 11.1), a temperature slightly below t_X on the phase-

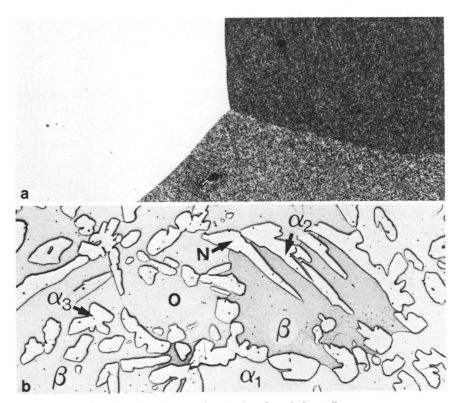

(a) Heated to 800 °C (1470 °F) and cooled rapidly to room temperature. Structure representative of point 1 in Fig. 11.1.

(b) Heated to 800 °C (1470 °F), cooled very slowly to 600 °C (1110 °F) and then cooled rapidly to room temperature. Structure representative of point 2 in Fig. 11.1.

Optical micrographs; 100×.

Fig. 11.2. Microstructures of the Cu–40Zn alloy labeled alloy X in Fig. 11.1.

boundary line for alloy X, the structure consists of a mixture of a small volume fraction of alpha phase containing e% zinc and beta phase containing f% zinc. The structure illustrated in Fig. 11.2(b) is representative of the phases and the proportions to be expected. We now have to inquire how this structure formed from the all-beta-phase structure of Fig. 11.2(a). First of all, a number of nuclei of alpha phase would have to appear throughout the beta-phase crystals; perhaps the point labeled N in Fig. 11.2(b) was one of these. The formation of each of these nuclei requires that the crystal structure of a volume of beta phase change from the body-centered cubic of the beta phase to face-centered cubic, which is the crystal structure of alpha phase. Moreover, it has to do this throughout a certain minimum volume of material to be stable; a potential nucleus which is too small will be reabsorbed, and there are well-established thermodynamic reasons for this.* The rearrangement of the atoms needed to effect this change in crystal structure has to occur essentially by statistical chance, and this takes time. Consequently, there is a delay before a nucleus becomes active and the transformation starts. The driving force for nucleus formation is small at temperatures just below the equilibrium transformation temperature, and the delay time is comparatively long. Moreover, comparatively few widely spaced nuclei become active. So only a few grains of alpha phase would be nucleated in our alloy X at 600 °C (Fig. 11.2b), and even then only after a comparatively long time delay.

Suppose, however, that things were arranged so that the beta phase began to transform at a temperature well below t_X. The driving force for nucleus formation would then be greater than at 600 °C (1110 °F), and hence stable nuclei would appear after a shorter time delay. Moreover, they would be more numerous and more closely spaced, increasingly so as the degree of undercooling increased. However, there is a limit to this trend, because the composition of a nucleus region has to change as well as its crystal structure. In alloy X at 600 °C (1100 °F), for example, zinc atoms would have to diffuse away from the nucleus site to reduce the local composition from X% to f% zinc. This diffusion inevitably occurs more slowly as the undercooling increases and the temperature is lowered (p. 235), and so a counteracting effect is introduced which tends to discourage the formation of nuclei. Eventually the diffusion-limiting effect dominates and the nucleation rate decreases with a further decrease in temperature.

The second step, following this first step of nucleation, is for each active nucleus to grow progressively into a larger crystal of the new phase, such as the

---

*For a particle of a new phase to be stable, the energy saved by the creation of the particle must exceed the energy needed to form the new interphase surface. The former is related to the volume of the phase, and the latter to its surface area. The ratio of surface area to volume is largest with smaller particles, and so this energy balance is unfavorable until a particle exceeds a certain size.

crystal of alpha phase surrounding the nucleus labeled N in Fig. 11.2(b). Once growth is under way, it is generally easier for more of the new phase to be added onto these crystals than for new nuclei to become active. So the first-formed crystals of the new phase tend to continue to grow until the required volume fraction has formed. Considerable local changes in composition have to occur to effect this growth process. In our example, zinc atoms have to diffuse away from the boundary of each crystal of alpha phase to reduce the zinc content from X% to e% before the boundary can move to enlarge a crystal of alpha phase. These zinc atoms have to diffuse throughout each remaining crystal of beta phase to increase its zinc content to f% everywhere, such as at point O in Fig. 11.2(b). Consequently, the growth phase of the transformation slows down exponentially with increased undercooling, because the diffusion rate slows down exponentially (p. 235).

The interplay between these factors can be summarized graphically in an *isothermal transformation diagram (IT diagram)*, also known as a *time-temperature-transformation diagram (TTT diagram)*. The times for the transformation to start and to finish are plotted against temperature, as indicated in Fig. 11.3(a), and intermediate curves for the various proportions of transformation can also be plotted if desired. The characteristic behavior is for the transformation-start curve to have a "C" shape. The start (i.e., nucleation) time is reduced at first because the increasing drive to form nuclei with increasing departure from equilibrium is dominant. The nose in the start curve occurs when this drive is more than counterbalanced by the effect of the reduction in diffusion rate on nucleus formation. The transformation-finish curve follows the general shape of the start curve, although the displacement between the two increases in absolute time as the temperature decreases (time is plotted on a logarithmic scale in IT diagrams, and this latter point is not immediately obvious). This is because of the increasing influence of the reduction in diffusion rate on the rate of growth of the new phase, and because larger volume fractions of the new phase must form at lower temperatures to maintain the equilibrium phase relationships.

Morphology of the Transformation Products

The factors that we have discussed so far bear only on the amount and composition of the new phase that forms. The other factor of importance is its morphology — that is, the size, shape, and distribution of the individual particles of the phase. This feature is determined by a different set of factors from those already discussed — namely, by the location of the nuclei and the crystallography of the growth process.

We have already noted that nucleation is infrequent with small degrees of undercooling, that comparatively few grains of the new phase are then able to form, and that each of them has to grow to a comparatively large size to make

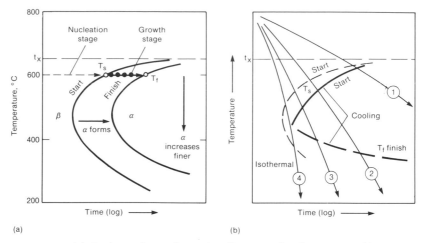

(a) Isothermal transformation diagram. This diagram would be determined by experiments of the type illustrated in Fig. 11.13. The curves drawn are the loci of the start and finish times determined for a range of temperatures. Curves for other fractions of transformation could also be constructed if desired. Note that time has to be plotted on a logarithmic scale to accommodate the range involved.

(b) A corresponding continuous cooling transformation diagram. The dashed curve at the left is the isothermal start time curve transposed from Fig. 11.3(a) for comparison. The light curves (numbered 1 through 4; see text) are cooling curves for cooling at linear rates. This diagram would be determined by cooling specimens at a linear rate for a known time, and then quenching them. The times along the cooling curve for transformation to start and to finish would be determined for each cooling rate, and loci of the points drawn would serve as the start and finish curves.

Fig. 11.3. Diagrammatic sketches of isothermal and continuous cooling transformation diagrams for alloy X in Fig. 11.1.

up the required volume fraction of transformation product. More nuclei become operative at lower transformation temperatures, and so the particles of the new phase become more numerous and each becomes smaller. This can be illustrated by the photomicrographs in Fig. 11.6; the specimens illustrated there were heat treated in a slightly different way from that being implied in the present discussion, but the end result is the same. A second point of significance is that the nuclei are not necessarily located randomly throughout the grains of the parent phase. On the contrary, they tend to develop at preferred sites, the most favored ones being the grain boundaries. It can well be imagined that it would be easier for the new arrangement of atoms that constitutes a nucleus to be established in this disordered region (p. 241) than throughout the more perfect core of the crystal. Next in favor as nucleus sites come certain crystal planes in the parent crystal. The preferred planes tend to be those along which interfaces can be developed between the old and new phases such that the atoms of the two crystals match with minimum disturbance (see p. 245). Random sites throughout the parent grains are usually the final choice and tend to become active only when more-preferred sites are remote or unavailable. Local disturbances such as the interfaces at foreign inclusions or constituents, or crystal defects such as dislocations, are particularly likely points of more random nucleation.

Now let us look at how the growth of these nuclei is influenced by the crystallography of both the parent and growing phases. Nuclei established at a grain boundary characteristically grow by extending along the boundary, thickening only a little because thickening is energetically more difficult. The growing films join up with one another, perhaps, to form a continuous film around the boundaries of the parent grain. Only a comparatively small total volume fraction of the phase may be needed to do this. *Grain-boundary allotriomorphs* is the grand name given to this type of arrangement. Grain-boundary allotriomorphs do thicken a little during growth, during which process they tend to develop faceted sides (e.g., grain α_1 in Fig. 11.2b; see also Fig. 6.3, p. 189). These facets are the result of the development of low-energy interfaces on which there is a favorable match between the adjoining crystal lattices. Long side plates may even extend well into the parent grain (grain α_2 in Fig. 11.2b).

Nuclei developed along preferred planes within the parent grains frequently grow as very elongated plates or needles to establish minimum-energy interfaces. It is then energetically much easier for a plate to grow by lengthening than by thickening. Groups of parallel plates tend to form in one region of a parent grain, and similar groups on different and intersecting sets of the preferred planes in other regions. Good examples are to be seen in the central grain in Fig. 11.4(b) and the grain at the top of Fig. 11.5(a). This builds up a sort of chevron pattern, which is the basis of the morphology known as a Widmanstätten pattern (cf. Fig. 11.4b with Fig. 6.1, p. 185, and Fig. 6.3, p. 189).

Although the grains of a new phase always grow in these modes, this is apparent in the final structure only when a limited volume fraction of the new phase has formed (Fig. 11.5a). Eventually, with larger amounts of the phase, the grains grow to meet one another and then interfere with their respective growth patterns. They fill the available space and appear in massed areas with, perhaps, isolated areas of residuals of the parent phase left in between (Fig. 11.5b). Each massed area typically is occupied by a number of grains. Most commonly they are equiaxed (Fig. 11.5b), although they may have an elongated shape inherited from Widmanstätten-type growth (Fig. 11.20a). We should also remember that the morphology of the new phase may be changed by subsequent cold working (see, for example, Fig. 9.12, p. 302), and even further by cold working and annealing.

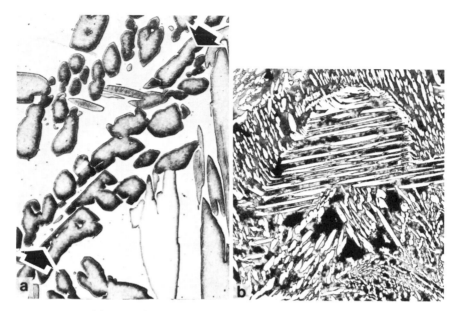

(a) Very slowly cooled. Large crystals of alpha phase (dark here) have formed along a grain boundary in the beta phase (arrows) and as plates within the beta grains.

(b) More rapidly cooled. The crystals of alpha phase (light here) are smaller, more numerous, and more closely spaced. Most have formed as Widmanstätten plates. Note that regions of a parent crystal are occupied by a family of parallel plates of beta phase, the families being oriented differently in different regions of the parent grain.

Optical micrographs; 100×.

Fig. 11.4. Microstructures of the copper-zinc alloy labeled X in Fig. 11.1. Specimens were cooled at different rates from a temperature at which they were comprised entirely of beta phase.

Transformation During Continuous Cooling

Transformation during cooling can be described graphically in a diagram, similar to an isothermal transformation diagram, which we can call a *cooling transformation diagram (CT diagram)*. A hypothetical example is sketched in Fig. 11.3(b), in which the relationship to the corresponding IT diagram in Fig. 11.3(a) is also indicated. The pertinent points illustrated in Fig. 11.3(b) concerning the start (i.e., the nucleation) of the transformation are:

1. At very low cooling rates (curve 1), transformation commences at a temperature below, but close to, the equilibrium transformation temperature.
2. With increased cooling rates, transformation commences at progressively lower temperatures and after progressively longer times (curves 2 and 3).

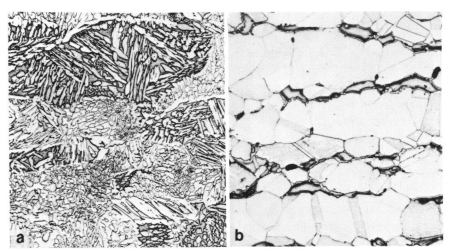

(a) An alloy where the balance of composition and cooling rate has been such that alpha phase (light) is in the minority. The Widmanstätten mode of growth of the alpha phase is apparent.

(b) A slightly different alloy in which the alpha phase is in the majority. The alpha phase grew as Widmanstätten plates, but the plates have joined together as massed areas. The remaining beta phase occupies the residual areas between these massed areas.

Optical micrographs: (a) 100×; (b) 500×.

Fig. 11.5. Representative microstructures of wrought Cu-40Zn alloys.

3. A cooling rate is eventually reached (called the *critical cooling rate*) which does not allow sufficient time at high temperatures for the transformation to start (curve 4); the transformation is suppressed. A phase which theoretically is stable only at high temperatures is then retained to room temperature (although it may transform in an entirely different non-equilibrium reaction at lower temperatures [see p. 389], but brasses do not do this).

The structure illustrated in Fig. 11.2(a) was obtained by cooling at a higher-than-critical rate, and the phase seen in this photomicrograph is just as it would have appeared at 800 °C (1470 °F).*

Further complications arise during continuous cooling after the transformation has started. Typically, the volume fraction of transformation product

---

*You may have noticed, however, that the beta fields are labeled β' in Fig. 11.1 at temperatures below about 450 °C (840 °F). A change occurs in the arrangement of the copper and zinc atoms in the crystal lattice of the β phase at this temperature, the zinc atoms taking up regularly ordered positions in the lattice (p. 228). The phase with the ordered crystal structure is called β' (*beta prime*). The ordering cannot be suppressed by rapid cooling, so the phase illustrated in Fig. 11.2(a) is β' and not β. This transformation is not without its importance, but we can ignore it here. The beta phase also undergoes a further phase transformation at about 250 °C (480 °F), but the probability of this reaction occurring in practice is so remote that it too can be disregarded here.

that forms should increase as the temperature falls because of the changing equilibrium phase relationships. The equilibrium volume fraction of alpha phase in our alloy X, for example, increases steadily from 0% at t_X to about 75% at 200 °C (390 °F) (Fig. 11.1). At the same time, the rate at which a grain of alpha phase can be expected to grow decreases because the migration of the boundary is diffusion-controlled (p. 244). Moreover, the two phases have to adjust their compositions continuously as the new phase forms. Thus the grains of alpha phase which form first in alloy X have to adjust their composition as they cool and grow by losing zinc as per line eg in Fig. 11.1; the untransformed remnants of the beta grains have to gain zinc as per line fh. The diffusion of zinc atoms required to make these adjustments becomes increasingly difficult as the temperature falls, and eventually becomes impossibly so. The net result is that the transformation slows down as the temperature falls and, to all intents and purposes, stops at some temperature such as that indicated by the heavy broken line in Fig. 11.3(b).

The volume fraction of the new phase that is present in the structure when material is continuously cooled to room temperature consequently is likely to be smaller than that predicted by the phase diagram. Moreover, the discrepancy is likely to increase as the cooling rate is increased. Material cooled at a very low rate (curve 1 in Fig. 11.3b) thus has the opportunity for most of the equilibrium volume fraction of alpha phase to form and will contain close to the equilibrium amount of alpha phase (Fig. 11.4a), but the opportunity is smaller in material cooled at the rate represented by curve 2 and so less alpha phase forms (Fig. 11.4b). Very little will form at the rate illustrated by curve 3 in Fig. 11.3(b) and, of course, none at all when the critical cooling rate represented by curve 4 is exceeded, which was so for the material illustrated in Fig. 11.2(a).

The depression in the temperature at which the transformation starts during continuous cooling also has an influence on the morphology of the transformation product. The number of nuclei of a phase of this type which become active during a cooling sequence is determined essentially by the number that become active when, or soon after, transformation starts. Thereafter, newly forming transformation product grows mostly on pre-existing grains rather than by the nucleation of new grains. Thus the number of particles of new phases increases with increasing cooling rate and the size of each particle decreases (Fig. 11.4). But this can continue only down to a limit set by the number that nucleate at the nose of the CT curve. This limit arises because, as we have just seen, transformation cannot be nucleated at a temperature below the nose temperature during continuous cooling. Thus a structure much finer than that illustrated in Fig. 11.4(b) cannot be obtained by continuous cooling in the brass being illustrated here.

Decomposition of Metastable Phases

A high-temperature phase retained at room temperature, such as the beta phase in Fig. 11.2(a), is stable only in the sense that it will remain unchanged at room temperature indefinitely by human standards. This is so only because the diffusion that is essential for decomposition occurs only at a vanishingly low rate. The metastable phase will decompose, however, when it is heated at a temperature at which the diffusion rate becomes finite (Fig. 11.6). The time required, the nature of the transformation product, and the morphology of the product vary with the temperature (Fig. 11.6). They are all essentially the same as for isothermal transformation after direct cooling to the temperature concerned. More finely dispersed structures can thus be formed by decomposing a metastable phase than are possible during continuous cooling, because nucleations can be achieved at a lower temperature (cf. Fig. 11.2a, 11.4b, and 11.6).

Sometimes a metastable phase decomposes in a different way from that predicted by the above equilibrium considerations. The equilibrium transfor-

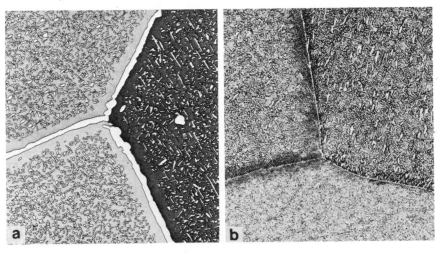

(a) Heated at 600 °C (1110 °F) for 1 h. Comparatively large particles of alpha phase (light) have formed as films along the grain boundaries and as plates within the grains of beta phase.

(b) Heated at 400 °C (750 °F) for 1 h. Particles of alpha phase have again precipitated, but the particles are much smaller and more closely spaced than those in (a). Note that these structures are finer than the finest that can be obtained by continuous cooling, which is about that illustrated in Fig. 11.4(b).

Optical micrographs; 100×.

Fig. 11.6. Photomicrographs illustrating the decomposition of metastable beta phase during heating at elevated temperatures. The starting material was alloy X in Fig. 11.1, with the microstructure illustrated in Fig. 11.2(a).

mation products are ultimately produced, but through a sequence of intermediate phases which have special morphologies. These phenomena will receive
special attention in the sections on precipitation hardening (p. 367) and on tempering of martensites (p. 408).

Phase Transformations on Heating to High Temperatures

Phase transformations also occur when an alloy is heated into a temperature
range in which a new phase becomes stable. For example, our alloy X with
structures such as those illustrated in Fig. 11.2(b) and 11.4 will transform completely to beta phase as in Fig. 11.2(a) when heated to a temperature above t_X in
Fig. 11.1.

In cases where none of the high-temperature phase is present in the parent
structure (e.g., in the steel illustrated in Fig. 11.7a), nuclei of the new high-
temperature phase have to form first and then these nuclei have to grow to
absorb the low-temperature phase (Fig. 11.7). The nucleation stage is obviated
when the high-temperature phase pre-exists in, say, a duplex structure such as
the alpha-beta brasses illustrated in Fig. 11.4 to 11.6, but the growth stage still
has to occur. The growth stage is diffusion-controlled, particularly when adjustments in composition are necessary, and so the transformation always requires a finite time (Fig. 11.7). The lower the diffusion rates of the elements
concerned, the longer this time will be. Nevertheless, the transformation times
tend to be short compared with those for cooling transformations because the
transformation is attempted only at comparatively high temperatures. Moreover, ample time frequently is available during a heating cycle and it may be
possible to hold the alloy at the transformation temperature, luxuries which
are not available during cooling. Thus it is comparatively easy to obtain an
equilibrium structure during heating.

A new family of grains of the high-temperature phase is generated during a
heating transformation whenever nucleation is needed, from which it follows
that an opportunity then arises to change the grain size. This is illustrated
diagrammatically in Fig. 11.8. The new grain size is determined by the number
of nuclei that become active and by the rate at which the new grains grow.
Nucleation sites typically are numerous, the most highly preferred sites being
phase interfaces in the parent structure (e.g., the cementite/ferrite interface in
Fig. 11.7) and the grain boundaries of the parent phase (as implied in Fig. 11.8b).
The base grain size establishes when the transformation has just been completed (that is, when the boundaries of the grains growing from these nuclei
have just met one another), at which stage the grain size typically is smaller
than that of the parent grains by a considerable margin (cf. Fig. 11.8a and c).
This is because a number of nuclei are likely to become active in each parent
grain. Thereafter, some of the new grains grow at the expense of others (cf.

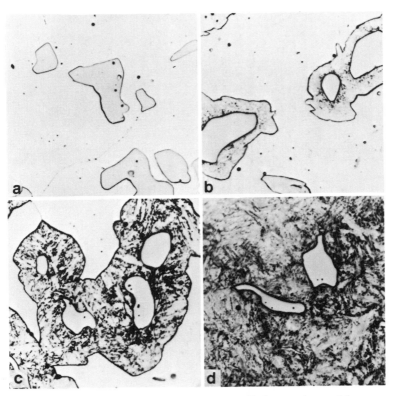

The material starting structure consisted of cementite particles (light gray) dispersed in a matrix of ferrite (white). The steel was heated to 745 °C for a measured time and then cooled rapidly.

(a) Original structure. (b) Heated for 5 s. The austenite that formed at temperature has transformed to martensite during cooling and is seen here as dark gray rims to the cementite particles. Only a small volume of ferrite has transformed to austenite in this time. (c) Heated for 15 s. More ferrite has transformed to austenite and some cementite has dissolved, the carbon so released diffusing into the austenite. (d) Heated for 60 s. All of the ferrite has now transformed and most of the cementite has dissolved. But this has taken a finite time.

Optical micrographs; 1500×.

Fig. 11.7. A series of photomicrographs illustrating the formation of austenite in steel by a transformation on heating.

Fig. 11.8c and d), the higher the superheat temperature the more rapid the grain growth and hence the larger the final grain size.

Heating conditions thus have to be well controlled if minimum grain size is to be obtained in the transformation product. The material has to be heated to a temperature as little as possible above the transformation temperature and to be held there for the minimum time necessary for complete transformation.

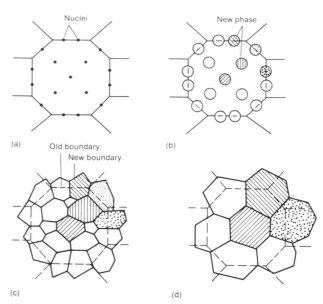

(a) Original grain structure, sites of potential nuclei for the formation of new grains being indicated.   (b) New grains have formed at the nuclei and grown radially.

(c) The new grains have grown until they have just contacted one another and so stopped their mutual growth. The resultant grain size is finer than the original grain size and is a minimum for the conditions.

(d) The grain size has increased due to the growth of some grains at the expense of others. Specific grains are shaded in the same way in (c) and (d) to illustrate this point.

Fig. 11.8. Sketches illustrating the possibility of changing the grain size during a heating transformation.

However, metallurgists sometimes are able to modify alloys so that these requirements can be relaxed. They do this by adding small amounts of alloying elements which appear to form finely dispersed and stable insoluble particles. The particles stop the migration of grain boundaries that is essential for grain growth. The grain-refining influence usually breaks down at a certain temperature, presumably because the particles coarsen, but all that is required even then is to ensure that the coarsening temperature is not exceeded. Steels can be treated in this way by adding very small amounts of elements such as aluminum, titanium, and zirconium. These elements are thought to form compounds with the nitrogen that is always present. Steels so treated are often referred to as *fine-grained steels*, which means in practice that the grains stay small during austenization until a certain temperature is exceeded.

The practical end result is that the presence in the parent phase of finely dispersed particles, a large grain-boundary area (fine parent grain size), and low superheat temperatures favor the development of a small grain size in a phase formed at a high temperature. Coarse grain structures developed during previous uncontrolled or poorly controlled excursions to high temperatures can be refined by reheating in this way under more closely controlled conditions, and the grain refinement often can be enhanced by successive cycles of transformation. But remember that this result is achieved only if the heating temperature and time are properly selected and controlled. The result is a desirable one on its own merits and because it encourages a fine morphology in any transformation product formed in the grains during subsequent cooling. One of the repeated themes of modern metallurgy is the beneficial influence of fine grain size on mechanical properties.

Residual Stresses Induced During Heat Treatment

We have noted on several occasions that residual tensile stresses may contribute to unexpected service failures because they supplement applied stresses. They may consequently tip the balance toward failure even when the stresses applied by the user apparently are within safe limits. The rapid cooling from a high temperature that is required in some heat treatment cycles is one important source of internal stresses. The stresses develop due to the interaction of four factors — namely, the temperature gradient that is set up through a section when heat is extracted from its surface, the contraction that occurs when the temperature of a metal is reduced, the volume change that occurs during a phase transformation (if any), and the decrease in yield strength with temperature that occurs in metals.

Consider a highly simplified model in which a cylinder is divided into a surface shell and an inner core in a manner similar to that sketched in Fig. 9.13 (p. 303), and consider first only the residual stresses that arise from the thermal contraction. The situation when the shell has cooled to room temperature and the core is still warm is that the shell has contracted onto a still-warm core which has a comparatively low yield strength at the time. The core is then able to deform plastically as it is squeezed by the contracting shell. But then the core's turn to contract comes, and in so doing it tries to pull the shell in with it. But by this time the cold shell may have become too strong to be easily deformed plastically. It is still compressed, but it is at least partly compressed elastically and is subsequently held in elastic restraint. Thus compressive residual stresses are induced in the surface layers, and they have to be balanced by tensile stresses in the core, giving rise to a stress distribution somewhat of the same general form as that sketched in Fig. 9.13 (p. 303). Likewise, systems of residual stresses may be developed between regions which are cooled at different rates either because of differences in section thickness or because of variations in the effectiveness of the cooling media. Other things being equal, the magnitude of the stresses increases with cooling rate.

Surface compressive stresses predicted by this model are advantageous in most situations, but regions of tensile stress may be exposed at the surface by, say, the surface skin being machined away. Localized stress systems developed by variations in cooling rate also have balancing regions of residual stresses which may be exposed somewhere at a surface. In any event, any localized removal of material disturbs the stress system, perhaps sufficiently to cause distortion of, or dimensional changes in, a component.

Phase transformations occurring during cooling may add to or subtract from the stresses induced by thermal contraction. Phase changes are accompanied by changes in crystal structure and atomic spacing, and hence in volume. The change may be one of either contraction or expansion. One involving contraction can enhance the effects of thermal contraction and so increase the magnitude of the residual stresses just discussed. One involving expansion, if small, can counteract the effects of thermal contraction and so reduce the residual stresses. On the other hand, a change involving expansion that is larger than the effects of the thermal expansion can cause the directions of the dimensional changes described earlier to be reversed and tensile stresses to be induced at the surface. This potentially can have serious consequences, most likely so when a considerable expansion occurs during the phase transformation and when the transformation occurs at low temperatures. There is, in fact, one such case, and a most important one it is. It involves the transformation of austenite to martensite, which is responsible for the quench hardening of steels (p. 401). The stresses induced in this case can be large enough to cause local fracture (i.e., cracking) of a component. At a lower level, the residual stresses can cause distortion of a component. Finally, they can enhance the detrimental effects of externally applied tensile stresses. The transformations in steels of higher carbon contents occur at comparatively low temperatures and involve larger volume expansions, and the likelihood of distortion and cracking is then greatest. Heating at a low temperature after quenching, as is done during tempering (p. 408), alleviates residual stresses.

Principles of Practical Heat Treatment Cycles

We are now in a position to outline the characteristic steps of heat treatment procedures which have the objective of changing the structures of metals, assuming that the metals are amenable to these procedures. The steps are:

Step 1. Heat to a temperature at which a different phase is stable. Hold there for a time sufficient to allow complete transformation to the new phase.

Step 2A. Cool at a rate chosen to do one of the following: (a) allow transformation to the old phase but with a different morphology; (b) suppress this transformation and retain the high-temperature phase as a metastable phase; or (c) same as (b), but allow the high-temperature phase to transform to a nonequilibrium metastable phase.

Step 2B. As an alternative to step 2A, cool the high-temperature phase rapidly to a chosen temperature and allow it to transform there isothermally, selecting the temperature so as to obtain a particular morphology of transformation products.

Step 3. Decompose a metastable phase by heating, selecting the temperature and time so as to produce a chosen structure.

The remarks to this stage have been directed specifically at heat treatments which are carried out deliberately with a definite end result in mind. They apply equally well, however, to any thermal cycle to which a metal is subjected. This includes cooling after solidification, heating for the purpose of fabrication (e.g., welding), and heating in service. The prior structure, the temperature attained, the time at temperature, and the subsequent cooling rate all have the same influence as in a deliberate heat treatment.

The Transformation in Practice

Although the transformations described in the discussion to Fig. 11.1 have been considered as generalized models, they are actually involved in many thermal treatments to which metals are subjected in practice.

The transformations described govern, of course, the structures developed in commercial brasses which have zinc contents within a certain range. These alloys are not normally heat treated deliberately but are subjected to a range of thermal cycles during fabrication. As a consequence, the transformations discussed govern to a large extent the microstructures of these alloys as they are supplied and to some extent their mechanical properties. They may also govern other critical properties. Copper-zinc alloys which have a fully-beta microstructure (as in Fig. 11.2a) are susceptible to devastating forms of intercrystalline fracture which can be avoided with certainty only by ensuring that a reasonable volume fraction of alpha phase is present to line the grain boundaries (as in Fig. 11.2b). This can be assured at higher zinc contents, which tend to be used to obtain higher strengths, only if final cooling from the beta field is carried out at an appropriately slow rate. It cannot be assured at all, however, if a certain zinc content is exceeded. This is, incidentally, a prime example of an alloy type becoming to be unacceptable in engineering practice because of only a single critical deficiency.

The titanium alloy of greatest commercial importance also relies on the type of phase relationship being discussed for its response to heat treatment. The alloy has a nominal composition of 6% aluminum and 4% vanadium (it is commonly referred to as a Ti-6Al-4V alloy), and the relationships between possible phases in the alloy can, as a starting point, be described in a pseudobinary phase diagram for aluminum-vanadium alloys having a constant 6% aluminum content (Fig. 11.8A). Note that this diagram is similar in principle to the portion of the copper–zinc diagram sketched in Fig. 11.1, but this time the phase

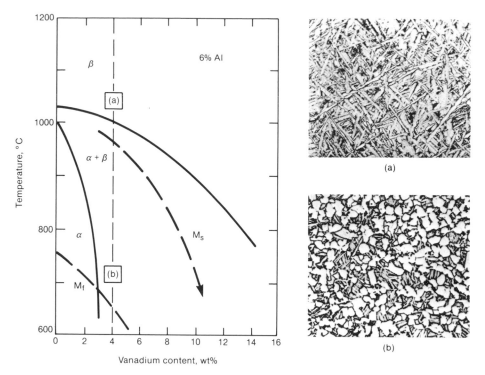

The micrographs illustrate typical microstructures produced by annealing an alloy containing 4% vanadium by cooling slowly after heating into (a) the beta field and (b) the (alpha-plus-beta) field. In (a), the symmetrically arranged plates are the alpha phase precipitated during cooling; the dark islands included between them are the beta phase that remained untransformed. In (b), the white equiaxed areas are the alpha phase that was not transformed to beta phase during heating; the dark equiaxed areas are the areas in which beta phase was formed during heating and which have transformed during cooling to a structure similar to but finer than that in (a). Optical micrographs; 500× (shown here at 75% of actual size).

Fig. 11.8A. A pseudobinary phase diagram for nickel-vanadium alloys containing a constant 6% aluminum.

transformations are initiated by an allotropic transformation (p. 226) which occurs in titanium.

Ti-6Al-4V alloys are most commonly used in an annealed condition, but they may be annealed by slow cooling after heating either into the beta field (heated to 1000 °C [1830 °F] or more) or into the alpha-plus-beta field (usually to about 700 °C [1290 °F]). In the first case (called a *beta anneal*), a fully-beta structure is established by heating and alpha phase precipitates from this beta phase during cooling. The alpha phase forms as thin plates with a Widmanstät-ten morphology [insert (a) in Fig. 11.8A], and the size of the plates is established essentially by the grain size of the parent beta phase. With the compositions concerned, only a small volume fraction of beta phase usually remains between the plates of alpha phase when room temperature has been reached. In the second case (*alpha-plus-beta anneal*), a duplex structure is obtained by heating — namely, a random distribution of equiaxed grains of the two phases. The phase diagram (Fig. 11.8A) tells us that the composition of these two phases will depend on the heating temperature and that the beta phase will contain more vanadium than the beta phase formed during a beta anneal. The grain size of the beta phase will also be smaller. The relative proportion of the two phases will depend on both composition and heating temperature. The primary alpha phase cannot be expected to undergo much change during subsequent cooling, because the revelant phase boundary is steep (Fig. 11.8A). The beta phase, on the other hand, can be expected to undergo the same type of transformation as occurs during a beta anneal. The final structure thus consists of equiaxed grains of alpha phase interspersed among equiaxed regions which have a duplex structure of Widmanstätten plates of alpha phase enclosing islands of beta phase [insert (b) in Fig. 11.8A]. The duplex structure typically is much finer than that produced in a beta anneal [as in insert (a) in Fig. 11.8A], because the grains of the parent beta phase were smaller. Note, moreover, that both annealing cycles provide an opportunity to refine the general grain size, as was described in connection with Fig. 11.8.

The mechanical properties of these alloys are largely determined by their structure, solid-solution hardening making only a small contribution to strengthening. We have just seen that a range of structures is possible. Consequently, control of structure by heat treatment becomes important in the alloys, which typically are used in critical applications. It turns out that structures obtained by a beta anneal have notably better toughness and resistance to the growth of fatigue cracks, both of these being characteristics which may be of greater importance than straight strength in the aerospace components in which the alloy is commonly used. Structures obtained by an alpha-plus-beta anneal, on the other hand, have slightly better yield strength and ductility, a combination which is desirable in other situations (e.g., bolts and fasteners). The yield strength of these alloys is actually 900 to 950 MPa (130 to 140 ksi), which is about three times that of unalloyed titanium, similar to that of medium- to high-strength heat treated steels, and two to three times that of the strongest aluminum alloys. The alloys also retain their strength to higher temperatures than aluminum alloys (p. 113). The mass density is only a little over half that of steel and about the same as that of aluminum alloys. These comparisons make very apparent the advantages of these structure-strengthened titanium alloys for components which are highly stressed in either tension or compression (see p. 78), particularly when weight is important. Unfortunately, they are very expensive, to the extent that cost is a major factor inhibiting their use.

An extra complication, the principles of which we have not yet dealt with, arises if these titanium alloys are cooled rapidly from the heating temperature. It might be thought from the above discussion (p. 355) that the beta phase formed during heating might be retained unchanged at room temperature when the cooling rate was fast enough to suppress the alpha phase transformation. It does do so with some alloy compositions, but with others the beta phase undergoes a so-called martensitic transformation at lower temperatures (p. 401), the significance of which will be mentioned later (p. 422).

The type of transformation being discussed is also relevant to the formation of Widmanstätten structures in metallic meteorites, the observation of which was such an important step in the foundation of metallography (p. 185). This may be of interest to you. All the meteorites found on Earth originated during the formation of the solar system, initially becoming part of the asteroid belt which is now located between Mars and Jupiter. Some of these asteroids were drawn into orbits which intersected the orbit of the Earth, and eventually entered the Earth's atmosphere. The larger ones survived entry to strike the Earth's surface, which still happens occasionally. Only a small fraction of meteorites are metallic, but all metallic meteorites are very similar in composition, being basically iron-nickel alloys containing about 8% nickel. The phase relationships to be expected in such an alloy are indicated in Fig. 11.8B. Meteorites have cooled extremely slowly over eons of time in interplanetary space before they reach the Earth, solidification being so slow that each mass solidified as a single crystal (p. 240) or at least very large grains. The crystals are of the phase which metallurgists call gamma phase or *austenite* and which geologists call *taenite*. Alpha phase (*ferrite* to metallurgists and *kamacite* to geologists) separated from the gamma phase during subsequent cooling (Fig. 11.8B) and, as we now know to expect, precipitated as plates on particular planes in the parent crystal of gamma phase. Moreover, in a typical meteorite, only a small volume fraction of gamma phase remained untransformed. The really unusual feature of a meteorite is that the cooling rate during the transformation was so slow, slower than anything that we could comprehend or duplicate, that equilibrium might actually have been achieved during cooling. More spectacularly, only comparatively few plates of alpha phase were nucleated and each had the opportunity to grow right across the parent crystal and to a considerable thickness (Fig. 6.1, p. 185). Hence an unusually coarse structure was developed, coarse enough to be discerned easily by the human eye, as Widmanstätten discovered. However, although the macrostructure is simple, the microstructure is not so, because transformations occur in both primary phases. Meteorites also contain inclusions of a number of types, some large enough to be seen in macrographs.

The proeutectoid transformation in steel, which we shall soon discuss in some detail, is also a case of the type of transformation under discussion. It is a limiting case in the sense that the equilibrium composition of the low-tempera-

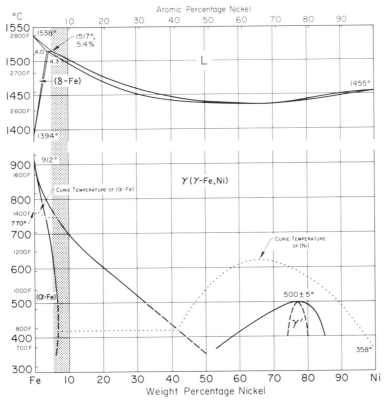

Fig. 11.8B. The iron-nickel phase diagram. The band between 5% and 10% Ni indicates the range of compositions of most metallic meteorites. Source: *Metals Handbook*, 8th Ed., Vol 7, American Society for Metals, 1973.

ture phase does not change with temperature — for most practical purposes, at any rate. Moreover, the transformation terminates at a certain temperature because it is supplanted by a different type of transformation — namely, a eutectoid transformation (p. 382).

PRECIPITATION HARDENING IN SUPERSATURATED SOLID SOLUTIONS

A German metallurgist named Alfred Wilm discovered in 1909 an entirely new method of strengthening metals, the first new method to be discovered since the ancients had worked out how to harden bronzes by hammering and steels by quenching from a high temperature. Wilm was in the midst of a program sponsored by the German government to develop an improved alloy for ammunition cartridge cases and happened to be experimenting with an aluminum alloy containing 3.5% copper and 0.5% magnesium. He had heated the alloy to a high temperature for annealing and had then quenched it in water. His schedule was such that he had to leave this material over a weekend before he was able to get around to testing it, and when he returned he found that the material had unexpectedly good mechanical properties. A very fortuitous set of events, you might say, but what was not fortuitous was Wilm's decision to follow up the observation. This was the mark of a perceptive investigator, who went to work and developed an alloy and a heat treatment procedure that could be handed over to industry for exploitation. This outcome was not the one that his government sponsor had hoped for, but was perhaps an even more important one. Modern sponsors of research would recognize the syndrome even if the outcome were not as fortunate. Successes of this magnitude are indeed rare.

It took a major effort by industry to turn Wilm's laboratory work into an industrial reality, but the eventual outcome was a range of usable products in a light, high-strength alloy. The alloy finally was called "Duralumin", and the heat treatment procedure was called *age hardening*. This alloy appeared at a propitious time. Design engineers at the Zeppelin company were assessing the practicability of producing a large powered airship and they were quick to recognize that the new alloy opened up revolutionary possibilities. The great success of their airship — unfortunately as, among other things, the first aircraft to bomb civilian populations — depended on the use of the alloy in major structural elements. This was, moreover, a beginning to the enormous usage of age-hardenable aluminum alloys in aircraft construction that we know today, a usage that has spread to general engineering applications where weight is limiting. It is also an example of the symbiotic relationship between advances in engineering design and in materials technology. The development of modern aircraft would not have been possible without these alloys; there would have been much less need for them without the development of aircraft.

It took some time after these alloys had become well established in engineering practice before an adequate understanding was developed of the mechanism by which the hardening occurred. This had to await, in fact, the development of advanced techniques of electron microscopy and improvements in the theoretical knowledge of the solid state. At the same time, new types of hardenable aluminum alloys, and also age-hardenable alloys based on other metals, were discovered. It turned out that hardening in all instances was due to the formation of a very fine precipitate, or a feature associated with precipitation, during the decomposition of a metastable supersaturated solid solution. Moreover, the supersaturated solution often had to be heated at a temperature somewhat above room temperature to achieve this result. Consequently, the phenomenon is perhaps more properly called *precipitation hardening*, although it is still known as *age hardening* or *aging*. It is called *natural aging* when hardening occurs during standing at ambient temperatures, as happened with Wilm's alloy, and *artificial aging* when heating to elevated temperatures is required.

General Characteristics of Precipitation-Hardening Systems

The basic requirement for an alloy to be amenable to precipitation hardening is that it should be possible to form a supersaturated solid solution of one or more alloying elements. Preferably, the solution should be highly supersaturated, and this requirement can be met only when the solubility of the alloying element concerned decreases markedly with temperature. A second requirement is that it should be possible subsequently to decompose the supersaturated solution to form precipitates in a particular distribution and spacing.

A precipitation-hardening heat treatment cycle thus characteristically involves three steps, namely:

1. A *solution treatment* in which the alloy is heated to dissolve the alloying element(s) concerned.
2. Cooling, usually to room temperature, rapidly enough to retain the alloying element(s) in solution and so obtain a supersaturated solid solution.
3. Heating at a temperature and time selected to decompose the supersaturated solid solution. The temperature may range upward from room temperature but is always lower than the solution treatment temperature. This is the aging treatment.

There are many alloy systems in which the phase relationships are such that it is possible to achieve steps 1 and 2. Some have already been discussed earlier in this chapter, and others will be discussed later. The characteristic requirement is that the alloy composition be in a region where a phase boundary slopes significantly, the more slope the better. But only in a few of these systems is the decomposition of the supersaturated solution accompanied by a worthwhile degree of hardening. The aluminum-copper-magnesium alloy which Wilm stumbled upon just happened to be one of the few, and one of the very few in which the hardening precipitation occurs at ambient temperature. It is more convenient, however, to use the binary aluminum-copper alloys to describe the principles of precipitation-hardening heat treatment. These alloys are simpler and have been thoroughly investigated.

The relevant portion of the aluminum-copper phase diagram is illustrated in Fig. 11.9(a), the alloys in this range constituting a eutectic system between an aluminum-rich terminal solid solution and the valence compound $CuAl_2$, which is also known as θ phase. The solubility of copper in the terminal solid solution changes considerably with temperature, decreasing from 5.7% at the eutectic temperature of 548 °C (1018 °F) to less than 0.5% at 300 °C (570 °F) and virtually to nil at room temperature.

Solution Treatment

To ensure maximum solution of the alloying elements, the alloy has to be heated to a temperature which is above the solid-solubility boundary (t_s in an Al-4Cu alloy, for example; see Fig. 11.9a) but below the temperature at which another phase transformation would commence. This temperature is the liquidus temperature in aluminum-copper alloys, t_l in Fig. 11.9(a), at which temperature melting would commence and ruin the material. Commercial alloys are likely to be designed for maximum effect by attempting to incorporate the maximum amount of alloying element. The gap between t_s and t_e is then likely to be small with the type of phase relationship illustrated in Fig. 11.9(a), and it follows that the solution-treatment temperature has to be accurately controlled.

The alloy has to be held at the solution-treatment temperature for a time sufficient to ensure that all of the alloying element is dissolved, as has occurred

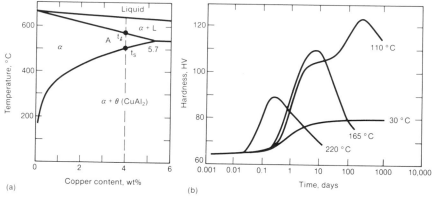

(a) Phase diagram representative of one type of alloy system which is amenable to precipitation hardening — in this case, the aluminum-copper system. At the solution treatment stage, an alloy containing, say, 4% copper would be heated to a temperature between t_s and t_l for a time sufficient to dissolve all of the CuAl$_2$ (θ) phase. For aging, the solution-treated material would be heated at a temperature below, and usually well below, t_s.

(b) Variation in hardness with time at some representative aging temperatures for a solution-treated Al-4Cu alloy. Some hardening occurs at room temperature in this particular alloy, but more occurs at higher aging temperatures, at least in practically acceptable times.

Fig. 11.9. Diagrams illustrating the characteristics of a precipitation-hardening system.

in the material illustrated in Fig. 8.7(b) (p. 260). Many hours at the solution-treatment temperature may be necessary, depending on the diffusivity of the alloying element in the base metal.

Cooling

Quite rapid cooling frequently is required to retain the alloying element in solid solution. Quenching in cold water may be desirable, particularly for thicker sections. However, rapid quenching distorts thinner irregular sections and introduces residual stresses in thicker ones (p. 361), and slower cooling becomes necessary when the consequences become troublesome. It may even become necessary to sacrifice some degree of efficiency in retaining a fully supersaturated solution, and hence in eventual strengthening.

The solution treatment results in a degree of solution strengthening which is always small compared with that obtained during subsequent precipitation hardening. It may, however, still make a useful contribution to the final strength, and an alloy may be designed to take advantage of this. A good level of ductility is retained in the solution-treated alloy, and advantage can be taken of this in practice by forming components while the material is in the solution-treated condition, the aging treatment being carried out after forming. The yield strength is usually too high and the ductility usually too low to allow much forming after aging. On the other hand, excessive distortion of shape might occur if the solution treatment were carried out after forming.

Precipitation

The hardening achieved during a precipitation treatment characteristically varies with the temperature and the time of heating in the manner illustrated in Fig. 11.9(b) and 11.10. For each temperature, the hardness starts to increase after a delay, rises to a maximum, and then falls gradually to that of the as-quenched material. In the latter event, the material is said to have been *overaged*. The peak hardness increases with decreasing precipitation temperature, but the time required to reach peak hardness then increases markedly. The time may even be infinitely long at very low temperatures (e.g., at 30 °C, or 85 °F, in Fig. 11.9b). The quantitative details of a family of aging curves are characteristic of a particular alloy type and of the solution treatment to which it has been subjected.

Decomposition of the supersaturated solution always occurs in two or more steps. Each intermediate step produces a metastable intermediate phase, and the equilibrium phase is formed during the last step in the sequence. The number of steps and the nature of the metastable phases vary with the alloy, but the common feature is that the intermediate phases are much more finely dispersed than the equilibrium phase. This is basically because they are easier to

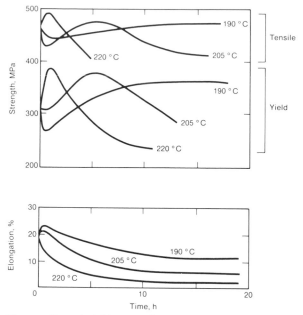

The tensile and yield strengths vary in the same way as for hardness. The tensile elongation decreases rapidly to a low level by the time that peak strength has been achieved.

Fig. 11.10. Variation with aging time of the tensile mechanical properties of a 4Cu-0.5Mg aluminum alloy which has been solution treated.

nucleate. The requirement is that one or more of the intermediate phases be dispersed finely enough to impede dislocation movement strongly by one of the mechanisms illustrated in Fig. 10.12 (p. 340) and 10.13 (p. 342).

Let us again take a simple aluminum-copper alloy as an example. The first step in the aging of such alloys produces a large number of closely spaced thin plates, each probably consisting of a single layer of copper atoms lying on simple cube planes of the aluminum crystal. Copper atoms do not have to diffuse very far to form such simple clusters and hence they can form easily at a large number of closely adjacent points. Moreover, the clusters are surrounded by a coherency strain field (Fig. 11.11a, and see p. 245) which effectively further reduces the spacing through which dislocation lines have to squeeze. There may be as many as 10^{17} of these zones per cubic centimetre of crystal. The clusters are known as Guinier-Preston (GP) zones after their co-discoverers, Guinier in France and Preston in England, and have all the characteristics required to impede the movement of dislocations effectively. At the next stage, the GP zones are replaced by coherent precipitates (Fig. 11.11b, and see p. 245) of a constituent called θ''. These precipitates are similar to but larger than the GP zones and probably consist of several layers of copper atoms separated by several layers of aluminum atoms. Next to form are small semicoherent precipitates (Fig. 11.11c, and see p. 245) of a constituent called θ' which has a compo-

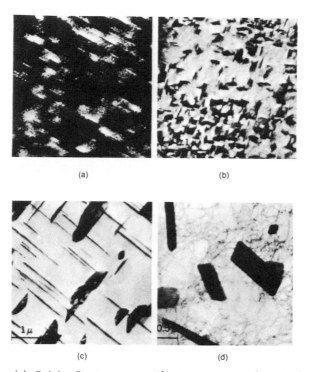

(a) Guinier-Preston zones. The image contrast here is due mainly to the coherency strain field around each zone.   (b) A precipitate known as θ'', which is also surrounded by a strain field. (c) The transition phase known as θ'.   (d) The equilibrium phase, θ (CuAl$_2$).

Note that the size of the precipitates decreases, and so their spacing increases, progressively. In making this comparison, allowance must be made for the differences in magnification of the micrographs. Transmission electron micrographs: (a) and (b) 150,000×; (c) 15,000×; (d) 7500×.

Source: A. G. Guy, *Introduction to Materials Science*, McGraw-Hill, New York, 1972.

Fig. 11.11. Representative examples of the precipitates formed during aging of a solution-treated Al-4.6Cu alloy. The alloy was aged for increasing times at constant temperature.

sition similar to that of CuAl$_2$ (θ) but a different crystal structure. The fourth and final step is the formation of the equilibrium phase (θ), by which time the precipitate particles are comparatively large (Fig. 11.11d) and the alloy has overaged. The particles of θ phase grow as overaging progresses, and it is only at some stage in this growth process that they become large enough to be resolved by optical microscopy. The important intermediate zones and precipitates can be detected only by transmission electron microscopy, which is only to be expected when the scale of phenomena needed to affect the movement of dislocations is considered. Little wonder, then, that metallurgists had difficulty early on in developing an adequate understanding of age hardening.

All age-hardenable alloys go through some sequence of this nature, although the number of steps and the nature of the intermediate phases are different and are unique to each alloy type. For example, GP zones form in all types of aluminum alloys but the nature of the zones is different in each. On the other hand, GP zones usually do not form at all in age-hardenable alloys based on other metals, but at least one closely spaced intermediate precipitate always does.

Limitations of Strengthening by Precipitation Hardening

So far we have mentioned mainly the effects of precipitation on hardness. Hardness is a measure of yield strength in compression (p. 80), but tensile mechanical properties are generally of more interest to a designer. Tensile properties follow the same general trends as hardness during aging, but there are differences in detail. In aluminum alloys, for example, yield strength increases more markedly than tensile strength (Fig. 11.10), so that the yield/tensile strength ratio increases. As might be expected, tensile elongation decreases, and it decreases steadily, the loss being significant by the time that maximum strengthening has been achieved. Aging treatments consequently have to be chosen to optimize some desirable combination of properties, commensurate with practically acceptable aging conditions. However, other factors which we shall now outline may require further modifications to a treatment arrived at on the basis of these simple considerations.

Strengthening that is due to the development of a particular distribution of precipitated particles can be expected to be affected if this distribution is altered or destroyed in service. It is obvious, for example, that strength will gradually be lost if the material is heated in use to a temperature at which it continues to age. There is consequently a limit on the temperature at which an age-hardened alloy can be operated in service, particularly under long-term creep conditions, and this temperature will in general be well below that at which the aging heat treatment was carried out. For example, the aluminum alloy used in the structure and skin of the supersonic "Concorde" aircraft is initially aged at 190 °C (375 °F), but its maximum safe operating temperature is 110 °C (230 °F). This sets a limit on the speed at which the aircraft can be flown, because the airframe structure heats up during the flight, and the higher the speed the higher the temperature reached. The Concorde is designed to operate at a cruising speed of about twice the speed of sound (called Mach 2 — 2200 km/h, or 1370 mph), at which speed the skin temperature remains within acceptable limits. Even so, the airframe will have a limited flight-time life because the alloy will eventually overage — and this is the most heat-resistant alloy that aluminum metallurgists have been able to develop so far. Much more expensive titanium alloys have to be used in aircraft which are required to operate at significantly higher speeds and skin temperatures.

Nevertheless, age-hardened alloys have been developed in which the hardening precipitates are stable for long periods at quite high temperatures, the prime example being the nickel-base "superalloys" used in the hot-end components of jet engines. These alloys retain their strength for practically acceptable times at temperatures close to those at which they are aged, temperatures that are not far below their melting points (p. 378).

The hardening precipitate may also be affected by cycling stresses. The slip processes that occur during cyclical stressing are concentrated in bands (p. 321), and it is found that the hardening precipitates tend to disappear from these bands with increasing numbers of stress cycles. The material in the bands then effectively becomes softer, slip becomes more strongly concentrated in the bands, and the fatigue cracking process is accelerated. Most probably, the precipitates are broken up into smaller and smaller fragments by repeated intersections with dislocations — that is, by the very process that is responsible for the strengthening. The fragments eventually become so small that they become thermodynamically unstable and then, in effect, redissolve. This is why the ratio of endurance limit to tensile strength for age-hardened aluminum alloys is smaller than might have been hoped for (Fig. 4.4, p. 109) — the stronger the alloy and the more it depends on age-hardening for its static strength, the lower the ratio.

There may be other limitations as well. Although the phenomena involved in the over-all process are complex, it is probable that the factors which determine the resistance to stress-corrosion cracking of alloys strengthened by precipitation hardening include details of the precipitation structures. This is so, of course, only in certain types of alloys that are susceptible to stress-corrosion cracking, but aluminum alloys are among these. Deterioration in fracture toughness (p. 95) is another factor that may limit the advantage that can be taken of the strengthening potential of precipitation hardening.

Precipitation-hardened alloys by their very nature are in a metastable condition. It takes some skilled fine tuning to get them balanced in a state which ensures fully acceptable behavior in service. But a whole new avenue for the development of alloys is opened when this is done, an avenue which has had a profound influence on progress in many fields of engineering.

Precipitation Hardening in Practice

Aluminum alloys are still by far the most widely used group of alloys which rely on precipitation hardening for strengthening. Copper, on which Wilm's original discovery was based, is, however, no longer the only alloying element that is used to achieve hardening. Zinc, magnesium, and silicon also have appropriate characteristics* — they are all highly soluble in aluminum, super-

---

*Lithium is also suitable, and has the advantage that its mass density is much lower than that of aluminum and hence reduces the density of the alloys. Lithium-containing alloys are, however, just entering commercial usage.

saturated solid solutions can be obtained by a solution heat treatment followed by rapid cooling, and the resultant solid solution can be decomposed to form hardening precipitates. The details of the precipitation processes differ considerably with each alloy type, but the principles are essentially the same as for aluminum-copper alloys. Finally, as alloying elements they are not too expensive and can be handled easily enough in production.

Commercial alloys, with a minor exception or two, all use combinations of two or more of the potentially useful alloying elements, and a very large number of combinations obviously are possible. To complicate matters further, additions of small amounts of elements such as titanium, zirconium, vanadium, and chromium, either singly or in combination, have been found to be useful to control grain size and sometimes to modify the aging behavior. Fortunately, suppliers and users have agreed to reduce some of the possible confusion by standardizing on a limited number of alloy types. Most are for wrought alloys, but some casting alloys are heat treatable. Details can be obtained from standard reference books, some of which are listed at the end of this chapter, and from various national standards.

The net result is that alloys covering a range of strengths (yield strengths from 200 to 450 MPa, or 30 to 65 ksi) are available. The alloys also cover a range of other attributes, some favorable and some unfavorable. For example, alloys of the highest strengths may have unacceptably poor toughness and perhaps be excessively susceptible to stress-corrosion cracking. The improvement in fatigue strength may be too small to make worthwhile the use of higher-strength alloys. Copper-containing alloys have, relatively, poor corrosion resistance and so may be avoided on this count. Some alloy types are more difficult to fabricate than others; in particular, only alloys in the lower strength ranges can be fabricated readily by welding. All of these things have to be taken into account as well as yield strength when selecting the optimum alloy for a particular application. For this a considerable depth of knowledge and experience is needed.

The most effective solution treatment temperature must be expected to vary considerably among such a wide range of alloy compositions, because both the solidus and solvus temperatures are affected by alloy content. It varies in fact from about 465 to 550 °C (870 to 1020 °F), data for specific alloys again being available in standard references. The time required at temperature to achieve full solution also varies a little, but the heating time in practice is largely governed by the time needed for a component to reach the intended temperature throughout. This varies from tens of minutes for sheet to tens of hours for forgings, depending on the section thickness of the component and the type of heating unit. Rapid cooling, such as that obtained by quenching in water, is desirable to ensure full retention of alloying elements in solution. But this may cause components to distort excessively or cause undesirably large internal stresses to be induced. Less severe cooling procedures then have to be employed, usually resulting in some loss in final strength. A compromise often has to be reached in practice.

Only a few of the currently used alloys can be age hardened satisfactorily by allowing them to stand at room temperature, as was possible with Wilm's original Duralumin. Most alloys age to some extent at room temperature, but to achieve maximum strength they have to be artificially aged at a temperature and for a time which again varies with each alloy type. The aging temperatures vary from about 120 to 205 °C (250 to 400 °F), and the time required is usually some tens of hours. However, the optimum aging treatment is sometimes modified to improve some critical property, a loss in strength being accepted. For example, overaging may be used to reduce internal stresses or to improve corrosion resistance. Two-stage aging treatments may be used to improve resistance to stress corrosion. Solution-treated material can be straightened and even cold worked immediately after quenching before natural aging becomes advanced, when it is in its softest condition. This is a useful attribute in production. Final strength is also increased, but internal stresses may be introduced.

Over all, precipitation hardening allows the yield strength of commercial aluminum alloys to be increased to some ten to fifteen times greater than that of unalloyed commercially pure aluminum. This is to about half the strength that can be obtained in medium- to high-strength heat treated steels, but at a third the mass density. Aluminum alloys thus have a clear advantage in many weight-critical components, particularly when buckling in compression is the determining failure mode (see p. 78). Their wide use in the transport and aerospace industries testifies to the usefulness of this attribute. A limitation may be that strength and creep properties are retained to only comparatively modest temperatures, as mentioned earlier. This is demonstrated semiquantitatively in Fig. 4.5B (p. 113). Moreover, there is often not a fully commensurate improvement in fatigue properties (p. 109), and the strengthening that can safely be used in engineering practice may have to be limited by inadequate toughness (p. 95) or, most particularly, resistance to stress-corrosion cracking (p. 164). The latter is the Achilles' heel of high-strength aluminum alloys, and is responsible for some loss in popularity for the strongest of them.

Alloys which respond to precipitation hardening have been developed for all the common metals, but only a few of them are used extensively. Several types of precipitation-hardening titanium alloys have had limited success, perhaps the most successful containing 2.5% copper. The advantage of this type of alloy is that it can be formed cold into complex shapes after solution treatment and before aging whereas conventional alloys have to be formed hot using special equipment. Useful strengthening can be obtained in several magnesium alloy systems; in fact, most of the commonly used casting alloys of magnesium are heat treated by a precipitation-hardening routine.

Only one copper alloy is used widely in the precipitation-hardened condition, but it is a spectacular example. The essential alloying element is 1.5 to

2.0% beryllium, and a hardness of more than 400 HV and a yield strength of 1200 MPa (175 ksi) can be obtained. The alloy is solution treated at about 800 °C (1470 °F) and has a yield strength of only about 220 MPa (30 ksi) after quenching. It can be cold formed and cold worked easily in this condition and any strain hardening is retained during aging because the aging temperature of about 350 °C (660 °F) is below the recrystallization temperature. The yield strength of these *beryllium bronzes*, as they are called, is comparable to that of the best heat treated spring steel. The alloy has, moreover, comparative advantages of high electrical and heat conductivities and good corrosion resistance. It has the disadvantages of comparatively low elastic modulus (p. 64) and high cost. The cost confines its use to applications in which its special characteristics are really needed in items such as springs and diaphragms, electrical contactor bridges, and spark-resistant tools.

The "superalloys" used in the hot-end components of jet engines are, however, undoubtedly the greatest triumph of the precipitation strengthening, apart from aluminum alloys. These alloys are very complex in detail and a number of factors contribute to their strength, but the main one is precipitation hardening in a nickel-base* alloy. The primary requirement of these alloys is, of course, that they maintain their mechanical strength at, and have good creep properties at, high temperature. But they also have to survive physically in an aggressive environment, and this dictates that they contain 15 to 20% chromium as an oxidation inhibitor (p. 176). The chromium does confer some solution-hardening strengthening, but the main source of strengthening is precipitation hardening resulting from the addition of titanium and aluminum in amounts of 1 to 3% each.

Nickel-base superalloys are solution treated at about 1100 °C (2010 °F) and aged at 700 to 850 °C (1290 to 1560 °F), sometimes in a complicated way, the objective being to precipitate a compound based on the formula $Ni_3(Al,Ti)$ as a strengthening phase. The precipitate is known as gamma prime (γ') phase and the nickel solid solution matrix phase as gamma (γ) phase. Gamma prime is an atypical hardening precipitate in many respects. It is the equilibrium phase, not an intermediate metastable phase; the individual particles are comparatively large, cubic in shape, and present in a large volume fraction. The precipitate particles are coherent with respect to the matrix gamma phase and so are surrounded by a strain field (p. 245) which effectively increases their size. The mechanisms of strengthening are dislocation bowing (p. 340) and particle shearing (p. 342). The most important characteristic of the precipitate, however, is that it is unusually stable at high temperatures, and this is the basic reason for the excellent high-temperature properties of the alloys.

Superalloys were originally developed in England by a team led by Walter

---

*Cobalt, a cousin element of nickel, is sometimes substituted partly for nickel. A similar alternative alloy system is based entirely on cobalt.

Betteridge. During the height of World War II, he was given the task of producing alloys which would turn the jet engine then being developed by Sir Frank Whittle into a practical reality. Betteridge and his colleagues decided to do this by improving the nickel-chromium alloys that had been used for many years in heating elements, including those in domestic radiators. By perception, persistence and inspired empiricism they succeeded brilliantly, although it must be admitted that it was some years before anyone fully worked out why. The alloy type has been improved out of sight since then (p. 119). Compositions have been adjusted, additional alloying elements (molybdenum, in particular) added, and heat treatment schedules varied, but this time more by design than by empiricism. Much of the improvement can be credited to the scientific knowledge which physical metallurgists have acquired in the meantime.

A range of precipitation hardening effects also occur in steels, some of them being in the normal run of steels. The phenomenon can be disadvantageous, so that precautions have to be taken to avoid it. It can be advantageous, in which event steels may be designed to take full advantage of the phenomenon.

First, let us consider the principal example of where age hardening is disadvantageous, the case being that of low-carbon steels which have a predominantly ferritic matrix. Ferrite can dissolve up to about 0.02% carbon at 720 °C (1330 °F), but the solubility falls to virtually zero at room temperature (Fig. 11.11A), the general nature of the relationship being similar to that for the aluminum-copper system (see Fig. 11.9a). A supersaturated solution of carbon thus tends to form in steels containing carbon in the range of about 0.01 to 0.1%, as many sheet steels do, if they are cooled at all rapidly from temperatures in the range of about 450 to 720 °C (840 to 1330 °F). The resultant solid solution

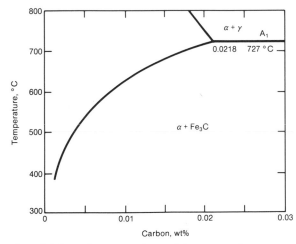

The solubility of carbon in ferrite (α) varies with temperature in a manner similar to that for copper in aluminum (cf. Fig. 11.9a).

Fig. 11.11A. Detail of portion of the iron-carbon phase diagram at low carbon contents.

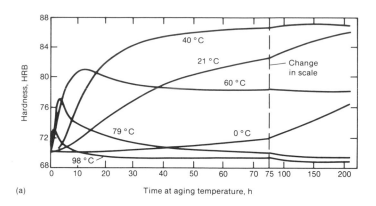

(a)

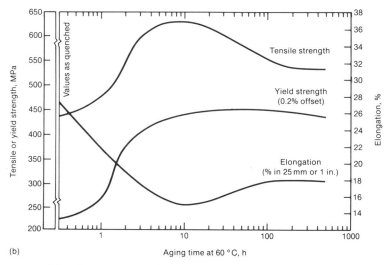

(b)

The characteristics of two different steels are illustrated, but the trends are representative and in this sense comparative. The steel for (a) contained 0.06% C and was quenched from 720 °C (1330 °F). The steel for (b) contained 0.03% C and was quenched from 725 °C (1340 °F).

Fig. 11.11B. Typical examples of the variation of hardness (a) and tensile properties (b) during the aging of ferrite which is supersaturated with carbon.

age hardens quite rapidly at room temperature [Fig. 11.11B(a)] and even more rapidly at slightly elevated temperatures, because a transition form of a metastable iron carbide is precipitated in a very fine form. The stable carbide, cementite, forms only at the overaging stage. The aging is rapid because interstitial diffusion of carbon in iron is relatively fast. This phenomenon is called *quench aging* and must be distinguished from strain aging (p. 330). A yield point phenomenon is not introduced by quench aging.

The difficulty is, however, that hardening in this case is accompanied by a serious loss in ductility [Fig. 11.11B(b)]. The treatment consequently is not used deliberately, but it may occur inadvertently. For example, steel sheet is coated with zinc to confer corrosion resistance by the process called galvanizing (p. 145). This is commonly done by passing low-carbon steel sheet continuously through a bath of molten zinc maintained at a temperature of about 480 °C (900 °F), the sheet being cooled with a blast of air as it emerges from the molten zinc. A solution of carbon is thereby produced which is sufficiently supersaturated to age harden subsequently at room temperature, and ductility is lost. A supplementary treatment at 315 °C (600 °F) is required when maximum ductility is desired in the galvanized product. Thus quench hardening is a phenomenon which has to be either avoided or circumvented in the processing of low-carbon steels.

Now to turn to advantages of precipitation hardening phenomena in steels. We shall see later (p. 413) that the decrease in hardness of martensite with increase in tempering temperature is delayed when certain alloying elements are added. The hardness may even increase (Fig. 11.28 and 11.29, p. 408). This phenomenon, known as *secondary hardening*, is due to the development of a dispersion of precipitates of alloy carbides. Advantage has been taken of this to design two groups of tool steels which can be operated at high temperatures while still retaining high hardness, at least for a reasonable time. One group is used in dies for hot-working operations (*hot-die steels*) and the other in metal-cutting tools (*high-speed tool steels*).

The concept has been taken a step further in several types of specialty steels. One type, *maraging steels*, is designed to induce precipitation hardening in a martensitic phase in a steel which is as free from carbon as possible (the characteristic of martensitic transformations will be discussed soon). The steels are of complex composition, containing between 15 and 25% nickel to produce the martensitic structure, and essential additions of molybdenum (about 5%) and titanium (0.2 to 0.6%) to induce the precipitation hardening. They also typically contain cobalt (8 to 12%) and aluminum (about 0.1%). Heat treatment is, however, comparatively simple. The steels have to be cooled only slowly from a high austenitizing temperature (900 to 1100 °C, or 1650 to 2010 °F) to transform to a massive form of martensite, and this possibility of slow cooling has obvious practical advantages. The martensite is soft because of its low carbon content, and components can be machined and formed in this condition. The components are then aged at a temperature in the range 450 to 500 °C (840 to 930 °F) for 3 to 12 hours when very large volume fractions of complex phases are precipitated. This causes considerable strengthening, with very high yield strengths in the range 1500 to 2500 MPa (215 to 360 ksi) being obtained. The important point is that these high strengths are combined with good ductility and toughness. The steels also are weldable, but they are difficult to produce, particularly with consistency, must be handled with special care, and are expensive. Nevertheless, they do find important applications in high-performance equipment.

Another special precipitation-hardening type of steel has a substantial addition of chromium to confer a level of corrosion resistance similar to that obtained in standard stainless steels. Nickel (4 to 7%) is added to enable an austenitic (sometimes a duplex austenite-martensite) structure to be produced by rapid cooling from a high solution temperature (1050 to 1150 °C, or 1920 to 2100 °F), and other elements such as 2% molybdenum and 1% aluminum are added to induce precipitation-hardening characteristics. They are called *precipitation-hardened stainless steels* and have yield strengths approaching those of maraging steels. Their special characteristic is that the high strength is combined with good corrosion resistance.

EUTECTOID TRANSFORMATIONS

Characteristics of the Transformation

The portion of the iron-carbon phase diagram shown in Fig. 11.12 represents a particular type of relationship that can develop among three phases in a crystalline solid. Although this relationship occurs in a number of other systems, the one of great practical importance is that in iron-carbon alloys, on which we shall concentrate here. The relationship involves the following phases:

1. *Gamma phase*, also called *austenite*, an allotrope (p. 226) of iron which is stable at intermediate-to-high temperatures, has a face-centered cubic crystal structure, and can dissolve considerable amounts of carbon. Austenite was named after Professor Sir William Robert-Austens, a pioneering English metallurgist.
2. *Alpha phase*, also called *ferrite*, an allotrope of iron which is stable at intermediate-to-low temperatures, has a body-centered cubic crystal structure, and can dissolve only very small amounts of carbon. The name is derived from *ferrum*, the Latin word for iron.
3. *Cementite*, a valence compound of iron and carbon which has the approximate composition Fe_3C and which has a complex crystal structure. Strictly, cementite is a metastable phase, the true stable phase of the iron-carbon system being graphite. But cementite can be considered to be the stable phase in steels for most practical purposes. Its name was proposed by Professor Marion Howe, one of the fathers of physical metallurgy in the United States. It is based on the term "cementation", which had been used since ancient times for a process in which carbon is diffused into solid iron to make steel.

The relationship among these three phases, which is illustrated in Fig. 11.12(a), arises basically because the relative stabilities of the two allotropes of iron are affected by carbon. We can see this at its simplest by looking at the effects of carbon in amounts less than 0.8%. The transformation from gamma

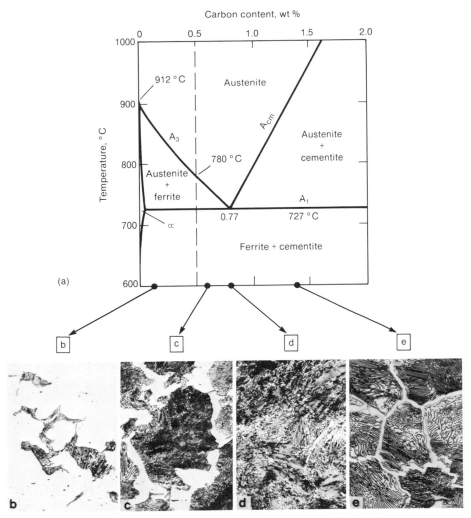

The steels illustrated have carbon contents of (b) 0.15%, (c) 0.6%, (d) 0.8%, and (e) 1.4%. These structures are near-equilibrium structures. The darker constituent is pearlite, the eutectoid constituent, in all cases. The light constituent is ferrite in (b) and (c) and cementite in (e). Optical micrographs; 500×; shown here at 79%.

Fig. 11.12. The eutectoid portion of the iron-carbon phase diagram, and structures of four steels having various carbon contents.

to alpha phase occurs at a specified temperature of 912 °C (1674 °F) in pure iron (p. 227), but occurs over a range of temperatures, and at lower temperatures, when carbon is present. The upper temperature of the range varies with carbon content as indicated by the line labeled A_3 in Fig. 11.12(a). The lower temperature of the transformation range is constant for all carbon contents (the A_1 temperature in Fig. 11.12a). The lower temperature of the transformation range also varies with carbon content, even more markedly so as indicated by the phase-boundary line close to the temperature axis in Fig. 11.12. The relationships to this point are similar to those for the titanium alloys discussed in connection with Fig. 11.8A (p. 364), but now an important difference is introduced in that the simple allotropic transformation ceases when a certain temperature is reached (the temperature labeled A_1 in Fig. 11.12a). The austenite remaining decomposes instead into a mixture of the low-temperature allotrope (ferrite) and cementite. This is called a *eutectoid transformation*, the temperature at which it occurs the *eutectoid temperature* (abbreviation A_1 temperature in the iron-carbon system), and the composition the *eutectoid composition* (0.77% C in the iron-carbon system). The transformation in high-carbon steels (C $>$ 0.77%) are different only in that austenite first precipitates cementite, starting at what is called the A_{cm} temperature, until its carbon content is reduced to the eutectoid composition. The eutectoid transformation then occurs as before.

The relationships which develop among these three phases, as described diagrammatically in Fig. 11.12(a), is similar in a general way to that of a eutectic system (p. 270), for reasons that we need not go into here. The system is consequently called a *eutectoid* system to acknowledge the similarity but recognize the difference. The difference is, of course, that all interactions among the phases occur inside a solid — between a solid crystal and a solid crystal rather than between a liquid and a solid crystal. This, as we have already seen, can be expected to have a determining influence on the kinetics of the reaction and on the morphology of the products. It is with these two aspects that we shall now mostly be concerned.

Let us consider first a steel of eutectoid composition* (0.77% C) that has been heated to a temperature at which austenite is the stable phase, in this case to slightly above the eutectoid temperature of 727 °C (1341 °F) (Fig. 11.12a). Assume that it is allowed to transform there completely to austenite into which is absorbed in solution all of the carbon present, as per the principles described on p. 358. This is called an *austenitizing treatment*, and the grain size produced in the new phase (p. 358) is called the *austenitic grain size*. Now let us see what happens when this austenite is cooled under conditions which maintain equilibrium. Figure 11.12(a) tells us that the austenite decomposes at the eutectoid temperature into two phases, ferrite and cementite, during a thermal arrest (cf. solidification of eutectics, p. 271). But this phase diagram does not tell us anything about how the two phases are arranged after transformation. Observations indicate that ideally they are arranged in colonies of thin, roughly parallel plates not unlike those of some lamellar eutectics. This is the structural constit-

---

*The composition and temperature quoted here are for pure iron-carbon alloys. The values are slightly different in commercial steels due to the presence of certain other elements, some inadvertently present and some deliberately added, but the principles are the same.

uent known as *pearlite*, a constituent to which we have referred on a number of occasions (see Fig. 6.4, p. 190, for examples). The use of the term "pearlite" for this constituent was also sponsored by Professor Howe, as a contraction of "pearly constituent", the descriptor used by Henry Sorby when he first reported its existence (p. 186). Sorby used this term because, when prepared for microscopic examination, surfaces at which the constituent is exposed have a pearly appearance when viewed in reflected light by the unaided eye. This effect is due to the diffraction of light by the closely spaced lines of the structure, which act as a diffracting grating. This is also the reason for the sheen of pearls, which have a lamellar microstructure.

Kinetics of the Transformation and Morphology of Products

The transformation of austenite to a mixture of cementite and ferrite does not occur instantaneously. Nuclei for the two phases have to form adjacent to one another, and this takes time (p. 350). The nuclei then have to grow, and this too takes time (Fig. 11.13) because it is diffusion-controlled (p. 351). These important features of the reaction kinetics can be summarized in an isothermal transformation diagram (cf. discussion on p. 352), and a schematic IT diagram for eutectoid austenite is sketched in Fig. 11.14(a). We may first confine our atten-

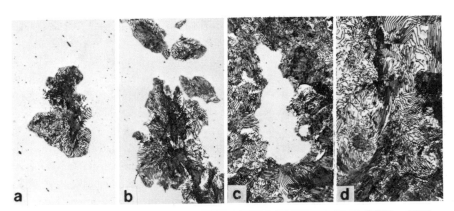

This is a eutectoid steel which has been austenitized and then cooled rapidly to a temperature below A_1 (705 °C, or 1300 °F, in this particular case). Specimens were allowed to stay at that temperature for times of: (a) 150 s; (b) 300 s; (c) 800 s; and (d) 2000 s (33 min). They were then cooled rapidly to room temperature. The darker areas are the pearlite constituent to which the austenite transformed. The times for the transformation to start and to finish can be determined in this way for a range of temperatures, and from this information an isothermal transformation diagram can be constructed. Optical micrographs; 250×.

Fig. 11.13. A series of micrographs illustrating the transformation of austenite at constant temperature.

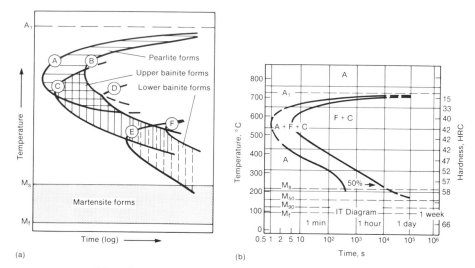

(a) A diagrammatic sketch separating individual transformations into different constituents. The key to the labeling of the curves is: A, pearlite starts to form; B, pearlite finishes forming; C, upper bainite starts to form; D, upper bainite finishes forming; E, lower bainite starts to form; F, lower bainite finishes forming. M_s and M_f indicate the temperatures at which transformation to martensite starts and finishes, respectively.   (b) An experimentally determined diagram. The various curves in (a) are smeared together in this diagram, which is the usual practice.

Fig. 11.14. Isothermal transformation diagrams for a eutectoid iron-carbon austenite.

tion to lines A and B in this diagram, which indicate the times necessary for the transformation to pearlite to start and to finish, respectively. The start curve has the standard "C" shape (p. 352) with the nose of the C being located well below the eutectoid temperature. The transformation nuclei form mostly at the austenitic grain boundaries, although some may also develop within the grains. Thus many colonies of pearlite characteristically develop in each parent grain of austenite (Fig. 11.15a), and a final colony size is then established which is smaller than the parent austenitic grain size. The number of nuclei increases with decreasing transformation temperature, which further reduces the colony size. Moreover, and more importantly, the spacing of the lamellae of the pearlite decreases with transformation temperature, over about the range illustrated in Fig. 10.9(a) to (c) (p. 336). We have seen that this has important influences on mechanical properties, some advantageous and some disadvantageous (p. 337).

Complete transformation to pearlite requires increasingly longer times as the temperature falls below the noses of the curves A and B in Fig. 11.14(a). It is then found that constituents with different morphological arrangements of ferrite and cementite (examples are illustrated in Fig. 11.15) form more quickly

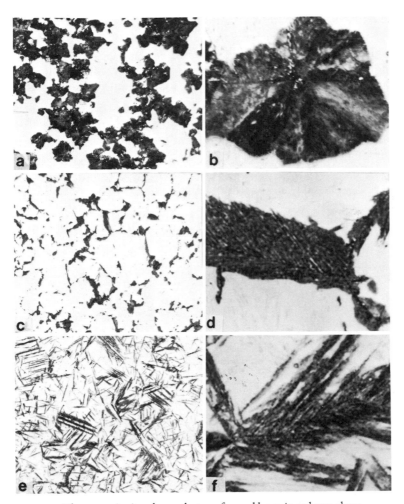

The austenite is only partly transformed here, in order to show the morphologies of the constituents more clearly. (a) and (b) Pearlite formed at 550 °C (1020 °F). (c) and (d) Upper bainite formed at 450 °C (840 °F). (e) and (f) Lower bainite formed at 300 °C (570 °F). Optical micrographs: (a, c, and e) 250×; (b, d, and f) 2000×.

Fig. 11.15. Micrographs illustrating the constituents which are characteristically formed during the isothermal transformation of a eutectoid iron-carbon austenite.

than, and hence in preference to, pearlite. This is indicated in Fig. 11.14(a) by curves C and D, which refer to the formation of a constituent known as *upper bainite* (Fig. 11.15c and d), and by curves E and F, which indicate that another constituent, known as *lower bainite* (Fig. 11.15e and f), forms preferentially at still lower temperatures. Mixtures of these constituents and pearlite may form over intermediate temperature ranges.

Both bainites are mixtures of ferrite and cementite, but the cementite is present as smaller, more closely spaced, and somewhat less regularly arranged particles than those in pearlite. The exact morphology depends on the temperature at which transformation occurs. Bainites have some interesting mechanical properties, but specialized texts need to be consulted for further details. The phases are named to recognize the work of Dr. Edgar C. Bain, the eminent American metallurgist who perhaps contributed more than anyone else to the development of a scientific understanding of the heat treatment of steels.

Practical IT diagrams for steels do not distinguish among the three reactions which we have sketched in Fig. 11.14(a) but smooth out the start and finish curves, as shown in Fig. 11.14(b). The phases existing in each transformation field are identified in these diagrams, but usually not the nature of the constituents which are produced. The hardnesses of the products usually are indicated, however, as they are in Fig. 11.14(b). IT diagrams of this nature are available for a wide range of commercial steels.

Transformations During Continuous Cooling

A continuous cooling transformation diagram (p. 354) corresponding to Fig. 11.14(a) is sketched in Fig. 11.16(a), the relationship with the IT curves being indicated (experimentally determined IT and CT diagrams for a commercial steel are also compared in Fig. 11.17). The CT curves in Fig. 11.16(a) are drawn alone for greater clarity in Fig. 11.16(b), on which some representative linear-rate cooling curves have also been drawn. These cooling curves represent the following situations:

1. Curve A represents slow cooling, such as might occur during cooling in a furnace (called *annealing* in the special terminology of steel technology). Transformation starts and finishes at comparatively high temperatures, and so comparatively coarse pearlite is formed.
2. Curve N represents somewhat faster cooling, such as might occur if the material were removed from the austenitizing furnace and cooled in air (called *normalizing*). Transformation occurs at lower temperatures and hence more finely spaced pearlite is produced, progressively so as the cooling rate approaches that of curve Q.
3. Curve Q represents cooling at a rate slightly higher than the critical rate (p. 355), such as might occur during quenching in a liquid (called *quench hardening*). The transformation to pearlite or bainite is suppressed, and the high-temperature phase could in principle be retained at room temperature. However, in most eutectoid systems, including iron-carbon alloys, the high-temperature phase transforms to a nonequilibrium phase before it reaches room temperature. This transformation product is called a *martensite*, which we shall discuss soon.

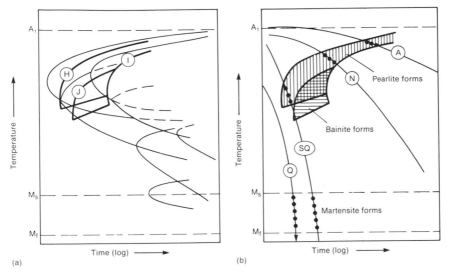

In (a), the heavy lines are the continuous cooling diagram. The light lines are the corresponding isothermal transformation diagram sketched in Fig. 11.14(a); they are included for comparison. The key to the labeling of the curves is: H, pearlite starts to form; J, upper bainite starts to form; I, transformations finish.

In (b), the isothermal transformation curves are shown alone, but superimposed are number of linear-rate cooling curves the import of which is described in text. The dots on these curves indicate periods during which transformation occurs.

Fig. 11.16. Diagrammatic continuous cooling transformation diagrams for a eutectoid steel.

4. Curve SQ represents cooling at a rate slightly lower than the critical rate (called *slack quenching*). Some of the austenite transforms to a fine pearlite, or a mixture of pearlite and bainite, and some persists either to be retained to room temperature or to transform at a lower temperature to a nonequilibrium constituent.

Note that a simple steel which has characteristics of the type described by the CT diagram sketched in Fig. 11.16 cannot transform completely to bainite during cooling at a reasonably steady rate.

The absolute values of the parameters in the IT and CT diagrams are affected to some extent by the carbon content of the steel and to a considerable extent by the addition of a number of alloying elements, such as manganese, nickel, chromium, and molybdenum. The first important change is that the noses of the transformation-start curves are shifted to longer times. You will appreciate, perhaps, from the values of the time scale in Fig. 11.14(b), that quite rapid cooling is required to suppress the high-temperature transformations in this unalloyed plain carbon steel. Cooling by immersion in water or brine is, in fact, necessary, and even then the critical cooling rate can be achieved through only a comparatively thin section; slightly thicker sections will not then harden throughout, but the center regions will be slack quenched. Thicker sections again will transform throughout to bainite or pearlite. The consequence of the delay of the high-temperature transformations resulting from the addition of appropriate alloying elements is that progressively gentler quenching procedures become adequate for quench hardening. The extent to which this is so depends on the types, amounts, and combination of alloying elements. Quenching in oil or even cooling in air may then suffice for full hardening, and hardening may also be achieved throughout much thicker sections. This can be most beneficial in practice. It enables larger components to be heat treated successfully and reduces internal stresses and the risks of distortion and cracking (p. 361). But this has to be balanced against the extra cost of incorporating the alloying elements.

The property of a steel which determines its behavior in this respect is referred to as its *hardenability*. It is a property which is well understood and can be adequately quantified, and it is a factor which often has to be taken into consideration in the selection of steels.

Alloying elements commonly also have the effect of slowing down the pearlite transformation more than the bainite transformation; this is so for the steel whose transformation characteristics are summarized in Fig. 11.17(a). It is then possible to obtain largely or fully bainitic structures during continuous cooling (curves C and D in Fig. 11.17b). It is also possible to obtain mixtures of bainite and martensite with no pearlite. In fact, the critical cooling rate is then determined by the rate required to suppress the bainite transformation (curve E in Fig. 11.17b). Nevertheless, mixtures of bainite and martensite obtained by cooling rates between D and E in Fig. 11.15(b) may be acceptable in practice.

PROEUTECTOID TRANSFORMATIONS

General Characteristics

Alloys of a eutectoid system which are off-eutectoid in composition contain, at equilibrium, one of the terminal phases of the system as well as the eutectoid constituent. The volume fraction of the terminal phases increases directly in proportion to the deviation in composition from eutectoid (cf. eutectic systems, p. 270). Thus in the iron-carbon eutectoid system described in Fig. 11.12(a), alloys which contain less than about 0.8% C and which are called *hypoeutectoid* steels, contain areas of ferrite intermingled with areas of pearlite, and the volume fraction of this ferrite decreases as carbon content increases (Fig. 11.12b to d). At the other end of the system, areas of cementite in increasing volume fraction are present in alloys which contain more than 0.8% C, and these are called *hypereutectoid* steels (Fig. 11.12e).

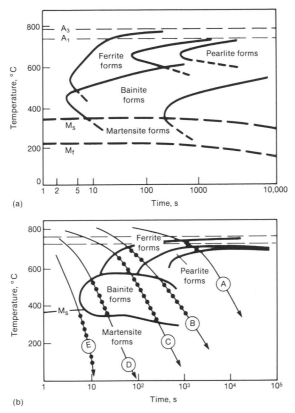

(a) Isothermal transformation diagram.   (b) Continuous cooling transformation diagram. Cooling curves are included, the significance of which are explained in the text. The dots on these curves indicate periods during which transformation occurs.

Fig. 11.17. Experimentally determined transformation diagrams for an alloy steel containing approximately 0.4% C, 1% Cr, and 0.15% Mo.

Transformations During Heating

Let us now consider what happens when an off-eutectoid steel is heated to a temperature at which austenite becomes the stable phase (i.e., into the field labeled "austenite" in Fig. 11.12a). Take, as an example, a 0.6% C steel which has an initial structure similar to that illustrated in Fig. 11.12(c). As described on p. 384, the pearlite areas of this structure transform to austenite when the eutectoid temperature, which is called the A_1* temperature, is reached. This does not occur instantaneously, however. The ferrite has first to undergo the allotropic transformation to austenite. The austenite has to dissolve the cemen-

---

*The origin of the use of the designations A_1, A_3, and A_{cm} is buried in the early history of the study of phase transformations in steels and need not concern us. Just regard their use as a convention.

tite. The carbon so dissolved has to diffuse to even-up the composition of the austenite. All in all, this takes several minutes to complete (Fig. 11.7). The areas of proeutectoid ferrite remain unchanged at this stage but are gradually absorbed into the austenite when the temperature is increased further. The carbon content of the austenite decreases during this stage from the eutectoid composition to that of the average composition of the steel, the carbon content at any temperature being indicated by the line labeled A_3 in Fig. 11.12(a). The absorption of ferrite is complete, and the carbon content of the austenite reaches average composition, when the A_3 temperature for the particular composition is reached (about 780 °C, or 1440 °F, for 0.6% C). Both steps in the ferrite absorption sequence require that carbon diffuse through the two phases and across the interface between them, and this takes time — perhaps ten minutes or so in this case.

Transformations During Cooling

Now let us outline the sequences of transformation events that occur when austenite obtained as above is cooled, assuming again that equilibrium is maintained and considering first a hypoeutectoid 0.6% C steel. Ferrite begins to precipitate from the austenite when the temperature falls below the A_3 temperature and continues to do so as the temperature falls toward A_1. At the same time, the carbon content of the untransformed austenite increases, its composition at any temperature being that indicated by the A_3 line. Precipitation of ferrite stops when the A_1 temperature is reached and the carbon content of the austenite remaining is about 0.8%. The same sequence of events occurs in hypereutectoid steel, except that cementite is the proeutectoid constituent instead of ferrite and the line labeled A_{cm} in Fig. 11.12(a) defines the temperature-composition relationship. In either event, the austenite remaining at the A_1 temperature transforms to pearlite during a thermal arrest in exactly the same way as that already described for austenite of eutectoid composition. So we now need to consider only the characteristics of the extra proeutectoid transformations.

Kinetics of the Transformations

The same general principles apply here as for the model system discussed on p. 347. The effect of temperature on the most important parameter of the transformation kinetics — namely, the time for the transformation to start — can be summarized by adding an appropriate line to the IT diagram in Fig. 11.14(a), as has been done in Fig. 11.18(a). Corresponding additional lines are present in the experimentally determined IT diagrams in Fig. 11.17(b) and 11.18(b). The proeutectoid transformation starts, of course, before the eutectoid transforma-

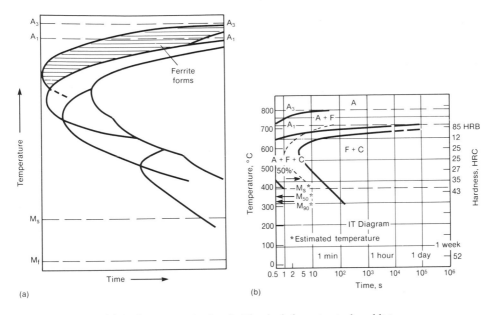

(a) A diagrammatic sketch. The shaded portion is the addition to Fig. 11.14(a) which is necessary to characterize the formation of the proeutectoid constituent.   (b) An experimentally determined diagram for a steel containing 0.35% C.

Fig. 11.18. Isothermal transformation diagrams for hypoeutectoid steels.

tion initiates, be it to pearlite or to bainite, and the start curve is of the normal C shape, the nose being located below the A_1 temperature. The precipitation of the proeutectoid phase stops at or about the time that the eutectoid transformation starts. Thus ferrite can, under practical conditions, continue to form at temperatures well below the A_1 temperature but in progressively decreasing amounts as the temperature is lowered, because the pearlite transformation intervenes. Eventually, none can form at all, and this occurs at about the nose of the pearlite (or bainite) start curve. The pearlite (or bainite) that is then formed has a different carbon content, and has to have a different structure, from those of an equilibrium transformation product (e.g., it contains less cementite than the equilibrium eutectoid amount in a hypoeutectoid steel).

The consequence during continuous cooling is that the volume fraction of proeutectoid phase that is formed decreases with increasing cooling rate until it becomes zero at the critical cooling rate, as defined previously, with corresponding changes in the eventual transformation product. Putting it in another way, the proeutectoid transformation can be suppressed at the same time as the eutectoid transformation. In this event, the carbon content of the austenite which survives to be transformed to other constituents is that of the average composition of the steel and not the 0.8% of the eutectoid composition.

Morphology of the Proeutectoid Phases

The most favored nucleation sites for the proeutectoid constituents by far are the grain boundaries of the parent austenite. The phases grow first as films around the grain boundaries of the parent austenite (that is, as grain-boundary allotriomorphs, p. 353; and see, for example, Fig. 11.12c and e), because they thicken only slowly and so soon extend into continuous envelopes around the austenite grains. The thickness of the films then depends, first, on the total volume fraction of the phase that forms, which in turn depends on the carbon content and cooling rate, and, secondly, on the austenitic grain size (thicker films form with larger grain sizes, which have smaller grain-boundary areas). Films of this nature, even when thin, can have serious consequences when the proeutectoid phase is brittle, as cementite is (p. 333), or when it affects some other property. They then provide a continuous path of easy fracture. They are not particularly detrimental, however, when the phase concerned is ductile (e.g., ferrite).

Both ferrite and cementite, as proeutectoid constituents, may grow as Widmanstätten plates in addition to grain-boundary allotriomorphs (Fig. 11.19b). However, such plates grow only over a restricted range of conditions, the factors involved including carbon content (Fig. 11.19), transformation temperature, and austenitic grain size. Occasionally, the proeutectoid phase may also grow as isolated grains within the parent austenite grains, then most likely initiating at the sites of foreign inclusions.

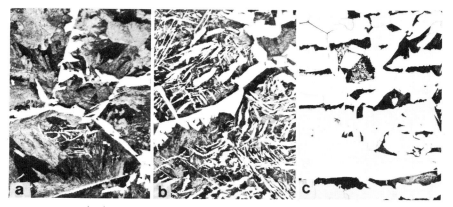

The heat treatments and compositions in all three cases are identical, only the carbon content varying in the areas illustrated. (a) Highest carbon content. The proeutectoid ferrite is present mostly as grain-boundary allotriomorphs. (b) Intermediate carbon content. Much of the ferrite is present as Widmanstätten plates. (c) Lowest carbon content. The ferrite is present in massed areas. Optical micrographs; 100×.

Fig. 11.19. An illustration of the influence of carbon content on the morphology of proeutectoid ferrite in steels.

Large volume fractions of proeutectoid ferrite grow in low-carbon hypoeutectoid steels, and, for reasons that were discussed earlier (p. 353), massed areas of ferrite then accumulate (Fig. 11.19c). Each area of *massive ferrite* consists of a number of grains, the size of these grains being referred to as the *ferritic grain size* (which is to be distinguished from the austenitic grain size). The individual ferrite grains most commonly are equiaxed (Fig. 11.20b), but sometimes they have an elongated shape (Fig. 11.20a) which has been inherited from a Widmanstätten mode of growth. In any event, a small austenitic grain size gives rise to smaller ferrite grains largely because the larger grain-boundary area provides more ferrite nucleation sites. So too does a higher cooling rate, which lowers the transformation temperature, activating more nuclei (cf. Fig. 11.20a and b). A small ferritic grain size is an essential need in ferritic steels which are required to be tough as well as strong (p. 95).

Eutectoid Transformations in Practice

We have mentioned on several occasions that the most important practical application of a eutectoid transformation is found in the many steels that are basically iron-carbon alloys. Significant variations in structure and properties become possible, most of which have been touched on at various points throughout earlier discussions. It remains only to draw the threads together and to summarize how use is made of the transformation in practice. It will not be the intention, however, to cover the heat treatment of steels comprehensively. This is the proper function of specialist books, several of which are listed at the end of this chapter.

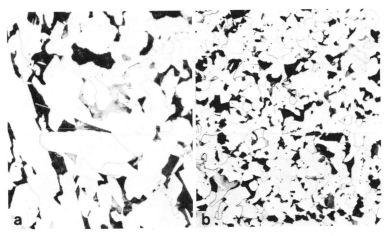

(a) The ferrite grains are elongated, giving some indication that they grow as Widmanstätten plates.   (b) The ferrite grains are equiaxed. Optical micrographs; 500X.

Fig. 11.20. Examples of the grain structure of ferrite in massed areas.

The usual objective of the first step of any full heat treatment schedule for steels is to form a fully austenitic structure (*austenitization*). An opportunity also presents itself here to achieve a reduction in grain size (p. 360), which is always desirable, and so this becomes a secondary objective of an austenitization treatment. Dealing first with hypoeutectoid steels, achieving austenitization requires that the material be heated to above the A_1 temperature throughout the component concerned. The A_1 temperature must in fact be exceeded by a margin to allow for variations in temperature throughout the furnace and for inaccuracies in the temperature control equipment. However, the second requirement of producing a fine grain size demands that this allowance be kept to the minimum. The optimum austenitizing temperature thus varies with carbon content, and it is generally recommended that it lie within the band drawn in Fig. 11.21. Temperatures toward the top of the band tend to be used when a pre-existing coarse structure may have to be refined, as may be required of an annealing treatment (see below). Temperatures toward the bottom of the band may be used when the aim is to produce as fine a final structure as possible, as is likely to be desirable during a typical normalizing treatment (see below).

Assuming that the desired temperature has been reached, the material must be left at temperature for a finite time (perhaps 10 minutes or so) to ensure that austenitization is completed. In practice, however, the factor that determines

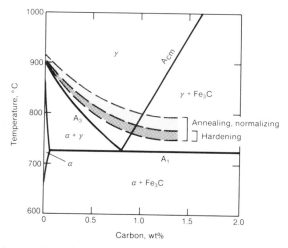

Austenitizing for annealing or normalizing typically is carried out at a temperature within the wider band, with normalizing usually being carried out at temperatures toward the bottom of the band. Austenitizing for quench hardening typically is carried out at temperatures in the lower band indicated.

Fig. 11.21. Diagrammatic illustration of the variation with carbon content of the temperatures used in practice to austenitize plain carbon steels. The band is superimposed on the relevant portion of the iron-carbon constituent diagram.

the minimum heating time usually is that required to ensure that the desired temperature is attained throughout a component. This is determined by the heating method used and the section thickness of the component.

The optimum austenitizing temperatures for alloy steels generally are somewhat different from plain-carbon steels of the same carbon content (i.e., different from those indicated in Fig. 11.21), because alloying elements alter both the eutectoid temperature and the eutectoid carbon content. Most alloying elements increase the eutectoid temperature, some more so than others. A few, principally nickel and manganese, reduce it. All alloying elements reduce the eutectoid carbon content, again some much more so than others. Standard texts and steelmakers' catalogs advise on recommended austenitizing temperatures for various steel compositions.

For hypereutectoid steels, on the other hand, austenitizing is typically carried out at temperatures only slightly above the A_1 temperature and well below the A_{cm} temperature. This also is indicated in Fig. 11.21. Some proeutectoid cementite is then left out of solution (Fig. 10.7, p. 333) because excessive grain growth would occur if this were not done. Moreover, an austenite is formed which contains only a little more than eutectoid carbon, and this produces a much more desirable structure when the material is quench hardened (see below). The cementite which remains enhances wear resistance, which again is a desirable characteristic in these steels.

Now to consider cooling schedules, which have the objective of producing structures of ferrite and cementite, including pearlite and similar high-temperature transformation products. As we have already explained, the rate of cooling determines the morphology of these transformation products, and, in principle, an infinite range of cooling rates is possible by varying the method by which heat is extracted and the section thickness of the component. In a most general way, however, two broad categories of cooling are recognized (p. 388), namely:

Annealing: Very slow cooling during which transformation occurs at temperatures not far below the A_1 temperature (represented by the cooling curves labeled A in Fig. 11.16 and 11.17b).

Normalizing: Faster cooling during which transformation occurs at lower temperatures close to the nose of the IT curve (represented by the cooling curve labeled N in Fig. 11.16 and that labeled B in Fig. 11.17b).

We have also seen that, compared to annealing, the structures produced by normalizing hypereutectoid steels: (*a*) contain less proeutectoid ferrite; (*b*) have a smaller ferritic grain size; (*c*) always in plain carbon steels contain pearlite as the major transformation product but in alloy steels may contain either pearlite or mixtures of the two; and (*d*) have a smaller lamellar spacing in the pearlitic transformation product. The structures obtained are quite strong but are not particularly tough (p. 334).

These annealing and normalizing heat treatments have, knowingly or unknowingly, been used from time immemorial. Indeed, they represented the full gamut of possibilities until the kinetics of eutectoid transformations and the existence of transformation products other than pearlite became to be understood. It then became apparent that an austenitized steel could be cooled rapidly at faster than a critical rate (e.g., rate Q in Fig. 11.16 or rate E in Fig. 11.17) to suppress the formation of proeutectoid ferrite and of pearlite, and that cooling could then be arrested at a temperature chosen to allow isothermal transformation to a different bainitic type of structure. This type of cooling cycle, which has been named *austempering*, is illustrated diagrammatically in Fig. 11.22. It is achieved in practice by quenching the austenitized steel into a molten salt bath held at the chosen temperature, where it is allowed to remain for an appropriate time, then removed and cooled in air. Austempered steels have unique structures and hence have a unique combination of properties. In practice, these properties can constitute a worthwhile improvement when the product is a lower bainite having a hardness in the range 350 to 550 HV. The austempered material is then tougher than material of comparable hardness

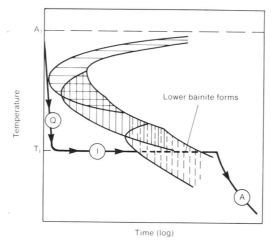

A curve representing the cooling schedule is superimposed on the isothermal transformation diagram that is sketched in Fig. 11.14(a). The austenetized steel is quenched at faster than the critical rate (Q) into a salt bath held at a temperature, T_i, which was chosen to form a particular transformation product. It is held there for sufficient time for the transformation to complete (I). It is then removed from the salt bath and cooled slowly (A) to room temperature. A lower bainite is usually chosen as the transformation product.

Fig. 11.22. Diagrammatic illustration of the cooling schedule used in an austempering heat treatment of steels.

produced by alternative heat treatment cycles. Also, components distort less during treatment, an important advantage in production. However, the austempering procedure usually is suited only to alloy steels of reasonably high hardenability. Plain carbon steels can be treated successfully but only in thin sections, because the high critical cooling rate of these steels cannot easily be achieved during quenching into a salt bath. Moreover, this is no blacksmith's heat treatment. Special equipment and sound knowledge of the transformation characteristics of steels are required.

Although eutectoid phase systems are not common, that of the iron-carbon system is not unique. A eutectoid region is present, for example, in the copper-aluminum system, the eutectoid composition being at 11.8% Al (Fig. 11.23a). A group of commercial alloys known as aluminum bronzes have aluminum contents ranging around 10%, and their equilibrium structure consists of proeutectoid alpha phase and areas of a eutectoid constituent (Fig. 11.23b). The eutectoid constituent turns out to have a lamellar structure very similar to that of pearlite (Fig. 11.23b), the lamellar plates being of a brittle gamma (γ) phase. Alloys containing close to 12% Al consist entirely of lamellar eutectoid (Fig. 11.23c).

These alloys are potentially amenable to the same types of structural manipulation by heat treatment as steels, but the possible transformations have not been explored in as much detail. It is known that the critical cooling rate to suppress the eutectoid transformation is slower than for steels but that the critical rate for the proeutectoid transformation is probably faster. Thus the eutectoid regions in the alloys illustrated in Fig. 11.23(b) and (c) are replaced by a martensitic phase (see next section of this chapter) when the material is moderately rapidly cooled (Fig. 11.23d and e); quenching is not usually required. The eutectoid transformation has been suppressed but the proeutectoid precipitation has not. It is usual in practice to ensure that the eutectoid transformation is suppressed because the eutectoid constituent is embrittling. The proeutectoid alpha phase, on the other hand, has adequate properties and can safely be left.

A group of aluminum bronzes is used as specialty alloys in practice, principally as castings and in marine applications. High-performance marine propellers, for example, are almost exclusively cast in one of these alloys. They have excellent corrosion resistance and a good combination of mechanical properties — similar, in fact, to those of medium-strength steels. The commercial alloys are, however, more complex than straight copper-aluminum alloys, usually containing substantial amounts of iron, nickel, and manganese. Nevertheless, their structure is still dominated by the eutectoid transformation even though deliberate efforts are not normally made to vary structure by heat treatments. It is sufficient to ensure that final cooling from high temperature is carried at a rate fast enough to suppress the eutectoid transformation.

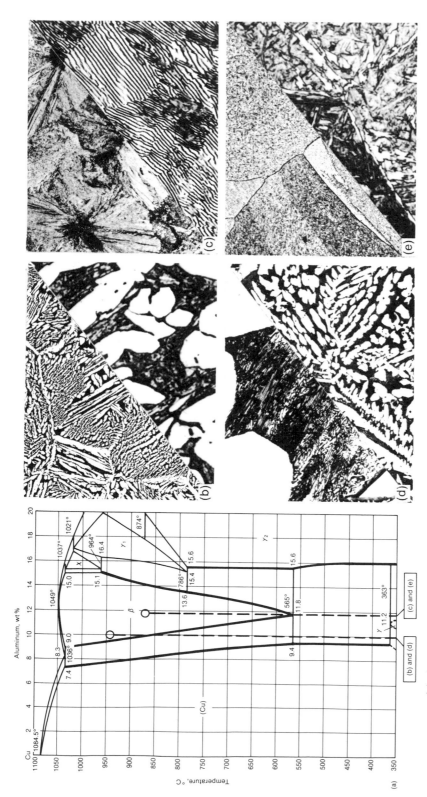

(a) The relevant constitutional diagram, the eutectoid part of the system being emphasized. Source: *Metals Handbook*, 8th Ed., Vol 7, American Society for Metals, 1973.

(b) and (d) Structures in an alloy containing 10% Al which has been heated to 925 °C (1700 °F) and then cooled very slowly for (b) and relatively quickly for (d). The needle-shaped constituents in both are proeutectoid alpha phase. The dark interleaving constituent is a lamellar eutectoid in (b) and an acicular martensitic form of beta phase (β') in (d).

(c) and (e) Structures in an alloy containing 11.8% Al which has been heated to 900 °C (1650 °F) and then cooled very slowly for (c) and relatively quickly for (e). The structure is composed entirely of lamellar eutectoid in (c) and almost entirely of the martensitic form of beta phase in (e).

Optical micrographs; upper left portions, 100×; lower right portions 1000×.

Fig. 11.23. Illustrations of the structures that can be produced by the eutectoid transformation in copper-aluminum alloys.

MARTENSITIC TRANSFORMATIONS

Characteristics of the Transformation

We have shown earlier in this chapter that phase transformations which by rights should occur during cooling from a high temperature may be prevented from doing so if the material is cooled rapidly enough. This occurs because insufficient time is made available for the long-range movement, or diffusion, of atoms that is required to effect the transformation. A high-temperature phase so retained is obviously metastable at best when it reaches low temperatures, and so it becomes a candidate for transformation by an alternative mechanism which does not require the long-range movement of atoms but which produces a more stable phase, although it may still be a metastable phase. *Martensitic* transformations provide one such possibility.

The term "martensite" was actually first applied to the constituent responsible for the high hardness of quench-hardened steels, and was chosen to honor Adolf Martens, a pioneering German metallurgist. Only later was it realized that this constituent was produced by a special type of transformation, and still later that the same type of transformation occurred in other alloys. Nevertheless, the term was retained and used generally for both the type of transformation and its product.

The essential characteristic of a martensitic transformation is that it occurs by the coordinated and simultaneous movement of all of the atoms in a volume of crystal, each atom being moved a short distance by a shearing displacement. The atoms take up positions in which they are arranged in a different type of crystal lattice. Some additional adjustments by slip or twinning are also necessary to preserve geometric continuity within the material, but these processes do not require long-range movements of atoms either. So all atoms retain their positions relative to one another, which means that there are no changes in composition even on the scale of a crystal cell. The movements required occur virtually instantaneously and cannot be suppressed no matter how fast the cooling. Martensitic transformations conforming to these principles occur in a number of metallic alloys and even in some nonmetals (e.g., silica and zirconia), although there are differences in detail. The transformation of greatest practical importance undoubtedly is the one that occurs in iron-carbon alloys (steels), and we shall concentrate on this transformation.

The martensitic transformation in steels involves the transformation of a high-temperature allotrope of iron, which is called gamma phase or austenite and which has a face-centered cubic crystal structure, to the low-temperature allotrope, which is called alpha phase or ferrite and which has a body-centered cubic crystal structure (see p. 226). As we have already discussed in connection with eutectoid and proeutectoid transformations, this transformation should start (in unalloyed steels) at some temperature between 910 and 727 °C (1670 and 1341 °F), depending on the carbon content of the austenite, and should be complete at 727 °C (Fig. 11.12). As we have also discussed, this requires diffusion-controlled transformations, which can be prevented if the austenite is cooled rapidly enough. It is this austenite which transforms by a martensitic transformation when it reaches lower temperatures.

Kinetics of Martensitic Transformations

Austenite, which we will use as a representative example, begins to transform when the temperature falls to an identifiable temperature called the M_s *temperature* (martensite start temperature). This is the temperature at which the high-temperature phase has become unstable enough to force a transformation by an alternative mechanism. Only a small volume of parent crystal is affected at first, the transformation propagating in a slab-shape volume across the crystal until it encounters an obstruction, such as a grain boundary or a previously transformed slab. The slab projects across the parent grain at nearly the speed of an elastic wave in iron (1000 m/s, or 3280 ft/s), which is a far cry from the speed at which diffusion-controlled transformations, such as the one illustrated in Fig. 11.13, procede. However, with a few exceptions in unusual alloys, the transformation then stops and everything is quiescent, no matter how long the holding time, until the temperature is lowered again.

Transformation resumes when the temperature is lowered, an additional volume fraction of martensite forming to produce a total amount that is characteristic of the temperature (Fig. 11.24). This process occurs continuously, of course, if the temperature is reduced continuously, discrete slabs of martensite being produced one by one to fill in the untransformed space in the parent grain (Fig. 11.24). A temperature can be identified at which, for all practical purposes, all of the austenite has transformed, and this is called the M_f temperature (martensite finish temperature).

A transformation with the above type of reaction kinetics is referred to as being *athermal*, in contrast to isothermal transformations, which occur over time at constant temperature. It is strictly not possible to mix athermal data with the isothermal information contained in an IT or CT diagram, but by convention this is done by adding markers on the temperature axis of the diagram. Lines may even be drawn parallel with the time axis from these markers. The M_s and M_f temperatures can be marked in this way, and perhaps also the temperatures at which intermediate amounts of martensite form (this has been done, for example, in Fig. 11.14, 11.16, and 11.17).

The M_s temperature of a steel is affected most strongly, weight for weight, by its carbon content, decreasing from about 500 °C (930 °F) for 0.1% C to about 100 °C (210 °F) for 1.4% C. All principal alloying elements further reduce the M_s temperature to varying degrees, and to extents which may be

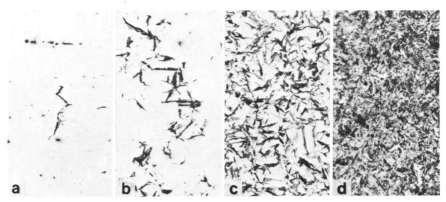

The dark needles in these photomicrographs are martensite plates that formed virtually instantaneously at the temperatures indicated. The temperatures were: (a) 210 °C (410 °F); (b) 200 °C (390 °F); (c) 180 °C (360 °F); and (d) 20 °C (70 °F). The M_s temperature of this 0.8% C steel is about 220 °C (430 °F), and the M_f temperature is about 10 °C (50 °F).

Fig. 11.24. An illustration of the kinetics of formation of martensite.

significant considering the amounts of alloying elements that are added in many cases. The M_s temperature can even be reduced to below room temperature in some highly alloyed steels, in which event the steel remains fully austenitic at room temperature. The 18Cr-8Ni steels commonly used in corrosion-resistant applications such as in kitchens and food-processing equipment are examples. Reliable formulas are available to calculate the M_s temperature of a steel from its composition.

The M_f temperature of a steel is usually about 120 °C (215 °F) below its M_s temperature. Thus, the M_f temperatures of steels containing less than about 0.5% C (also allowing for its equivalence in alloying elements) are above, or at least close to, ambient. It follows that virtually all of the austenite transforms to martensite when these steels are quench hardened, although a few volume per cent of untransformed austenite may remain. Most steels used in stress-bearing engineering components are in this category. The M_f temperatures of steels containing more than 0.6% C, however, fall progressively below room temperature, and now some of the austenite must remain untransformed after normal quench hardening (the steel illustrated in Fig. 11.26a is an example). The lower the M_f temperature, the larger the volume fraction of this *retained austenite*. Most tool and wear-resistant steels are in this category. As much as 25 vol % austenite could be retained in steels with the highest carbon contents, although precautions are taken in practice to avoid this. For example, the steel may be chilled after quench hardening, the aim being to cool it to below the M_f temperature to permit transformation of all of the retained austenite. This is called a *subzero treatment*.

Crystallography of Martensite

Austenite, the parent phase of martensite, can hold comfortably in interstitial solid solution (p. 228) large numbers of carbon atoms. Ferrite, on which the product martensite is based, can however hold virtually none at all at equilibrium, yet all of the carbon atoms that were present in the parent austenite are frozen in place in the martensite. The body-centered cubic lattice of ferrite obviously has to adjust itself considerably to accommodate them. It actually does this by expanding one axis of the cube markedly and contracting the other two a little and equally, which results in a tetragonal lattice cell instead of a cubic cell, the tetragonality increasing with carbon content. One direct consequence is that there is an over-all expansion of the material during a martensitic transformation which can result in the development of tensile internal surface stresses (p. 361), distortion, and cracking. The likelihood of these undesirable results increases with carbon content because the lattice expansion is greater, the martensite is more brittle, and the transformation temperature is lower.

These phenomena are, in degree but not in kind, special characteristics of the martensitic transformation in iron-carbon alloys. There are no other common cases where such a highly supersaturated solid solution can be obtained, especially in an interstitial solution.

Morphology of Martensite

We noted earlier that a martensitic transformation takes place in a series of discrete events each of which occurs in a confined slab of the parent crystal. The slab is unlimited in length, until it encounters an external obstruction, but is limited in thickness. There is a good reason for this. The slab of crystal shears in the manner illustrated diagrammatically in Fig. 11.25 and in the process plastically deforms the adjoining regions of the parent austenite to an extent which increases as the slab thickens. This in turn imposes a restraint on the slab of martensite, which eventually halts the thickening process. Further transformation then has to occur by the nucleation of new slabs. The constraint is greatest at the ends of the slabs, and so they are reduced in thickness there and end up with a pointed, or lenticular, shape. Plates of this shape have a needle-like outline when viewed in a two-dimensional section (Fig. 11.26a) and hence are sometimes described as being acicular.

The martensite slabs form as very thin *laths* (Fig. 11.27a, c, and e) in low- and medium-carbon steels containing up to about 0.6% C, which encompasses all of the steels used for stress-bearing structural purposes. The laths are 0.1 μm (3.9 μin.) or less in thickness and are grouped in *packets* which are on the order of micrometres in thickness. The packets are the features visible in the optical photomicrograph in Fig. 11.27c. The thickness of the packets decreases with

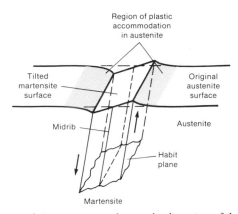

Shearing of the martensite plate in the direction of the arrows induces the plastic strains indicated in the adjoining regions of the austenite. The resistance of the austenite to deformation opposes widening of the plate.

Fig. 11.25. A sketch illustrating how the shear that occurs during the formation of a martensite plate plastically deforms the adjacent regions of the parent austenite.

austenitic grain size. They are aligned on a simple plane of high symmetry in the parent crystal of austenite, a plane for which there are only four alternatives in each crystal. This allows the packets to fill in each austenite grain with symmetrically arranged arrays (Fig. 11.27a and c). The microstructures of this type of martensite consequently appear orderly. Individual laths contain high densities of dislocations produced by plastic deformation during formation, which occurs, remember, at comparatively high temperatures. Carbon atoms tend to diffuse to these dislocations and lock them in place (p. 328). Also, some tempering (see below) can occur during cooling from the M_s temperature. All in all, the structure is complex.

Martensite starts to form in a different morphology at a carbon content of about 0.6%, and forms entirely in this different morphology at 1.0% C, steels with carbon contents between these values having mixtures of the two types. This range encompasses the steels used in tools and in wear-resisting applications. The martensite now forms in units which are several micrometers wide and which have an easily resolved lenticular shape (Fig. 11.26a and 11.27d). It was consequently called *plate martensite* in the early days of optical microscopy, when it could easily be studied, and is still known as such when being distinguished from the *lath martensite* previously discussed. The plates of this morphology develop along one of two possible complex planes in the austenite crystals, for both of which there are large numbers of variants in each crystal. Thus the plates appear to be rather haphazardly arranged when viewed in section (Fig. 11.27d) because each one that develops as the parent grain is filled can initiate on any one of the possible variant planes (Fig. 11.27b). This has an

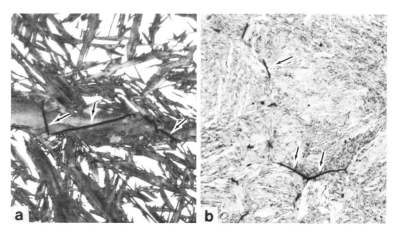

(a) Transverse cracks (arrows) produced when a developing plate impinged against a pre-existing plate. The white background is retained austenite; the dark acicular regions are the martensite plates.   (b) Cracks (arrows) produced at the austenitic grain boundaries. Optical micrographs: (a) 1000X; (b) 250X.

Fig. 11.26. Examples of cracks produced during the formation of martensite.

important practical consequence in that newly forming plates frequently collide with pre-existing plates and may then crack them (Fig. 11.26a). Fortunately, the victim plates are not always cracked, but large, thick plates produced in coarse-grain austenite of comparatively high carbon content are at risk (Fig. 11.26a). The presence of these *microcracks* severely increases brittleness. Cracks may also develop at the austenitic grain boundaries (Fig. 11.27b), in which event they may extend to produce a continuous *hardening crack*.

Plate martensite also has a fine substructure, but a substructure which is characteristically different from that of lath martensite. The plates contain dislocations, but these dislocations are introduced by plastic deformation at lower temperatures commensurate with the lower M_s temperatures of high-carbon austenites. The plates also characteristically contain numerous very thin parallel twins (Fig. 11.27f). The substructures of both plate and lath martensites are caused, incidentally, by the strains necessary to meet the geometrical requirements of the shape changes that occur during transformation.

Properties of Iron-Carbon Martensites

The high hardness of iron-carbon martensites, which increases with carbon content, is their one really unique feature (Fig. 10.6, p. 332). The reason for this hardness has long been a matter of intense speculation because it is so unusual and because it is the basis for many of the useful characteristics of steels. It is now recognized that a number of factors contribute, namely:

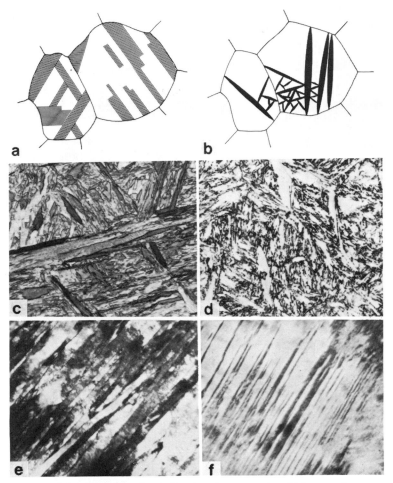

(a) and (b) Sketches indicating the manner in which austenite grains are filled by the two types of martensite, lath martensite (a) and plate martensite (b). (c) and (d) Optical micrographs illustrating (c) the regular geometrical arrangement of the packets in which martensite laths are grouped, and (d) the irregular arrangement of plate martensite. Optical micrographs; 1000X.

(e) and (f) Transmission electron micrographs showing (e) individual laths of which the packets seen in optical micrographs are composed, and (f) transverse parallel twins which characteristically are present in martensite plates. Transmission electron micrographs: (e) 10,000X; (f) 24,000X. Micrograph (f) is from *Metals Handbook*, 9th Ed., Vol 9, *Metallography and Microstructures*, American Society for Metals, Metals Park, OH, 1985.

Fig. 11.27. Illustrations of the two types of martensite formed in steels.

1. The high degree of supersaturation of the interstitial solid solution of carbon in iron (p. 331). This is the most important factor, particularly in steels of higher carbon contents.
2. The fine grain structure of martensite. This is particularly important in low-carbon lath martensites where strength increases with decreasing lath-packet size, which in turn decreases with the grain size of the parent austenite.
3. The dislocation and twin substructure in the laths and plates, particularly when the dislocations are locked by diffused carbon, as they are in low-carbon martensites.
4. The presence of alloying elements in substitutional solid solution in the martensite, although this is usually of minor importance.

A high price is paid for this hardness at its higher levels in the form of impaired toughness; high-carbon martensites can fairly be described as being brittle (p. 88). These martensites consequently cannot be used in their virgin states but have to be *tempered* to achieve an acceptable combination of mechanical properties. This modifying treatment is discussed in the immediately following section.

Tempering of Iron-Carbon Martensites

Martensitic transformations are in principle reversible on reheating, but iron-carbon martensites are exceptions because they decompose before the transformation gets a chance to reverse. The special martensitic crystal structure is destroyed in the process and the special mechanical properties of martensite are modified. Hardness and strength are reduced (Fig. 11.28 and 11.29), but with

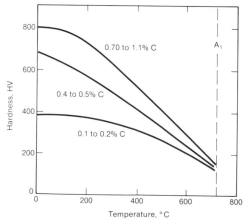

Fig. 11.28. Variation of hardness with tempering temperature for three representative plain carbon steels.

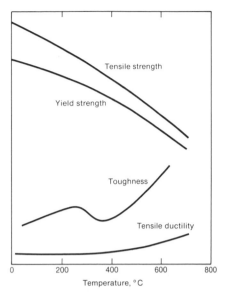

Strength and ductility are as assessed in a tensile test. Toughness is as assessed in a notched impact test.

Fig. 11.29. Diagrammatic representation of the variation with tempering temperature of the mechanical properties of a 0.4% C steel.

the bonus that ductility and toughness are improved (Fig. 11.29). This opens a new and profitable avenue for producing particular combinations of properties in steels. A heat treatment which decomposes martensite in this way is known as *tempering** and is unique to steels among engineering alloys.

All of this seems to be straightforward enough. The martensite might well be expected to decompose into the equilibrium phases (ferrite and cementite) at any temperature at which the diffusion of carbon is reasonably rapid. The temperature would not have to be very high for this to occur, because the interstitial solute carbon can diffuse with unusual ease. The mechanical properties would then be expected to revert gradually to those of the base phase, strengthened only a little by the dispersion of the second phase (p. 334). However, the actuality is more complicated. The ultimate result is indeed the production of the equilibrium phases, but this state is arrived at through a sequence of overlapping intermediate stages each of which has its own particular effects on properties. It will be most convenient to consider these effects in steels grouped by carbon content.

Low-Carbon Steels (Less Than 0.25% C). As-quenched martensites with carbon contents in this range consist of packets of laths which contain high densities of dislocations (p. 405). Most of the carbon segregates to these disloca-

---

*This term is used in metallurgy with several meanings (p. 472). Here it means heating of an iron-carbon martensite at a temperature between ambient and the A_1 temperature.

tions and lath boundaries during cooling from the martensite-formation temperature range, a process which certainly will be completed (i.e., carbon will have segregated to all available crystal defects) if the steel is tempered at low temperatures, say up to 200 °C (390 °F). Sufficient numbers of crystal defects typically are present to accommodate up to a little more than 0.2% C. The "martensite" then is not detectably tetragonal because no carbon is left in supersaturated solution. Strengthening is now due to locking of the dislocations and boundaries by the segregated carbon (p. 328). The segregated carbon is precipitated as carbides at higher tempering temperatures in the same way as we shall soon describe for steels of higher carbon contents.

The net result is that the hardness of low-carbon steels does not decrease until quite high tempering temperatures are reached, and then only a little (Fig. 11.28). There is, however, no real need to temper them at high temperatures, and this is not typically done in practice. These steels are reasonably tough, even in the as-quenched condition, and certainly so after tempering at low temperatures. Such steels consequently have good combinations of ductility, toughness, and strength and are the basis for a generation of high-strength weldable steels which have excellent characteristics (p. 95). The fact that they can be used untempered is a further advantage in a welding steel. A zone in the parent plate adjacent to a weld inevitably is austenitized and then quench hardened because it is cooled rapidly by the adjacent still-cold parent metal. It is clearly an advantage not to have to heat a whole structure after welding just to temper this heat-affected zone.

Medium-Carbon Steels (0.25 to 0.6% C). The general expectation for tempering of steels in this range of carbon contents is that the hardness should decrease gradually, as illustrated by the middle curve in Fig. 11.28. Tensile and yield strengths decrease roughly in proportion to the decrease in hardness, as illustrated diagrammatically in Fig. 11.29. Both ductility and toughness increase at the same time, but they do not increase significantly until softening is well advanced (Fig. 11.29). Moreover and more importantly, there is a minimum cusp in the toughness relationship in the range of tempering temperatures from 250 to 375 °C (480 to 710 °F), the decrease being so significant that tempering in this range has to be avoided in practice.

Toughness increases rapidly when the tempering temperature exceeds about 400 °C (750 °F), to very acceptable levels when the carbon content is less than about 0.5%. A regime is entered in which great versatility is available in the choice of combinations of strength and toughness. It is from this range of compositions (probably with alloying elements added to improve hardenability and perhaps other properties) and heat treatments that the typical steels used in stressed machinery components and the like are drawn. Usable tensile strengths ranging from 800 to 1800 MPa (115 to 260 ksi) can be obtained. These steels can also be used after tempering in the low-temperature range of 150 to

200 °C (300 to 390 °F). The increase in toughness is then only modest, but is still worthwhile, and there are applications where this toughness is adequate because the component is stressed largely in compression and where the higher yield strength is advantageous. Bearings and gears are two examples.

In steels with carbon contents ranging from 0.5 to 0.6%, toughness does improve with tempering at higher temperatures but not enough to reach levels acceptable for most components which are highly stressed in tension. These steels consequently can be used only in applications where the poor toughness can be coped with and where the extra yield strength is of significant advantage. Examples include screwdrivers, wrenches, and springs (p. 67). Steels in this range can still, however, be used in a lightly tempered condition in applications where their high hardness is of itself useful. Low-duty cutting blades such as those on metal chisels, and hand tools such as pliers and hammers, are examples.

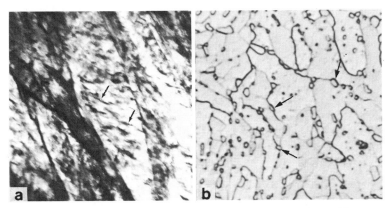

(a) Extremely small carbide particles (some are indicated by arrows) precipitated by tempering at 300 °C (570 °F). Shown at an early stage of tempering. Transmission electron micrograph; 20,000×.

(b) Large cementite particles in a matrix of ferrite, many of the particles (some are indicated by arrows) being located at the ferrite grain boundaries. Tempering temperature, 600 °C (1110 °F). Shown at a late stage of tempering. Optical micrograph; 2000×.

Fig. 11.30. Illustration of the range of structures developed during tempering of the martensite in a 0.4% C steel.

Let us now consider briefly the structural changes that occur during the tempering of these martensites. Note first that 0.2% of the carbon present again segregates to dislocations and boundaries during cooling, and some small carbides may also precipitate in these cases if the cooling rate is on the low side — that is, some *autotempering* may occur during quench hardening. The first deliberate tempering step involves the precipitation of very small particles of carbide (Fig. 11.30a) and occurs at temperatures up to about 300 °C (570 °F).

These particles are not the equilibrium phase cementite but rather a carbide, called ϵ (epsilon) carbide, which has a different crystal structure and a slightly different composition from those of cementite. Nevertheless, the tetragonality of the parent martensite is reduced as the carbon is removed from the supersaturated solution, and this is the major softening mechanism. However, the ϵ carbide is precipitated as very small particles, and this can induce dispersion hardening (p. 340), which delays softening (middle curve, Fig. 11.28). Next in order of occurrence, any retained austenite, of which there will usually be very little at these carbon contents, decomposes in the temperature range of 200 to 300 °C (390 to 570 °F), constituents not unlike bainites being formed. This too may delay softening if a sufficient amount of retained austenite is present. At the same time, the carbide is gradually replaced by somewhat larger rod- or plate-shape Widmanstätten particles of true cementite (Fig. 11.30a). Above 350 °C (660 °F), the cementite particles spheroidize and grow (p. 423), and the dislocations produced with the martensite are rearranged and reduced in number. Eventually, the now-ferritic matrix recrystallizes and these ferrite grains grow larger as temperature increases.

The steel has now reached a basic state in which large, approximately spherical particles of cementite are distributed uniformly through strain-free grains of ferrite (Fig. 11.27b). This structure is, however, not a very strong one because particles of a second phase of this size do not have a great strengthening effect (p. 335). The volume fraction of cementite increases with carbon content, but this has only a small additional effect on strength. Hence the hardness of fully tempered steels does not vary much with carbon content. This is indicated by the way in which the three curves in Fig. 11.28 converge at higher tempering temperatures. The structures are, however, ductile and may be produced deliberately to facilitate a fabrication process, such as forming or machining, or subsequent heat treatment.

It is of some interest that many of the carbide particles end up at the ferritic grain boundaries (Fig. 11.30b). Actually, it is the other way around. The grain boundaries move around after the ferrite has recrystallized, are captured by cementite particles, and then tend to be held in place. The presence of the carbide particles restrains the growth of the ferrite grains, the more so the more numerous the particles, and the growth of the ferrite grains has to follow the growth of the carbide particles.

The reduction in toughness which results from tempering in the range from 250 to 350 °C (480 to 660 °F) is due to different phenomena from those which we have so far discussed. It occurs during the cementite-formation stage of tempering when the cementite tends to form as films around the austenitic grain boundaries. However, the loss in toughness probably is not directly due to the formation of these brittle films, but more likely to the concurrent segregation of certain impurities at the same boundaries. These impurities include elements such as phosphorus, antimony, tin, and arsenic. Only extremely small amounts of impurities are required, which can scarcely be avoided in commercial steels. The problem has yet to be overcome and imposes an unfortunate limitation on steel technology by excluding from use a range of strengths that otherwise would be useful. Several other forms of embrittlement can arise during particular tempering circumstances, but we need not go into them other than to note that methods of coping with them satisfactorily have been developed.

All of the changes that occur during tempering involve diffusion and hence are time- as well as temperature-dependent. A structure and hardness may be obtained by tempering for a short time at a high temperature or for a longer time at a lower temperature, the relationship being an exponential one. In heat treatment practice, however, temperature usually is chosen as the controlling variable, with some constant time (commonly 1h) being used. The time parameter becomes important, however, during long-term exposure at temperature in service. A heat treated steel loses its strength if its structure changes (p. 113), the temperature and time at which this occurs being determined by the resistance of the steel to the progression of the tempering structural changes (see below).

High-Carbon Steels (More Than 0.6% C). Hypereutectoid steels normally are austenitized at a temperature which leaves most of the proeutectoid cementite out of solution (p. 333). The matrix martensite obtained during quench hardening then contains a maximum of about 0.9% C, so we are concerned here only with the tempering of martensites containing between, say, 0.6 and 0.9% C. This martensite will generally have associated with it a significant amount of retained austenite, and the transformation of this austenite becomes a more important part of the tempering process. The contribution of dispersion hardening at low tempering temperatures is also greater because a large volume fraction of carbides forms. Softening is delayed noticeably, and slight hardening may even occur (upper curve in Fig. 11.28).

A high degree of toughness cannot be achieved in these steels, and so they are used in the lightly tempered condition at maximum hardness. Their applications are in fields requiring hardness and wear resistance *per se* and in which poor toughness is not a disastrous impediment. Examples include tools such as wood chisels and high-performance cutting edges which require little toughness; punches; dies; gages; and, at the higher end of the carbon range, metal-cutting files and drills.

Effects of Alloying Elements. Alloying elements can have either indirect or direct influences on both the kinetics of the tempering reactions and the relative importance of the reactions that occur. The influence of a particular alloying element depends primarily on whether it forms a compound with carbon (a carbide) more strongly than does iron. Elements such as chromium, molybdenum, tungsten, and vanadium do so; manganese is about neutral; and nickel and silicon do not.

The main indirect effect is that strong carbide forming elements delay the onset of softening by delaying the decomposition of the martensitic supersaturated solution. Weak carbide-formers have the reverse effect. Elements of both types can, however, slow down the coarsening of cementite, the reorganization of dislocations, and the recrystallization and grain growth of ferrite. All of this delays softening at higher tempering temperatures also. Alloyed steels usually also contain more retained austenite than plain carbon steels, and the transformation of this austenite is another factor which tends to delay softening.

The direct effects of alloying elements are associated with the formation of alloy-rich carbides instead of iron-base carbides. These elements diffuse substitutionally and much more slowly than carbon, so much so that they cannot diffuse to form alloy-rich carbides until a tempering temperature of 500 to 600 °C (930 to 1110 °F) is reached, and so it is only at these temperatures that alloying elements have any direct effects. The benefit is that they then form as a finer dispersion which does not coarsen as rapidly as cementite. Higher strengths can consequently be obtained at comparable levels of toughness than for plain carbon steels, and this strength is retained at higher temperatures and/or times.

Thus the general result is that alloy steels tend to be stronger than equivalent plain carbon steels when tempered at the same low temperatures as would be used to obtain the highest strengths (Fig. 11.31a). Moreover, they also tend to be significantly tougher, particularly so for some alloying elements (Fig. 11.31b).

With high concentrations of carbide-forming elements, hardness may even increase during tempering in the appropriate range, even to the extent that hardnesses greater than that of the as-quenched martensite are obtained (Fig. 11.32). This phenomenon, which is known as secondary hardening, is a type of precipitation hardening (p. 371) that is dependent on the attainment of a critical dispersion of carbide particles. Consequently, it takes a certain tempering time for maximum hardness to be attained, the hardness decreasing with longer times. But hardness is retained for long periods in service at high temperatures, at least at those below the tempering temperature. Hardening is now caused, it should be noted, by an entirely different mechanism from that for the parent martensite, and so a different combination of properties is to be expected.

Secondary hardening occurs in steels containing strong carbide-forming elements such as molybdenum, vanadium, tungsten, and chromium when present in higher concentrations. These elements are such strong carbide-formers that they tend to be located largely in the carbides in the softened steel. It is

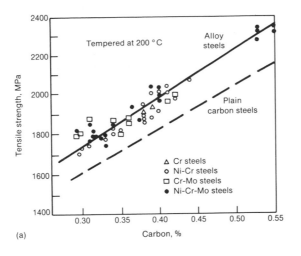

(a)

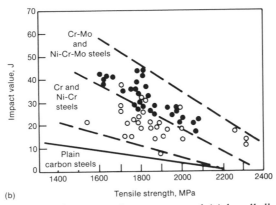

(b)

Carbon has the dominant influence on strength (a), but all alloying elements increase strength somewhat. Moreover, alloy steels are significantly tougher than plain carbon steels at the same strength (b), molybdenum being particularly effective in this respect. Source: K. J. Irving and F. B. Pickering, *J. Iron Steel Inst.*, 1960, *194*, 137.

Fig. 11.31. Comparison of the mechanical properties of plain carbon and alloy steels which have been austenitized, quenched, and tempered at 200 °C (390 °F).

then only the portions of the elements that are dissolved from these carbides during austenitization that contribute to secondary hardening. So austenitizing treatments have to be designed to dissolve adequate amounts of carbides and typically are carried out at temperatures close to the solidus temperature of the alloy rather than just above the eutectoid temperature. Moreover, austenitization has to be carried out reproducibly, because otherwise the response to tempering varies (Fig. 11.33). These highly alloyed steels, incidentally, tend to contain particularly large amounts of retained austenite after quench harden-

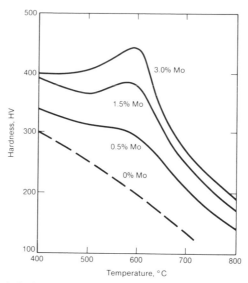

The dashed curve is the corresponding curve for a molybdenum-free steel, transposed from Fig. 11.28. Note that the presence of molybdenum has retarded softening at all tempering temperatures. Its important effect, however, has been to cause the hardness actually to increase during tempering in the range from 550 to 650 °C (1020 to 1200 °F). Adapted from K. J. Irving and F. B. Pickering, *J. Iron Steel Inst.*, 1960, *194*, 137.

Fig. 11.32. Examples of the effects of a strong carbide-forming alloying element (molybdenum) on the variation of hardness with tempering temperature for quench-hardened 0.4% C steels.

ing. This austenite does not transform during tempering but is changed in some way so that this time around most of it does transform to martensite during cooling after tempering. The newly formed martensite then has to be treated by a second tempering stage.

A series of steel compositions has been developed on these principles for use in components which are required to operate at high temperatures, each steel being designed to meet a needed combination of operating temperature, wear resistance, strength, and toughness. The best known of these steels perhaps are those commonly called high speed steels. These are the steels from which, for example, the twist drills familiar to both professional and amateur metal machinists are manufactured. They were first developed by empirical methods but subsequently were refined greatly by a more scientific approach, and they constituted at the time a major breakthrough in the technology of metal machining. The cutting edge of a metal-cutting tool attains a high temperature in use, the higher the cutting speed the higher the temperature. Plain carbon steels can be used satisfactorily only at comparatively low speeds because they are limited by the temperature at which the edge material softens and becomes

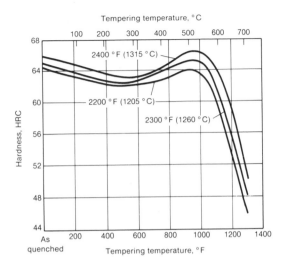

This type of steel retains its hardness until a temperature of about 550 °C (1020 °F) is exceeded. The three curves show that the exact response to tempering varies with the austenitizing treatment. This is because the amount of alloying element taken into solution is affected. Source: *Heat Treater's Guide*, P. M. Unterweiser, H. E. Boyer, and J. J. Kubbs, eds., American Society for Metals, Metals Park, OH, 1982.

Fig. 11.33. Variation of hardness with temperature during tempering of a high speed tool steel of nominal composition (0.75% C, 18% W, 4% Cr, 1% V).

useless. Tools made of high speed steel can be operated at much higher speeds before softening becomes serious, thereby affording a great improvement in the economy of the operation.* High speed steels in turn have been superseded in many machining applications by other more refractory materials (principally carbides of heavy metals such as tungsten), but they still have a niche of usefulness. A range of somewhat similar but less highly alloyed steels has also been developed for use in dies which have to operate hot — for example, forging and extrusion dies.

Martensitic Transformations in Practice

The great flexibility and wide usage of steels as engineering materials, past and present, is essentially based on the potential provided by the martensitic transformation. At one end of the spectrum, it enables hard structures suitable for the ever–essential cutting tools. At the other end, and combined with a tempering heat treatment, it enables a variety of combinations of strength and toughness to be obtained by balancing carbon content and tempering temperature.

*These steels are also designed to contain a goodly volume of alloy carbides which confer wear resistance superior to that of cementite–containing steels.

In principle, three steps are involved in quench hardening heat treatments in which the formation of martensite is the key, namely:

(a) *Austenitize*, usually with the objective of obtaining the finest possible grain size. The same principles apply as for austenitizing during annealing or normalizing heat treatments (p. 396), but temperatures in the lower band in Fig. 11.21 are used (for plain carbon steels) to ensure minimum grain size. Heat treaters' guides should be consulted, however, for recommendations on the optimum austenitizing conditions for specific steels, plain carbon and alloy.

(b) *Cool*, preferably immediately after removal from the austenitizing furnace, at a rate faster than the critical rate which is necessary to suppress transformations at high temperatures to pearlitic and/or bainitic structures. Quenching in a liquid medium is usually required.

(c) *Temper*, preferably as soon as possible after quenching, at a temperature chosen to produce the desired combination of properties. Some of the factors that enter into the choice of tempering conditions were outlined on p. 408, but again a large body of knowledge has been built up on which the choice of optimum tempering conditions for particular circumstances can be based. This type of information is also available in standard references on the heat treatment of steels. Remember always that martensite itself is next to useless as an engineering material. It becomes useful only when it has been tempered, so that tempering is an integral and essential part of a quench-hardening heat treatment.

Cooling is the crucial and, in practice, difficult stage of the sequence, the critical cooling rate of the steel in the condition in which it has been austenitized being the controlling parameter. We have seen that the critical cooling rates of plain carbon steels are quite high and can usually be attained only by vigorous quenching methods, such as by immersion in water. Even then it can be achieved only in comparatively thin sections (1 or 2 cm or less). The first and foremost advantage of steels alloyed with a few weight percent of elements such as nickel, chromium and molybdenum is that the critical cooling rate is reduced, and substantially so in the more highly alloyed steels. This allows less vigorous cooling methods to be used. Quenching in an oil becomes adequate (oil-hardening steels) and sometimes even cooling in air (air-hardening steels). Moreover, full hardening can be achieved in thicker sections (5 cm at least).

Several other problems are inherent to the quenching stage and are magnified when fast cooling is required. The first occurs because the cooling rate inevitably is slower toward the center of a section than at the surface which is exposed directly to the cooling medium. Moreover, this difference is greater the greater the cooling rate attempted and the thicker the section of the workpiece. The normal laws of heat transfer ensure this. Let us now assume that the critical cooling rate is just achieved at the surface of a workpiece. The critical rate is then not achieved in the central regions of its section, and so the central regions are slack quenched, as is illustrated diagrammatically in Fig. 11.34(a) (compare also with curve Q in Fig. 11.16b). The structure of the workpiece then varies from surface to center. The outer shell is composed entirely of martensite, but with increasing depth a layer is soon met which contains some areas of high-temperature transformation product as well. The structure there would be similar to that illustrated in Fig. 11.15(a). The volume fraction of high-temperature transformation product then increases rapidly with further depth and soon becomes 100%; the structure then is similar to that which would be obtained by normalizing.* On all surface tests, the workpiece would appear to have been satisfactorily hardened, but it hasn't been, a substantial proportion of the cross section having a structure with inferior mechanical properties after tempering.

A surface cooling rate which considerably exceeds the critical rate must be achieved before the critical rate is reached in the central regions. This stage, which is described diagrammatically in Fig. 11.34(b), is the one at which full quench hardening is first achieved. A corollary is that full hardening cannot be achieved at all if this state is not reached when the most vigorous cooling method that it is practicable or safe to use has been used. So it sets the limit to the thickness of section that can be successfully quench hardened. Obviously, the higher the hardenability of the steel and the lower the critical cooling rate, the thicker will be the limiting section under conditions which are achievable.

A second major problem arises during quench hardening because of the volume expansion that occurs during the transformation of austenite to martensite. This expansion tends to induce tensile residual stresses at the surface of a workpiece, because the transformation is likely to occur at the surface of a section before it does in its central regions (p. 361). These residual stresses at their lowest level add to service stresses, then at the next level cause undesirable distortion of components, and finally at their most serious level give rise to destructive cracking. The effect becomes more severe the higher the carbon content of the martensite, because the volume expansion during transformation increases. For a given composition, it also becomes more severe with increasing severity of quench, because the time difference between transformation in the surface and central regions is increased. Consequently, steels with high hardenability have an advantage from this point of view as well, provided of course that advantage is taken of their hardenability by using a less severe quenching procedure.

But this problem with residual stresses can also be alleviated by a cooling schedule called *martempering*, the design of the procedure having been based on our understanding of the mechanism of stress development and of the harden-

---

*This sequence of structures also develops at the boundary of a region which is quenched locally (*selective quenching*).

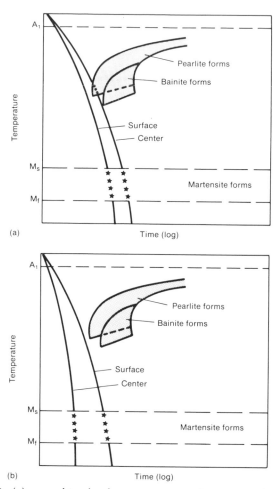

In (a), quenching has been severe enough to ensure that the surface regions have cooled fast enough to avoid transformation to pearlite or bainite, the austenite transforming completely to martensite (stars on cooling curve). The central regions have cooled more slowly, however, and some of the austenite has transformed to pearlite (dots on cooling curve). Only the austenite remaining has transformed to martensite (stars).

In (b), the quenching has been more severe and the cooling rates at both the surface and central regions have been fast enough to ensure that no transformation to pearlite could occur. All of the austenite in both regions has transformed to martensite (stars on cooling curves). Full through hardening has occurred. However, in both (a) and (b) the transformation to martensite has occurred at different times in the surface and central regions. This can result in the development of tensile residual stresses at the surface.

Fig. 11.34. Diagrammatic illustration of the effect of section thickness on the quench-hardening of steels. Cooling curves for the surface and central regions, respectively, of a workpiece are shown here superimposed on a continuous cooling transformation diagram for a plain carbon eutectoid austenite.

ing transformations. These considerations indicate that: (*a*) the objective should be to ensure that the martensite forms at the same time throughout the section of the workpiece, (*b*) fast cooling is really necessary only at higher temperatures at which pearlites and bainites could form, and (*c*) cooling through the martensitic transformation temperature range can be carried at any slow rate. Consequently, a workpiece can be quenched into a bath held at a temperature above the M_s temperature of the steel in the expectation that high-temperature transformations will be suppressed. It can then be allowed to stay there until the temperature equalizes throughout the section of the workpiece, provided that it is not left for so long that isothermal transformation to bainite could occur as in austempering. It can finally be removed from the quench bath and allowed to cool as slowly as is convenient, typically in still air, to ambient temperature. The cooling schedule is illustrated diagrammatically in Fig. 11.35. Martensite now forms at approximately the same time throughout the section.

Note that martempering is a quenching procedure, not a tempering procedure. The product is normal martensite which has to be tempered in the same way as for conventional quenching. The procedure perhaps might more prop-

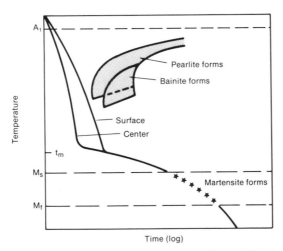

The workpiece is quenched into a bath held at temperature t_m, which is above the M_s temperature. It is allowed to stay there until the temperatures of the surface and central regions equalize, and is then withdrawn and allowed to cool slowly to ambient. The martensitic transformation now occurs at approximately the same time throughout the section of the workpiece, thus reducing the tendency for residual stresses to develop.

Fig. 11.35. Diagrammatic illustration of the cooling cycle used in a martempering heat treatment procedure. The cooling curve here is superimposed on a continuous cooling transformation diagram for a plain carbon eutectoid martensite. Compare with Fig. 11.34.

erly be called "marquenching". As for austempering, the procedure is really applicable, particularly in reasonably thick sections, only to steels which have good hardenability. This is because only comparatively modest cooling rates can be obtained by quenching into a bath of hot liquid, cooling rates comparable to those obtained during oil quenching at best.

Quite a number of types of commercial alloys undergo martensitic transformations, but none of these achieve the degree of supersaturation or the degree of crystal-lattice distortion that is attained in iron-carbon martensites. Consequently, they do not develop particularly unusual properties, as do iron-carbon martensites. Perhaps the most important of these nonferrous martensitic transformations are those which occur in titanium alloys and which follow because titanium undergoes an allotropic transformation from a low-temperature form (designated alpha phase) with a close-packed hexagonal crystal structure to a high-temperature form (designated beta phase) with a body-centered cubic crystal structure, at 882 °C (1620 °F) in pure titanium. The resulting transformations in alloys occur in some cases as eutectoid systems and in others as simpler isomorphous systems of the type illustrated in Fig. 11.8A (p. 364) for the commercially important titanium-vanadium-aluminum alloys.

An alloy of the latter type, say the one containing 6% Al and 4% V which is illustrated in Fig. 11.8A and discussed in the associated text, does not precipitate alpha phase when cooled even moderately rapidly from the beta field to room temperature. The beta phase persists over a range of temperatures but eventually transforms martensitically to a form of alpha phase which is designated alpha prime (α'). The transformation starts and finishes at characteristic temperatures (M_s and M_f temperatures) which vary with alloy content. Approximate M_s and M_f temperature lines are drawn in Fig. 11.8A. Note that the M_s temperature for these alloys falls to ambient at about 12% Al and the M_f temperature at about 7% Al. It follows that only alloys containing less than 7% Al can be transformed completely to alpha prime and those containing more than 12% Al do not transform at all, beta phase being retained as a metastable phase. Decreasing amounts of alpha prime form, of course, in alloys containing 7% and 12% Al.

Some strengthening above that which would be found in a corresponding alpha solid solution does occur as a result of this martensitic transformation, but not nearly as much as for iron-carbon alloys. On the other hand, there is no significant loss in ductility or toughness. Maximum strengthening occurs in alloys which have just sufficient aluminum to ensure complete transformation to martensitic alpha prime (i.e., the composition at which M_f is still slightly above room temperature). Little change in properties occurs, moreover, when the martensitic phase is tempered. The phase decomposes directly to a dispersion of beta phase in alpha phase, these being the equilibrium phases, which is fine but not fine enough to induce precipitation hardening. Retained metastable beta does precipitation harden a little during such a tempering treatment, and this in fact contributes more to strengthening than the martensitic phase. The formation of several other types of martensitic phase is also possible in titanium alloys, but none are of great strengthening significance. It is easy to see, even from this very brief outline, that the metallography of titanium alloys can become very complex and that the effects of heat treatments on properties can become very subtle.

The beta phase in eutectoid copper-aluminum alloys also transforms martensitically during moderately rapid cooling which suppresses the eutectoid transformation (p. 399 and Fig. 11.23). The martenistic form of beta phase is designated as beta prime (β') and decomposed directly during a tempering treatment to a fine distribution of a phase (designated as γ_2) in alpha phase. These again are the equilibrium phases, which is in fact more typical of the tempering of martensites than the complex effects that occur during the tempering of iron-carbon martensites. This martensitic phase also does not contribute useful strengthening, either before or after tempering, compared with an equivalent equilibrium phase. However, the presence of the martensitic phase instead of the eutectoid constituent is desirable, because the eutectoid is embrittling whereas the martensitic beta prime is not. Precautions are taken during the production of these alloys to ensure that they are cooled fast enough to produce the martensitic phase.

SPHEROIDIZATION AND GROWTH OF PHASE PARTICLES

We have seen that the individual particles of a phase may form initially in a variety of shapes, but often they are extended like needles or plates. Their shapes are determined primarily by the criterion that the interface between the phase and its matrix should have minimum energy, but it still has energy. Sometimes this energy is very small, but at other times it is comparatively large. The laws of nature, as codified in the laws of thermodynamics, tell us that there will always be a drive to reduce this energy, which can be achieved by reducing the interfacial area. The shape which has the minimum surface area per unit volume is, of course, a sphere. So we can expect the particles of a phase to be driven to adopt a spherical shape regardless of their original shapes. Moreover, we can expect these spheres to try to grow to as large a diameter as possible, smaller particles tending to disappear in the process, because this too reduces the surface area per unit volume.

The rate at which the desired end result is achieved is, however, another thing — as we have seen on a number of occasions for other reactions. A change in either the shape or the size of a particle requires that the atoms at one point on its surface enter into solution in the matrix, then diffuse through the matrix to another point on either the original particle or another particle, and deposit there. The diffusion step is the rate-determining step in this sequence. This means that the rate at which *spheroidization* or *growth* can occur, other things being equal, is determined by the rate at which the slowest of the elements concerned diffuses through the matrix. This rate is likely to be significant only at comparatively high temperatures (p. 235) and will increase exponentially with temperature. So we have to keep in mind here two factors: the interfacial energy between the two phases, and the rate of diffusion.

Now let us look at the spheroidization of lamellar pearlite, considering it both as a model system and as one in which *spheroidizing heat treatments* are carried out deliberately in practice. In this case, the maximum temperature at which spheroidization can be attempted is the A_1 temperature, above which a phase transformation intervenes (p. 391). There is a limiting temperature of this or some similar nature in all systems even if it is the solidus temperature. When heated at temperatures approaching the A_1 temperature, the cementite plates in pearlite break up, colony by colony, into a series of fragments each of which gradually assumes a spherical shape (Fig. 11.36a to c). The interleaving ferrite joins up between these particles to form a continuous matrix of equiaxed grains (Fig. 11.37). The cementite particles often are found at ferritic grain boundaries (Fig. 11.37), producing a structure which is essentially the same as that finally developed by quench hardening and full tempering (cf. Fig. 11.30b), and for the same reasons. The cementite spheres grow with time, allowing the ferrite grains also to grow.

The spheroidization process in lamellar pearlite is slow, some hundreds of hours being required even at the highest practicable temperature to achieve the complete spheroidization illustrated in Fig. 11.37. This is because the crystal structures of cementite and ferrite are well matched at the interface and so the interface in pearlite has low surface energy. The spheroidization rate decreases with increasing pearlite spacing, because the diffusion distances are increased, and when alloying elements are present in the carbide, because they diffuse much more slowly than carbon.

However, the spheroidization rate is greatly increased if the pearlite is plastically deformed before or during heating, the increase being as great as several orders of magnitude. The primary reason is that the orientations of the ferrite and cementite are changed by the plastic deformation, the crystallographic match at the interface is destroyed, and so the interface becomes one of higher surface energy. The deformation structures introduced into the two phases also make it easier for the cementite plates to break up, and they do so into much smaller fragments than before. The particles thereafter grow, of course, in the same way as before.

Phases for which there is no special initial crystallographic match at the interface with the matrix spheroidize comparatively rapidly. Grain-boundary allotriomorphs of cementite in annealed high-carbon steels are an example (Fig. 11.36d and e), and they spheroidize much more rapidly than cementite

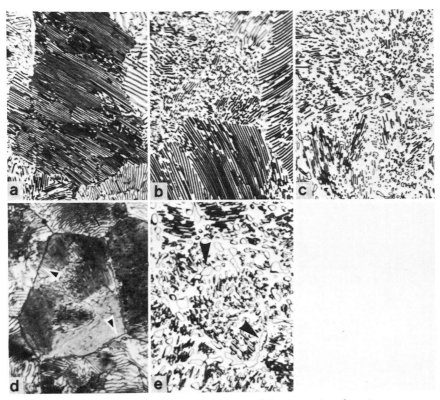

(a) to (c) Gradual spheroidization of the cementite plates in pearlite during heating at 650 °C (1200 °F). Heated for (a) 0 h; (b) 16 h; and (c) 240 h. Spheroidization is not complete even after several hundred hours at temperature.

(d) and (e) Spheroidization of proeutectoid cementite during heating at 700 °C (1290 °F). Heated for (d) 0 h and (e) 16 h. Unaffected proeutectoid cementite is indicated by the arrow in (d), and the spheroidized product of this cementite is indicated by the arrow in (e). The proeutectoid cementite has spheroidized much more rapidly than the cementite plates in pearlite.

Optical micrographs; 1000×.

Fig. 11.36. Illustrations of the spheroidization of lamellar pearlite and proeutectoid cementite in steels.

plates in pearlite (Fig. 11.36d and e; cf. Fig. 11.36b). On the other hand, there are phases for which the interfacial energy is so low that spheroidization or growth does not occur at all, at least for all practical purposes.

Spheroidized structures are low-strength structures (p. 334) and hence are not normally sought in final products. They are, however, ductile structures and so may be produced deliberately to facilitate intermediate fabrication stages, such as pressing and some types of machining. They may also be produced to facilitate final heat treatment and to overcome the deleterious effects of brittle phases when they are present as large particles (p. 333). For example,

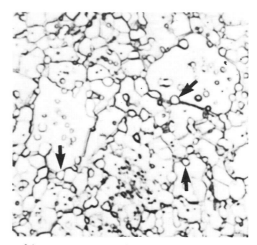

Many of the cementite particles (arrows) are located at the ferrite grain boundaries (network of dark lines) because they have held the migrating boundaries in place. This structure is very similar to that illustrated in Fig. 11.30(b) but has been obtained via an entirely different route.

Fig. 11.37. An example of the structure that exists when lamellar pearlite has been fully spheroidized.

the proeutectoid cementite in high-carbon steels is spheroidized in the manner illustrated in Fig. 11.36(e) before the material is quench hardened, so that a final structure of the type illustrated in Fig. 10.7(b) is obtained instead of that illustrated in Fig. 10.7(a).

We have already noted that particle spheroidization and growth occur during the final stages of heat treatments such as precipitation hardening and tempering. These heat treatments usually are designed to avoid the phenomena but their subsequent occurrence may set a limit on the temperatures at which, and times for which, alloys which have been hardened in this way can be used in service. The alloys in effect overage or overtemper, being likely to be exposed for much longer times in service than during heat treatment. These effects are even more likely to occur under creep conditions because of the accelerating effect of the simultaneous deformation. The useful operating temperature (about 180 °C, or 360 °F) of precipitation-hardened aluminum alloys (p. 374) is determined by factors of this nature, for example. One of the challenges facing metallurgists interested in aluminum alloys is to develop alloys in which the strengthening precipitates are just as effective as those used in current alloys but which are stable at higher temperatures. There are many situations where even small increases in the permissible operating temperatures of aluminum alloys would be of considerable engineering and economic advantage. At the other extreme, the dispersed phase which is primarily responsible for the strengthening of nickel-base superalloys has a very low interfacial energy and is extraordinarily stable at high temperatures. This is the basic reason for the outstanding creep resistance of these alloys (p. 119). Metallurgists seek ways of strengthening alloys by development of phases which are as stable as this. One approach is to incorporate fine particles of a refractory oxide in the alloy. Oxides which would be stable can be chosen easily enough, but the trick is to incorporate enough of them in a fine enough dispersion to achieve the desired strengthening. Such alloys also have to be fabricated by unusual methods, such as by the powder metallurgy route, and these methods present problems involving both cost and reliability.

FURTHER READING

General

A. D. Pelton, "Phase Diagrams", Chapter 7, p. 327, in *Physical Metallurgy*, 3rd Ed., edited by R. W. Cahn and P. Haasen, North Holland, Amsterdam, 1983.

Binary Alloy Phase Diagrams, edited by T. B. Massalski, American Society for Metals, Metals Park, OH, 1976.

ASM Metals Reference Book, 2nd Ed., American Society for Metals, Metals Park, OH, 1982.

Metals Handbook Desk Edition, edited by H. E. Boyer and T. L. Gall, American Society for Metals, Metals Park, OH, 1985.

Metals Handbook, 9th Ed., Vol 4, *Heat Treating*, American Society for Metals, Metals Park, OH, 1981.

H. E. Boyer, *Practical Heat Treating*, American Society for Metals, Metals Park, OH, 1984.

Nonferrous Metals

C. R. Brooks, *Heat Treatment, Structures and Properties of Nonferrous Alloys*, American Society for Metals, Metals Park, OH, 1981.

I. J. Polmear, *Light Alloys. Metallurgy of Light Metals*, 2nd Ed., Edward Arnold, London, 1988.

A. Donachie, *Titanium. A Technical Guide*, ASM International, Metals Park, OH, 1988.

E. Bradley, *Superalloys. A Technical Guide*, ASM International, Metals Park, OH, 1988.

Ferrous Metals

G. Krauss, *Principles of Heat Treatment of Steel*, American Society for Metals, Metals Park, OH, 1980.

R. W. K. Honeycombe, *Steels, Microstructures and Properties*, Edward Arnold, London, 1981.

Atlas of Isothermal Transformation and Continuous Cooling Transformation Diagrams, American Society for Metals, Metals Park, OH, 1977.

L. E. Samuels, *Optical Microscopy of Carbon Steels*, American Society for Metals, Metals Park, OH, 1980.

Heat Treater's Guide. Standard Practices and Procedures for Steels, edited by P. M. Unterweiser, H. E. Boyer, and J. J. Kubbs, American Society for Metals, Metals Park, OH, 1982.

Appendix 1
Glossary of Metallurgical and Metalworking Terms

This glossary covers a number of specialized technical terms pertinent to the text of this book. They have been selected from a basic list compiled and collected by the ASM Committee on Definitions of Metallurgical Terms and printed in full in other ASM publications such as *Metals Handbook Desk Edition* and *ASM Metals Reference Book*.

Many terms can have more than one meaning in metallurgical literature; alternative meanings are identified here by parenthetical numbers preceding each alternative. Wherever possible, a general or generic meaning is given before a specialized meaning, but there is no significance in the order in which alternative meanings are given. Terms whose meanings are the same in both technical and nontechnical contexts are largely excluded.

A_{cm}, A_1, A_3. Defined under *transformation temperature*.

abrasion. A roughening or scratching of a surface due to *abrasive wear*. On aluminum parts, also known as a rub mark or traffic mark.

abrasive. (1) A hard substance used for grinding, honing, lapping, superfinishing, *polishing*, pressure blasting or barrel finishing. It includes natural materials such as garnet, emery, corundum and diamond, and electric-furnace products like aluminum oxide, silicon carbide and boron carbide. (2) Hard particles, such as rocks, sand or fragments of certain hard metals, that wear away a surface when they move across it under pressure.

abrasive wear. The removal of material from a surface when hard particles slide or roll across the surface under pressure. The particles may be loose or may be part of another surface in contact with the surface being worn. Contrast with *adhesive wear*.

Ac_{cm}, Ac_1, Ac_3. Defined under *transformation temperature*.

accuracy. The closeness of approach of a measurement to the true value of the quantity measured. Since the true value cannot actually be measured, the most probable value from the available data, critically considered for sources of error, is used as "the truth". Contrast with *precision*.

acicular ferrite. A highly substructured nonequiaxed ferrite that forms upon continuous cooling by a mixed diffusion and shear mode of transformation that begins at a temperature slightly higher than the temperature transformation range for upper bainite. It is distinguished from bainite in that it has a limited amount of carbon available; thus, there is only a small amount of carbide present.

acicular ferrite steels. Those steels having a microstructure consisting of either acicular ferrite or a mixture of acicular and equiaxed ferrite.

activation. The changing of a passive surface of a metal to a chemically active state. Contrast with *passivation*.

activation energy. The energy required for initiating a metallurgical reaction; for example, plastic flow, diffusion, chemical reaction. The activation energy may be calculated from the slope of the line obtained by plotting the natural log of the reaction rate versus the reciprocal of the absolute temperature.

adhesion. Force of attraction between the molecules (or atoms) of two different phases. Contrast with *cohesion*.

adhesive wear. The removal of material from a surface by the welding together and subsequent shearing of minute areas of two surfaces that slide across each other under pressure. In advanced stages, may lead to *galling* or seizing. Contrast with *abrasive wear*.

Ae$_{cm}$, Ae$_1$, Ae$_3$. Defined under *transformation temperature*.

age hardening. Hardening by aging, usually after rapid cooling or cold working. See *aging*.

age softening. Spontaneous decrease of strength and hardness that takes place at room temperature in certain strain hardened alloys, especially those of aluminum.

aging. A change in the properties of certain metals and alloys that occurs at ambient or moderately elevated temperatures after hot working or a heat treatment (quench aging in ferrous alloys, natural or artificial aging in ferrous and nonferrous alloys) or after a cold working operation (strain aging). The change in properties is often, but not always, due to a phase change (precipitation), but never involves a change in chemical composition of the metal or alloy. See also *age hardening, artificial aging, interrupted aging, natural aging, overaging, precipitation hardening, precipitation heat treatment, progressive aging, quench aging, step aging, strain aging.*

air-hardening steel. A steel containing sufficient carbon and other alloying elements to harden fully during cooling in air or other gaseous mediums from a temperature above its transformation range. The term should be restricted to steels that are capable of being hardened by cooling in air in fairly large sections, about 2 in. or more in diameter. Same as self-hardening steel.

alclad. Composite wrought product comprised of an aluminum alloy core having on one or both surfaces a metallurgically bonded aluminum or aluminum alloy coating that is anodic to the core and thus electrically protects the core against corrosion.

alkali metal. A metal in group IA of the periodic system — namely, lithium, sodium, potassium, rubidium, cesium and francium.

They form strongly alkaline hydroxides; hence, the name.

alkaline earth metal. A metal in group IIA of the periodic system — namely, beryllium, magnesium, calcium, strontium, barium and radium — so called because the oxides or "earths" of calcium, strontium and barium were found by the early chemists to be alkaline in reaction.

allotriomorphic crystal. A crystal whose lattice structure is normal but whose external surfaces are not bounded by regular crystal faces; rather, the external surfaces are impressed by contact with other crystals or another surface such as a mold wall, or are irregularly shaped because of nonuniform growth. Compare with *idiomorphic crystal*.

allotropy. A near synonym for *polymorphism*. Allotropy is generally restricted to describing polymorphic behavior in elements, terminal phases, and alloys whose behavior closely parallels that of the predominant constituent element.

allowance. The specified difference in limiting sizes (minimum clearance or maximum interference) between mating parts, as computed arithmetically from the specified dimensions and tolerances of each part.

alloy. A substance having metallic properties and being composed of two or more chemical elements of which at least one is a *metal*.

alloying element. An element added to a metal to effect changes in properties and which remains within the metal.

alloy steel. Steel containing specified quantities of alloying elements (other than carbon and the commonly accepted amounts of manganese, copper, silicon, sulfur and phosphorus) within the limits recognized for constructional alloy steels, added to effect changes in mechanical or physical properties.

alpha ferrite. See *ferrite.*

alpha iron. The body-centered cubic form of pure iron, stable below 910 °C (1670 °F).

amorphous. Not having a crystal structure; noncrystalline.

angstrom (unit). A unit of linear measurement equal to 10^{-10} m, or 0.1 nm, sometimes used to express small distances such as interatomic distances and some wavelengths.

anion. A negatively charged ion; it flows to the anode in electrolysis.

anisotropy. The characteristic of exhibiting different values of a property in different directions with respect to a fixed reference system in the material.

annealing. A generic term denoting a treatment, consisting of heating to and holding at a suitable temperature followed by cooling at a suitable rate, used primarily to soften metallic materials, but also to simultaneously produce desired changes in other properties or in microstructure. The purpose of such changes may be, but is not confined to: improvement of machinability, facilitation of cold work, improvement of mechanical or electrical properties, and/or increase in stability of dimensions. When the term is used without qualification, full annealing is implied. When applied only for the relief of stress, the process is properly called stress relieving or stress-relief annealing.

In ferrous alloys, annealing usually is done above the upper critical temperature, but the time-temperature cycles vary widely in both maximum temperature attained and in cooling rate employed, depending on composition, material condition, and results desired. When applicable, the following commercial process names should be used: black annealing, blue annealing, box annealing, bright annealing, cycle annealing, flame annealing, full annealing, graphitizing, in-process annealing, isothermal annealing, malleablizing, orientation annealing, process annealing, quench annealing, spheroidizing, subcritical annealing.

In nonferrous alloys, annealing cycles are designed to: (*a*) remove part or all of the effects of cold working (recrystallization may or may not be involved); (*b*) cause substantially complete coalescence of precipitates from solid solution in relatively coarse form; or (*c*) both, depending on composition and material condition. Specific process names in commercial use are *final annealing, full annealing, intermediate annealing, partial annealing, recrystallization annealing*, stress-relief annealing, *anneal to temper*.

annealing carbon. Fine, apparently amorphous carbon particles formed in white cast iron and certain steels during prolonged annealing. Also called *temper carbon.*

annealing twin. A *twin* formed in a crystal during recrystallization.

anneal to temper. A final partial anneal that softens a cold worked nonferrous alloy to a specified level of hardness or tensile strength.

anode. The electrode where electrons leave an operating system such as a battery, an electrolytic cell, an x-ray tube or a vacuum tube. In the first of these, it is negative; in the other three, positive. In a battery or electrolytic cell, it is the electrode where oxidation occurs. Contrast with *cathode.*

anodic protection. Imposing an external electrical potential to protect a metal from corrosive attack. (Applicable only to metals that show active-passive behavior.) Contrast with *cathodic protection.*

anodizing. Forming a conversion coating on a metal surface by anodic oxidation; most frequently applied to aluminum.

anvil. (1) In drop forging, the base of the hammer into which the sow block and lower die part are set. (2) A block of steel upon which metal is forged.

apparent density. (1) The weight per unit volume of a metal powder, in contrast to the weight per unit volume of the individual particles. (2) The weight per unit volume of a porous solid, where the unit volume is determined from external dimensions of the mass. Apparent density is always less than the true density of the material itself.

Ar_{cm}, Ar_1, Ar_3. Defined under *transformation temperature.*

artifact. A feature of artificial character (such as a scratch or a piece of dust on a metallographic specimen) that can be erroneously interpreted as a real feature. In inspection, an artifact often produces a *false indication.*

artificial aging. Aging above room temperature. See *aging.* Compare with *natural aging.*

athermal transformation. A reaction that proceeds without benefit of thermal fluctuations; that is, thermal activation is not required. Such reactions are diffusionless and can take place with great speed when the driving force is sufficiently high. For example, many martensitic transformations occur athermally on cooling, even at relatively low temperatures, because of the progressively increasing driving force. In contrast, a reaction that occurs at constant temperature is an *isothermal transformation;* thermal activation is necessary in this case and the reaction proceeds as a function of time.

atomic number. The number of protons in an atomic nucleus; determines the individuality of the atom as a chemical element.

atomic percent. The number of atoms of an

element in a total of 100 representative atoms of a substance.

attritious wear. Wear of abrasive grains in grinding such that the sharp edges gradually become rounded. A grinding wheel that has undergone such wear usually has a glazed appearance.

austempering. A heat treatment for ferrous alloys in which a part is quenched from the austenitizing temperature at a rate fast enough to avoid formation of ferrite or pearlite and then held at a temperature just above M_s until transformation to bainite is complete.

austenite. A solid solution of one or more elements in face-centered cubic iron. Unless otherwise designated (such as nickel austenite), the solute is generally assumed to be carbon.

austenitic grain size. The size attained by the grains of steel when heated to the austenitic region; may be revealed by appropriate etching of cross sections after cooling to room temperature.

austenitic steel. An alloy steel whose structure is normally austenitic at room temperature.

austenitizing. Forming austenite by heating a ferrous alloy into the transformation range (partial austenitizing) or above the transformation range (complete austenitizing). When used without qualification, the term implies complete austenitizing.

autoradiography. An inspection technique in which radiation spontaneously emitted by a material is recorded photographically. The radiation is emitted by radioisotopes that are (a) produced in a metal by bombarding it with neutrons, (b) added to a metal such as by alloying, or (c) contained within a cavity in a metal part. The technique serves to locate the position of the radioactive element or compound.

B

back draft. A reverse taper on a casting pattern or a forging die that prevents the pattern or forged stock from being removed from the cavity.

bainite. A metastable aggregate of ferrite and cementite resulting from the transformation of austenite at temperatures below the pearlite range but above M_s. Its appearance is feathery if formed in the upper part of the bainite transformation range; acicular, resembling tempered martensite, if formed in the lower part.

banded structure. A segregated structure consisting of alternating nearly parallel bands of different composition, typically aligned in the direction of primary hot working.

base metal. (1) The metal present in the largest proportion in an alloy; brass, for example, is a copper-base alloy. (2) The metal to be brazed, cut, soldered or welded. (3) After welding, that part of the metal which was not melted. (4) A metal that readily oxidizes, or that dissolves to form ions. Contrast with *noble metal* (2).

beach marks. Progression marks on a fatigue fracture surface that indicate successive positions of the advancing crack front. The classic appearance is of irregular elliptical or semielliptical rings, radiating outward from one or more origins. Beach marks (also known as clamshell marks or tide marks) are typically found on service fractures where the part is loaded randomly, intermittently, or with periodic variations in mean stress or alternating stress.

beta structure. A Hume-Rothery designation for structurally analogous body-centered cubic phases (similar to beta brass) or electron compounds that have ratios of three valence electrons to two atoms. Not to be confused with a beta phase on a constitution diagram.

billet. (1) A solid semifinished round or square product that has been hot worked by forging, rolling or extrusion; usually smaller than a *bloom*. (2) A general term for wrought starting stock used in making forgings or extrusions.

billet mill. A primary rolling mill used to make steel billets.

binary alloy. An alloy containing only two component elements.

biscuit. (1) An upset blank for drop forging. (2) A small cake of primary metal (such as uranium made from uranium tetrafluoride and magnesium by bomb reduction).

black annealing. Box annealing or pot annealing ferrous alloy sheet, strip or wire. See *box annealing*.

blank. (1) In forming, a piece of sheet material, produced in cutting dies, that is usually subjected to further press operations. (2) A pressed, presintered or fully sintered powder metallurgy compact, usually in the unfinished condition and requiring cutting, machining or some other operation to produce the final shape. (3) A piece of stock from which a forging is made; often called a slug or multiple.

blank holder. The part of a drawing or forming die that holds the workpiece against the draw ring to control metal flow.

blanking. Producing desired shapes from metal to be used for forming or other operations, usually by punching.

blemish. A nonspecific quality control term designating an imperfection that mars the appearance of a part but does not detract from its ability to perform its intended function.

blind riser. A *riser* that does not extend through the top of the mold.

blister. A raised area, often dome shaped, resulting from (*a*) loss of adhesion between a coating or deposit and the basis metal or (*b*) delamination under the pressure of expanding gas trapped in a metal in a near subsurface zone. Very small blisters may be called pinheads or pepper blisters.

blocker. The impression in the dies (often one of a series of impressions in a single die set) that imparts to the forging an intermediate shape, preparatory to forging of the final shape. Also called blocking impression.

blocker-type forging. A forging that approximates the general shape of the final part with relatively generous finish allowance and radii. Such forgings are sometimes specified to reduce die costs where only a small number of forgings is desired and the cost of machining each part to its final shape is not excessive.

bloom. (1) A semifinished hot rolled product, rectangular in cross section, produced on a blooming mill. See also *billet*. For steel, the width of a bloom is not more than twice the thickness, and the cross-sectional area is usually not less than about 230 cm^2 (36 in.^{2}). Steel blooms are sometimes made by forging. (2) A visible exudation or efflorescence on the surface of an electroplating bath. (3) A bluish fluorescent cast to a painted surface caused by deposition of a thin film of smoke, dust or oil. (4) A loose, flowerlike corrosion product that forms when certain metals are exposed to a moist environment.

bloomer. The mill or other equipment used in reducing steel ingots to blooms.

blooming mill. A primary rolling mill used to make blooms.

blowhole. A hole in a casting or a weld caused by gas entrapped during solidification.

blue annealing. Heating hot rolled ferrous sheet in an open furnace to a temperature within the transformation range and then cooling in air, in order to soften the metal. The formation of a bluish oxide on the surface is incidental.

bond. (1) In grinding wheels and other relatively rigid abrasive products, the material that holds the abrasive grains together. (2) In welding, brazing or soldering, the junction of joined parts. Where filler metal is used, it is the junction of the fused metal and the heat-affected base metal. (3) In an adhesive bonded or diffusion bonded joint, the line along which the faying surfaces are joined together.

box annealing. Annealing a metal or alloy in a sealed container under conditions that minimize oxidation. In box annealing a ferrous alloy, the charge is usually heated slowly to a temperature below the transformation range, but sometimes above or within it, and is then cooled slowly; this process is also called close annealing or pot annealing. See *black annealing*.

brake. A device for bending sheet metal to a desired angle.

brass. An alloy consisting mainly of copper (over 50%) and zinc, to which smaller amounts of other elements may be added.

braze welding. A method of welding by using a filler metal having a liquidus above 450 °C (840 °F) and below the solidus of the base metals. Unlike *brazing*, in braze welding, the filler metal is not distributed in the joint by capillary attraction.

brazing. A group of welding processes that join solid materials together by heating them to a suitable temperature and by using a filler metal having a liquidus above 450 °C (840 °F) and below the solidus of the base materials. The filler metal is distributed between the closely fitted surfaces of the joint by capillary attraction.

bright annealing. Annealing in a protective medium to prevent discoloration of the bright surface.

Brinell hardness test. A test for determining the hardness of a material by forcing a hard steel or carbide ball of specified diameter into it under a specified load. The result is expressed as the Brinell hardness number, which is the value obtained by dividing the applied load in kilograms by the surface area of the resulting impression in square millimetres.

brittle crack propagation. A very sudden propagation of a crack with the absorption of no energy except that stored elastically in the body. Microscopic examination may reveal some deformation even though it is not noticeable to the unaided eye.

brittle fracture. Separation of a solid accompanied by little or no macroscopic plastic deformation. Typically, brittle fracture occurs by rapid crack propagation with less expenditure of energy than for ductile fracture.

brittleness. The quality of a material that leads to crack propagation without appreciable plastic deformation.

bronze. A copper-rich copper-tin alloy with or without small proportions of other elements such as zinc and phosphorus. By extension, certain copper-base alloys containing considerably less tin than other alloying elements, such as manganese bronze (copper-zinc plus manganese, tin and iron) and leaded tin bronze (copper-lead plus tin and sometimes zinc). Also, certain other essentially binary copper-base alloys containing no tin, such as aluminum bronze (copper-aluminum), silicon bronze (copper-silicon) and beryllium bronze (copper-beryllium). Also, trade designations for certain specific copper-base alloys that are actually brasses, such as architectural bronze (57 Cu, 40 Zn, 3 Pb) and commercial bronze (90 Cu, 10 Zn).

buckle. (A) A local waviness in metal bar or sheet, usually transverse to the direction of rolling. (2) An indentation in a casting resulting from expansion of molding sand into the mold cavity.

burned-on sand. A mixture of sand and cast metal adhering to the surface of a casting.

burning. (1) Permanently damaging a metal or alloy by heating to cause either incipient melting or intergranular oxidation. See *overheating.* (2) In grinding, getting the work hot enough to cause discoloration or to change the microstructure by tempering or hardening.

C

camber. (1) Deviation from edge straightness, usually referring to the greatest deviation of side edge from a straight line. (2) Sometimes used to denote crown in rolls where the cen-

ter diameter has been increased to compensate for deflection caused by the rolling pressure.

capped steel. A type of steel similar to rimmed steel, usually cast in a bottle-top ingot mold, in which the application of a mechanical or a chemical cap renders the rimming action incomplete by causing the top metal to solidify. The surface condition of capped steel is much like that of rimmed steel, but certain other characteristics are intermediate between those of *rimmed steel* and those of *semikilled steel.*

carbide. A compound of carbon with one or more metallic elements.

carbon steel. Steel having no specified minimum quantity for any alloying element (other than the commonly accepted amounts of manganese, silicon and copper) and that contains only an incidental amount of any element other than carbon, silicon, manganese, copper, sulfur and phosphorus.

carburizing. Absorption and diffusion of carbon into solid ferrous alloys by heating, to a temperature usually above Ac_3, in contact with a suitable carbonaceous material. A form of *case hardening* that produces a carbon gradient extending inward from the surface, enabling the surface layer to be hardened either by quenching directly from the carburizing temperature or by cooling to room temperature, then reaustenitizing and quenching.

case hardening. A generic term covering several processes applicable to steel that change the chemical composition of the surface layer by absorption of carbon, nitrogen, or a mixture of the two and, by diffusion, create a concentration gradient. The processes commonly used are carburizing and quench hardening; cyaniding; nitriding; and carbonitriding. The use of the applicable specific process name is preferred.

casting. (1) An object at or near finished shape obtained by solidification of a substance in a mold. (2) Pouring molten metal into a mold to produce an object of desired shape.

casting shrinkage. (1) Liquid shrinkage — the reduction in volume of liquid metal as it cools to the liquidus. (2) Solidification shrinkage — the reduction in volume of metal from the beginning to ending of solidification. (3) Solid shrinkage — the reduction in volume of metal from the solidus to room temperature.

casting strains. Strains in a casting caused by *casting stresses* that develop as the casting cools.

casting stresses. Residual stresses set up when the shape of a casting impedes contraction of the solidified casting during cooling.

cast iron. A generic term for a large family of cast ferrous alloys in which the carbon content exceeds the solubility of carbon in austenite at the eutectic temperature. Most cast irons contain at least 2% carbon, plus silicon and sulfur, and may or may not contain other alloying elements. For the various forms *gray cast iron*, white cast iron, malleable cast iron and ductile cast iron, the word "cast" is often left out, resulting in "gray iron", "white iron", "malleable iron" and "ductile iron", respectively.

cast steel. Steel in the form of *castings.*

cast structure. The metallographic structure of a *casting* evidenced by shape and orientation of grains and by segregation of impurities.

catastrophic failure. Sudden failure of a component or assembly that frequently results in extensive secondary damage to adjacent components or assemblies.

cathode. The electrode where electrons enter an operating system such as a battery, an electrolytic cell, an x-ray tube or a vacuum tube. In the first of these, it is positive; in the other three, negative. In a battery or electrolytic cell, it is the electrode where reduction occurs. Contrast with *anode.*

cathode film. The portion of solution in immediate contact with the cathode during electrolysis.

cathodic protection. Partial or complete protection of a metal from corrosion by making it a cathode, using either a galvanic or impressed current. Contrast with *anodic protection.*

cation. A positively charged ion; it flows to the cathode in electrolysis.

caustic cracking. A form of *stress-corrosion cracking* most frequently encountered in carbon steels or iron-chromium-nickel alloys that are exposed to concentrated hydroxide solutions at temperatures of 200 to 250 °C (400 to 480 °F).

cavitation. The formation and instantaneous collapse of innumerable tiny voids or cavities within a liquid subjected to rapid and intense pressure changes. Cavitation produced by ultrasonic radiation is sometimes used to give violent localized agitation. That caused by severe turbulent flow often leads to *cavitation damage.*

cavitation damage. Erosion of a solid surface through the formation and collapse of cavities in an adjacent liquid.

cavitation erosion. See preferred term, *cavitation damage.*

cementation. The introduction of one or more elements into the outer portion of a metal object by means of diffusion at high temperature.

cemented carbide. A solid and coherent mass made by pressing and sintering a mixture of powders of one or more metallic carbides and a much smaller amount of a metal, such as cobalt, to serve as a binder.

cementite. A compound of iron and carbon, known chemically as iron carbide and having the approximate chemical formula Fe_3C. It is characterized by an orthorhombic crystal structure. When it occurs as a phase in steel, the chemical composition will be altered by the presence of manganese and other carbide-forming elements.

chafing fatigue. Fatigue initiated in a surface damaged by rubbing against another body. See *fretting.*

chaplet. Metal support that holds a core in place within a mold; molten metal solidifies around a chaplet and fuses it into the finished casting.

Charpy test. A pendulum-type single-blow impact test in which the specimen, usually notched, is supported at both ends as a simple beam and broken by a falling pendulum. The energy absorbed, as determined by the subsequent rise of the pendulum, is a measure of impact strength or notch toughness. Contrast with *Izod test.*

chevron pattern. A fractographic pattern of radial marks (shear ledges) that looks like nested letters "V"; sometimes called a herringbone pattern. Chevron patterns are typically found on brittle fracture surfaces in parts whose widths are considerably greater than their thicknesses. The points of the chevrons can be traced back to the fracture origin.

chill. (1) A metal or graphite insert embedded in the surface of a sand mold or core or placed in a mold cavity to increase the cooling rate at that point. (2) White iron occurring on a gray or ductile iron casting, such as the chill in the wedge test.

Chinese script. The angular microstructural form suggestive of Chinese writing and characteristic of the constituents α(Al-Fe-Si) and α(Al-Fe-Mn-Si) in cast aluminum alloys. A similar microstructure is found in cast magnesium alloys containing silicon as Mg_2Si.

chipping. (1) Removing seams and other surface imperfections in metals manually with a chisel or gouge, or by a continuous machine, before further processing. (2) Similarly, removing excessive metal.

chips. Pieces of material removed from a workpiece by cutting tools or an abrasive medium.

clad metal. A composite metal containing two or three layers that have been bonded together. The bonding may have been accomplished by co-rolling, welding, casting, heavy chemical deposition or heavy electroplating.

cleavage. The splitting (fracture) of a crystal on a crystallographic plane of low index.

cleavage fracture. A fracture, usually of a polycrystalline metal, in which most of the grains have failed by cleavage, resulting in bright reflecting facets. It is one type of *crystalline fracture* and is associated with low-energy brittle fracture. Contrast with *shear fracture.*

cleavage plane. A characteristic crystallographic plane or set of planes on which cleavage fracture easily occurs.

close annealing. Same as *box annealing.*

closed die forging. See *impression die forging.*

closed dies. Forging or forming impression dies designed to restrict the flow of metal to the cavity within the die set, as opposed to open dies, in which there is little or no restriction to lateral flow.

close tolerance forging. A forging held to unusually close dimensional tolerances. Often, little or no machining is required after forging.

cluster mill. A rolling mill where each of the two working rolls of small diameter is supported by two or more backup rolls.

coarsening. An increase in the grain size, usually, but not necessarily, by *grain growth.*

coefficient of elasticity. Same as *modulus of elasticity.*

cogging mill. A *blooming mill.*

coherency. The continuity of lattice of precipitate and parent phase (solvent) maintained by mutual strain and not separated by a phase boundary.

coherent precipitate. A crystalline precipitate that forms from solid solution with an orientation that maintains continuity between the crystal lattice of the precipitate and the lattice of the matrix, usually accompanied by some strain in both lattices. Because the lattices fit at the interface between precipitate and matrix, there is no discernible phase boundary.

cohesion. Force of attraction between the molecules (or atoms) within a single phase. Contrast with *adhesion.*

cohesive strength. (1) The hypothetical stress causing tensile fracture without plastic deformation. (2) The stress corresponding to the forces between atoms. (3) Same as *disruptive strength.*

coining. (1) A closed-die squeezing operation, usually performed cold, in which all surfaces of the work are confined or restrained, resulting in a well-defined imprint of the die upon the work. (2) A *restriking* operation used to sharpen or change an existing radius or profile. (3) The final pressing of a sintered powder metallurgy compact to obtain a definite surface configuration (not to be confused with *re-pressing* or sizing).

cold chamber machine. A *die-casting* machine where the metal chamber and plunger are not heated.

cold extrusion. See *impact; extrusion.*

cold heading. Working metal at room temperature in such a manner that the cross-sectional area of a portion or all of the stock is increased.

cold lap. Wrinkled markings on the surface of an ingot, caused by incipient freezing of the surface while the liquid is still in motion; results from too low a pouring temperature. See also *cold shut* (1).

cold mill. A mill for cold rolling sheet or strip.

cold pressing. Forming a powder metallurgy *compact* at a temperature low enough to avoid *sintering,* usually room temperature. Contrast with *hot pressing.*

cold rolled sheets. A mill product produced from a hot rolled pickled coil that has been given substantial cold reduction at room temperature. The resulting product usually requires further processing to make it suitable for most common applications. The usual end product is characterized by improved surface, greater uniformity in thickness and improved mechanical properties compared to hot rolled sheet.

cold shortness. Brittleness that exists in some metals at temperatures below the recrystallization temperature.

cold shut. (1) A discontinuity that appears on the surface of cast metal as a result of two streams of liquid meeting and failing to unite. (2) A lap on the surface of a forging or billet that was closed without fusion during deformation. (3) Freezing of the top surface of an ingot before the mold is full.

cold treatment. Exposing to suitable subzero temperatures for the purpose of obtaining desired conditions or properties such as dimensional or microstructural stability. When the treatment involves the transformation of retained austenite, it is usually followed by tempering.

cold welding. A solid state welding process in which pressure is used at room temperature to produce coalescence of metals with substantial deformation at the weld. Compare *diffusion welding* and *forge welding.*

cold work. Permanent strain in a metal accompanied by strain hardening.

cold working. Deforming metal plastically under conditions of temperature and strain rate that induce strain hardening. Usually, but not necessarily, conducted at room temperature. Contrast with *hot working.*

columnar structure. A coarse structure of parallel elongated grains formed by unidirectional growth, most often observed in castings, but sometimes in structures resulting from diffusional growth accompanied by a solid-state transformation.

combined carbon. The part of the total carbon in steel or cast iron that is present as other than *free carbon.*

compact. An object produced by the compression of metal powder, generally while confined in a die, with or without the inclusion of nonmetallic constituents.

component. (1) One of the elements or compounds used to define a chemical (or alloy) system, including all phases, in terms of the fewest substances possible. (2) One of the individual parts of a vector as referred to a system of coordinates.

composite material. A heterogeneous, solid structural material consisting of two or more distinct components that are mechanically or metallurgically bonded together (such as a *cermet,* or boron wire embedded in a matrix of epoxy resin).

composite structure. A structural member (such as a panel, plate, pipe or other shape) that is built up by bonding together two or more distinct components, each of which may be made of a metal, alloy, nonmetal or *composite material.* Examples of composite structures include: honeycomb panels, clad plate, electrical contacts, sleeve bearings, carbide-tipped drills or lathe tools, and weldments constructed of two or more different alloys.

compound die. Any die so designed that it performs more than one operation on a part with one stroke of the press, such as blanking and piercing, where all functions are performed simultaneously within the confines of the particular blank size being worked.

compressibility. In powder metallurgy, the reciprocal of the compression ratio where a compact is made following a procedure in which the die, the pressure and the pressing speed are specified.

compressive strength. The maximum compressive stress that a material is capable of developing, based on original area of cross section. If a material fails in compression by a shattering fracture, the compressive strength has a very definite value. If a material does not fail in compression by a shattering fracture, the value obtained for compressive strength is an arbitrary value depending upon the degree of distortion that is regarded as indicating complete failure of the material.

conditioning heat treatment. A preliminary heat treatment used to prepare a material for a desired reaction to a subsequent heat treatment. For the term to be meaningful, the exact heat treatment must be specified.

constituent. (1) One of the ingredients that make up a chemical system. (2) A phase or combination of phases that occur in a characteristic configuration in an alloy microstructure.

constitution diagram. A graphical representation of the temperature and composition limits of phase fields in an alloy system as they actually exist under the specific conditions of heating or cooling (synonymous with phase diagram). A constitution diagram may be an equilibrium diagram, an approximation to an equilibrium diagram or a representation of metastable conditions or phases. Compare with *equilibrium diagram.*

constraint. Any restriction that occurs to the transverse contraction normally associated with a longitudinal tension, and that hence causes a secondary tension in the transverse direction; usually used in connection with welding. Contrast with *restraint*.

contact fatigue. Cracking and subsequent pitting of a surface subjected to alternating Hertzian stresses such as those produced under rolling contact or combined rolling and sliding. The phenomenon of contact fatigue is encountered most often in rolling-element bearings or in gears, where the surface stresses are high due to the concentrated loads and are repeated many times during normal operation.

continuous casting. A casting technique in which a cast shape is continuously withdrawn through the bottom of the mold as it solidifies, so that its length is not determined by mold dimensions. Used chiefly to produce semifinished mill products such as billets, blooms, ingots, slabs and tubes. See also *strand casting*.

continuous mill. A rolling mill consisting of a number of stands of synchronized rolls (in tandem) in which metal undergoes successive reductions as it passes through the various stands.

continuous phase. In an alloy or portion of an alloy containing more than one phase, the phase that forms the matrix in which the other phase or phases are present as isolated units.

continuous precipitation. Precipitation from a supersaturated solid solution in which the precipitate particles grow by long-range diffusion without recrystallization of the matrix. Continuous precipitates grow from nuclei distributed more or less uniformly throughout the matrix. They usually are randomly oriented, but may form a *Widmanstätten structure*. Also called general precipitation. Compare with *discontinuous precipitation, localized precipitation*.

controlled cooling. Cooling from an elevated temperature in a predetermined manner, to avoid hardening, cracking, or internal damage, or to produce desired microstructure or mechanical properties.

conventional forging. A forging characterized by design complexity and tolerances that fall within the broad range of general forging practice.

conventional strain. See *strain*.

conventional stress. See *stress*.

cooling curve. A curve showing the relation between time and temperature during the cooling of a material.

cooling stresses. Residual stresses resulting from nonuniform distribution of temperature during cooling.

cope. The upper or topmost section of a flask, mold or pattern.

copper brazing. A term improperly used to denote joining with a copperbase filler metal. See preferred terms *brazing* and *braze welding*.

core. (1) A specially formed material inserted in a mold to shape the interior or other part of a casting that cannot be shaped as easily by the pattern. (2) In a ferrous alloy prepared for *case hardening*, that portion of the alloy that is not part of the case. Typically considered to be the portion that (*a*) appears light on an etched cross section, (*b*) has an essentially unaltered chemical composition, or (*c*) has a hardness, after hardening, less than a specified value.

coring. (1) A condition of variable composition between the center and surface of a unit of microstructure (such as a dendrite, grain, carbide particle); results from nonequilibrium solidification, which occurs over a range of temperature. (2) A central cavity at the butt end of rod extrusions, sometimes called extrusion pipe.

corrosion. The deterioration of a metal by chemical or electrochemical reaction with its environment.

corrosion embrittlement. The severe loss of ductility of a metal resulting from corrosive attack, usually intergranular and often not visually apparent.

corrosion fatigue. Cracking produced by the combined action of repeated or fluctuating stress and a corrosive environment.

counterblow hammer. A forging hammer in which both the *ram* and *anvil* are driven simultaneously toward each other by air or steam pistons.

covalent bond. A bond between two or more atoms resulting from the completion of shells by the sharing of electrons.

crank press. A mechanical press, the slides of which are actuated by a crankshaft.

creep. Time-dependent strain occurring under stress. The creep strain occurring at a diminishing rate is called primary creep; that occurring at a minimum and almost constant rate, secondary creep; that occurring at an accelerating rate, tertiary creep.

creep limit. (1) The maximum stress that will cause less than a specified quantity of creep in a given time. (2) The maximum nominal stress under which the creep strain rate decreases continuously with time under constant load and at constant temperature. Sometimes used synonymously with *creep strength*.

creep recovery. Time-dependent strain after release of load in a creep test.

creep-rupture test. Same as *stress-rupture test*.

creep strength. (1) The constant nominal stress that will cause a specified quantity of creep in a given time at constant temperature. (2) The constant nominal stress that will cause a specified rate of secondary creep at constant temperature.

crevice corrosion. A type of concentration-cell corrosion; corrosion caused by the concentration or depletion of dissolved salts, metal ions, oxygen or other gases, and such, in crevices or pockets remote from the principal fluid stream, with a resultant building up of differential cells that ultimately cause deep pitting.

critical cooling rate. The rate of continuous cooling required to prevent undesirable transformation. For steel, it is the minimum rate at which austenite must be continuously cooled to suppress transformations above the M$_s$ temperature.

critical current density. In an electrolytic process, a current density at which an abrupt change occurs in an operating variable or in the nature of an electrodeposit or electrode film.

critical point. (1) The temperature or pressure at which a change in crystal structure, phase or physical properties occurs. Same as *transformation temperature*. (2) In an equilibrium diagram, that specific value of composition, temperature and pressure, or combinations thereof, at which the phases of a heterogeneous system are in equilibrium.

critical shear stress. The shearing stress required to cause slip in a designated slip direction on a given slip plane. It is called the critical resolved shear stress if the shearing stress is induced by tension or compression forces acting on the crystal.

critical strain. The strain just sufficient to cause *recrystallization*; because the strain is small, usually only a few percent, recrystallization takes place from only a few nuclei, which produces a recrystallized structure consisting of very large grains.

critical temperature. (1) Synonymous with *critical point* if the pressure is constant. (2) The temperature above which the vapor phase cannot be condensed to liquid by an increase in pressure.

critical temperature ranges. Synonymous with *transformation ranges*, which is the preferred term.

crop. (1) An end portion of an ingot that is cut off as scrap. (2) To shear a bar or billet.

crop end. See *end loss*.

cross-country mill. A rolling mill in which the mill stands are so arranged that their tables are parallel with a transfer (or crossover) table connecting them. They are used for rolling structural shapes, rails and any special form of bar stock not rolled in the ordinary bar mill.

cross forging. Preliminary working of forging stock in flat dies to develop mechanical properties, particularly in the center portions of heavy sections.

cross rolling. The rolling of sheet or plate so that the direction of rolling is about 90° from the direction of a previous rolling.

crucible. A vessel or pot, made of a refractory substance or of a metal with a high melting point, used for melting metals or other substances.

crushing test. (1) A radial compressive test applied to tubing, sintered-metal bearings or other similar products for determining radial crushing strength (maximum load in compression). (2) An axial compressive test for determining quality of tubing, such as soundness of weld in welded tubing.

crystal. A solid composed of atoms, ions or molecules arranged in a pattrn that is repetitive in three dimensions.

crystalline fracture. A pattern of brightly reflecting crystal facets on the fracture surface of a polycrystalline metal and resulting from cleavage fracture of many individual crystals. Contrast with *fibrous fracture, silky fracture*.

crystallization. (1) The separation, usually from a liquid phase on cooling, of a solid crystalline phase. (2) Sometimes erroneously used to explain fracturing that actually has occurred by fatigue.

crystal orientation. See *orientation*.

cubic plane. A plane perpendicular to any one of the three crystallographic axes of the cubic (isometric) system.

cup. (1) Sheet-metal part, the product of the first step in deep drawing. (2) Any cylindrical part or shell closed at one end.

cup fracture (cup-and-cone fracture). A mixed-mode fracture, often seen in tensile test specimens of a ductile material, where the central portion undergoes *plane-strain* fracture and the surrounding region undergoes *plane-stress* fracture. It is called a cup fracture (or cup-and-cone fracture) because one of the mating fracture surfaces looks like a miniature cup—that is, it has a central depressed flat-face region surrounded by a shear lip; the other fracture surface looks like a miniature truncated cone.

cupping. (1) The first step in deep drawing (2) The fracture of severely worked rods or wire where one end has the appearance of a cup and the other that of a cone.

current efficiency. The proportion of current used in a given process to accomplish a desired result; in electroplating, the proportion used in depositing or dissolving metal.

cycle annealing. An annealing process employing a predetermined and closely controlled time-temperature cycle to produce specific properties or microstructures.

D

damping capacity. The ability of a material to absorb vibration (cyclical stresses) by internal friction, converting the mechanical energy into heat.

dead soft. A *temper* of nonferrous alloys and some ferrous alloys corresponding to the condition of minimum hardness and tensile strength produced by *full annealing*.

decarburization. Loss of carbon from the surface layer of a carbon-containing alloy due to reaction with one or more chemical substances in a medium that contacts the surface.

deep drawing. Forming deeply recessed parts by forcing sheet metal to undergo plastic flow between dies, usually without substantial thinning of the sheet.

deep etching. Severe *macroetching*.

defect. A departure of any *quality* characteristic from its intended (usually specified) level that is severe enough to cause the product or service not to fulfill its anticipated function. According to ANSI standards, defects are classified according to severity:

Very serious defects lead directly to severe injury or catastrophic economic loss.

Serious defects lead directly to significant injury or significant economic loss.

Major defects are related to major problems with respect to anticipated use.

Minor defects are related to minor problems with respect to anticipated use.

defective. A quality control term describing a unit of product or service containing at least one *defect*, or having several lesser imperfections that, in combination, cause the unit not to fulfill its anticipated function.
NOTE: The term *defective* is not synonymous with noncomforming (or rejectable) and should be applied only to those units incapable of performing their anticipated functions.

deformation bands. Parts of a crystal that have rotated differently during deformation to produce bands of varied orientation within individual grains.

degasifier. A substance that can be added to molten metal to remove soluble gases that might otherwise be occluded or entrapped in the metal during solidification.

degassing. Removing gases from liquids or solids.

delayed yield. A phenomenon involving a delay in time between the application of a stress and the occurrence of the corresponding yield point strain.

delta ferrite. See *ferrite*.

dendrite. A crystal that has a treelike branching pattern, being most evident in cast metals slowly cooled through the solidification range.

depolarization. A decrease in the *polarization* of an electrode.

descaling. Removing the thick layer of oxides formed on some metals at elevated temperatures.

deseaming. Analogous to *chipping*, the surface imperfections being removed by gas cutting.

dezincification. Corrosion in which zinc is selectively leached from zinc-containing alloys. Most commonly found in copper-zinc alloys containing less than 85% copper after extended service in water containing dissolved oxygen.

diamond pyramid hardness test. See *Vickers hardness test*.

die. A tool, usually containing a cavity, that imparts shape to solid, molten or powdered metal primarily because of the shape of the tool itself. Used in many press operations (including blanking, drawing, forging and forming), in die casting and in forming green powder metallurgy compacts. Die-casting and powder-metallurgy dies are sometimes referred to as *molds*.

die block. A block, usually of tool steel, into which the desired impressions are sunk, formed, or machined and from which forgings or die castings are made.

die body. The stationary or fixed part of a powder pressing die.

die casting. (1) A casting made in a die. (2) A casting process where molten metal is forced under high pressure into the cavity of a metal mold.

die forging. A forging whose shape is determined by impressions in specially prepared dies.

die forming. The shaping of solid or powdered metal by forcing it into or through the cavity in a die.

die lubricant. A lubricant applied to working surfaces of dies and punches to facilitate drawing, pressing, stamping and/or ejection. In powder metallurgy, the die lubricant is sometimes mixed into the powder before pressing into a compact.

die set. A tool or tool holder consisting of a die base and punch plate for the attachment of a die and punch, respectively.

die sinking. Forming or machining a depressed pattern in a die.

differential heating. Heating that intentionally produces a temperature gradient within an object such that, after cooling, a desired stress distribution or variation in properties is present within the object.

diffusion. (1) Spreading of a constituent in a gas, liquid or solid, tending to make the composition of all parts uniform. (2) The spontaneous movement of atoms or molecules to new sites within a material.

diffusion bonding. See preferred term *diffusion welding.*

diffusion coefficient. A factor of proportionality representing the amount of substance diffusing across a unit area through a unit concentration gradient in unit time.

diffusion welding. A high-temperature solid state welding process that permanently joins faying surfaces by the simultaneous application of pressure and heat. The process does not involve macroscopic deformation, melting, or relative motion of parts. A solid filler metal (diffusion aid) may or may not be inserted between the faying surfaces.

dimple rupture. A fractographic term describing ductile fracture that occurred through the formation and coalescence of microvoids along the fracture path. The fracture surface of such a ductile fracture appears dimpled when observed at high magnification and usually is most clearly resolved when viewed in a scanning electron microscope.

dimpling. (1) Stretching a relatively small shallow indentation into sheet metal. (2) In aircraft, stretching thin metal into a conical flange for use with a countersunk head rivet.

direct chill casting. A continuous method of making ingots for rolling or extrusion by pouring the metal into a short mold. The base of the mold is a platform that is gradually lowered while the metal solidifies, the frozen shell of metal acting as a retainer for the liquid metal below the wall of the mold. The ingot is usually cooled by the impingement of water directly on the mold or on the walls of the solid metal as it is lowered. The length of the ingot is limited by the depth to which the platform can be lowered; therefore, it is often called semicontinuous casting.

directional property. Property whose magnitude varies depending on the relation of the test axis to a specific direction within the metal. The variation results from preferred orientation or from fibering of constituents or inclusions.

directional solidification. The solidification of molten metal in such a manner that feed metal is always available for that portion that is just solidifying.

discontinuity. Any interruption in the normal physical structure or configuration of a part, such as cracks, laps, seams, inclusions or porosity. A discontinuity may or may not affect the usefulness of a part.

discontinuous precipitation. Precipitation from a supersaturated solid solution in which the precipitate particles grow by short-range diffusion, accompanied by recrystallization of the matrix in the region of precipitation. Discontinuous precipitates grow into the matrix from nuclei near grain boundaries, forming cells of alternate lamellae of precipitate and depleted (and recrystallized) matrix. Often referred to as cellular or nodular precipitation. Compare with *continuous precipitation, localized precipitation.*

discontinuous yielding. The nonuniform plastic flow of a metal exhibiting a yield point in which plastic deformation is inhomogeneously distributed along the gage length. Under some circumstances, it may occur in metals not exhibiting a distinct yield point, either at the onset of or during plastic flow.

dishing. Forming a shallow concave surface, the area being large compared to the depth.

dislocation. A linear imperfection in a crystalline array of atoms. Two basic types are recognized: an edge dislocation corresponds to the row of mismatched atoms along the edge formed by an extra, partial plane of atoms within the body of a crystal; a screw dislocation corresponds to the axis of a spiral structure in a crystal, characterized by a distortion that joins normally parallel planes together to form a continuous helical ramp (with a pitch of one interplanar distance) winding about the dislocation. Most prevalent is the so-called mixed dislocation, which is the name given to any combination of an edge dislocation and a screw dislocation.

disordering. Forming a lattice arrangement in which the solute and solvent atoms of a solid solution occupy lattice sites at random. See also *ordering, superlattice.*

disruptive strength. The stress at which a metal fractures under hydrostatic tension.

distortion. Any deviation from an original size, shape or contour that occurs because of the application of stress or the release of residual stress.

divorced eutectic. A metallographic appearance in which the two constituents of a eutectic structure appear as massive phases rather than the finely divided mixture characteristic of normal eutectics. Often, one of the constituents of the eutectic is continuous with and indistinguishable from an accompanying proeutectic constituent.

double aging. Employment of two different aging treatments to control the type of precipitate formed from a supersaturated matrix in order to obtain the desired properties. The first aging treatment, sometimes referred to as intermediate or stabilizing, is usually carried out at higher temperature than the second.

double tempering. A treatment in which a quench-hardened ferrous metal is subjected to two complete tempering cycles, usually at substantially the same temperature, for the purpose of ensuring completion of the tempering reaction and promoting stability of the resulting microstructure.

draft. (1) An angle or taper on the surface of a pattern, core box, punch or die (or of the parts made with them) that makes it easier to remove the parts from a mold or die cavity, or to remove a core from a casting. (2) The change in cross section that occurs during rolling or cold drawing.

drag. The bottom section of a flask, mold or pattern.

drawability. A measure of the workability of a metal subject to a drawing process. A term usually expressed to indicate a metal's ability to be deep drawn.

drawbench. The stand that holds the die and draw head used in the drawing of wire, rod and tubing.

drawing. (1) Forming recessed parts by forcing the plastic flow of metal in dies. (2) Reducing the cross section of bar stock, wire or tubing by pulling it through a die. (3) A misnomer for *tempering.*

drawing compound. A substance applied to prevent pickup and scoring during drawing or pressing operations by preventing metal-to-metal contact of the work and die. Also known as *die lubricant.*

draw plate. A circular plate with a hole in the center contoured to fit a forming punch, used to support the blank during the forming cycle.

draw radius. The radius at the edge of a die or punch over which the work is drawn.

draw ring. A ring-shaped die part over the inner edge of which the metal is drawn by the punch.

drop forging. A shallow forging made in impression dies; usually with a drop hammer.

drop hammer. A forging hammer that depends on gravity for its force.

dry sand mold. A casting mold made of sand and then dried at 100°C (212°F) or above before using. Contrast with *green sand mold.*

ductile crack propagation. Slow crack propagation that is accompanied by noticeable plastic deformation and requires energy to be supplied from outside the body.

ductile fracture. Fracture characterized by tearing of metal accompanied by appreciable gross plastic deformation and expenditure of considerable energy.

ductility. The ability of a material to deform plastically without fracturing, being measured by elongation or reduction of area in a tensile test; by height of cupping in an Erichsen test or by other means.

dummy block. In extrusion, a thick unattached disk placed between the ram and billet to prevent overheating of the ram.

duralumin. (obsolete) A term formerly applied to the class of age-hardenable aluminum-copper alloys containing manganese, magnesium or silicon.

E

earing. The formation of scallops (ears) around the top edge of a drawn part caused by directional differences in the properties of the sheet metal used.

eddy-current testing. An electromagnetic nondestructive testing method in which eddy-current flow is induced in the test object. Changes in the flow caused by variations in the object are reflected into a nearby coil or coils where they are detected and measured by suitable instrumentation.

edge dislocation. See *dislocation.*

edge strain. Transverse strain lines or Lüders lines ranging from 25 to 300 mm (1 to 12 in.) in from the edges of cold rolled steel sheet or strip.

ejector. A device mounted in such a way that it removes or assists in removing a formed part from a die.

elastic constants. Factors of proportionality that describe elastic response of a material to applied forces; includes *modulus of elasticity* (either in tension, compression or shear), *Poisson's ratio, compressibility* and bulk modulus.

elastic deformation. A change in dimensions directly proportional to and in phase with an increase or decrease in applied force.

elasticity. Ability of a solid to deform in direct proportion to and in phase with increases or decreases in applied force.

elastic limit. The maximum stress to which a material may be subjected without any permanent strain remaining upon complete release of stress.

elastic modulus. Same as *modulus of elasticity.*

elastic ratio. *Yield point* divided by *tensile strength.*

elastic strain. Same as *elastic deformation.*

elastic strain energy. See *strain energy.*

electrochemical corrosion. Corrosion that is accompanied by a flow of electrons between cathodic and anodic areas on metallic surfaces.

electrochemical equivalent. The weight of an element, compound, radical, or ion involved in a specified electrochemical reaction during the passage of a unit quantity of electricity.

electrode. (1) In arc welding, a current-carrying rod that supports the arc between the rod and work, or between two rods as in twin carbon-arc welding. It may or may not furnish filler metal. (2) In resistance welding, a part of a resistance welding machine through which current and, in most instances, pressure are applied directly to the work. The electrode may be in the form of a rotating wheel, rotating roll, bar, cylinder, plate, clamp, chuck or modification thereof. (3) An electrical conductor for leading current into or out of a medium.

electrodeposition. The deposition of a substance upon an electrode by passing electric current through an electrolyte. Electroplating (plating), electroforming, electrorefining and electrowinning result from electrodeposition.

electrogalvanizing. The electroplating of zinc upon iron or steel.

electrolysis. Chemical change resulting from the passage of an electric current through an electrolyte.

electrolyte. (1) An ionic conductor. (2) A liquid, most often a solution, that will conduct an electric current.

electrolytic cell. An assembly, consisting of a vessel, electrodes and an electrolyte, in which electrolysis can be carried out.

electrolytic protection. See the preferred term, *cathodic protection.*

electron beam microprobe analyzer. An instrument for selective analysis of a microscopic component or feature in which an electron beam bombards the point of interest in a vacuum at a given energy level. Scanning of a larger area permits determination of the distribution of selected elements. The analysis is made by measuring the wavelengths and intensities of secondary electromagnetic radiation resulting from the bombardment.

electron compound. An intermediate phase on a *constitution diagram,* usually a binary phase, that has the same crystal structure and the same ratio of valence electrons to atoms as intermediate phases in several other systems. An electron compound is often a solid solution of variable composition and good metallic properties. Occasionally, an ordered arrangement of atoms is characteristic of the compound, in which case the range of composition is usually small. Phase stability depends essentially on electron concentration and crystal structure and has been observed at valence-electron-to-atom ratios of 3/2, 21/13 and 7/4.

electroplating. Electrodepositing a metal or alloy in an adherent form on an object serving as a cathode.

elongation. In tensile testing, the increase in the gage length, measured after fracture of the specimen within the gage length, usually expressed as a percentage of the original gage length.

embossing. Raising a design in relief against a surface.

embrittlement. Reduction in the normal ductility of a metal due to a physical or chemical change. Examples include blue brittleness, *hydrogen embrittlement* and *temper brittleness*.

emf. An abbreviation for electromotive force (electrical potential; voltage).

endurance limit. The maximum stress below which a material can presumably endure an infinite number of stress cycles. If the stress is not completely reversed, the value of the mean stress, the minimum stress or the stress ratio also should be stated. Compare with *fatigue limit*.

endurance ratio. The ratio of the *endurance limit* for completely reversed flexural stress to the tensile strength of a given material.

equiaxed grain structure. A structure in which the grains have approximately the same dimensions in all directions.

equilibrium. A dynamic condition of physical, chemical, mechanical or atomic balance, where the condition appears to be one of rest rather than change.

equilibrium diagram. A graphical representation of the temperature, pressure and composition limits of phase fields in an alloy system as they exist under conditions of complete equilibrium. In metal systems, pressure is usually considered constant.

erosion. Destruction of metals or other materials by the abrasive action of moving fluids, usually accelerated by the presence of solid particles or matter in suspension. When *corrosion* occurs simultaneously, the term *erosion-corrosion* is often used.

erosion-corrosion. See *erosion*.

etchant. A chemical substance or mixture used for *etching*.

etching. (1) Subjecting the surface of a metal to preferential chemical or electrolytic attack in order to reveal structural details for metallographic examination. (2) Chemically or electrochemically removing tenacious films from a metal surface to condition the surface for a subsequent treatment, such as painting or electroplating.

eutectic. (1) An isothermal reversible reaction in which a liquid solution is converted into two or more intimately mixed solids on cooling, the number of solids formed being the same as the number of components in the system. (2) An alloy having the composition indicated by the eutectic point on an equilibrium diagram. (3) An alloy structure of intermixed solid constituents formed by a eutectic reaction.

eutectoid. (1) An isothermal reversible reaction in which a solid solution is converted into two or more intimately mixed solids on cooling, the number of solids formed being the same as the number of components in the system. (2) An alloy having the composition indicated by the eutectoid point on an equilibrium diagram. (3) An alloy structure of intermixed solid constituents formed by a eutectoid reaction.

exfoliation. A type of corrosion that progresses approximately parallel to the outer surface of the metal, causing layers of the metal to be elevated by the formation of corrosion product.

extractive metallurgy. The branch of process metallurgy dealing with the winning of metals from their ores.

extra hard. A *temper* of nonferrous alloys and some ferrous alloys characterized by tensile strength and hardness about one-third of the way from *full hard* to *extra spring* temper.

extra spring. A *temper* of nonferrous alloys and some ferrous alloys corresponding approximately to a cold worked state above *full hard* beyond which further cold work will not measurably increase the strength and hardness.

extrusion. Conversion of an ingot or billet into lengths of uniform cross section by forcing metal to flow plastically through a die orifice. In direct extrusion (forward extrusion), the die and ram are at opposite ends of the extrusion stock, and the product and ram travel in the same direction. Also, there is relative motion between the extrusion stock and the container. In indirect extrusion (backward extrusion), the die is at the ram end of the stock and the product travels in the opposite direction as the ram, either around the ram (as in the impact extrusion of cylinders such as cases for dry cell batteries) or up through the center of a hollow ram.

Impact extrusion is the process (or resultant product) in which a punch strikes a slug (usually unheated) in a confining die. The metal flow may be either between punch and die or through another opening. Impact extrusion of unheated slugs is often called cold extrusion.

A stepped extrusion is a single product having one or more abrupt changes in cross section. It is produced by stopping extrusion to change dies. Often, such an extrusion is made in a complex die having a die section that can be freed from the main die and allowed to ride out with the product when extrusion is resumed.

extrusion billet. A metal slug used as *extrusion stock*.

extrusion ingot. A cast metal slug used as *extrusion stock*.

extrusion stock. A rod, bar or other section used to make extrusions.

F

failure. A general term used to imply that a part in service (*a*) has become completely inoperable, (*b*) is still operable but is incapable of satisfactorily performing its intended function, or (*c*) has deteriorated seriously, to the point that it has become unreliable or unsafe for continued use.

false indication. In nondestructive inspection, an *indication* that may be interpreted erroneously as an *imperfection*. See also *artifact*.

fatigue. The phenomenon leading to fracture under repeated or fluctuating stresses having a maximum value less than the tensile strength of the material. Fatigue fractures are progressive, beginning as minute cracks that grow under the action of the fluctuating stress.

fatigue life. The number of cycles of stress that can be sustained prior to failure for a stated test condition.

fatigue limit. The maximum stress that presumably leads to fatigue fracture in a specified number of stress cycles. If the stress is not completely reversed, the value of the mean stress, the minimum stress or the stress ratio also should be stated. Compare with *endurance limit*.

fatigue notch factor (K_f). The ratio of the fatigue strength of an unnotched specimen to the fatigue strength of a notched specimen of the same material and condition; both strengths are determined at the same number of stress cycles.

fatigue notch sensitivity (q). An estimate of the effect of a notch or hole on the fatigue properties of a material; measured by $q = (K_f-1)/(K_t-1)$. A material is said to be fully notch sensitive if q approaches a value of 1.0; it is not notch sensitive if the ratio approaches 0. K_f is the *fatigue notch factor*, and K_t is the *stress-concentration factor*, for a specimen of the material containing a notch or hole of a given size and shape.

fatigue ratio. The *fatigue limit* under completely reversed flexural stress divided by the tensile strength for the same alloy and condition.

fatigue strength. The maximum stress that can be sustained for a specified number of cycles without failure, the stress being completely reversed within each cycle unless otherwise stated.

fatigue-strength reduction factor (K_f). The ratio of the fatigue strength of a member or specimen with no stress concentration to the fatigue strength with stress concentration. K_f has no meaning unless the stress range and the shape, size and material of the member or specimen are stated.

fatigue striations. Parallel lines frequently observed in electron microscope fractographs of fatigue fracture surfaces. The lines are transverse to the direction of local crack propagation; the distance between successive lines represents the advance of the crack front during one cycle of stress variation.

feeding. (1) Conveying metal stock or workpieces to a location for use or processing, such as wire to a consumable electrode, strip to a die, or workpieces to an assembler. (2) In casting, providing molten metal to a region undergoing solidification, usually at a rate sufficient to fill the mold cavity ahead of the solidification front and to make up for any shrinkage accompanying solidification.

ferrite. (1) A solid solution of one or more elements in body-centered cubic iron. Unless otherwise designated (for instance, as chromium ferrite), the solute is generally assumed to be carbon. On some equilibrium diagrams, there are two ferrite regions separated by an austenite area. The lower area is alpha ferrite; the upper, delta ferrite. If there is no designation, alpha ferrite is assumed. (2) In the field of magnetics, substances having the general formula:

$$M^{++}O \cdot M_2^{+++}O_3$$

the trivalent metal often being iron.

ferrite banding. Parallel bands of free ferrite aligned in the direction of working. Sometimes referred to as ferrite streaks.

ferrite streaks. Same as *ferrite banding.*

fiber. (1) The characteristic of wrought metal that indicates *directional properties* and is revealed by the etching of a longitudinal section or is manifested by the fibrous or woody appearance of a fracture. It is caused chiefly by the extension of the constituents of the metal, both metallic and nonmetallic, in the direction of working. (2) The pattern of preferred orientation of metal crystals after a given deformation process, usually wiredrawing. See *preferred orientation.*

fibrous fracture. A fracture where the surface is characterized by a dull gray or silky appearance. Contrast with *crystalline fracture.*

fibrous structure. (1) In forgings, a structure revealed as laminations, not necessarily detrimental, on an etched section or as a ropy appearance on a fracture. It is not to be confused with the silky or ductile fracture of a clean metal. (2) In wrought iron, a structure consisting of slag fibers embedded in ferrite. (3) In rolled steel plate stock, a uniform, fine-grained structure on a fractured surface, free of laminations or shale-type discontinuities. As contrasted with part (1) above, it is virtually synonymous with silky or ductile fracture.

filamentary shrinkage. A fine network of shrinkage cavities, occasionally found in steel castings, that produces a radiographic image resembling lace.

file hardness. Hardness as determined by the use of a file of standardized hardness on the assumption that a material that cannot be cut with the file is as hard as, or harder than, the file. Files covering a range of hardnesses may be employed.

final annealing. An imprecise term used to denote the last anneal given to a nonferrous alloy prior to shipment.

finish. (1) Surface condition, quality or appearance of a metal. (2) Stock on a forging or casting to be removed when finish machined.

finishing die. The die used to make the final impression on a forging. Sometimes called finisher.

finishing temperature. The temperature at which *hot working* is completed.

flame annealing. Annealing in which the heat is applied directly by a flame.

flame hardening. A process for hardening the surfaces of hardenable ferrous alloys in which an intense flame is used to heat the surface layers above the upper transformation temperature, whereupon the workpiece is immediately quenched.

flash. (1) In forging, excess metal forced out between the upper and lower dies. (2) In casting, a fin of metal that results from leakage between mating mold surfaces. (3) In resistance butt welding, a fin formed perpendicular to the direction of applied pressure.

flash extension. Portion of flash remaining after trimming. Flash extension is measured from the intersection of the draft and flash at the body of the forging to the trimmed edge of the stock.

flash line. The line of location of flash formed around a forging or casting.

flask. A metal or wood frame used for making and holding a sand mold. The upper part is called the cope; the lower, the drag.

flat die forging. Forging metal between flat or simple contour dies by repeated strokes and manipulation of the workpiece. Also known as open die forging, hand forging or smith forging.

flattening. (1) A preliminary operation performed on forging stock so as to position the metal for a subsequent forging operation. (2) Removing irregularities or distortion in sheets or plates by a method such as roller leveling or *stretcher leveling.*

flat wire. A roughly rectangular or square mill product, narrower than *strip,* in which all surfaces are rolled or drawn without any previous slitting, shearing or sawing.

flaw. A nonspecific term often used to imply a cracklike discontinuity. See preferred terms *discontinuity, imperfection, defect.*

flospinning. Forming cylindrical, conical and curvilinear shaped parts by power spinning over a rotating mandrel.

flowability. A characteristic of a foundry sand mixture that enables it to move under pressure or vibration so that it makes intimate contact with all surfaces of the pattern or core box.

flow lines. (1) Texture showing the direction of metal flow during hot or cold working. Flow lines often can be revealed by etching the surface or a section of a metal part (see macrograph on this page). (2) In mechanical metallurgy, paths followed by minute volumes of metal during deformation.

flow stress. The uniaxial true stress at the onset of plastic deformation in a metal.

fluidity. The ability of liquid metal to run into and fill a mold cavity.

fluorescence. The emission of characteristic electromagnetic radiation by a substance as a result of the absorption of electromagnetic or corpuscular radiation having a greater unit energy than that of the fluorescent radiation. It occurs only so long as the stimulus responsible for it is maintained.

fluorescent magnetic-particle inspection. Inspection with either dry magnetic particles or those in a liquid suspension, the particles being coated with a fluorescent substance to increase the visibility of the indications.

fluorescent penetrant inspection. Inspection using a fluorescent liquid that will penetrate any surface opening; after wiping the surface clean, the location of any surface flaws may be detected by the fluorescence, under ultraviolet light, of back-seepage of the fluid.

fluoroscopy. An inspection procedure in which the radiographic image of the subject is viewed on a fluorescent screen, normally limited to low-density materials or thin sections of metals because of the low light output of the fluorescent screen at safe levels of radiation.

flux. (1) In metal refining, a material used to remove undesirable substances, like sand, ash or dirt, as a molten mixture. It is also used as a protective covering for certain molten metal baths. Lime or limestone is generally used to remove sand, as in iron smelting; sand, to remove iron oxide in copper refining. (2) In brazing, cutting, soldering or welding, material used to prevent the formation of, or to dissolve and facilitate removal of, oxides and other undesirable substances.

flux lines. Imaginary lines used as a means of explaining the behavior of magnetic and other fields. Their concept is based on the pattern of lines produced when magnetic particles are sprinkled over a permanent magnet. Sometimes called magnetic lines of force.

flying shear. A machine for cutting continuous rolled products to length that does not require a halt in rolling, but rather moves along the runout table at the same speed as the product while performing the cutting, then returns to the starting point in time to cut the next piece.

foil. Metal in sheet form less than 0.15 mm (0.006 in.) in thickness.

fold. Same as *lap*.

forgeability. Term used to describe the relative ability of material to flow under a compressive load without rupture.

forge welding. Solid state welding in which metals are heated in a forge (in air) then welded together by applying pressure or blows sufficient to cause permanent deformation at the interface.

forging. Plastically deforming metal, usually hot, into desired shapes with compressive force, with or without dies.

forging billet. A wrought metal slug used as *forging stock*.

forging ingot. A cast metal slug used as *forging stock*.

forging machine. A type of forging equipment, related to the mechanical press, in which the main forming energy is applied horizontally to the workpiece, which is held by dies. Commonly called upsetter or header.

forging plane. In forging, the plane that includes the principal die face and that is perpendicular to the direction of ram travel. When parting surfaces of the dies are flat, the forging plane coincides with the parting line. Contrast *parting plane*.

forging range. Temperature range in which a metal can be forged successfully.

forging rolls. A machine used in *roll forging*. Also called gap rolls.

forging stock. A rod, bar or other section used to make forgings.

formability. The relative ease with which a metal can be shaped through plastic deformation. See *drawability*.

form block. Tooling, usually the male part, used for forming sheet-metal contours, being generally employed in the rubber-pad process.

form die. A die used to change the shape of a blank with minimum plastic flow.

forming. Making a change, with the exception of shearing or blanking, in the shape or contour of a metal part without intentionally altering the thickness.

form rolling. Hot rolling to produce bars having contoured cross sections; not to be confused with roll forming of sheet metal or with roll forging.

forward extrusion. Same as direct extrusion. See *extrusion.*

foundry. A commercial establishment or building where metal castings are produced.

four-high mill. A type of rolling mill, commonly used for flat-rolled mill products, in which two large-diameter backup rolls are employed to reinforce two smaller working rolls, which are in contact with the product. Either the working rolls or the backup rolls may be driven. Compare with *two-high mill, cluster mill.*

fractography. Descriptive treatment of fracture, especially in metals, with specific reference to photographs of the fracture surface. Macrofractography involves photographs at low magnification; microfractography, at high magnification.

fracture mechanics. See *linear elastic fracture mechanics.*

fracture stress. (1) The maximum principal true stress at fracture. Usually refers to unnotched tensile specimens. (2) The (hypothetical) true stress that will cause fracture without further deformation at any given strain.

fracture test. Breaking a specimen and examining the fractured surface with the unaided eye or with a low-power microscope to determine such things as composition, grain size, case depth or soundness.

fracture toughness. See *stress-intensity factor.*

fragmentation. The subdivision of a grain into small discrete crystallites outlined by a heavily deformed network of intersecting slip as a result of cold working. These small crystals or fragments differ from one another in orientation and tend to rotate to a stable orientation determined by the slip systems.

free carbon. The part of the total carbon in steel or cast iron that is present in elemental form as graphite or temper carbon. Contrast with *combined carbon.*

free ferrite. Ferrite that is formed directly from the decomposition of hypoeutectoid austenite during cooling, without the simultaneous formation of cementite. Also proeutectoid ferrite.

freezing range. That temperature range between *liquidus* and *solidus* temperatures in which molten and solid constituents coexist.

fretting. A type of wear that occurs between tight-fitting surfaces subjected to cyclic relative motion of extremely small amplitude. Usually, fretting is accompanied by corrosion, especially of the very fine wear debris. Also referred to as fretting corrosion, false brinelling (in rolling-element bearings), friction oxidation, chafing fatigue, molecular attrition and wear oxidation.

fretting fatigue. Fatigue fracture that initiates at a surface area where fretting has occurred.

full annealing. An imprecise term that denotes an annealing cycle to produce minimum strength and hardness. For the term to be meaningful, the composition and starting condition of the material and the time-temperature cycle used must be stated.

fuller. In preliminary forging, the portion of a die that reduces the cross-sectional area between the ends of the stock and permits the metal to move outward.

full hard. A *temper* of nonferrous alloys and some ferrous alloys corresponding approximately to a cold worked state beyond which the material can no longer be formed by bending. In specifications, a full hard temper is commonly defined in terms of minimum hardness or minimum tensile strength (or, alternatively, a range of hardness or strength) corresponding to a specific percentage of cold reduction following a full anneal. For aluminum, a full hard temper is equivalent to a reduction of 75% from *dead soft;* for austenitic stainless steels, a reduction of about 50 to 55%.

fusion. A change of state from solid to liquid; melting.

fusion face. A surface of the base metal that will be melted during welding.

fusion welding. Any welding process in which filler metal and base metal (substrate), or base metal only, are melted together to complete the weld.

fusion zone. In a weldment, the area of base metal melted as determined on a cross section through the weld.

G

gage. (1) The thickness (or diameter) of sheet or wire. The various standards are arbitrary and differ, ferrous from nonferrous products and sheet from wire. (2) An instrument used to measure thickness or length. (3) An aid for visual inspection that enables the inspector to determine more reliably whether the size or contour of a formed part meets dimensional requirements.

gage length. The original length of that portion of the specimen over which strain, change of length and other characteristics are measured.

galling. A condition whereby excessive friction between high spots results in localized welding with subsequent spalling and a further roughening of the rubbing surfaces of one or both of two mating parts.

galvanic cell. A cell in which chemical change is the source of electrical energy. It usually consists of two dissimilar conductors in contact with each other and with an electrolyte, or of two similar conductors in contact with each other and with dissimilar electrolytes.

galvanic corrosion. Corrosion associated with the current of a galvanic cell consisting of two dissimilar conductors in an electrolyte or two similar conductors in dissimilar electrolytes. Where the two dissimilar metals are in contact, the resulting reaction is referred to as couple action.

galvanic series. A series of metals and alloys arranged according to their relative electrode potentials in a specified environment. Compare with *electromotive series*.

galvanize. To coat a metal surface with zinc using any of various processes.

galvanneal. To produce a zinc-iron alloy coating on iron or steel by keeping the coating molten after hot-dip galvanizing until the zinc alloys completely with the basis metal.

gamma iron. The face-centered cubic form of pure iron, stable from 910 to 1400 °C (1670 to 2550 °F).

gamma ray. Short wavelength electromagnetic radiation, similar to x-rays but of nuclear origin, with a range of wavelengths from about 10^{-14} to 10^{-10} m.

gas holes. Holes in castings or welds that are formed by gas escaping from molten metal as it solidifies. Gas holes may occur individually, in clusters, or distributed throughout the solidified metal.

gas pocket. A cavity caused by entrapped gas.

gas porosity. Fine holes or pores within a metal that are caused by entrapped gas or by evolution of dissolved gas during solidification.

gassing. (1) Absorption of gas by a metal. (2) Evolution of gas from a metal during melting operations or on solidification. (3) The evolution of gas from an electrode during electrolysis.

gate. The portion of the runner in a mold through which molten metal enters the mold cavity. Sometimes the generic term is applied to the entire network of connecting channels that conduct metal into the mold cavity.

gated pattern. A *pattern* that includes not only the contours of the part to be cast, but also the *gates*.

gathering. A forging operation that increases the cross section of part of the stock; usually a preliminary operation.

ghost lines. Lines running parallel to the rolling direction that appear in a panel when it is stretched. These lines may not be evident unless the panel has been sanded or painted. (Not to be confused with leveler lines.)

glide. (1) Same as *slip*. (2) A noncrystallographic shearing movement, as one grain over another.

G-P zone. A *Guinier-Preston zone*.

grain. An individual crystal in a polycrystalline metal or alloy; it may or may not contain twinned regions and subgrains.

grain boundary corrosion. Same as *intergranular corrosion*. See also *interdendritic corrosion*.

grain fineness number. A weighted average grain size of a granular material. The AFS grain fineness number is calculated with prescribed weighting factors from the standard screen analysis.

grain flow. Fiberlike lines appearing on polished and etched sections of forgings, which are caused by orientation of the constituents of the metal in the direction of working during forging. Grain flow produced by proper die design can improve required mechanical properties of forgings.

grain growth. An increase in the average size of the grains in polycrystalline metal, usually as a result of heating at elevated temperature.

grain refiner. A material added to a molten metal to induce a finer than normal grain size in the final structure.

grain size. A measure of the areas or volumes of grains in a polycrystalline material, usually expressed as an average when the individual sizes are fairly uniform. In metals containing two or more phases, the grain size refers to that of the matrix unless otherwise specified. Grain sizes are reported in terms of number of grains per unit area or volume, average diameter, or as a grain-size number derived from area measurements.

granular fracture. A type of irregular surface produced when metal is broken that is characterized by a rough, grainlike appearance as differentiated from a smooth silky, or fibrous, type. It can be subclassified into transgranular and intergranular forms. This type of fracture is frequently called crystalline fracture, but the inference that the metal broke because it "crystallized" is not justified because all metals are crystalline when in the solid state. Contrast with *fibrous fracture, silky fracture.*

graphitic corrosion. Corrosion of gray iron in which the iron matrix is selectively leached away, leaving a porous mass of graphite behind; it occurs in relatively mild aqueous solutions and on buried pipe and fittings.

graphitization. Formation of graphite in iron or steel. Where graphite is formed during solidification, the phenomenon is called primary graphitization; where formed later by heat treatment, secondary graphitization.

gray cast iron. A *cast iron* that gives a gray fracture due to the presence of flake graphite. Often called gray iron.

green sand. A naturally bonded sand, or a compounded molding sand mixture, that has been "tempered" with water and used while still moist.

green sand core. (1) A *core* made of *green sand* and used as rammed. (2) A sand core that is used in the unbaked condition.

green sand mold. A casting mold composed of moist prepared molding sand. Contrast with *dry sand mold.*

grinding stress. *Residual stress,* generated by grinding, in the surface layer of work. It may be tensile, compressive or both.

gross porosity. In weld metal or in a casting, pores, gas holes or globular voids that are larger and in much greater number than obtained in good practice.

Guinier-Preston (G-P) zone. A small precipitation domain in a supersaturated metallic solid solution. A G-P zone has no well-defined crystalline structure of its own and contains an abnormally high concentration of solute atoms. The formation of G-P zones constitutes the first stage of precipitation and is usually accompanied by a change in

properties of the solid solution in which they occur.

gutter. The clearance around the land of a forging die providing space for the flash without trapping it in the dies.

H

habit plane. The plane or system of planes of a crystalline phase along which some phenomenon such as twinning or transformation occurs.

half hard. A *temper* of nonferrous alloys and some ferrous alloys characterized by tensile strength about midway between that of *dead soft* and *full hard* tempers.

hammer forging. Forging in which the work is deformed by repeated blows. Compare with *press forging.*

hammering. Beating metal sheet into a desired shape either over a form or on a high-speed mechanical hammer and a similar anvil to produce the required dishing or thinning.

hammer welding. *Forge welding* by hammering.

hand forging. See *flat die forging*

hard drawn. An imprecise term applied to drawn products, such as wire and tubing, that indicates substantial cold reduction without subsequent annealing. Compare with *light drawn.*

hardenability. The relative ability of a ferrous alloy to form martensite when quenched from a temperature above the upper critical temperature. Hardenability is commonly measured as the distance below a quenched surface where the metal exhibits a specific hardness (50 HRC, for example) or a specific percentage of martensite in the microstructure.

hardener. An alloy, rich in one or more alloying elements, added to a melt to permit closer composition control than possible by addition of pure metals or to introduce refractory elements not readily alloyed with the base metal. Sometimes called master alloy or rich alloy.

hardening. Increasing hardness by suitable treatment, usually involving heating and cooling. When applicable, the following more specific terms should be used: *age hardening, case hardening, flame hardening, induction hardening, precipitation hardening* and *quench hardening.*

hardness. Resistance of metal to plastic deformation, usually by indentation. However, the term may also refer to stiffness or temper, or to resistance to scratching, abrasion or cutting. Indentation hardness may be measured by various hardness tests, such as *Brinell, Rockwell* and *Vickers*.

hard temper. Same as *full hard* temper.

heading. Upsetting wire, rod or bar stock in dies to form parts that usually have some of the cross-sectional area larger than the original.

heat-affected zone. That portion of the base metal that was not melted during brazing, cutting or welding, but whose microstructure and mechanical properties were altered by the heat.

heat check. A pattern of parallel surface cracks that are formed by alternate rapid heating and cooling of the extreme surface metal, sometimes found on forging dies and piercing punches. There may be two sets of parallel cracks, one set perpendicular to the other.

heat-resisting alloy. An alloy developed for very high temperature service where relatively high stresses (tensile, thermal, vibratory or shock) are encountered and where oxidation resistance is frequently required.

heat tinting. Coloration of a metal surface through oxidation by heating to reveal details of the microstructure.

heat treatable alloy. An alloy that can be hardened by heat treatment.

heat treating film. A thin coating or film, usually an oxide, formed on the surface of metals during heat treatment.

heat treatment. Heating and cooling a solid metal or alloy in such a way as to obtain desired conditions or properties. Heating for the sole purpose of hot working is excluded from the meaning of this definition.

herringbone pattern. Same as *chevron pattern*.

homogenizing. Holding at high temperature to eliminate or decrease chemical segregation by diffusion.

Hooke's law. Stress is proportional to strain. The law holds only up to the proportional limit.

hot chamber machine. A *die casting* machine in which the metal chamber under pressure is immersed in the molten metal in a furnace. The chamber is sometimes called a gooseneck and the machine, a gooseneck machine.

hot crack. A crack formed in a cast metal because of internal stress developed on cooling following solidification. A hot crack is less open than a *hot tear* and usually exhibits less oxidation and decarburization along the fracture surface.

hot extrusion. Extrusion at elevated temperature that does not cause strain hardening. See also *extrusion*.

hot forming. See *hot working*.

hot mill. A production line or facility for hot rolling metals.

hot press forging. Plastically deforming metals between dies in presses at temperatures high enough to avoid strain hardening.

hot pressing. Forming a powder metallurgy compact at a temperature high enough to have concurrent *sintering*.

hot shortness. A tendency for some alloys to separate along grain boundaries when stressed or deformed at temperatures near the melting point. Hot shortness is caused by a low-melting constituent, often present only in minute amounts, that is segregated at grain boundaries.

hot tear. A fracture formed in a metal during solidification because of *hindered contraction*. Compare with *hot crack*.

hot top. (1) A reservoir, thermally insulated or heated, to hold molten metal on top of a mold to feed the ingot or casting as it contracts on solidifying to avoid having pipe or voids. See accompanying sketch. (2) A refractory-lined steel or iron casting that is inserted into the tip of the mold and is supported at various heights to feed the ingot as it solidifies.

hot working. Deforming metal plastically at such a temperature and strain rate that recrystallization takes place simultaneously with the deformation, thus avoiding any strain hardening.

hydraulic press. A press in which fluid pressure is used to actuate and control the ram.

hydrogen damage. A general term for the embrittlement, cracking, blistering and hydride formation that can occur when hydrogen is present in some metals.

hydrogen embrittlement. A condition of low ductility in metals resulting from the absorption of hydrogen.

hydrostatic tension. Three equal and mutually perpendicular tensile stresses.

hypereutectic alloy. In an alloy system exhibiting a *eutectic*, any alloy whose composition has an excess of alloying element compared to the eutectic composition, and whose equilibrium microstructure contains some eutectic structure.

hypereutectoid alloy. In an alloy system exhibiting a *eutectoid*, any alloy whose composition has an excess of alloying element compared to the eutectoid composition, and whose equilibrium microstructure contains some eutectoid structure.

hypoeutectic alloy. In an alloy system exhibiting a *eutectic*, any alloy whose composition has an excess of base metal compared to the eutectic composition, and whose equilibrium microstructure contains some eutectic structure.

hypoeutectoid alloy. In an alloy system exhibiting a *eutectoid*, any alloy whose composition has an excess of base metal compared to the eutectoid composition, and whose equilibrium microstructure contains some eutectoid structure.

I

idiomorphic crystal. An individual crystal that has grown without restraint so that the habit planes are clearly developed. Compare with *allotriomorphic crystal*.

impact energy. The amount of energy required to fracture a material, usually measured by means of an *Izod* or *Charpy* test. The type of specimen and test conditions affect the values and therefore should be specified.

impact extrusion. See *extrusion*.

impact strength. Same as *impact energy*.

impact test. A test to determine the behavior of materials when subjected to high rates of loading, usually in bending, tension or torsion. The quantity measured is the energy absorbed in breaking the specimen by a single blow, as in the *Charpy* or *Izod* tests.

imperfection. (1) When referring to the physical condition of a part or metal product, any departure of a quality characteristic from its intended level or state. The existence of an imperfection does not imply *nonconformance*, nor does it have any implication as to the usability of a product or service. An imperfection must be rated on a scale of severity, in accordance with applicable specifications, to establish whether or not the part or metal product is of acceptable quality. (2) Generally, any departure from an ideal design, state or condition. (3) In crystallography, any deviation from an ideal space lattice.

impregnation. (1) The treatment of porous castings with a sealing medium to stop pressure leaks. (2) The process of filling the pores of a sintered compact, usually with a liquid such as a lubricant. (3) The process of mixing particles of a nonmetallic substance in a matrix of metal powder, as in diamond-impregnated tools.

impression die forging. A forging that is formed to the required shape and size by machined impressions in specially prepared dies that exert three-dimensional control on the workpiece.

impurities. Elements or compounds whose presence in a material is undesired.

inclusions. Particles of foreign material in a metallic matrix. The particles are usually compounds (such as oxides, sulfides or silicates), but may be of any substance that is foreign to (and essentially insoluble in) the matrix.

indentation hardness. The resistance of a material to indentation. This is the usual type of hardness test, in which a pointed or rounded indenter is pressed into a surface under a substantially static load.

indication. In inspection, a response to a nondestructive stimulus that implies the presence of an *imperfection*. The indication must be interpreted to determine if (*a*) it is a true indication or a *false indication* and (*b*) whether or not a true indication represents an unacceptable deviation.

induction hardening. A surface-hardening process in which only the surface layer of a suitable ferrous workpiece is heated by electromagnetic induction to above the upper critical temperature and immediately quenched.

ingate. Same as *gate*.

ingot. A casting of simple shape, suitable for hot working or remelting.

inhibitor. A substance that retards some specific chemical reaction. Pickling inhibitors retard the dissolution of metal without hindering the removal of scale from steel.

inoculation. The addition of a material to molten metal to form nuclei for crystallization.

intercrystalline. Between the crystals, or grains, of a metal.

interdendritic corrosion. Corrosive attack that progresses preferentially along interdendritic paths. This type of attack results from local differences in composition, such as coring commonly encountered in alloy castings.

interface. A surface that forms the boundary between phases or systems.

interfacial tension. The contractile force of an interface between two phases.

intergranular corrosion. Corrosion occurring preferentially at grain boundaries, usually with slight or negligible attack on the adjacent grains. See also *interdendritic corrosion*.

intermediate annealing. Annealing wrought metals at one or more stages during manufacture and before final treatment.

intermediate phase. In an alloy or a chemical system, a distinguishable homogeneous phase whose composition range does not extend to any of the pure components of the system.

intermetallic compound. An intermediate phase in an alloy system, having a narrow range of homogeneity and relatively simple stoichiometric proportions; the nature of the atomic binding can be of various types, ranging from metallic to ionic.

internal oxidation. Preferential in situ oxidation of certain components or phases within the bulk of a solid alloy accomplished by diffusion of oxygen into the body; a form of *subsurface corrosion*.

internal stress. See preferred term, *residual stress*.

interpass temperature. In a multipass weld, the lowest temperature of a *pass* before the succeeding one is commenced.

interrupted aging. Aging at two or more temperatures, by steps, and cooling to room temperature after each step. See *aging*, and compare with *progressive aging* and *step aging*.

interrupted quenching. A quenching procedure in which the workpiece is removed from the first quench at a temperature substantially higher than that of the quenchant and is then subjected to a second quenching system having a different cooling rate than the first.

interstitial solid solution. A solid solution in which the solute atoms occupy positions that do not correspond to lattice points of the solvent. Contrast with *substitutional solid solution*.

intracrystalline. Within or across the crystals or grains of a metal; same as transcrystalline and transgranular.

investment casting. (1) Casting metal into a mold produced by surrounding (investing) an expendable pattern with a refractory slurry that sets at room temperature after which the wax, plastic or frozen mercury pattern is removed through the use of heat. Also called precision casting, or lost-wax process. (2) A part made by the investment casting process.

ion. An atom, or group of atoms, that has gained or lost one or more outer electrons and thus carries an electric charge. Positive ions, or cations, are deficient in outer electrons. Negative ions, or anions, have an excess of outer electrons.

ionic bond. A bond between two or more atoms that is the result of electrostatic attractive forces between positively and negatively charged ions.

ionic crystal. A crystal in which atomic bonds are *ionic bonds*. This type of atomic linkage, also known as (hetero) polar bonding, is characteristic of many compounds (sodium chloride, for instance).

iron casting. A part made of *cast iron*.

ironing. Thinning the walls of hollow articles by drawing them between a punch and a die.

irradiation. The exposure of a material in a field of radiation; the cumulative exposure.

isothermal annealing. Austenitizing a ferrous alloy and then cooling to and holding at a temperature at which austenite transforms to a relatively soft ferrite carbide aggregate.

isothermal transformation. A change in phase that takes place at a constant temperature. The time required for transformation to be completed, and in some instances the time delay before transformation begins, depends on the amount of supercooling below (or superheating above) the equilibrium temperature for the same transformation.

isotope. One of several different nuclides of an element having the same number of protons in their nuclei and therefore the same atomic number, but differing in the number of neutrons and therefore in atomic weight.

isotropy. Quality of having identical properties in all directions.

Izod test. A pendulum-type single-blow impact test in which the specimen, usually notched, is fixed at one end and broken by a falling pendulum. The energy absorbed, as measured by the subsequent rise of the pendulum, is a measure of impact strength or notch toughness. Contrast with *Charpy test*.

J

joint. The location where two or more members are to be or have been fastened together mechanically or by brazing or welding.

K

keyhole specimen. A type of specimen containing a hole-and-slot notch, shaped like a keyhole, usually used in impact bend tests. See *Charpy* and *Izod tests*.

killed steel. Steel treated with a strong deoxidizing agent such as silicon or aluminum in order to reduce the oxygen content to such a level that no reaction occurs between carbon and oxygen during solidification.

knockout. (1) A mechanism for freeing formed parts from a die used for stamping, blanking, drawing, forging or heading operations. (2) A partly pierced hole in a sheet metal part, where the slug remains in the hole and can be forced out by hand if a hole actually is needed. (3) Removing sand cores from a casting. (4) Jarring an investment casting mold to remove the casting and investment from the flask.

Knoop hardness. Microhardness determined from the resistance of metal to indentation by a pyramidal diamond indenter, having edge angles of 172° 30′ and 130°, making a rhombohedral impression with one long and one short diagonal.

L

ladle. A receptacle used for transferring and pouring molten metal.

laminate. (1) A composite metal, usually in the form of sheet or bar, composed of two or more metal layers so bonded that the composite metal forms a structural member. (2) To form a metallic product of two or more bonded layers.

lamination. (1) A type of discontinuity with separation or weakness generally aligned parallel to the worked surface of a metal. May be the result of pipe, blisters, seams, inclusions or segregation elongated and made directional by working. Laminations may also occur in metal-powder compacts. (2) In electrical products such as motors, a blanked piece of electrical sheet that is stacked up with several other identical pieces to make a stator or rotor.

lap. A surface imperfection, appearing as a seam, caused by folding over hot metal, fins or sharp corners and then rolling or forging them into the surface, but not welding them.

lattice constant. See *lattice parameter*.

lattice parameter. The length of any side of a unit cell of a given crystal structure; if the lengths are unequal, all unequal lengths must be given.

leaching. Extracting an element or compound from a solid alloy or mixture by preferential dissolution in a suitable liquid.

leveling. Flattening of rolled sheet, strip or plate by reducing or eliminating distortions. See also *stretcher leveling*.

light drawn. An imprecise term applied to drawn products, such as wire and tubing, that indicates a lesser amount of cold reduction than for *hard drawn* products.

light metal. One of the low-density metals such as aluminum, magnesium, titanium, beryllium or their alloys.

linear elastic fracture mechanics. A method of fracture analysis that can determine the stress (or load) required to induce fracture instability in a structure containing a cracklike flaw of known size and shape. See *stress-intensity factor*.

linear strain. See *strain*.

liquation. The partial melting of an alloy, usually as a result of coring or other compositional heterogeneities.

liquation temperature. The lowest temperature at which partial melting can occur in an alloy that exhibits the greatest possible degree of segregation.

liquid penetrant inspection. A type of nondestructive inspection that locates discontinuities that are open to the surface of a metal by first allowing a penetrating dye or fluorescent liquid to infiltrate the discontinuity, removing the excess penetrant, and then applying a developing agent that causes the penetrant to seep back out of the discontinuity and register as an indication. Liquid penetrant inspection is suitable for both ferrous and nonferrous materials, but is limited to the detection of open surface discontinuities in nonporous solids.

liquid shrinkage. See *casting shrinkage*.

liquidus. In a constitution or equilibrium diagram, the locus of points representing the temperatures at which the various compositions in the system begin to freeze on cooling or finish melting on heating. See also *solidus*.

local action. Corrosion due to the action of "local cells"; that is, *galvanic cells* resulting from inhomogeneities between adjacent areas on a metal surface exposed to an electrolyte.

local current density. Current density at a point or on a small area.

localized precipitation. Precipitation from a supersaturated solid solution similar to *continuous precipitation*, except that the precipitate particles form at preferred locations, such as along slip planes, grain boundaries or incoherent twin boundaries.

lost-wax process. An *investment casting* process in which a wax pattern is used.

lot. A finite quantity of a given product manufactured under production conditions that are considered uniform. Often used to describe a finite quantity of product submitted for inspection as a single group. For a bulk product (such as a chemical or powdered metal), the term "batch" is often used synonymously with lot.

lubricant. Any substance used to reduce friction between two surfaces in contact.

Lüders lines. Elongated surface markings or depressions caused by localized plastic deformation that results from discontinuous (inhomogeneous) yielding. Also known as Lüders bands, Hartmann lines, Piobert lines or *stretcher strains*.

M

machinability. The relative ease of machining a metal.

machinability index. A relative measure of the machinability of an engineering material under specified standard conditions.

machining. Removing material from a metal part, usually using a cutting tool, and usually using a power-driven machine.

machining stress. *Residual stress* caused by machining.

macroetching. *Etching* a metal surface to accentuate gross structural details (such as grain flow, segregation, porosity or cracks) for observation by the unaided eye or at a magnification of ten diameters or less.

macrograph. A graphic reproduction of the surface of a prepared specimen at a magnification not exceeding ten diameters. When photographed, the reproduction is known as a photomacrograph.

macroscopic. Visible at magnifications up to ten diameters.

macroscopic stresses. Residual stresses that vary from tension to compression in a distance (presumably many times the grain size) that is comparable to the gage length in ordinary strain measurements, hence, detectable by x-ray or dissection methods.

macroshrinkage. Isolated, clustered or interconnected voids in a casting that are detectable macroscopically. Such voids are usually associated with abrupt changes in section size and are caused by a lack of adequate feeding to compensate for solidification shrinkage.

macrostress. Same as *macroscopic stress*.

macrostructure. The structure of metals as revealed by macroscopic examination of the etched surface of a polished specimen.

magnetic-particle inspection. A nondestructive method of inspection for determining the existence and extent of surface cracks and similar imperfections in ferromagnetic materials. Finely divided magnetic particles, applied to the magnetized part, are attracted to and outline the pattern of any magnetic-leakage fields created by discontinuities.

malleability. The characteristic of metals that permits plastic deformation in compression without rupture.

maraging. A precipitation-hardening treatment applied to a special group of iron-base alloys to precipitate one or more intermetallic compounds in a matrix of essentially carbon-free martensite. NOTE: The first developed series of maraging steels contained, in addition to iron, more than 10% nickel and one or more supplemental hardening elements. In this series, aging is done at 480 °C (900 °F).

marquenching. See *martempering*.

martempering. (1) A hardening procedure in which an austenitized ferrous workpiece is quenched into an appropriate medium whose temperature is maintained substantially at the M_s of the workpiece, held in the medium until its temperature is uniform throughout — but not long enough to permit bainite to form — and then cooled in air. The treatment is frequently followed by tempering. (2) When the process is applied to carburized material, the controlling M_s temperature is that of the case. This variation of the process is frequently called mar-quenching.

martensite. A generic term for microstruc-tures formed by diffusionless phase trans-formation in which the parent and product phases have a specific crystallographic rela-tionship. Martensite is characterized by an acicular pattern in the microstructure in both ferrous and nonferrous alloys. In alloys where the solute atoms occupy interstitial positions in the martensitic lattice (such as carbon in iron), the structure is hard and highly strained; but where the solute atoms occupy substitutional positions (such as nickel in iron), the martensite is soft and ductile. The amount of high temperature phase that transforms to martensite on cool-ing depends to a large extent on the lowest temperature attained, there being a rather distinct beginning temperature (M_s) and a temperature at which the transformation is essentially complete (M_f).

martensite range. The temperature interval between M_s and M_f.

martensitic transformation. A reaction that takes place in some metals on cooling, with the formation of an acicular structure called *martensite*.

master alloy. An alloy, rich in one or more desired addition elements, that can be added to a melt to raise the percentage of a desired constituent.

matrix. (1) The principal phase or aggregate in which another constituent is embedded. (2) In electroforming, a form used as a cathode.

mean stress. (1) In fatigue loading, the alge-braic mean of the maximum and minimum stress in one cycle. Also called the steady stress component. (2) In any multiaxial stress system, the algebraic mean of three principal stresses; more correctly called mean normal stress.

mechanical metallurgy. The science and technology dealing with the behavior of metals when subjected to applied forces; often considered restricted to the plastic working or shaping of metals.

mechanical properties. The properties of a material that reveal its elastic and inelastic behavior when force is applied, thereby in-dicating its suitability for mechanical appli-cations; for example, modulus of elasticity, tensile strength, elongation, hardness, and

fatigue limit. Compare with *physical prop-erties*.

mechanical testing. Determination of *me-chanical properties*.

mechanical twin. A *twin* formed in a crystal by simple shear under external loading.

mechanical working. Subjecting metal to pressure, exerted by rolls, hammers or presses, in order to change the metal's shape or physical properties.

melting point. The temperature at which a pure metal, compound or eutectic changes from solid to liquid; the temperature at which the liquid and the solid are in equilib-rium.

melting range. The range of temperature over which an alloy other than a compound or eutectic changes from solid to liquid; the range of temperature from *solidus* to *liquidus* at any given composition on a constitution diagram.

mesh. The screen number of the finest screen of a specified standard screen scale through which almost all the particles of a powder sample will pass. Also called mesh size.

metal. (1) An opaque lustrous elemental chem-ical substance that is a good conductor of heat and electricity and, when polished, a good reflector of light. Most elemental met-als are malleable and ductile and are, in gen-eral, denser than the other elemental sub-stances. (2) As to structure, metals may be distinguished from non-metals by their atomic binding and electron availability. Metallic atoms tend to lose electrons from the outer shells, the positive ions thus formed being held together by the electron gas produced by the separation. The ability of these "free electrons" to carry an electric current, and the fact that this ability de-creases as temperature increases, establish the prime distinctions of a metallic solid. (3) From the chemical viewpoint, an elemental substance whose hydroxide is alkaline. (4) An *alloy*.

metal leaf. Thin metal sheet, usually thinner than foil, and traditionally produced by beating rather than by rolling.

metallic bond. The principal bond between metal atoms, which arises from the in-creased spatial extension of valence-elec-tron wave functions when an aggregate of metal atoms is brought close together. See *covalent bond, ionic bond.*

metallic glass. A noncrystalline metal or alloy, commonly produced by drastic supercooling of a molten alloy, by electrodeposition, or by vapor deposition. Also called amorphous alloy.

metallograph. An optical instrument designed for both visual observation and photomicrography of prepared surfaces of opaque materials at magnifications ranging from about 25 to about 2000 diameters. The instrument consists of a high-intensity illuminating source, a microscope and a camera bellows. On some instruments, provisions are made for examination of specimen surfaces with polarized light, phase contrast, oblique illumination, darkfield illumination and customary brightfield illumination.

metallography. The science dealing with the constitution and structure of metals and alloys as revealed by the unaided eye or by such tools as low-powered magnification, optical microscope, electron microscope and diffraction or x-ray techniques.

metallurgy. The science and technology of metals and alloys. Process metallurgy is concerned with the extraction of metals from their ores and with the refining of metals; physical metallurgy, with the physical and mechanical properties of metals as affected by composition, processing and environmental conditions; and mechanical metallurgy, with the response of metals to applied forces.

metastable. Refers to a state of pseudoequilibrium that has a higher free energy than the true equilibrium state.

M_f temperature. For any alloy system, the temperature at which martensite formation on cooling is essentially finished. See *transformation temperature* for the definition applicable to ferrous alloys.

microfissure. A crack of microscopic proportions.

micrograph. A graphic reproduction of the surface of a prepared specimen, usually etched, at a magnification greater than ten diameters. If produced by photographic means it is called a photomicrograph (not a microphotograph).

microhardness. The hardness of a material as determined by forcing an indenter such as a Vickers or Knoop indenter into the surface of a material under very light load; usually, the indentations are so small that they must be measured with a microscope. Capable of determining hardnesses of different microconstituents within a structure, or of measuring steep hardness gradients such as those encountered in case hardening.

microprobe. See preferred term, *electron beam microprobe analyzer*.

microradiography. The technique of passing x-rays through a thin section of an alloy in contact with a fine-grained photographic film and then viewing the radiograph at 50 to 100X to observe the distribution of alloying constituents and voids.

microscopic. Visible at magnifications greater than ten diameters.

microscopic stresses. Residual stresses that vary from tension to compression in a distance (presumably approximating the grain size) that is small compared to the gage length in ordinary strain measurements. They are not detectable by dissection methods, but can sometimes be measured from line shift or line broadening in an x-ray diffraction pattern.

microsegregation. Segregation within a grain, crystal or small particle. See *coring*.

microshrinkage. A casting imperfection, not detectable microscopically, consisting of interdendritic voids. Microshrinkage results from contraction during solidification where there is not an adequate opportunity to supply filler material to compensate for shrinkage. Alloys with a wide range in solidification temperature are particularly susceptible.

microstress. Same as *microscopic stress*.

microstructure. The structure of metals as revealed by microscopic examination of the etched surface of a polished specimen.

migration. Movement of entities (such as electrons, ions, atoms, molecules, vacancies and grain boundaries) from one place to another under the influence of a driving force (such as an electrical potential or a concentration gradient).

mild steel. *Carbon steel* with a maximum of about 0.25% C.

mill. (1) A factory where metals are hot worked, cold worked, or melted and cast into standard shapes suitable for secondary fabrication into commercial products. (2) A production line, usually of four or more *stands*, for hot rolling metal into standard shapes such as bar, rod, plate, sheet or strip. (3) A single machine for hot rolling, cold rolling or extruding metal; examples include blooming mill, *cluster mill*, *four-high mill*, and Sendzimer mill. (4) A shop term for milling cutter. (5) A machine or group of machines for grinding or crushing ores and other minerals.

mill product. Any commercial product of a *mill*.

mill scale. The heavy oxide layer formed during hot fabrication or heat treatment of metals.

minimum bend radius. The minimum radius over which metal products can be bent to a given angle without fracture.

mismatch. Error in register between forged surfaces formed by opposing dies.

misrun. A casting not fully formed, resulting from the metal solidifying before the mold is filled.

modification. Treatment of molten hypoeutectic (8 to 13% Si) or hypereutectic (13 to 19% Si) aluminum-silicon alloys to improve mechanical properties of the solid alloy by refinement of the size and distribution of the silicon phase. Involves additions of small percentages of sodium or strontium (hypoeutectic alloys) or phosphorus (hypereutectic alloys).

modulus of elasticity. A measure of the rigidity of metal. Ratio of stress, below the proportional limit, to corresponding strain. Specifically, the modulus obtained in tension or compression is Young's modulus, stretch modulus or modulus of extensibility; the modulus obtained in torsion or shear is modulus of rigidity, shear modulus or modulus of torsion; the modulus covering the ratio of the mean normal stress to the change in volume per unit volume is the bulk modulus. The tangent modulus and secant modulus are not restricted within the proportional limit; the former is the slope of the stress-strain curve at a specified point; the latter is the slope of a line from the origin to a specified point on the stress-strain curve. Also called elastic modulus and coefficient of elasticity.

modulus of rigidity. See *modulus of elasticity*.

modulus of rupture. Nominal stress at fracture in a bend test or torsion test. In bending, modulus of rupture is the bending moment at fracture divided by the section modulus. In torsion, modulus of rupture is the torque at fracture divided by the polar section modulus.

modulus of strain hardening. See preferred term, *rate of strain hardening*.

mold. (1) A form made of sand, metal or other material that contains the cavity into which molten metal is poured to produce a casting of definite shape and outline. (2) Same as *die*.

molding machine. A machine for making sand molds by mechanically compacting sand around a pattern.

molding press. A press used to form powder metallurgy *compacts*.

mold jacket. Wood or metal form that is slipped over a sand mold for support during pouring.

mold wash. An aqueous or alcoholic emulsion or suspension of various materials used to coat the surface of a mold cavity.

monotropism. The ability of a solid to exist in two or more forms (crystal structures), but in which one form is the stable modification at all temperatures and pressures. Ferrite and martensite are a monotropic pair below Ac_1 in steels, for example. May also be spelled monotrophism.

M_s temperature. For any alloy system, the temperature at which martensite starts to form on cooling. See *transformation temperature* for the definition applicable to ferrous alloys.

multiaxial stresses. Any stress state in which two or three principal stresses are not zero.

N

native metal. (1) Any deposit in the earth's crust consisting of uncombined metal. (2) The metal in such a deposit.

natural aging. Spontaneous aging of a supersaturated solid solution at room temperature. See *aging*, and compare with *artificial aging*.

natural strain. See *strain*.

necking. (1) Reducing the cross-sectional area of metal in a localized area by stretching. (2) Reducing the diameter of a portion of the length of a cylindrical shell or tube.

necking down. Localized reduction in area of a specimen during tensile deformation.

necking strain. Same as *uniform strain*.

network structure. A structure in which one constituent occurs primarily at the grain boundaries, thus partially or completely enveloping the grains of the other constituents.

Neumann band. *Mechanical twin* in ferrite.

noble metal. (1) A metal whose potential is highly positive relative to the hydrogen electrode. (2) A metal with marked resistance to chemical reaction, particularly to oxidation and to solution by inorganic acids. The term as often used is synonymous with *precious metal*. Contrast with *base metal* (4).

nominal stress. See *stress*.

nondestructive inspection. Inspection by methods that do not destroy the part nor impair its serviceability.

nondestructive testing. Same as *nondestructive inspection*, but implying use of a method in which the part is stimulated and its response measured quantitatively or semiquantitatively.

nonmetallic inclusions. See *inclusions*.

normalizing. Heating a ferrous alloy to a suitable temperature above the transformation range and then cooling in air to a temperature substantially below the transformation range.

normal stress. See *stress*.

notch acuity. Relates to the severity of the stress concentration produced by a given notch in a particular structure. If the depth of the notch is very small compared with the width (or diameter) of the narrowest cross section, the acuity may be expressed as the ratio of the notch depth to the notch root radius. Otherwise, the acuity is defined as the ratio of one-half the width (or diameter) of the narrowest cross seciton to the notch root radius.

notch brittleness. Susceptibility of a material to brittle fracture at points of stress concentration. For example, in a notch tensile test, the material is said to be notch brittle if the *notch strength* is less than the tensile strength of an unnotched specimen. Otherwise, it is said to be notch ductile.

notch depth. The distance from the surface of a test specimen to the bottom of the notch. In a cylindrical test specimen, the percentage of the original cross-sectional area removed by machining an annular groove.

notch ductile. See *notch brittleness*.

notch ductility. The percentage reduction in area after complete separation of the metal in a tensile test of a notched specimen.

notch rupture strength. The ratio of applied load to original area of the minimum cross section in a stress-rupture test of a notched specimen.

notch sensitivity. A measure of the reduction in strength of a metal caused by the presence of stress concentration. Values can be obtained from static, impact or fatigue tests.

notch sharpness. See *notch acuity*.

notch strength. The maximum load on a notched tensile-test specimen divided by the minimum cross-sectional area (the area at the root of the notch). Also called notch tensile strength.

nucleation. The initiation of a phase transformation at discrete sites, the new phase growing on nuclei. See *nucleus*, (1).

nucleus. (1) The first structurally stable particle capable of initiating recrystallization of a phase or the growth of a new phase, and possessing an interface with the parent matrix. The term is also applied to a foreign particle that initiates such action. (2) The heavy central core of an atom, in which most of the mass and the total positive electric charge are concentrated.

O

octahedral plane. In cubic crystals, a plane with equal intercepts on all three axes.

offset. The distance along the strain coordinate between the initial portion of a stress-strain curve and a parallel line that intersects the stress-strain curve at a value of stress that is used as a measure of the *yield strength*. It is used for materials that have no obvious *yield point*. A value of 0.2% is commonly used.

open die forging. Same as *flat die forging*.

open dies. See *closed dies*.

operating stress. The stress to which a structural unit is subjected in service.

orange peel. A surface roughening in the form of a pebble-grained pattern where a metal of unusually coarse grain is stressed beyond its elastic limit. Also called pebbles and alligator skin.

ordering. Forming a *superlattice*.

orientation. Arrangement in space of the axes of a crystal with respect to a chosen reference or coordinate system. See also *preferred orientation*.

overaging. Aging under conditions of time and temperature greater than those required to obtain maximum change in a certain property, so that the property is altered in the direction of the initial value. See *aging*.

overheating. Heating a metal or alloy to such a high temperature that its properties are impaired. When the original properties cannot be restored by further heat treating, by mechanical working or by a combination of working and heat treating, the overheating is known as *burning*.

overstressing. (1) In fatigue testing, cycling at a stress level higher than that used at the end of the test.

oxidation. (1) A reaction in which there is an increase in valence resulting from a loss of electrons. Contrast with *reduction*. (2) A corrosion reaction in which the corroded metal forms an oxide; usually applied to reaction with a gas containing elemental oxygen, such as air.

oxidized surface (on steel). Surface having a thin, tightly adhering, oxidized skin (from straw to blue in color), extending in from the edge of a coil or sheet. Sometimes called annealing border.

oxidizing agent. A compound that causes oxidation, thereby itself becoming reduced.

oxidizing flame. A gas flame produced with excess oxygen in the inner flame.

P

pack rolling. Hot rolling a pack of two or more sheets of metal; scale prevents their being welded together.

pancake forging. A rough forged shape, usually flat, that may be obtained quickly with a minimum of tooling. It usually requires considerable machining to attain finish size.

partial annealing. An imprecise term used to denote a treatment given cold worked material to reduce the strength to a controlled level or to effect stress relief. To be meaningful, the type of material, the degree of cold work, and the time-temperature schedule must be stated.

parting line. (1) The intersection of the parting plane of a casting mold or the parting plane between forging dies with the mold or die cavity. (2) A raised line or projection on the surface of a casting or forging that corresponds to said intersection.

parting plane. (1) In forging, the dividing plane between dies. Contrast with *forging plane*. (2) In casting, the dividing plane between mold halves.

pass. (1) A single transfer of metal through a *stand* of rolls. (2) The open space between two grooved rolls through which metal is processed. (3) The weld metal deposited in one trip along the axis of a weld.

passivation. The changing of a chemically active surface of a metal to a much less reactive state. Contrast with *activation*.

passivity. A condition in which a piece of metal, because of an impervious covering of oxide or other compound, has a potential much more positive than when the metal is in the active state.

patenting. In wiremaking, a heat treatment applied to medium-carbon or high-carbon steel before the drawing of wire or between drafts. This process consists of heating to a temperature above the transformation range and then cooling to a temperature below Ae_1 in air or in a bath of molten lead or salt.

pattern. (1) A form of wood, metal or other material, around which molding material is placed to make a mold for casting metals. (2) A full-scale reproduction of a part used as a guide in cutting.

pearlite. A metastable lamellar aggregate of ferrite and cementite resulting from the transformation of austenite at temperatures above the bainite range.

penetrant. A liquid with low surface tension used in *liquid penetrant inspection* to flow into surface openings of parts being inspected.

penetrant inspection. See preferred term, *liquid penetrant inspection*.

penetration. (1) In founding, an *imperfection* on a casting surface caused by metal running into voids between sand grains; usually referred to as metal penetration. (2) In welding, the distance from the original surface of the base metal to that point at which fusion ceased. See *joint penetration*.

penetration hardness. Same as *indentation hardness*.

peritectic. An isothermal reversible reaction in which a liquid phase reacts with a solid phase to produce a single (and different) solid phase on cooling.

peritectoid. An isothermal reversible reaction in which a solid phase reacts with a second solid phase to produce a single (and different) solid phase on cooling.

permanent mold. A metal, graphite or ceramic mold (other than an ingot mold) of two or more parts that is used repeatedly for the production of many *castings* of the same form. Liquid metal is poured in by gravity.

permanent set. Plastic deformation that remains upon releasing the stress that produces the deformation.

pH. The negative logarithm of the hydrogen ion activity; it denotes the degree of acidity or basicity of a solution. At 25 °C (76 °F), 7.0 is the neutral value. Decreasing values below 7.0 indicate increasing acidity; increasing values above 7.0, increasing basicity.

phase. A physically homogeneous and distinct portion of a material system.

phase diagram. Same as *constitution diagram*.

photomacrograph. See *macrograph*.

photomicrograph. See *micrograph*.

physical metallurgy. The science and technology dealing with the properties of metals and alloys, and of the effects of composition, processing and environment on those properties.

physical properties. Properties of a metal or alloy that are relatively insensitive to structure and can be measured without the application of force; for example, density, electrical conductivity, coefficient of thermal expansion, magnetic permeability and lattice parameter. Does not include chemical reactivity. Compare with *mechanical properties*.

physical testing. Determination of *physical properties*.

pickling. Removing surface oxides from metals by chemical or electrochemical reaction.

pig. A metal casting used in remelting.

pig iron. (1) High-carbon iron made by reduction of iron ore in the blast furnace. (2) Cast iron in the form of *pigs*.

pinhole porosity. Porosity consisting of numerous small gas holes distributed throughout the metal; found in weld metal, castings or electrodeposited metal.

pipe. (1) The central cavity formed by contraction in metal, especially ingots, during solidification. See accompanying sketch. (2) An imperfection in wrought or cast products resulting from such a cavity. (3) A tubular metal product, cast or wrought.

pitting. Forming small sharp cavities in a metal surface by nonuniform electrodeposition or by corrosion.

plane strain. The stress condition in linear elastic fracture mechanics in which there is zero strain in a direction normal to both the axis of applied tensile stress and the direction of crack growth (i.e., parallel to the crack front); most nearly achieved in loading thick plates along a direction parallel to the plate surface. Under plane-strain conditions, the plane of fracture instability is normal to the axis of the principal tensile stress.

plane stress. The stress condition in linear elastic fracture mechanics in which the stress in the thickness direction is zero; most nearly achieved in loading very thin sheet along a direction parallel to the surface of the sheet. Under plane-stress conditions, the plane of fracture instability is inclined 45° to the axis of the principal tensile stress.

plastic deformation. Deformation that does or will remain permanent after removal of the load that caused it.

plastic flow. Same as *plastic deformation*.

plasticity. The ability of a metal to deform nonelastically without rupture.

plate. A flat-rolled metal product of some minimum thickness and width arbitrarily dependent on the type of metal.

platen. (1) Face of a bolster, slide or ram to which a tool assembly is attached. (2) A part of a resistance welding, mechanical testing or other machine with a flat surface to which dies, fixtures, backups or electrode holders are attached and that transmits pressure or force.

plates. Flat particles of metal powder having considerable thickness.

plating. Forming an adherent layer of metal upon an object; often used as a shop term for *electroplating*.

plug. (1) A rod or mandrel over which a pierced tube is forced. (2) A rod or mandrel that fills a tube as it is drawn through a die. (3) A punch or mandrel over which a cup is drawn. (4) A protruding portion of a die impression for forming a corresponding recess in the forging. (5) A false bottom in a die. Also called a "peg".

pointing. (1) Reducing the diameter of wire, rod or tubing over a short length at the end by swaging or hammer forging, turning or squeezing to facilitate entry into a drawing die and gripping in the drawhead. (2) The operation in automatic machines of chamfering or rounding the threaded end or the head of a bolt.

Poisson's ratio. The absolute value of the ratio of the transverse strain to the corresponding axial strain, in a body subjected to uniaxial stress; usually applied to elastic conditions.

polarization. A change in the potential of an electrode during electrolysis, such that the potential of an anode becomes more noble and that of a cathode more active than their respective reversible potentials. Often accomplished by the formation of a film on the electrode surface.

polishing. Smoothing metal surfaces, often to a high luster, by rubbing the surface with a fine abrasive, usually contained in a cloth or other soft lap. May be extended to include electropolishing.

polycrystalline. Pertaining to a solid composed of many crystals.

polymorphism. A general term for the ability of a solid to exist in more than one form. In metals, alloys and similar substances, this usually means the ability to exist in two or more crystal structures, or an amorphous state and at least one crystal structure. See also *allotropy, monotropism.*

pores. (1) Small voids in the body of a metal. (2) Minute cavities in a powder metallurgy compact, sometimes intentional. (3) Minute perforations in an electroplated coating.

porosity. Fine holes or pores within a metal.

pot annealing. Same as *box annealing.*

pouring. Transferring molten metal from a furnace or a ladle to a mold.

pouring basin. A basin on top of a mold to receive the molten metal before it enters the sprue or downgate.

powder metallurgy. The art of producing metal powders and of utilizing metal powders for the production of massive materials and shaped objects.

precious metal. One of the relatively scarce and valuable metals: gold, silver and the platinum-group metals.

precipitation hardening. Hardening caused by the precipitation of a constituent from a supersaturated solid solution. See also *age hardening* and *aging.*

precipitation heat treatment. *Artificial aging* in which a constituent precipitates from a supersaturated solid solution.

precision. The closeness of approach of each of a number of similar measurements to the arithmetic mean, the sources of error not necessarily being considered critically. *Accuracy* demands precision, but precision does not ensure accuracy.

precision casting. A metal casting of repro-

ducible accurate dimensions regardless of how it is made.

preferred orientation. A condition of a polycrystalline aggregate in which the crystal orientations are not random, but rather exhibit a tendency for alignment with a specific direction in the bulk material, commonly related to the direction of working; also called *texture.*

preforming. (1) The initial pressing of a metal powder to form a compact that is to be subjected to a subsequent pressing operation other than coining or sizing. Also, the preliminary shaping of a refractory metal compact after presintering and before the final sintering. (2) Preliminary forming operations, especially for impression die forging.

preheating. Heating before some further thermal or mechanical treatment. For tool steel, heating to an intermediate temperature immediately before final austenitizing. For some nonferrous alloys, heating to a high temperature for a long time, in order to homogenize the structure before working. In welding and related processes, heating to an intermediate temperature for a short time immediately before welding, brazing, soldering, cutting or thermal spraying.

press. A machine tool having a stationary bed and a slide or ram that has reciprocating motion at right angles to the bed surface, the slide being guided in the frame of the machine.

pressing. (1) In metalworking, the product or process of shallow drawing sheet or plate. (2) Forming a powder-metal part with compressive force.

pressure casting. (1) Making castings with pressure on the molten or plastic metal, as in injection molding, *die casting*, centrifugal casting, and cold chamber pressure casting. (2) A casting made with pressure applied to the molten or plastic metal.

primary creep. See *creep.*

primary crystal. The first type of crystal that separates from a melt on cooling.

primary metal. Metal extracted from minerals and free of reclaimed metal scrap. Compare with *native metal.*

primary mill. A mill for rolling ingots or the rolled products of ingots to blooms, billets or slabs. This type of mill is often called a *blooming mill* and sometimes a *cogging mill.*

process annealing. An imprecise term denoting various treatments used to improve workability. For the term to be meaningful, the condition of the material and the time-temperature cycle used must be stated.

process metallurgy. The science and technology of winning metals from their ores and purifying metals; sometimes referred to as chemical metallurgy. Its two chief branches are *extractive metallurgy* and refining.

progressive aging. Aging by increasing the temperature in steps or continuously during the aging cycle. See *aging* and compare with *interrupted aging* and *step aging*.

proof. Any reproduction of a die impression in any material, frequently a lead or plaster cast.

proofload. A predetermined load, generally some multiple of the service load, to which a specimen or structure is submitted before acceptance for use.

proof stress. (1) The stress that will cause a specified small permanent set in a material. (2) A specified stress to be applied to a member or structure to indicate its ability to withstand service loads.

proportional limit. The maximum stress at which strain remains directly proportional to stress.

pseudobinary system. (1) A three-component or ternary alloy system in which an intermediate phase acts as a component. (2) A vertical section through a ternary diagram.

punch. (1) The movable tool that forces material into the die in powder molding and most forming operations. (2) The movable die in a trimming press or a forging machine. (3) The tool that forces the stock through the die in rod and tube extrusion and forms the internal surface in can or cup extrusion.

punching. Producing a hole by die shearing, in which the shape of the hole is controlled by the shape of the punch and its mating die; piercing. Multiple punching of small holes is called perforating.

push bench. Equipment used for drawing moderately heavy-gage tubes by cupping sheet and forcing it through a die by pressure exerted against the inside bottom of the cup.

pusher furnace. A type of continuous furnace in which parts to be heated are periodically charged into the furnace in containers, which are pushed along the hearth against a line of previously charged containers thus advancing the containers toward the discharge end of the furnace, where they are removed.

pyrometer. A device for measuring temperatures above the range of liquid thermometers.

Q

quarter hard. A *temper* of nonferrous alloys and some ferrous alloys characterized by tensile strength about midway between that of *dead soft* and *half hard* tempers.

quality. (1) The totality of features and characteristics of a product or service that bear on its ability to satisfy a given need (fitness-for-use concept of quality). (2) Degree of excellence of a product or service (comparative concept). Often determined subjectively by comparison against an ideal standard or against similar products or services available from other sources. (3) A quantitative evaluation of the features and characteristics of a product or service (quantitative concept).

quench-age embrittlement. Embrittlement of low-carbon steel evidenced by a loss of ductility on aging at room temperature following rapid cooling from a temperature below the lower critical temperature.

quench aging. Aging induced by rapid cooling after *solution heat treatment*.

quench annealing. Annealing an austenitic ferrous alloy by *solution heat treatment* followed by rapid quenching.

quench cracking. Fracture of a metal during quenching from elevated temperature. Most frequently observed in hardened carbon steel, alloy steel or tool steel parts of high hardness and low toughness. Cracks often emanate from fillets, holes, corners or other stress raisers and result from high stresses due to the volume changes accompanying transformation to martensite.

quench hardening. (1) Hardening suitable alpha-beta alloys (most often certain copper or titanium alloys) by solution treating and quenching to develop a martensite-like structure. (2) In ferrous alloys, hardening by austenitizing and then cooling at a rate such that a substantial amount of austenite transforms to martensite.

quenching. Rapid cooling. When applicable, the following more specific terms should be used: direct quenching, fog quenching, hot quenching, *interrupted quenching, selective quenching*, spray quenching and *time quenching*.

R

radiograph. A photographic shadow image resulting from uneven absorption of penetrating radiation in a test object.

radiography. A method of nondestructive inspection in which a test object is exposed to a beam of x-rays or gamma rays and the resulting shadow image of the object is recorded on photographic film placed behind the object. Internal discontinuities are detected by observing and interpreting variations in the image caused by differences in thickness, density or absorption within the test object. Variations of radiography include electron radiography, *fluoroscopy*, neutron radiography.

ram. The moving member of a hammer, machine, or press to which a tool is fastened.

ramming. Packing sand, refractory or other material into a compact mass.

range. In inspection, the difference between the highest and lowest values of a given quality characteristic within a single *sample*.

rare earth metal. One of the group of 15 chemically similar metals with atomic numbers 57 through 71, commonly referred to as the lanthanides.

rate of strain hardening. Rate of change of true *stress* with respect to true *strain* in the plastic range.

recovery. (1) Reduction or removal of work-hardening effects, without motion of large-angle grain boundaries. (2) The proportion of the desired component obtained by processing an ore, usually expressed as a percentage.

recrystallization. (1) The formation of a new, strain-free grain structure from that existing in cold worked metal, usually accomplished by heating. (2) The change from one crystal structure to another, as occurs on heating or cooling through a critical temperature.

recrystallization annealing. Annealing cold worked metal to produce a new grain structure without phase change.

recrystallization temperature. The approximate minimum temperature at which complete recrystallization of a cold worked metal occurs within a specified time.

redrawing. Drawing metal after a previous cupping or drawing operation.

reduction. (1) In cupping and deep drawing, a measure of the percentage decrease from blank diameter to cup diameter, or of diameter reduction in redraws. (2) In forging, rolling and drawing, either the ratio of the original to final cross-sectional area or the percentage decrease in cross-sectional area. (3) A reaction in which there is a decrease in valence resulting from a gain in electrons. Contrast with *oxidation*.

reduction in area. (1) Commonly, the difference, expressed as a percentage of original area, between the original cross-sectional area of a tensile test specimen and the minimum cross-sectional area measured after complete separation. (2) The difference, expressed as a percentage of original area, between original cross-sectional area and that after straining the specimen.

refractory. (1) A material of very high melting point with properties that make it suitable for such uses as furnace linings and kiln construction. (2) The quality of resisting heat.

refractory alloy. (1) A heat-resistant alloy. (2) An alloy having an extremely high melting point. See *refractory metal*. (3) An alloy difficult to work at elevated temperatures.

refractory metal. A metal having an extremely high melting point; for example, tungsten, molybdenum, tantalum, niobium (columbium), chromium, vanadium and rhenium. In the broad sense, it refers to metals having melting points above the range of iron, cobalt and nickel.

reliability. A quantitative measure of the ability of a product or service to fulfill its intended function for a specified period of time.

re-pressing. The application of pressure to a previously pressed and sintered powder metallurgy compact, usually for the purpose of improving some physical property.

residual elements. Elements present in an alloy in small quantities, but not added intentionally.

residual stress. Stress present in a body that is free of external forces or thermal gradients.

resilience. (1) The amount of energy per unit volume released upon unloading. (2) The capacity of a metal, by virtue of high yield strength and low elastic modulus, to exhibit considerable elastic recovery upon release of load.

resolution. The ability to separate closely related items of data or physical features using a given test method; also a quantitative measure of the degree to which they can be discriminated.

restraint. Any external mechanical force that prevents a part from moving to accommodate changes in dimensions due to thermal expansion or contraction. Often applied to weldments made while clamped in a fixture. Compare with *constraint*.

restriking. (1) Striking a trimmed but slightly misaligned or otherwise faulty forging one or more blows to improve alignment, improve surface, maintain close tolerance, increase hardness or effect other improvements. (2) A sizing operation in which coining or stretching is utilized to correct or alter profiles and to counteract distortion.

reverse drawing. *Redrawing* in a direction opposite to that of the original drawing.

reverse redrawing. A second drawing operation in a direction opposite to that of the original drawing.

rimmed steel. A low-carbon steel containing sufficient iron oxide to give a continuous evolution of carbon monoxide while the ingot is solidifying, resulting in a case or rim of metal virtually free of voids. Sheet and strip products made from the ingot have very good surface quality.

ringing. The audible or ultrasonic tone produced in a mechanical part by shock, and having the natural frequency or frequencies of the part. The quality, amplitude or decay rate of the tone may sometimes be used to indicate quality or soundness. See also *sonic testing, ultrasonic testing*.

riser. A reservoir of molten metal connected to the casting to provide additional metal to the casting, required as the result of shrinkage before and during solidification.

river pattern. A term used in fractography to describe a characteristic pattern of cleavage steps that run parallel to the local direction of crack propagation on the fracture surface of grains that have separated by cleavage.

riveting. Joining of two or more members of a structure by means of metal rivets, the unheaded end being upset after the rivet is in place.

rock candy fracture. A fracture that exhibits separated-grain facets, most often used to describe intergranular fractures in large-grained metals.

Rockwell hardness test. An indentation hardness test based on the depth of penetration of a specified penetrator into the specimen under certain arbitrarily fixed conditions.

rod mill. (1) A *hot mill* for rolling rod. (2) A mill for fine grinding, somewhat similar to a *ball mill*, but employing long steel rods instead of balls to effect the grinding.

roll flattening. Flattening of sheets that have been rolled in packs by passing them separately through a two-high cold mill, there being virtually no deformation.

rolling. Reducing the cross-sectional area of metal stock, or otherwise shaping metal products, through the use of rotating rolls.

rolling mills. Machines used to decrease the cross-sectional area of metal stock and produce certain desired shapes as the metal passes between rotating rolls mounted in a framework comprising a basic unit called a *stand*. Cylindrical rolls produce flat shapes; grooved rolls produce rounds, squares and structural shapes. Among rolling mills may be listed the *billet mill, blooming mill,* breakdown mill, plate mill, sheet mill, *slabbing mill,* strip mill and temper mill.

roll straightening. Straightening of metal stock of various shapes by (1) passing it through a series of staggered rolls, the rolls usually being in horizontal and vertical planes; or (2) by reeling in two-roll straightening machines.

rotary forging. A process subjecting the workpiece to pressing between a flat anvil and a swiveling die with a conical working face; the platens move toward each other during forging.

rotary furnace. A circular furnace constructed so that the hearth and workpieces rotate around the axis of the furnace during heating.

rotary swager. A swaging machine consisting of a power-driven ring that revolves at high speed causing rollers to engage cam surfaces and force the dies to deliver hammerlike blows upon the work at high frequency. Both straight and tapered sections can be produced.

roughing stand. The first stand of rolls through which the reheated billet passes, or the last stand in front of the finishing rolls.

runner. (1) A channel through which molten metal flows from one receptacle to another. (2) The portion of the gate assembly of a casting that connects the sprue with the gate(s). (3) Parts of patterns and finished castings corresponding to the portion of the gate assembly described in definition (2).

runner box. A distribution box that divides molten metal into several streams before it enters the mold cavity.

rust. A corrosion product consisting of hydrated oxides of iron. Applied only to ferrous alloys.

S

sacrificial protection. Reducing the extent of corrosion of a metal in an electrolyte by coupling it to another metal that is electrochemically more active in the environment.

sample. One or more units of product (or a relatively small quantity of a bulk material) that is withdrawn from a *lot* or process stream, and that is tested or inspected to provide information about the properties, dimensions or other quality characteristics of the lot or process stream. Not to be confused with *specimen*.

sand. A granular material, naturally or artificially produced by the disintegration or crushing of rocks or mineral deposits. In casting, the term denotes an aggregate whose individual particle (grain) size is 0.06 to 2mm (1/400 to 1/12 in.) in diameter, and largely free of finer constituents such as silt and clay, which are often present in natural sand deposits. The most commonly used foundry sand is silica; however, zircon, olivine, chromite, alumina and other crushed ceramics are used for special applications.

scaling. (1) Forming a thick layer of oxidation products on metals at high temperature. (2) Depositing water-insoluble constituents on a metal surface, as in cooling tubes and water boilers.

scalping. Removing surface layers from ingots, billets or slabs. See *die scalping*.

scrap. (1) Products that are discarded because they are defective or otherwise unsuitable for sale. (2) Discarded metallic material, from whatever source, that may be reclaimed through melting and refining.

screw dislocation. See *dislocation*.

screw press. A press whose slide is operated by a screw rather than by a crank or other means.

scuffing. A form of *adhesive wear* that produces superficial scratches or a high polish on the rubbing surfaces. It is observed most often on inadequately lubricated parts.

sealing. (1) Closing pores in anodic coatings to render them less absorbent. (2) Plugging leaks in a casting by introducing thermosetting plastics into porous areas and subsequently setting the plastic with heat.

selective leaching. Corrosion in which one element is preferentially removed from an alloy, leaving a residue (often porous) of the elements that are more resistant to the particular environment. See also *decarburization, denickelification, dezincification, graphitic corrosion*.

selective quenching. Quenching only certain portions of an object.

self-diffusion. Thermally activated movement of an atom to a new site in a crystal of its own species, as, for example, a copper atom within a crystal of copper.

self-hardening steel. See preferred term, *air-hardening steel*.

semiconductor. An electronic conductor whose resistivity at room temperature is in the range of 10^{-7} to 1 Ω·m and in which the conductivity increases with increasing temperature over some temperature range.

semikilled steel. Steel that is incompletely deoxidized and contains sufficient dissolved oxygen to react with the carbon to form carbon monoxide to offset solidification shrinkage.

semipermanent mold. A permanent mold in which sand cores are used.

sensitivity. The smallest difference in values that can be detected reliably with a given measuring instrument.

sensitization. In austenitic stainless steels, the precipitation of chromium carbides, usually at grain boundaries, upon exposure to temperatures of about 550 to 850 °C (1000 to 1550 °F), leaving the grain boundaries depleted of chromium and therefore susceptible to preferential attack by a corroding (oxidizing) medium.

shakeout. Removing castings from a sand mold.

shaping. Producing flat surfaces using single-point tools. The work is held in a vise or fixture, or clamped directly to the table. The ram supporting the tool is reciprocated in a linear motion past the work.

shear. (1) That type of force that causes or tends to cause two contiguous parts of the same body to slide relative to each other in a direction parallel to their plane of contact. (2) A type of cutting tool with which a material in the form of wire, sheet, plate or rod is cut between two opposing blades. (3) The type of cutting action produced by rake so that the direction of chip flow is other than at right angles to the cutting edge.

shear fracture. A ductile fracture in which a crystal (or a polycrystalline mass) has separated by sliding or tearing under the action of shear stresses.

shearing strain. See *strain.*

shear lip. A narrow, slanting ridge along the edge of a fracture surface. The term sometimes also denotes a narrow, often crescent-shaped, fibrous region at the edge of a fracture that is otherwise of the cleavage type, even though this fibrous region is in the same plane as the rest of the fracture surface.

shear modulus. See *modulus of elasticity.*

shear strain. Same as shearing strain; see *strain.*

shear strength. The stress required to produce fracture in the plane of cross section, the conditions of loading being such that the directions of force and of resistance are parallel and opposite although their paths are offset a specified minimum amount. The maximum load divided by the original cross-sectional area of a section separated by shear.

shear stress. See *stress.*

sheet. A flat-rolled metal product of some maximum thickness and minimum width arbitrarily dependent on the type of metal. It is thinner than plate, and has a width-to-thickness ratio greater than about 50.

shell. (1) A hollow structure or vessel. (2) An article formed by deep drawing. (3) The metal sleeve remaining when a billet is extruded with a dummy block of somewhat smaller diameter. (4) In shell molding, a hard layer of sand and thermosetting plastic or resin formed over a pattern and used as the mold wall. (5) A tubular casting used in making seamless drawn tube. (6) A pierced forging.

shell molding. Forming a mold from thermosetting resin-bonded sand mixtures brought in contact with preheated (150 to 260 °C, or 300 to 500 °F) metal patterns, resulting in a firm shell with a cavity corresponding to the outline of the pattern. Also called Croning process.

shrinkage. See *casting shrinkage.*

shrinkage cavity. A void left in cast metals as a result of solidification shrinkage. See *casting shrinkage.*

shrinkage cracks. Hot tears associated with shrinkage cavities.

silky fracture. A metal fracture in which the broken metal surface has a fine texture, usu-ally dull in appearance. Characteristic of tough and strong metals. Contrast with *crystalline fracture, granular fracture.*

silver soldering. Nonpreferred term used to denote brazing with a silverbase filler metal; preferred terms furnace brazing, induction brazing, and torch brazing.

single-stand mill. A rolling mill of such design that the product contacts only two rolls at a given moment. Contrast with *tandem mill.*

sinkhead. Same as *riser.*

sinter. To heat a mass of fine particles for a prolonged time below the melting point, usually to cause agglomeration.

sintering. The bonding of adjacent surfaces in a mass of particles by molecular or atomic attraction on heating at high temperatures below the melting temperature of any constituent in the material. Sintering strengthens a powder mass and normally produces densification and, in powdered metals, recrystallization.

size effect. Effect of the dimensions of a piece of metal upon its mechanical and other properties and upon manufacturing variables such as forging reduction and heat treatment. In general, the mechanical properties are lower for a larger size.

skim gate. A gating arrangement designed to prevent the passage of slag and other undesirable materials into a casting.

skin pass. See *temper rolling.*

slab. A piece of metal, intermediate between ingot and plate, with the width at least twice the thickness.

slabbing mill. A primary mill that produces slabs.

slack quenching. The incomplete hardening of steel due to quenching from the austenitizing temperature at a rate slower than the critical cooling rate for the particular steel, resulting in the formation of one or more transformation products in addition to martensite.

slag. A nonmetallic product resulting from the mutual dissolution of flux and nonmetallic impurities in smelting, refining, and certain welding operations.

slag inclusion. Slag or dross entrapped in a metal.

slip. Plastic deformation by the irreversible shear displacement (translation) of one part of a crystal relative to another in a definite crystallographic direction and usually on a specific crystallographic plane. Sometimes called glide.

slip band. A group of parallel slip lines so closely spaced as to appear as a single line when observed under an optical microscope. See *slip line*.

slip direction. The crystallographic direction in which the translation of slip takes place.

slip line. The trace of the slip plane on the viewing surface; the trace is (usually) observable only if the surface has been polished before deformation. The usual observation on metal crystals (under the light microscope) is of a cluster of slip lines known as a slip band.

slip plane. The crystallographic plane in which slip occurs in a crystal.

slush casting. A hollow casting usually made of an alloy with a low but wide melting temperature range. After the desired thickness of metal has solidified in the mold, the remaining liquid is poured out.

smith forging. Manual forging with flat or simple-shaped dies that never completely confine the work.

S-N diagram. A plot showing the relationship of stress, *S*, and the number of cycles, *N*, before fracture in fatigue testing.

soaking. Prolonged holding at a selected temperature to effect homogenization of structure or composition.

soft temper. Same as *dead soft* temper.

soldering. A group of processes that join metals by heating them to a suitable temperature below the solidus of the base metals and applying a filler metal having a liquidus not exceeding 450 °C (840 °F). Molten filler metal is distributed between the closely fitted surfaces of the joint by capillary action.

solidification. The change in state from liquid to solid on cooling through the melting temperature or melting range.

solidification shrinkage. See *casting shrinkage*.

solid shrinkage. See *casting shrinkage*.

solid solution. A single solid homogeneous crystalline phase containing two or more chemical species.

solid state welding. A group of welding processes that join metals at temperatures essentially below the melting point of the base materials, without the addition of a brazing or soldering filler metal. Pressure may or may not be applied to the joint.

solidus. In a constitution or equilibrium diagram, the locus of points representing the temperatures at which various compositions finish freezing on cooling or begin to melt on heating. See also *liquidus*.

solute. The component of either a liquid or solid solution that is present to a lesser or minor extent; the component that is dissolved in the *solvent*.

solution heat treatment. Heating an alloy to a suitable temperature, holding at that temperature long enough to cause one or more constituents to enter into solid solution, and then cooling rapidly enough to hold these constituents in solution.

solvent. The component of either a liquid or solid solution that is present to a greater or major extent; the component that dissolves the *solute*.

solvus. In a constitution or equilibrium diagram, the locus of points representing the temperatures at which the various compositions of the solid phases coexist with other solid phases, that is, the limits of solid solubility.

sonic testing. Any inspection method that uses sound waves (in the audible frequency range, about 20 to 20 000 Hz) to induce a response from a part or test specimen. Sometimes used, but inadvisedly, as a synonym for *ultrasonic testing*.

space lattice. A regular, periodic array of points (lattice points) in space that represents the locations of atoms of the same kind in a perfect crystal. The concept may be extended, where appropriate, to crystalline compounds and other substances, in which case the lattice points often represent locations of groups of atoms of identical composition, arrangement and orientation.

specimen. A test object, often of standard dimensions or configuration, that is used for destructive or nondestructive testing. One or more specimens may be cut from each unit of a *sample*.

spheroidizing. Heating and cooling to produce a spheroidal or globular form of carbide in steel. Spheroidizing methods frequently used are

1 Prolonged holding at a temperature just below A_1.

2 Heating and cooling alternately between temperatures that are just above and just below A_1.

3 Heating to a temperature above A_1 or A_3 and then cooling very slowly in the furnace or holding at a temperature just below A_1.

4 Cooling at a suitable rate from the minimum temperature at which all carbide is dissolved, to prevent the reformation of a carbide network, and then reheating in ac-

cordance with method 1 or 2 above. (Applicable to hypereutectoid steel containing a carbide network.)

spinning. Forming a seamless hollow metal part by forcing a rotating blank to conform to a shaped mandrel that rotates concentrically with the blank. In the usual application, a flat-rolled metal blank is forced against the mandrel by a blunt, rounded tool; however, other stock (notably welded or seamless tubing) can be formed, and sometimes the working end of the tool is a roller.

springback. (1) The elastic recovery of metal after cold forming. (2) The degree to which metal tends to return to its original shape or contour after undergoing a forming operation. (3) In flash, upset or pressure welding, the deflection in the welding machine caused by the upset pressure.

spring temper. A *temper* of nonferrous alloys and some ferrous alloys characterized by tensile strength and hardness about two-thirds of the way from *full hard* to *extra spring* temper.

sprue. (1) The mold channel that connects the *pouring basin* with the runner or, in the absence of a pouring basin, directly into which molten metal is poured. Sometimes referred to as downsprue or downgate. (2) Sometimes used to mean all gates, risers, runners and similar scrap that are removed from castings after shakeout.

stabilizing treatment. (1) Before finishing to final dimensions, repeatedly heating a ferrous or nonferrous part to or slightly above its normal operating temperature and then cooling to room temperature to ensure dimensional stability in service. (2) Transforming retained austenite in quenched hardenable steels, usually by *cold treatment*. (3) Heating a solution-treated stabilized grade of austenitic stainless steel to 870 to 900 °C (1600 to 1650 °F) to precipitate all carbon as TiC, NbC, or TaC so that *sensitization* is avoided on subsequent exposure to elevated temperature.

stainless steel. Any of several steels containing 12 to 30% chromium as the principle alloying element; they usually exhibit *passivity* in aqueous environments.

stamping. A general term covering almost all press operations. It includes blanking, shearing, hot or cold forming, drawing, bending, coining.

stand. A piece of rolling mill equipment containing one set of working rolls. In the usual sense, any pass of a continuous, looping or cross-country hot rolling mill.

state of strain. A complete description of the deformation within a homogeneously deformed volume or at a point. The description requires, in general, the knowledge of six independent components of *strain*.

state of stress. A complete description of the stresses within a homogeneously stressed volume or at a point. The description requires, in general, the knowledge of six independent components of *stress*.

static fatigue. A term sometimes used to identify a form of hydrogen embrittlement in which a metal appears to fracture spontaneously under a steady stress less than the yield stress. There almost always is a delay between the application of stress (or exposure of the stressed metal to hydrogen) and the onset of cracking. More properly referred to as hydrogen-induced delayed cracking.

steel. An iron-base alloy, malleable in some temperature ranges as initially cast, containing manganese, usually carbon, and often other alloying elements. In carbon steel and low-alloy steel, the maximum carbon is about 2.0%; in high-alloy steel, about 2.5%. The dividing line between low-alloy and high-alloy steels is generally regarded as being at about 5% metallic alloying elements.

Steel is to be differentiated from two general classes of "irons": the cast irons, on the high-carbon side, and the relatively pure irons such as ingot iron, carbonyl iron, and electrolytic iron, on the low-carbon side. In some steels containing extremely low carbon, the manganese content is the principal differentiating factor, steel usually containing at least 0.25%; ingot iron, considerably less.

step aging. Aging at two or more temperatures, by steps, without cooling to room temperature after each step. See *aging*, and compare with *interrupted aging* and *progressive aging*.

stepped extrusion. A product with one or more abrupt cross-sectional changes usually obtained by interrupting the extrusion operation and exchanging dies.

stereoradiography. A technique for producing paired radiographs that may be viewed with a stereoscope to exhibit a shadowgraph in three dimensions with various sections in perspective and spatial relation.

stiffness. The ability of a metal or shape to resist elastic deflection. For identical shapes, the stiffness is proportional to the modulus of elasticity. For a given material, the stiffness increases with increasing moment of inertia, which is computed from cross-sectional dimensions.

stock. A general term for solid starting material that is formed, forged or machined to make parts.

strain. A measure of the relative change in the size or shape of a body. Linear strain is the change per unit length of a linear dimension. True strain (or natural strain) is the natural logarithm of the ratio of the length at the moment of observation to the original gage length. Conventional strain is the linear strain over the original gage length. Shearing strain (or shear strain) is the change in angle (expressed in radians) between two lines originally at right angles. When the term "strain" is used alone it usually refers to the linear strain in the direction of applied stress. See also *state of strain*.

strain-age embrittlement. A loss in ductility accompanied by an increase in hardness and strength that occurs with low-carbon steel (especially rimmed or capped steel) is aged following plastic deformation. The degree of embrittlement is a function of aging time and temperature, occurring in a matter of minutes at about 200 °C (400 °F) but requiring a few hours to a year at room temperature.

strain aging. Aging induced by cold working. See *aging*.

strain energy. (1) The work done in deforming a body. (2) The work done in deforming a body within the elastic limit of the material. It is more properly termed elastic strain energy and can be recovered as work rather than heat.

strain hardening. An increase in hardness and strength caused by plastic deformation at temperatures below the recrystallization range.

strain-hardening exponent. A measure of rate of strain hardening. The constant n in the expression:

$$\sigma = \sigma_0 \delta^n$$

where σ is true stress, σ_0 is true stress at unit strain, and δ is true strain.

strain rate. The time rate of straining for the usual tensile test. Strain as measured directly on the specimen gage length is used for determining strain rate. Because strain is dimensionless, the units of strain rate are reciprocal time.

strain-rate sensitivity. Qualitatively, the increase in stress (s) needed to cause a certain increase in plastic strain rate ($\dot{\epsilon}$) at a given level of plastic strain (ϵ) and a given temperature (T).

Strain-rate sensitivity =

$$m = \frac{\Delta \log s}{\Delta \log \dot{\epsilon}_{\epsilon,T}}$$

strain state. See *state of strain*.

strand casting. A generic term describing *continuous casting* of one or more elongated shapes such as billets, blooms or slabs; if two or more strands are cast simultaneously, they are often of identical cross section.

stress. Force per unit area, often thought of as force acting through a small area within a plane. It can be divided into components, normal and parallel to the plane, called normal stress and shear stress, respectively. True stress denotes the stress where force and area are measured at the same time. Conventional stress, as applied to tension and compression tests, is force divided by the original area. Nominal stress is the stress computed by simple elasticity formulas, ignoring stress raisers and disregarding plastic flow; in a notch bend test, for example, it is bending moment divided by minimum section modulus. See also *state of stress*.

stress amplitude. One-half the algebraic difference between the maximum and minimum stress in one cycle of a repetitively varying stress.

stress-concentration factor (K_t). A multiplying factor for applied stress that allows for the presence of a structural discontinuity such as a notch or hole; K_t equals the ratio of the greatest stress in the region of the discontinuity to the nominal stress for the entire section.

stress-corrosion cracking. Failure by cracking under combined action of corrosion and stress, either external (applied) stress or internal (residual) stress. Cracking may be either intergranular or transgranular, depending on metal and corrosive medium.

stress-intensity factor. A scaling factor, usually denoted by the symbol K, used in linear-elastic fracture mechanics to describe the intensification of applied stress at the tip of a crack of known size and shape. At the onset of rapid crack propagation in any structure containing a crack, the factor is called the critical stress-intensity factor, or the *fracture toughness.*

stress raisers. Changes in contour or discontinuities in structure that cause local increases in stress.

stress range. The algebraic difference between the maximum and minimum stress in one cycle of a repetitively varying stress.

stress ratio. In fatigue, the ratio of the minimum stress to the maximum stress in one cycle, considering tensile stresses as positive, compressive stresses as negative.

stress relieving. Heating to a suitable temperature, holding long enough to reduce residual stresses and then cooling slowly enough to minimize the development of new residual stresses.

stress-rupture test. A method of evaluating elevated-temperature durability in which a tension-test specimen is stressed under constant load until it breaks. Data recorded commonly include: initial stress, time to rupture, initial extension, creep extension, reduction of area at fracture. Also known as creep-rupture test.

stretcher leveling. Leveling a piece of metal (that is, removing warp and distortion) by gripping it at both ends and subjecting it to a stress higher than its yield strength. Sometimes called patent leveling.

stretcher straightening. Straightening rod, tubing or shapes by gripping the stock at both ends and applying tension. The products are elongated a definite amount to remove warpage.

stretcher strains. Elongated markings that appear on the surface of some materials when deformed just past the yield point. These markings lie approximately parallel to the direction of maximum shear stress and are the result of localized yielding. See also *Lüders lines.*

stretch former. (1) A machine used to perform *stretch forming* operations. (2) A device adaptable to a conventional press for accomplishing stretch forming.

stretch forming. Shaping of a sheet or part, usually of uniform cross section, by first applying suitable tension or stretch and then wrapping it around a die of the desired shape.

striation. A fatigue fracture feature, often observed in electron micrographs, that indicates the position of the crack front after each succeeding cycle of stress. The distance between striations indicates the advance of the crack front across that crystal during one stress cycle, and a line normal to the striations indicates the direction of local crack propagation.

striking surface. Those areas on the faces of a set of dies that are designed to meet when the upper and lower dies are brought together. Striking surface helps protect impressions from impact shock and aids in maintaining longer die life. Also called beating area.

stringer. In wrought materials, an elongated configuration of microconstituents or foreign material aligned in the direction of working. Commonly, the term is associated with elongated oxide or sulfide inclusions in steel.

strip. A flat-rolled metal product of some maximum thickness and width arbitrarily dependent on the type of metal. It is narrower than sheet.

stripper punch. A punch that serves as top or bottom of the die cavity and later moves farther into the die to eject the part or compact. See also *ejector, knockout* (1).

stripping. Removing a coating from a metal surface.

structural shape. Piece of metal of any of several designs accepted as standard by the structural branch of the iron and steel industries.

subboundary structure. A network of low-angle boundaries (usually less than one degree) within the main crystals of a metallographic structure.

subcritical annealing. A process anneal performed on ferrous alloys at a temperature below A_1.

subgrain. A portion of a crystal or grain, with an orientation slightly different from the orientation of neighboring portions of the same crystal. Generally, neighboring subgrains are separated by low-angle boundaries such as *tilt boundaries* and *twist boundaries.*

substitutional solid solution. A solid solution in which the solute atoms are located at some of the lattice points of the solvent, the distribution being random. Contrast with *interstitial solid solution.*

substrate. Layer of metal underlying a coating, regardless of whether the layer is basis metal.

substructure. Same as *subboundary structure.*

subsurface corrosion. Formation of isolated

particles of corrosion products beneath a metal surface. This results from the preferential reaction of certain alloy constituents by inward diffusion of oxygen, nitrogen or sulfur.

superalloy. See *heat-resisting alloy.*

supercooling. Cooling below the temperature at which an equilibrium phase transformation can take place, without actually obtaining the transformation.

superficial Rockwell hardness test. Form of Rockwell hardness test using relatively light loads that produce minimum penetration by the indenter. Used for determining surface hardness or hardness of thin sections or small parts, or where a large hardness impression might be harmful.

superheating. (1) Heating above the temperature at which an equilibrium phase transformation should occur without actually obtaining the transformation. (2) Heating molten metal above the normal casting temperature so as to obtain more complete refining or greater fluidity.

superlattice. A lattice arrangement in which solute and solvent atoms of a solid solution occupy different preferred sites in the array. Contrast with *disordering.*

superplasticity. The ability of certain metals to undergo unusually large amounts of plastic deformation before local necking occurs.

supersonic. Pertains to phenomena in which the speed is higher than that of sound. Not synonymous with ultrasonic; see *ultrasonic frequency.*

surface finish. (1) Condition of a surface as a result of a final treatment. (2) Measured surface profile characteristics, the preferred term being roughness.

surface hardening. A generic term covering several processes applicable to a suitable ferrous alloy that produces, by quench hardening only, a surface layer that is harder or more wear resistant than the core. There is no significant alteration of the chemical composition of the surface layer. The processes commonly used are induction hardening, flame hardening and shell hardening. Use of the applicable specific process name is preferred.

surface tension. Interfacial tension between two phases, one of which is a gas.

swaging. Tapering bar, rod, wire or tubing by forging, hammering or squeezing; reducing a section by progressively tapering lengthwise until the entire section attains the smaller dimension of the taper.

T

tandem mill. A rolling mill consisting of two or more stands arranged so that the metal being processed travels in a straight line from stand to stand. In continuous rolling, the various stands are synchronized so that the strip may be rolled in all stands simultaneously. Contrast with *single-stand mill.*

tarnish. Surface discoloration of a metal caused by formation of a thin film of corrosion product.

teeming. Pouring molten metal from a ladle into ingot molds. The term applies particularly to the specific operation of pouring either iron or steel into ingot molds.

temper. (1) In heat treatment, reheating hardened steel or hardened cast iron to some temperature below the eutectoid temperature for the purpose of decreasing hardness and increasing toughness. The process also is sometimes applied to normalized steel. (2) In tool steels, temper is sometimes used, but inadvisedly, to denote the carbon content. (3) In nonferrous alloys and in some ferrous alloys (steels that cannot be hardened by heat treatment), the hardness and strength produced by mechanical or thermal treatment, or both, and characterized by a certain structure, mechanical properties, or reduction in area during cold working. (4) To moisten sand for casting molds with water.

temper brittleness. Brittleness that results when certain steels are held within, or are cooled slowly through, a certain range of temperature below the transformation range. The brittleness is manifested as an upward shift in ductile-to-brittle transition temperature, but only rarely produces a low value of reduction of area in a smooth-bar tension test of the embrittled material.

temper carbon. Same as *annealing carbon.*

temper color. A thin, tightly adhering oxide skin (only a few molecules thick) that forms when steel is tempered at a low temperature, or for a short time, in air or a mildly oxidizing atmosphere. The color, which ranges from straw to blue depending on the thickness of the oxide skin, varies with both tempering time and temperature.

temper rolling. Light cold rolling of sheet steel. This operation is performed to improve flatness, minimize the tendency to stretcher strain and flute, and obtain the desired texture and mechanical properties.

tensile strength. In tensile testing, the ratio of maximum load to original cross-sectional area. Also called *ultimate strength.* Compare with *yield strength.*

terminal phase. A solid solution having a restricted range of compositions, one end of the range being a pure component of an alloy system.

ternary alloy. An alloy that contains three principal elements.

terne. An alloy of lead containing 3 to 15% tin, used as a hot dip coating for steel sheet or plate. Terne coatings, which are smooth and dull in appearance, give the steel better corrosion resistance and enhance its ability to be formed, soldered or painted.

tertiary creep. See *creep.*

texture. In a polycrystalline aggregate, the state of distribution of crystal orientations. In the usual sense, it is synonymous with *preferred orientation.*

thermal fatigue. Fracture resulting from the presence of temperature gradients that vary with time in such a manner as to produce cyclic stresses in a structure.

thermal shock. The development of a steep temperature gradient and accompanying high stresses within a structure.

thermal stresses. Stresses in metal resulting from nonuniform temperature distribution.

thermocouple. A device for measuring temperatures, consisting of lengths of two dissimilar metals or alloys that are electrically joined at one end and connected to a voltage-measuring instrument at the other end. When one junction is hotter than the other, a thermal electromotive force is produced that is roughly proportional to the difference in temperature between the hot and cold junctions.

thermomechanical working. A general term covering a variety of processes combining controlled thermal and deformation treatments to obtain synergistic effects such as improvement in strength without loss of toughness. Same as thermal-mechanical treatment.

three-quarters hard. A *temper* of nonferrous alloys and some ferrous alloys characterized by tensile strength and hardness about mid-

way between those of *half hard* and *full hard* tempers.

tilt boundary. A subgrain boundary consisting of an array of edge *dislocations.*

time quenching. Interrupted quenching in which the time in the quenching medium is controlled.

tin pest. A polymorphic modification of tin that causes it to crumble into a powder known as gray tin. It is generally accepted that the maximum rate of transformation occurs at about –40 ° C (–40 °F), but transformation can occur as high as about 13 °C (55 °F).

tolerance. The specified permissible deviation from a specified nominal dimension, or the permissible variation in size or other quality characteristic of a part.

tolerance limits. The boundaries that define the range of permissible variation in size or other quality characteristic of a part.

tool steel. Any of a class of carbon and alloy steels commonly used to make tools. Tool steels are characterized by high hardness and resistance to abrasion, often accompanied by high toughness and resistance to softening at elevated temperature. These attributes are generally attained with high carbon and alloy contents.

torsion. A twisting action resulting in shear stresses and strains.

total carbon. The sum of the free and combined carbon (including carbon in solution) in a ferrous alloy.

toughness. Ability of a metal to absorb energy and deform plastically before fracturing. It is usually measured by the energy absorbed in a notch impact test, but the area under the stress-strain curve in tensile testing is also a measure of toughness.

tramp alloys. Residual alloying elements that are introduced into steel when unidentified alloy steel is present in the scrap charge to a steelmaking furnace.

transcrystalline. Same as *intracrystalline.*

transformation ranges. Those ranges of temperature within which a phase forms during heating and transforms during cooling. The two ranges are distinct, sometimes overlapping but never coinciding. The limiting temperatures of the ranges depend on the composition of the alloy and on the rate of change of temperature, particularly during cooling. See *transformation temperature.*

transformation temperature. The temperature at which a change in phase occurs. The term is sometimes used to denote the limiting temperature of a transformation range. The following symbols are used for iron and steels:

Ac$_{cm}$. In hypereutectoid steel, the temperature at which the solution of cementite in austenite is completed during heating.

Ac$_1$. The temperature at which austenite begins to form during heating.

Ac$_3$. The temperature at which transformation of ferrite to austenite is completed during heating.

Ae$_{cm}$, Ae$_1$, Ae$_3$, (or A$_{cm}$, A$_1$, A$_3$). The temperatures of phase changes at equilibrium.

Ar$_{cm}$. In hypereutectoid steel, the temperature at which precipitation of cementite starts during cooling.

Ar$_1$. The temperature at which transformation of austenite to ferrite or to ferrite plus cementite is completed during cooling.

Ar$_3$. The temperature at which austenite begins to transform to ferrite during cooling.

Ar′. The temperature at which transformation of austenite to pearlite starts during cooling.

M$_f$. The temperature at which transformation of austenite to martensite finishes during cooling.

M$_s$. The temperature at which transformation of austenite to martensite starts during cooling.

transgranular. Same as *intracrystalline*.

transition lattice. An unstable crystallographic configuration that forms as an intermediate step in a solid-state reaction such as precipitation from solid solution or eutectoid decomposition.

transition point. At a stated pressure, the temperature (or at a stated temperature, the pressure) at which two solid phases exist in equilibrium; that is, an allotropic transformation temperature (or pressure).

transition temperature. (1) An arbitrarily defined temperature that lies within the temperature range in which metal fracture characteristics (as usually determined by tests of notched specimens) change rapidly, such as from primarily fibrous (shear) to primarily crystalline (cleavage) fracture. Commonly used definitions are "transition temperature for 50% cleavage fracture", "10 ft·lb transition temperature", and "transition temperature for half maximum

energy". (2) Sometimes used to denote an arbitrarily defined temperature within a range in which the ductility changes rapidly with temperature.

transverse. Literally, "across", usually signifying a direction or plane perpendicular to the direction of working. In rolled plate or sheet, the direction across the width is often called long transverse, and the direction through the thickness, short transverse.

tribology. The science and art concerned with the design, friction, lubrication and wear of contacting surfaces that move relative to each other (as in bearings, cams or gears, for example).

trimmers. The combination of trimmer punch, trimmer blades and perhaps trimmer shoe used to remove the flash from the forging.

trimming. (1) In drawing, shearing the irregular edge of the drawn part. (2) In forging or die casting, removing any parting-line flash and gates from the part by shearing. (3) In casting, the removal of gates, risers and fins.

true strain. See *strain*.

true stress. See *stress*.

tube sinking. Drawing tubing through a die or passing it through rolls without the use of an interior tool (such as a mandrel or plug) to control inside diameter; sinking generally produces a tube of increased wall thickness and length.

twin. Two portions of a crystal having a definite crystallographic relationship; one may be regarded as the parent, the other as the twin. The orientation of the twin is either a mirror image of the orientation of the parent about a "twinning plane" or an orientation that can be derived by rotating the twin portion about a "twinning axis". See also *annealing twin, mechanical twin*.

twist boundary. A subgrain boundary consisting of an array of screw *dislocations*.

two-high mill. A type of rolling mill in which only two rolls, the working rolls, are contained in a single housing. Compare with *four-high mill, cluster mill*.

U

ultimate strength. The maximum conventional stress (tensile, compressive or shear) that a material can withstand.

ultrasonic beam. A beam of acoustical radiation with a frequency higher than the frequency range for audible sound — that is, above about 20 kHz.

ultrasonic frequency. A frequency, associated with elastic waves, that is greater than the highest audible frequency, generally regarded as being higher than 15 kHz.

ultrasonic testing. A nondestructive test applied to sound-conductive materials having elastic properties for the purpose of locating inhomogeneities or structural discontinuities within a material by means of an *ultrasonic beam*.

undercooling. Same as *supercooling*.

understressing. Applying a cyclic stress lower than the *endurance limit*. This may improve fatigue life if the member is later cyclically stressed at levels above the endurance limit.

uniform strain. The strain occurring prior to the beginning of localization of strain (necking); the strain to maximum load in the tension test.

unit cell. In crystallography, the fundamental building block of a space lattice. Space lattices are constructed by stacking identical unit cells — that is, parallelepipeds of identical size, shape and orientation, each having a lattice point at every corner — face to face in perfect three-dimensional alignment.

universal mill. A rolling mill in which rolls with a vertical axis roll the edges of the metal stock between some of the passes through the horizontal rolls.

upset. (1) The localized increase in cross-sectional area of a workpiece or weldment resulting from the application of pressure during mechanical fabrication or welding. (2) That portion of a welding cycle during which the cross-sectional area is increased by the application of pressure.

upset forging. A forging obtained by upset of a suitable length of bar, billet or bloom.

upsetter. A horizontal mechanical press used to make parts from bar stock or tubing by upset forging, piercing, bending or otherwise forming in dies. Also known as a header.

upsetting. Working metal so that the cross-sectional area of a portion or all of the stock is increased. See also *heading*.

V

vacancy. A type of lattice imperfection in which an individual atom site is temporarily unoccupied. Diffusion (of other than inter-

stitial solutes) is generally visualized as the shifting of vacancies.

veining. A type of subboundary structure that can be delineated because of the presence of a greater than average concentration of precipitate or possibly solute atoms.

vent. A small opening in a mold for the escape of gases.

Vickers hardness test. An indentation hardness test employing a 136° diamond pyramid indenter (Vickers) and variable loads enabling the use of one hardness scale for all ranges of hardness from very soft lead to tungsten carbide.

virgin metal. Same as *primary metal*.

W

warm working. Plastically deforming metal at a temperature above ambient (room) temperature but below the temperature at which the material undergoes recrystallization.

weld. A union made by *welding*.

weldability. A specific or relative measure of the ability of a material to be welded under a given set of conditions. Implicit in this definition is the ability of the completed weldment to fulfill all service designed into the part.

weld bead. A deposit of filler metal from a single welding *pass*.

welder. A person who makes welds using manual or semiautomatic equipment. Formerly used as a synonym for *welding machine*.

welding. (1) Joining two or more pieces of material by applying heat or pressure, or both, with or without filler material, to produce a localized union through fusion or recrystallization across the interface. The thickness of the filler material is much greater than the capillary dimensions encountered in *brazing*. (2) May also be extended to include brazing and soldering.

welding machine. Equipment used to perform the welding operation — for example, spot welding machine, arc welding machine, seam welding machine.

welding procedure. The detailed methods and practices, including joint preparation and welding procedures, involved in the production of a *weldment*.

welding rod. Welding or brazing filler metal, usually in rod or wire form, but not a consumable electrode. Welding rod does not conduct the electric current to an arc, and may be either fed into the weld puddle or preplaced in the joint.

welding sequence. The order in which the various component parts of a weldment or structure are welded.

welding stress. *Residual stress* caused by localized heating and cooling during welding.

welding technique. The details of a welding operation that, within the limitations of a welding procedure, are performed by the *welder.*

weld line. The junction of the weld metal and the base metal, or the junction of the base-metal parts when filler metal is not used.

weldment. An assembly whose component parts are joined by welding.

weld metal. That portion of a weld that has been melted during welding.

weld nugget. The weld metal in spot, seam or projection welding.

whiskers. Metallic filamentary growths, often microscopic, sometimes formed during electrodeposition and sometimes spontaneously during storage or service, after finishing.

white rust. Zinc oxide; the powdery product of corrosion of zinc or zinc-coated surfaces.

Widmanstätten structure. A structure characterized by a geometrical pattern resulting from the formation of a new phase along certain crystallographic planes of the parent solid solution. The orientation of the lattice in the new phase is related crystallographically to the orientation of the lattice in the parent phase. The structure was originally observed in meteorites, but is readily produced in many other alloys by appropriate heat treatment.

wire. (1) A thin, flexible, continuous length of metal, usually of circular cross section, and usually produced by drawing through a die. See also *flat wire.* (2) A length of single metallic electrical conductor; it may be of solid, stranded or tinsel construction, and may be either bare or insulated.

wire bar. A cast shape, particularly of tough pitch copper, that has a cross section approximately square with tapered ends, designed for hot rolling to rod for subsequent drawing into wire.

wiredrawing. Reducing the cross section of wire by pulling it through a die.

wire rod. Hot rolled coiled stock that is to be cold drawn into wire.

work hardening. Same as *strain hardening.*

wrought iron. A commercial iron consisting of slag (iron silicate) fibers entrained in a ferrite matrix.

X

x-ray. Electromagnetic radiation, of wavelength less than about 50 nm, emitted as the result of deceleration of fast-moving electrons (bremsstrahlung, continuous spectrum) or decay of atomic electrons from excited orbital states (*characteristic radiation*); specifically, the radiation produced when an electron beam of sufficient energy impinges on a target of suitable material.

Y

yield point. The first stress in a material, usually less than the maximum attainable stress, at which an increase in strain occurs without an increase in stress. Only certain metals exhibit a yield point. If there is a decrease in stress after yielding, a distinction may be made between upper and lower yield points.

yield strength. The stress at which a material exhibits a specified deviation from proportionality of stress and strain. An offset of 0.2% is used for many metals. Compare with *tensile strength.*

Young's modulus. See *modulus of elasticity.*

Appendix 2
SI Units and the
Metals Engineer

Scientists and engineers throughout the world agreed some years ago to present data in a new system of metric units known as the Système International d'Unités (SI). Some countries have also adopted SI as their legal system of units, including many, such as my own, that have made a full transition from nonmetric systems. In other countries, SI has not achieved public recognition but has been adopted by technical societies and associations as a preferred system, a procedure which has increasingly become necessary to make their publications intelligible internationally. Nevertheless, many engineers, particularly those trained in earlier systems, usually based on the English *Imperial* system, still think in terms of a non-SI system of units and convert from one to the other. But there is no getting away from the fact that all engineers have now to face up to acquiring a sound understanding of what SI is all about.

GENERAL CONCEPTS

The first and essential thing to realize is that SI is more than a metric decimal system. It is a rational system which has been developed to be internally consistent and unambiguous, and to correct some deficiencies which experience had shown to exist in the long-established metric system. Expert members of the Conférence Générale de Poids and Mesures labored long and valiantly to eliminate ambiguities, to reduce the probability that errors might be made in calculations, and to define units which were convenient in size for both science and engineering. Their final recommendations are detailed and are to be trifled with only at the risk of introducing ambiguities or of being misunderstood.

Base Units and Symbols

The system is based on six *fundamental* or *basic* physical quantities, of which five are of concern to metals engineers. They are:

Physical quantity	Name of unit	Symbol
length	metre	m
mass	kilogram	kg
time	second	s
electric current	ampere	A
thermodynamic temperature	kelvin	K

This list can be used additionally to illustrate the fundamental concepts on which names of, and the symbols for, the units are based. Names are spelled with a small first letter even when they were derived from a person's name. A capital letter is used for the symbol, however, when it is derived from a person's name — but never otherwise and even then only for the first letter of the symbol if more than one is used. As examples, the capital K is used for the kelvin, the unit of temperature, and Pa for the pascal, the unit of pressure or stress, but the small letter k is used for the kilogram. Symbols are the same in the singular as the plural and are written without a period (unless, of course, they are at the end of a sentence). Thus:

$$1 \text{ kg and } 5 \text{ kg}$$

but not   1 kg. or 5 kgs

Decimal multiples and submultiples of units are formed by adding the following prefixes:

Factor by which the unit is multiplied	Prefix	Symbol	Factor by which the unit is multiplied	Prefix	Symbol
10^{12}	tera	T	10^{-2}	centi	c
10^{9}	giga	G	10^{-3}	milli	m
10^{6}	mega	M	10^{-6}	micro	μ
10^{3}	kilo	k	10^{-9}	nano	n
10^{2}	hecto	h	10^{-12}	pico	p
10	deca	da	10^{-15}	femto	f
10^{-1}	deci	d	10^{-18}	atto	a

The principle is to use multipliers and submultipliers to provide steps of a thousand times, with a concession for convenience in the first to provide steps of ten and a hundred times. The use of the last two is discouraged but the use of "centi" is so well established that its continued use must be expected. Small letters are used for all of these prefix symbols except for tera (T), giga (G) and mega (M). There are good reasons for these rules. For example, a small m used for mega might be confused with metre (m); a large K used for kilo might be confused with K for kelvin. In combination, the prefix and unit name are written as one word. Thus: micrometre (μm) and not micro metre.* There are also agreed rules on the pronunciation of these prefixes and their combinations. Thus kil'o-metre and not kil-om'-etre. But this is probably a lost cause already, and perhaps is of lesser importance anyway.

Derived Units

Many quantities are derived from a combination of fundamental units. In principle, the symbols for such units are built up from combinations of the symbols of the fundamental units involved. Thus the symbol for speed, which is distance moved in unit time, is m/s (or m $\cdot$ s^{-1}, whichever you prefer). Some derived units have been given special names, typically the name of the scientist who was instrumental for the development of the area of science in which they are used. Thus the unit of stress (newtons per square metre; see below) is the pascal (Pa).

Some rules are necessary to ensure that no ambiguity arises when a number of symbols are strung together in a derived unit. First of all, as a general rule, a dot needs to be introduced between the symbols of units which are multiplied or divided. Thus a dot is needed between m and s in m $\cdot$ s^{-1} as indicating metres divided by seconds; ms without the dot might be taken to be the reciprocal of a millisecond. But the dot may be omitted when there is no risk of ambiguity, as in m^{-1} s^{-1}. But when in doubt, use a dot. In the case of division, either a solidus (oblique line, /), or a horizontal line, or a negative index may be used, as is convenient. For example, the unit for thermal conductivity is watts per metre kelvin. The following symbols are equally acceptable, although the last one is preferred:

$$\text{W/(m} \cdot \text{K)} \qquad \frac{\text{W}}{\text{m} \cdot \text{K}} \qquad \text{W} \cdot \text{m}^{-1}\,\text{K}^{-1}$$

The parentheses around m $\cdot$ K are necessary in the first example because there would be two possibilities if it were omitted — namely: W divided by m $\cdot$ K or W/m multiplied by K. The brackets can be omitted in the second and third examples because there is now no possibility of confusion. Finally, never use more than one solidus. For example: W/m/K should not be used, because it could be taken to mean quite a number of things.

---

*Certainly not "micron", a name previously used for this unit of length, or its old associated symbol, μ. Anyone who has not heard previously of a "micron" possibly will not be able to work out what you mean — a millionth of nothing, perhaps.

UNITS OF SPECIAL INTEREST TO METALS ENGINEERS

Metals engineers have to deal regularly with only a limited range of the very large number of units that are dealt with in SI, but some of these units are a little tricky. It is only those which are of direct concern to metals engineers that we shall discuss here, but full details on others are available in authoritative documents such as those listed at the end of this appendix.

Length, Area, and Volume

The basic unit of length is the metre (m), the preferred first multiple being the kilometre (km) and first submultiple the millimetre (mm). The submultiple of centimetre (cm) for one hundredth of a metre (ten millimetres), although not preferred, is in common usage, as also is decimetre in Europe. Engineers also commonly use the micrometre (μm) as a submultiple. It is often thought of as a thousandth of a millimetre although the term actually means a millionth of a metre (10^{-6} m). This length was once called a micron (symbol simply μ). Metals engineers are also likely to encounter a strange length unit known as the Angstrom ($\text{Å} = 10^{-10}$ m). This unit has been used for many years by crystallographers because it is convenient in their scheme of things. The dimensions of the unit cell of crystals are of the order of an Angstrom but unfortunately the unit does not fit in with the SI concept of submultiplier steps of 10^{-3}. The nearest smaller unit is the picometre, an Angstrom being 100 pm. Thus, the lattice constant of copper should in SI be quoted as 361.53 pm instead of the 3.6153 Å that crystallographers have always used. This change to strict SI was too much for them and they have stuck to their guns, with a degree of acceptance provided that the use of Å is confined to their field.

The quantity for area is not likely to cause confusion, except to note that a hundred metres square (10^2 m) is called a hectare (ha), which is not to be confused with an acre. It is, in fact, about two and a half acres. The SI unit for volume is the cubic metre (m^3) with the usual multiples and submultiples. The name litre (ℓ) is acceptable for liquid volumes in general usage, being synonymous for a cubic decimetre ($dm^3 = 10^{-3} m^3$).

Time

The SI unit is the second (s), which in some ways is the most unsatisfactory of the SI units. It is inconveniently small in magnitude and lacks decimal multiples in common usage. The units day (d), hour (h) and minute (min) are therefore still used and would indeed be difficult to proscribe. Note also that the symbol for hour is h, not hr, and that the unit of frequency is the hertz (Hz), not cycles per second (cps). Hertz warrants the honor as being the first to demonstrate the spreading of electromagnetic waves in space.

Temperature

The fundamental temperature scale is based on a thermodynamic concept, the unit being the kelvin (K), honoring the founder of the science of thermodynamics. The scale sets absolute zero at 0 K and the triple point of water (approximately its freezing point) at 273.16 K. This is a device chosen to make the scale compatible with the earlier and widely used (but less precisely defined) scale which fixed points of 0 and 100 for the freezing and boiling temperatures of water at normal atmospheric temperature. This was formerly known as the centigrade scale but is now named Celsius after the man who defined it. The temperature interval, the degree celsius (°C), is the same as for the kelvin. Thus:

$$0 \,°C = 273.15 \,K \qquad \text{and} \qquad 100 \,°C = 373.15 \,K$$

It is the notation for temperature to which attention must be paid. The practice in the past has been to refer to a temperature of so many "degrees" and to write this with a superscript (°). The same notation has been used to signify both temperature and temperature difference. SI practice is that temperatures kelvin should be indicated simply by K (not degrees kelvin or °K), but that the traditional notation °C be retained for customary temperature. The superscript used with the °C is an exception in the SI notational system and must be used because C alone is the symbol for quantity of electricity, the coulomb. The new notation makes no distinction between temperature and temperature interval. It was once regarded as desirable to use the abbreviation "deg" to express temperature interval. This is now regarded as obsolescent, partly because it is unacceptable to many countries in which the abbreviation has no significance in its language. Nevertheless, it may be used in textual matter written in English; e.g., "increase the temperature by 20 deg C".

Mass, Force and Stress

Mass is a fundamental unit which defines quantity of matter, and the unit is the kilogram (kg). A multiple is used as the unit in this instance because the gram itself is too small to construct accurately as a fundamental standard. The definition of the unit of mass was originally to have been based on the mass of a certain volume of water at prescribed conditions but was soon, and still is, taken to be that of a certain cylinder of platinum. A thousand kilograms (1000 kg) is called a tonne (t), which is conveniently close to the Imperial long ton of 2240 lb (= 1016 kg).

The unit of force, on the other hand, is derived from Newton's first law of motion, which he stated in a most general way, namely:

> "The alteration in motion is ever proportional to the motive force impressed, and is made in the direction of the right line in which the force is impressed."

Applied to the restricted case of defining force, this means that:

$$\text{force} = \text{mass} \times \text{acceleration}$$

Simple enough in principle, but confusion arises on several fronts. The first problem arises because the word "weight" is often used interchangeably with and indistinguishably from mass in common usage. Weight is in fact quite different from mass, being a force of a particular type — namely, the attractive force exerted by the mass of the Earth on any other mass. We call this the gravitational attraction of the Earth. It is a force which we may perceive through some measuring device be it the muscles of your legs as you hold yourself upright or the bathroom scales when you weigh yourself. The magnitude of the gravitational attraction varies a little (by up to 1%) from place to place on the Earth. So you may find that you have lost a little weight when you try out the scales at some foreign place, but this does not necessarily mean that you have lost any of your mass of fat. It may mean only that gravitational attraction is less at the new location than at home. A trivial matter in this context but not in that of calculating the force that will be required to launch a rocket. It would be better perhaps if the term weight were dropped from our vocabulary, but that is unlikely, to say the least. So we have to stop each time it is used and decide whether a mass or a force is being discussed.

The distinction between mass and weight as a force was blurred even further by the original definition of units of force, which was arranged to make the absolute values of force and mass approximately the same. This was done by defining the unit as the force required to accelerate unit mass by a value known as "g" for which a value of $9.80665 \, \text{m} \cdot \text{s}^{-2}$ was agreed upon in the metric system. The value was taken as the value of the acceleration due to gravity in Newton's equation. The resultant unit of force was called a "kilogram force" (kgf) to distinguish it from the unit of mass the kilogram (kg) — or at least it should have been. Unfortunately, however, the word "force" and the abbreviation "f" are all too often omitted in this context, leaving one to guess once more whether weight or mass is meant. The corresponding unit in the Imperial system is the "pound force" (lbf). You may think that these distinctions are examples of pure pedantry, but, to repeat, they do matter in measurements which pretend to any precision. The value of g used in these definitions of force is actually an estimate of a world average, and adjustments are required for local variations from average. *Load* is another engineering term which is often the cause of confusion between mass and force. Frequently, load has been measured in units of mass. From an SI standpoint it should be measured in N, and there is a difference. For example, if the load is 1 t (1000 kg), the force, according to Newton's gravitational equation, is:

$$1000 \times 9.80 = 9.80 \, \text{kN}$$

Thus, if you load a structure with 1 t you produce a force of 9.8 kN.

A second source of confusion has arisen with this system because conventions have been adopted which require that the value g should be introduced into some types of calculations and not into others. Many's the time that the wrong choice has been made.

The SI solution to these dilemmas was to define an absolute unit as being the force required to accelerate unit mass (1 kg) by $1 \text{ m} \cdot \text{s}^{-2}$. This is called the newton (N). It is appropriate, but quite coincidental of course, that a newton is about the value of the force on the apple which is said to have fallen beside Newton and to have been responsible for the flash of inspiration which spawned the laws of motion. Metals engineers are particularly interested in stress and pressure, which are both force per unit area. The SI unit is a newton per square metre $(\text{N} \cdot \text{m}^{-2})$ which is now almost universally called the pascal (Pa) in engineering circles. This is convenient, if nothing else, and appropriate because Pascal, as you may remember, was the first to realize that the Earth's atmosphere exerted a pressure on its surface. Nevertheless, a pascal (imagine an apple spread over a metre square) is a very small unit by engineering standards and so a multiplier is normally used. Most commonly, the megapascal (MPa) is used when considering the strength of metals.

You will often see other units of stress used in engineering practice. Examples are pounds force per square inch (lbf/in.^2), thousands of pounds force per square inch (ksi), and kilograms force per square millimetre (kgf/mm^2). Conversions between a number of these less precise legacies, which it is hoped will disappear in time (my grandchildren do not know what a pound or an inch is because they have been taught exclusively in SI, so there *is* hope), are given in Appendix 3. The unit kgf/mm^2 warrants some further attention here because the hardness numbers obtained in Vickers and Brinell hardness tests (p. 80) are actually calculated as a pressure in this unit. You might therefore ask whether these hardness numbers should not now be calculated in units of pascals. The various bodies responsible for national standards say no. Indenting loads have been retained at the same values of kg as of yore but the force is quoted in newtons, using the soft conversion described above of 1 kgf = 9.8 N. So leave well enough alone, and regard hardness just as a number.

There is one other area of special interest to metals engineers, and this concerns the unit for stress intensity used in evaluating fracture toughness (p. 90). The unit that should be used strictly is $\text{MN} \cdot \text{m}^{-3/2}$, but many practitioners prefer to use $\text{MPa} \cdot \text{m}^{-1/2}$, which may even be written as $\text{MPa} \sqrt{\text{m}}$. The concepts of fracture toughness were developed in pre-SI days, and stress intensity was then quoted by most of the pioneering investigators in units of $\text{ksi} \sqrt{\text{in.}}$, a unit which is still widely used. Hence, the desire to use a similar form in the SI unit without disrupting the system too much. By good luck, the absolute values of fracture toughness are the same within about 10% in the two units.

The units for energy and power are also derived from the newton. The work done (mechanical energy used) when a force of one newton moves something over a distance of one metre (N · m) is called a joule (J). In metals engineering this is the SI unit that is used to quantify the energy absorbed in impact tests, the old units being ft · lbf. Power is a measure of the rate at which energy is expanded, the rate of one joule per second being called a watt (W). Joule was the first to realize the equivalence of thermal and mechanical energy, and Watt invented and developed the condensing steam engine — one of the revolutionary inventions in the development of civilization.

FURTHER READING

ISO Recommendation R1000, *Rules for the Use of Units of the International System of Units*, International Organization for Standardization, 1969.

The Use of SI Units, PD 5686, British Standards Institution, London, 1969.

The Modernized Metric System, SP 304, National Bureau of Standards, Washington, DC, 1968.

D. R. Blackman, *SI Units in Engineering*, Macmillan, Melbourne, 1969.

Appendix 3
Conversion Factors and
Conversion Tables

Table A3.1. Conversion Multipliers for Some Units Commonly Used by Metals Engineers
Multiply by factor listed to convert in direction of arrow.
Example: a kg = (a × 2.204) lb
b lb = (b × 0.454) kg

Quantity	SI unit	Multiply by (to convert in the direction of the arrow)		Second unit
Mass	kilogram (kg)	× 2.204 → ←	0.454 ×	pound (lb)
	tonne (t) (1Mg)	× 1.103 → ←	0.907 ×	short ton (2000 lb)
		× 0.984 → ←	1.016 ×	long ton (2240 lb)
Force	newton (N) (kg · m/s^2)	× 0.102 → ←	9.807 ×	kilograms force (kgf)
		× 0.224 → ←	4.448 ×	pounds force (lbf)
Stress or pressure	pascal (Pa) (N/m^2)	× (0.102 × 10^{-4}) → ← (9.807 × 10^4) ×		kilograms force per square centimetre (kgf/cm^2)
		× (1.450 × 10^{-4}) → ← (6.985 × 10^3) ×		pounds force per square inch (lbf/in.^{2})
	megapascal (MPa)	× 0.145 → ←	6.895 ×	thousands of pounds force per square inch (ksi)
		× (6.475 × 10^{-2}) → ←	15.44 ×	long tons force per square inch (tsi)
Energy	joule (J)	× 0.239 → ←	4.187 ×	calories (cal)
		× 0.738 → ←	1.356 ×	foot-pounds force (ft·lbf)

(continued)

Table A3.1. Conversion Multipliers (continued)
Multiply by factor listed to convert in direction of arrow.
Example: a kg = (a × 2.204) lb
b lb = (b × 0.454) kg

Quantity	SI unit	Multiply by (to convert in the direction of the arrow)	Second unit
Length	metre (m)	× 1.094 → ← 0.914 ×	yard (yd)
	kilometre (km)	× 0.622 → ← 1.608 ×	mile
	millimetre (mm)	× (0.394 × 10⁻¹) → ← 25.4 ×	inch (in.)
	micrometre (μm)	× 39.4 → ← (2.54 × 10⁻²) ×	microinch (μin.)
		× 10⁴ → ← 10⁻⁴ ×	Angstrom (Å) (10⁻¹⁰m)
	picometre (pm)	× 10⁻² → ← 10² ×	Angstrom (Å)
Area	square metre (m²)	× (1.53 × 10³) → ← (6.451 × 10⁻⁴) ×	square inch (in.²)
		× 1.196 → ← 0.836 ×	square yard (yd²)
		× (2.47 × 10⁻⁴) → ← (4.047 × 10³) ×	acre
	hectare (ha) (10⁴ m²)	× 2.47 → ← 0.405 ×	acre
Volume	cubic metre (m³)	× (1 × 10³) → ← (1 × 10⁻³) ×	litre (ℓ)
		× 35.3 → ← (2.831 × 10⁻²) ×	cubic foot (ft³)
Density	kilograms per cubic metre (kg/m³)	× (1 × 10⁻³) → ← (1 × 10³) ×	grams per cubic centimetre (g/cm³)
		× (3.614 × 10⁻⁵) → ← (2.768 × 10⁴) ×	pounds per cubic inch (lb/in.³)

Table A3.2. Stress or Pressure Conversions

The middle column of figures (in bold-faced type) contains the reading (in MPa or ksi) to be converted. If converting from ksi to MPa, read the MPa equivalent in the column headed "MPa". If converting from MPa to ksi, read the ksi equivalent in the column headed "ksi". 1 ksi = 6.894757 MPa. 1 psi = 6.894757 kPa.

ksi		MPa	ksi		MPa	ksi		MPa	ksi		MPa
0.14504	**1**	6.895	8.2672	**57**	393.00	33.359	**230**	1585.8	114.58	**790**	...
0.29008	**2**	13.790	8.4122	**58**	399.90	34.809	**240**	1654.7	116.03	**800**	...
0.43511	**3**	20.684	8.5572	**59**	406.79	36.259	**250**	1723.7	117.48	**810**	...
0.58015	**4**	27.579	8.7023	**60**	413.69	37.710	**260**	1792.6	118.93	**820**	...
0.72519	**5**	34.474	8.8473	**61**	420.58	39.160	**270**	1861.6	120.38	**830**	...
0.87023	**6**	41.369	8.9923	**62**	427.47	40.611	**280**	1930.5	121.83	**840**	...
1.0153	**7**	48.263	9.1374	**63**	434.37	42.061	**290**	1999.5	123.28	**850**	...
1.1603	**8**	55.158	9.2824	**64**	441.26	43.511	**300**	2068.4	124.73	**860**	...
1.3053	**9**	62.053	9.4275	**65**	448.16	44.962	**310**	2137.4	126.18	**870**	...
1.4504	**10**	68.948	9.5725	**66**	455.05	46.412	**320**	2206.3	127.63	**880**	...
1.5954	**11**	75.842	9.7175	**67**	461.95	47.862	**330**	2275.3	129.08	**890**	...
1.7405	**12**	82.737	9.8626	**68**	468.84	49.313	**340**	2344.2	130.53	**900**	...
1.8855	**13**	89.632	10.008	**69**	475.74	50.763	**350**	2413.2	131.98	**910**	...
2.0305	**14**	96.527	10.153	**70**	482.63	52.214	**360**	2482.1	133.43	**920**	...
2.1756	**15**	103.42	10.298	**71**	489.53	53.664	**370**	2551.1	134.89	**930**	...
2.3206	**16**	110.32	10.443	**72**	496.42	55.114	**380**	2620.0	136.34	**940**	...
2.4656	**17**	117.21	10.588	**73**	503.32	56.565	**390**	2689.0	137.79	**950**	...
2.6107	**18**	124.11	10.733	**74**	510.21	58.015	**400**	2757.9	139.24	**960**	...
2.7557	**19**	131.00	10.878	**75**	517.11	59.465	**410**	2826.9	140.69	**970**	...
2.9008	**20**	137.90	11.023	**76**	524.00	60.916	**420**	2895.8	142.14	**980**	...
3.0458	**21**	144.79	11.168	**77**	530.90	62.366	**430**	2964.7	143.59	**990**	...
3.1908	**22**	151.68	11.313	**78**	537.79	63.817	**440**	3033.7	145.04	**1000**	...
3.3359	**23**	158.58	11.458	**79**	544.69	65.267	**450**	3102.6	147.94	**1020**	...
3.4809	**24**	165.47	11.603	**80**	551.58	66.717	**460**	3171.6	150.84	**1040**	...
3.6259	**25**	172.37	11.748	**81**	558.48	66.168	**470**	3240.5	153.74	**1060**	...
3.7710	**26**	179.26	11.893	**82**	565.37	69.618	**480**	3309.5	156.64	**1080**	...
3.9160	**27**	186.16	12.038	**83**	572.26	71.068	**490**	3378.4	159.54	**1100**	...
4.0611	**28**	193.05	12.183	**84**	579.16	72.519	**500**	3447.4	162.44	**1120**	...
4.2061	**29**	199.95	12.328	**85**	586.05	73.969	**510**	...	165.34	**1140**	...
4.3511	**30**	206.84	12.473	**86**	592.95	75.420	**520**	...	168.24	**1160**	...
4.4962	**31**	213.74	12.618	**87**	599.84	76.870	**530**	...	171.14	**1180**	...
4.6412	**32**	220.63	12.763	**88**	606.74	78.320	**540**	...	174.05	**1200**	...
4.7862	**33**	227.53	12.909	**89**	613.63	79.771	**550**	...	176.95	**1220**	...
4.9313	**34**	234.42	13.053	**90**	620.53	81.221	**560**	...	179.85	**1240**	...
5.0763	**35**	241.32	13.198	**91**	627.42	82.672	**570**	...	182.75	**1260**	...
5.2214	**36**	248.21	13.343	**92**	634.32	84.122	**580**	...	185.65	**1280**	...
5.3664	**37**	255.11	13.489	**93**	641.21	85.572	**590**	...	188.55	**1300**	...
5.5114	**38**	262.00	13.634	**94**	648.11	87.023	**600**	...	191.45	**1320**	...
5.6565	**39**	268.90	13.779	**95**	655.00	88.473	**610**	...	194.35	**1340**	...
5.8015	**40**	275.79	13.924	**96**	661.90	89.923	**620**	...	197.25	**1360**	...
5.9465	**41**	282.69	14.069	**97**	668.79	91.374	**630**	...	200.15	**1380**	...
6.0916	**42**	289.58	14.214	**98**	675.69	92.824	**640**	...	203.05	**1400**	...
6.2366	**43**	296.47	14.359	**99**	682.58	94.275	**650**	...	205.95	**1420**	...
6.3817	**44**	303.37	14.504	**100**	689.48	95.725	**660**	...	208.85	**1440**	...
6.5267	**45**	310.26	15.954	**110**	758.42	97.175	**670**	...	211.76	**1460**	...
6.6717	**46**	317.16	17.405	**120**	827.37	98.626	**680**	...	214.66	**1480**	...
6.8168	**47**	324.05	18.855	**130**	896.32	100.08	**690**	...	217.56	**1500**	...
6.9618	**48**	330.95	20.305	**140**	965.27	101.53	**700**	...	220.46	**1520**	...
7.1068	**49**	337.84	21.756	**150**	1034.2	102.98	**710**	...	223.36	**1540**	...
7.2519	**50**	344.74	23.206	**160**	1103.2	104.43	**720**	...	226.26	**1560**	...
7.3969	**51**	351.63	24.656	**170**	1172.1	105.88	**730**	...	229.16	**1580**	...
7.5420	**52**	358.53	26.107	**180**	1241.1	107.33	**740**	...	232.06	**1600**	...
7.6870	**53**	365.42	27.557	**190**	1310.0	108.78	**750**	...	234.96	**1620**	...
7.8320	**54**	372.32	29.008	**200**	1379.0	110.23	**760**	...	237.86	**1640**	...
7.9771	**55**	379.21	30.458	**210**	1447.9	111.68	**770**	...	240.76	**1660**	...
8.1221	**56**	386.11	31.908	**220**	1516.8	113.13	**780**	...	243.66	**1680**	...

Table A3.3 Stress Conversion Table, kg/mm² to psi

kg/mm₂	psi	kg/mm²	psi	kg/mm²	psi	kg/mm²	psi	kg/mm²	psi	kg/mm²	psi
10	14 223	35	49 782	60	85 340	85	120 899	110	156 457	135	192 016
11	15 646	36	51 204	61	86 763	86	122 321	111	157 880	136	193 438
12	17 068	37	52 627	62	88 185	87	123 744	112	159 302	137	194 861
13	18 490	38	54 049	63	89 607	88	125 166	113	160 724	138	196 283
14	19 913	39	55 471	64	91 030	89	126 588	114	162 147	139	197 705
15	21 335	40	56 894	65	92 452	90	128 011	115	163 569	140	199 128
16	22 757	41	58 316	66	93 874	91	129 433	116	164 991	141	200 550
17	24 180	42	59 738	67	95 297	92	130 855	117	166 414	142	201 972
18	25 602	43	61 161	68	96 719	93	132 278	118	167 836	143	203 395
19	27 024	44	62 583	69	98 141	94	133 700	119	169 258	144	204 817
20	28 447	45	64 005	70	99 564	95	135 122	120	170 681	145	206 239
21	29 869	46	65 428	71	100 986	96	136 545	121	172 103	146	207 662
22	31 291	47	66 850	72	102 408	97	137 967	122	173 525	147	209 084
23	32 714	48	68 272	73	103 831	98	139 389	123	174 948	148	210 506
24	34 136	49	69 695	74	105 253	99	140 812	124	176 370	149	211 929
25	35 558	50	71 117	75	106 675	100	142 234	125	177 792	150	213 351
26	36 981	51	72 539	76	108 098	101	143 656	126	179 215	151	214 773
27	38 403	52	73 962	77	109 520	102	145 079	127	180 637	152	216 196
28	39 826	53	75 384	78	110 943	103	146 501	128	182 059	153	217 618
29	41 248	54	76 806	79	112 365	104	147 923	129	183 482	154	219 040
30	42 670	55	78 229	80	113 787	105	149 346	130	184 904	155	220 463
31	44 093	56	79 651	81	115 210	106	150 768	131	186 327	156	221 885
32	45 515	57	81 073	82	116 632	107	152 190	132	187 749	157	223 307
33	46 937	58	82 496	83	118 054	108	153 613	133	189 171	158	224 730
34	48 360	59	83 918	84	119 477	109	155 035	134	190 594	159	226 152

Table A3.4. Stress-Intensity Conversions

The middle column of figures (in bold-faced type) contains the reading (in MPa$\sqrt{m}$ or ksi$\sqrt{in}$.) to be converted. If converting from ksi$\sqrt{in}$. to MPa$\sqrt{m}$, read the MPa$\sqrt{m}$ equivalent in the column headed "MPa$\sqrt{m}$". If converting from MPa$\sqrt{m}$ to ksi$\sqrt{in}$., read the ksi$\sqrt{in}$. equivalent in the column headed "ksi$\sqrt{in}$.". 1 ksi$\sqrt{in}$. = 1.098845 MPa$\sqrt{m}$.

ksi, $\sqrt{in}$.		MPa, $\sqrt{m}$	ksi, $\sqrt{in}$.		MPa, $\sqrt{m}$	ksi, $\sqrt{in}$.		MPa, $\sqrt{m}$	ksi, $\sqrt{in}$.		MPa, $\sqrt{m}$	ksi, $\sqrt{in}$.		MPa, $\sqrt{m}$
0.91005	**1**	1.0988	37.312	**41**	45.051	73.714	**81**	89.003	110.12	**121**	132.95	146.52	**161**	176.91
1.8201	**2**	2.1976	38.222	**42**	46.150	74.624	**82**	90.102	111.03	**122**	134.05	147.43	**162**	178.01
2.7301	**3**	3.2964	39.132	**43**	47.248	75.534	**83**	91.200	111.94	**123**	135.15	148.34	**163**	179.10
3.6402	**4**	4.3952	40.042	**44**	48.347	76.444	**84**	92.300	112.85	**124**	136.25	149.25	**164**	180.20
4.5502	**5**	5.4940	40.952	**45**	49.446	77.354	**85**	93.398	113.76	**125**	137.35	150.16	**165**	181.30
5.4603	**6**	6.5928	41.862	**46**	50.545	78.264	**86**	94.497	114.67	**126**	138.45	151.07	**166**	182.40
6.3703	**7**	7.6916	42.772	**47**	51.644	79.174	**87**	95.596	115.58	**127**	139.55	151.98	**167**	183.50
7.2804	**8**	8.7904	43.682	**48**	52.742	80.084	**88**	96.694	116.49	**128**	140.65	152.89	**168**	184.60
8.1904	**9**	9.8892	44.592	**49**	53.841	80.994	**89**	97.793	117.40	**129**	141.75	153.80	**169**	185.70
9.1005	**10**	10.988	45.502	**50**	54.940	81.904	**90**	98.892	118.31	**130**	142.84	154.71	**170**	186.80
10.011	**11**	12.087	46.412	**51**	56.039	82.814	**91**	99.991	119.22	**131**	143.94	155.62	**171**	187.90
10.921	**12**	13.186	47.322	**52**	57.138	83.724	**92**	101.09	120.13	**132**	145.04	156.53	**172**	189.00
11.831	**13**	14.284	48.232	**53**	58.236	84.634	**93**	102.19	121.04	**133**	146.14	157.44	**173**	190.10
12.741	**14**	15.383	49.143	**54**	59.335	85.544	**94**	103.29	121.95	**134**	147.24	158.35	**174**	191.19
13.651	**15**	16.482	50.053	**55**	60.434	86.454	**95**	104.39	122.86	**135**	148.34	159.26	**175**	192.29
14.561	**16**	17.581	50.963	**56**	61.533	87.364	**96**	105.48	123.77	**136**	149.44	160.17	**176**	193.39
15.471	**17**	18.680	51.873	**57**	62.632	88.275	**97**	106.58	124.68	**137**	150.54	161.08	**177**	194.49
16.381	**18**	19.778	52.783	**58**	63.730	89.185	**98**	107.68	125.59	**138**	151.63	161.99	**178**	195.59
17.291	**19**	20.877	53.693	**59**	64.829	90.095	**99**	108.78	126.50	**139**	152.73	162.90	**179**	196.69
18.201	**20**	21.976	54.603	**60**	65.928	91.005	**100**	109.88	127.41	**140**	153.83	163.81	**180**	197.78
19.111	**21**	23.075	55.513	**61**	67.027	91.915	**101**	110.98	128.32	**141**	154.93	164.72	**181**	198.88
20.021	**22**	24.174	56.423	**62**	68.126	92.825	**102**	112.08	129.23	**142**	156.03	165.63	**182**	199.98
20.931	**23**	25.272	57.333	**63**	69.224	93.735	**103**	113.18	130.14	**143**	157.13	166.54	**183**	201.08
21.841	**24**	26.371	58.243	**64**	70.323	94.645	**104**	114.28	131.05	**144**	158.23	167.45	**184**	202.18
22.751	**25**	27.470	59.153	**65**	71.422	95.555	**105**	115.37	131.96	**145**	159.33	168.36	**185**	203.28
23.661	**26**	28.569	60.063	**66**	72.521	96.465	**106**	116.47	132.87	**146**	160.42	169.27	**186**	204.38
24.571	**27**	29.668	60.973	**67**	73.620	97.375	**107**	117.57	133.78	**147**	161.52	170.18	**187**	205.48
25.481	**28**	30.766	61.883	**68**	74.718	98.285	**108**	118.67	134.69	**148**	162.62	171.09	**188**	206.57
26.391	**29**	31.865	62.793	**69**	75.817	99.195	**109**	119.77	135.60	**149**	163.72	172.00	**189**	207.67
27.301	**30**	32.964	63.703	**70**	76.916	100.11	**110**	120.87	136.51	**150**	164.82	172.91	**190**	208.77
28.211	**31**	34.063	64.613	**71**	78.015	101.02	**111**	121.97	137.42	**151**	165.92	173.82	**191**	209.87
29.121	**32**	35.162	65.523	**72**	79.114	101.93	**112**	123.07	138.33	**152**	167.02	174.73	**192**	210.97
30.032	**33**	36.260	66.433	**73**	80.212	102.84	**113**	124.16	139.24	**153**	168.12	175.64	**193**	212.07
30.942	**34**	37.359	67.343	**74**	81.311	103.75	**114**	125.26	140.15	**154**	169.22	176.55	**194**	213.17
31.852	**35**	38.458	68.253	**75**	82.410	104.66	**115**	126.36	141.06	**155**	170.31	177.46	**195**	214.27
32.762	**36**	39.557	69.164	**76**	83.509	105.57	**116**	127.46	141.97	**156**	171.41	178.37	**196**	215.36
33.672	**37**	40.656	70.074	**77**	84.608	106.48	**117**	128.56	142.88	**157**	172.51	179.28	**197**	216.46
34.582	**38**	41.754	70.984	**78**	85.706	107.39	**118**	129.66	143.79	**158**	173.61	180.19	**198**	217.56
35.492	**39**	42.853	71.893	**79**	86.805	108.30	**119**	130.76	144.70	**159**	174.71	181.10	**199**	218.66
36.402	**40**	43.952	72.804	**80**	87.904	109.21	**120**	131.86	145.61	**160**	175.81	182.01	**200**	219.76

Table A3.5. Approximate Equivalent Hardness Numbers and Tensile Strengths for Vickers Hardness Numbers for Steel(a)

Vickers hardness No.	Brinell hardness No., 3000-kg load, 10-mm ball — Standard ball	Tungsten carbide ball	A scale, 60-kg load, Brale indenter	B scale, 100-kg load, 1/16-in.-diam ball	C scale, 150-kg load, Brale indenter	D scale, 100-kg load, Brale indenter	15N scale, 15-kg load	30N scale, 30-kg load	45N scale, 45-kg load	Tensile strength, (approx), MPa
940	...	...	85.6	...	68.0	76.9	93.2	84.4	75.4	...
920	...	...	85.3	...	67.5	76.5	93.0	84.0	74.8	...
900	...	...	85.0	...	67.0	76.1	92.9	83.6	74.2	...
880	...	(767)	84.7	...	66.4	75.7	92.7	83.1	73.6	...
860	...	(757)	84.4	...	65.9	75.3	92.5	82.7	73.1	...
840	...	(745)	84.1	...	65.3	74.8	92.3	82.2	72.2	...
820	...	(733)	83.8	...	64.7	74.3	92.1	81.7	71.8	...
800	...	(722)	83.4	...	64.0	73.8	91.8	81.1	71.0	...
780	...	(710)	83.0	...	63.3	73.3	91.5	80.4	70.2	...
760	...	(698)	82.6	...	62.5	72.6	91.2	79.7	69.4	...
740	...	(684)	82.2	...	61.8	72.1	91.0	79.1	68.6	...
720	...	(670)	81.8	...	61.0	71.5	90.7	78.4	67.7	...
700	...	(656)	81.3	...	60.1	70.8	90.3	77.6	66.7	...
690	...	(647)	81.1	...	59.7	70.5	90.1	77.2	66.2	...
680	...	(638)	80.8	...	59.2	70.1	89.8	76.8	65.7	2448
670	...	(630)	80.6	...	58.8	69.8	89.7	76.4	65.3	2399
660	...	620	80.3	...	58.3	69.4	89.5	75.9	64.7	2358
650	...	611	80.0	...	57.8	69.0	89.2	75.5	64.1	2317
640	...	601	79.8	...	57.3	68.7	89.0	75.1	63.5	2261
630	...	591	79.5	...	56.8	68.3	88.8	74.6	63.0	2227
620	...	582	79.2	...	56.3	67.9	88.5	74.2	62.4	2186
610	...	573	78.9	...	55.7	67.5	88.2	73.6	61.7	2137
600	...	564	78.6	...	55.2	67.0	88.0	73.2	61.2	2089
590	...	554	78.4	...	54.7	66.7	87.8	72.7	60.5	2055
580	...	545	78.0	...	54.1	66.2	87.5	72.1	59.9	2020
570	...	535	77.8	...	53.6	65.8	87.2	71.7	59.3	1986
560	...	525	77.4	...	53.0	65.4	86.9	71.2	58.6	1951
550	(505)	517	77.0	...	52.3	64.8	86.6	70.5	57.8	1903
540	(496)	507	76.7	...	51.7	64.4	86.3	70.0	57.0	1862
530	(488)	497	76.4	...	51.1	63.9	86.0	69.5	56.2	1827

(continued)

Table A3.5. Approximate Equivalent Hardness Numbers and Tensile Strengths for Vickers Hardness Numbers for Steel(a) (contd.)

Vickers hardness No.	Brinell hardness No., 3000-kg load, Standard ball (10-mm ball)	Tungsten carbide ball	Rockwell hardness No. — A scale, 60-kg load, Brale Indenter	B scale, 100-kg load, 1/16-in.-diam ball	C scale, 150-kg load, Brale Indenter	D scale, 100-kg load, Brale Indenter	Rockwell superficial hardness No., superficial Brale indenter — 15N scale, 15-kg load	30N scale, 30-kg load	45N scale, 45-kg load	Tensile strength (approx), MPa
520	(480)	488	76.1	...	50.5	63.5	85.7	69.0	55.6	1793
510	(473)	479	75.7	...	49.8	62.9	85.4	68.3	54.7	1751
500	(465)	471	75.3	...	49.1	62.2	85.0	67.7	53.9	1703
490	(456)	460	74.9	...	48.4	61.6	84.7	67.1	53.1	1662
480	(448)	452	74.5	...	47.7	61.3	84.3	66.4	52.2	1620
470	441	442	74.1	...	46.9	60.7	83.9	65.7	51.3	1572
460	433	433	73.6	...	46.1	60.1	83.6	64.9	50.4	1538
450	425	425	73.3	...	45.3	59.4	83.2	64.3	49.4	1496
440	415	415	72.8	...	44.5	58.8	82.8	63.5	48.4	1462
430	405	405	72.3	...	43.6	58.2	82.3	62.7	47.4	1413
420	397	397	71.8	...	42.7	57.5	81.8	61.9	46.4	1372
410	388	388	71.4	...	41.8	56.8	81.4	61.1	45.3	1331
400	379	379	70.8	...	40.8	56.0	80.8	60.2	44.1	1289
390	369	369	70.3	...	39.8	55.2	80.3	59.3	42.9	1248
380	360	360	69.8	(110.0)	38.8	54.4	79.8	58.4	41.7	1207
370	350	350	69.2	...	37.7	53.6	79.2	57.4	40.4	1172
360	341	341	68.7	(109.0)	36.6	52.8	78.6	56.4	39.1	1131
350	331	331	68.1	...	35.5	51.9	78.0	55.4	37.8	1096
340	322	322	67.6	(108.0)	34.4	51.1	77.4	54.4	36.5	1069
330	313	313	67.0	...	33.3	50.2	76.8	53.6	35.2	1034
320	303	303	66.4	(107.0)	32.2	49.4	76.2	52.3	33.9	1007
310	294	294	65.8	...	31.0	48.4	75.6	51.3	32.5	979
300	284	284	65.2	(105.5)	29.8	47.5	74.9	50.2	31.1	951
295	280	280	64.8	...	29.2	47.1	74.6	49.7	30.4	938
290	275	275	64.5	(104.5)	28.5	46.5	74.2	49.0	29.5	917
285	270	270	64.2	...	27.8	46.0	73.8	48.4	28.7	903
280	265	265	63.8	(103.5)	27.1	45.3	73.4	47.8	27.9	899
275	261	261	63.5	...	26.4	44.9	73.0	47.2	27.1	876
270	256	256	63.1	(102.0)	25.6	44.3	72.6	46.4	26.2	855
265	252	252	62.7	..	24.8	43.7	72.1	45.7	25.2	841
260	247	247	62.4	(101.0)	24.0	43.1	71.6	45.0	24.3	827
255	243	243	62.0	...	23.1	42.2	71.1	44.2	23.2	807
250	238	238	61.6	99.5	22.2	41.7	70.6	43.4	22.2	793
245	233	233	61.2	...	21.3	41.1	70.1	42.5	21.1	779
240	228	228	60.7	98.1	20.3	40.3	69.6	41.7	19.9	765
230	219	219	...	96.7	(18.0)	...	...	...	...	731
220	209	209	...	95.0	(15.7)	...	...	...	...	696
210	200	200	...	93.4	(13.4)	...	...	...	...	669
200	190	190	...	91.5	(11.0)	...	...	...	...	634
190	181	181	...	89.5	(8.5)	...	...	...	...	607
180	171	171	...	87.1	(6.0)	...	...	...	...	579
170	162	162	...	85.0	(3.0)	...	...	...	...	545
160	152	152	...	81.7	(0.0)	...	...	...	...	517
150	143	143	...	78.7	...	...	...	...	...	490
140	133	133	...	75.0	...	...	...	...	...	455
130	124	124	...	71.2	...	...	...	...	...	427
120	114	114	...	66.7	...	...	...	...	...	393
110	105	105	...	62.3	...	...	...	...	...	...
100	95	95	...	56.2	...	...	...	...	...	...
95	90	90	...	52.0	...	...	...	...	...	...
90	86	86	...	48.0	...	...	...	...	...	...
85	81	81	...	41.0	...	...	...	...	...	...

(a) For carbon and alloy steels in the annealed, normalized, and quenched-and-tempered conditions; less accurate for cold worked condition and for austenitic steels. The values in **boldface type** correspond to the values in the joint SAE-ASM-ASTM hardness conversions as printed in ASTM E140, Table 2. The values in parentheses are beyond normal range and are given for information only.

Table A3.6. Energy Conversions

The middle column of figures (in bold-faced type) contains the reading (in J or ft·lb) to be converted. If converting from ft·lb to J, read the J equivalent in the column headed "J". If converting from J to ft·lb, read the equivalent in the column headed "ft·lb". 1 ft·lb = 1.355818 J.

ft·lb		J	ft·lb		J	ft·lb		J	ft·lb		J
0.7376	1	1.3558	28.7649	39	52.8769	56.7923	77	104.3980	129.0734	175	237.2681
1.4751	2	2.7116	29.5025	40	54.2327	57.5298	78	105.7538	132.7612	180	244.0472
2.2127	3	4.0675	30.2400	41	55.5885	58.2674	79	107.1096	136.4490	185	250.8263
2.9502	4	5.4233	30.9776	42	56.9444	59.0050	80	108.4654	140.1368	190	257.6054
3.6878	5	6.7791	31.7152	43	58.3002	59.7425	81	109.8212	143.8246	195	264.3845
4.4254	6	8.1349	32.4527	44	59.6560	60.4801	82	111.1771	147.5124	200	271.1636
5.1629	7	9.4907	33.1903	45	61.0118	61.2177	83	112.5329	154.8880	210	284.7218
5.9005	8	10.8465	33.9279	46	62.3676	61.9552	84	113.8887	162.2637	220	298.2799
6.6381	9	12.2024	34.6654	47	63.7234	62.6928	85	115.2445	169.6393	230	311.8381
7.3756	10	13.5582	35.4030	48	65.0793	63.4303	86	116.6003	177.0149	240	325.3963
8.1132	11	14.9140	36.1405	49	66.4351	64.1679	87	117.9562	184.3905	250	338.9545
8.8507	12	16.2698	36.8781	50	67.7909	64.9055	88	119.3120	191.7661	260	352.5126
9.5883	13	17.6256	37.6157	51	69.1467	65.6430	89	120.6678	199.1418	270	366.0708
10.3259	14	18.9815	38.3532	52	70.5025	66.3806	90	122.0236	206.5174	280	379.6290
11.0634	15	20.3373	39.0908	53	71.8583	67.1182	91	123.3794	213.8930	290	393.1872
11.8010	16	21.6931	39.8284	54	73.2142	67.8557	92	124.7452	221.2686	300	406.7454
12.5386	17	23.0489	40.5659	55	74.5700	68.5933	93	126.0911	228.6442	310	420.3036
13.2761	18	24.4047	41.3035	56	75.9258	69.3308	94	127.4469	236.0199	320	433.8617
14.0137	19	25.7605	42.0410	57	77.2816	70.0684	95	128.8027	243.3955	330	447.4199
14.7512	20	27.1164	42.7786	58	78.6374	70.8060	96	130.1585	250.7711	340	460.9781
15.4888	21	28.4722	43.5162	59	79.9933	71.5435	97	131.5143	258.1467	350	474.5363
16.2264	22	29.8280	44.2537	60	81.3491	72.2811	98	132.8702	265.5224	360	488.0944
16.9639	23	31.1838	44.9913	61	82.7049	73.0186	99	134.2260	272.8980	370	501.6526
17.7015	24	32.5396	45.7288	62	84.0607	73.7562	100	135.5818	280.2736	380	515.2108
18.4390	25	33.8954	46.4664	63	85.4165	77.4440	105	142.3609	287.6492	390	528.7690
19.1766	26	35.2513	47.2040	64	86.7723	81.1318	110	149.1400	295.0248	400	542.3272
19.9142	27	36.6071	47.9415	65	88.1282	84.8196	115	155.9191	302.4005	410	555.8854
20.6517	28	37.9629	48.6791	66	89.4840	88.5075	120	162.6982	309.7761	420	569.4435
21.3893	29	39.3187	49.4167	67	90.8398	92.1953	125	169.4772	317.1517	430	583.0017
22.1269	30	40.6745	50.1542	68	92.1956	95.8831	130	176.2563	324.5273	440	596.5599
22.8644	31	42.0304	50.8918	69	93.5514	99.5709	135	183.0354	331.9029	450	610.1181
23.6020	32	43.3862	51.6293	70	94.9073	103.2587	140	189.8145	339.2786	460	623.6762
24.3395	33	44.7420	52.3669	71	96.2631	106.9465	145	196.5936	346.6542	470	637.2344
25.0771	34	46.0978	53.1045	72	97.6189	110.6343	150	203.3727	354.0298	480	650.7926
25.8147	35	47.4536	53.8420	73	98.9747	114.3221	155	210.1518	361.4054	490	664.3508
26.5522	36	48.8094	54.5796	74	100.3305	118.0099	160	216.9308	368.7811	500	677.9090
27.2898	37	50.1653	55.3172	75	101.6863	121.6977	165	223.7099			
28.0274	38	51.5211	56.0547	76	103.0422	125.3856	170	230.4890			

Table A3.7. Temperature Conversions

The general arrangement of this conversion table was devised by Sauveur and Boylston. The middle columns of numbers (in **boldface** type) contain the temperature readings (°F or °C) to be converted. When converting from degrees Fahrenheit to degrees Celsius, read the Celsius equivalent in the column headed "C". When converting from Celsius to Fahrenheit, read the Fahrenheit equivalent in the column headed "F".

F		C	F		C	F		C	F		C
	-458	-272.22		**-378**	-227.78		**-298**	-183.33	-360.4	**-218**	-138.89
	-456	-271.11		**-376**	-226.67		**-296**	-182.22	-356.8	**-216**	-137.78
	-454	-270.00		**-374**	-225.56		**-294**	-181.11	-353.2	**-214**	-136.67
	-452	-268.89		**-372**	-224.44		**-292**	-180.00	-349.6	**-212**	-135.56
	-450	-267.78		**-370**	-223.33		**-290**	-178.89	-346.0	**-210**	-134.44
	-448	-266.67		**-368**	-222.22		**-288**	-177.78	-342.4	**-208**	-133.33
	-446	-265.56		**-366**	-221.11		**-286**	-176.67	-338.8	**-206**	-132.22
	-444	-264.44		**-364**	-220.00		**-284**	-175.56	-335.2	**-204**	-131.11
	-442	-263.33		**-362**	-218.89		**-282**	-174.44	-331.6	**-202**	-130.00
	-440	-262.22		**-360**	-217.78		**-280**	-173.33	-328.0	**-200**	-128.89
	-438	-261.11		**-358**	-216.67		**-278**	-172.22	-324.4	**-198**	-127.78
	-436	-260.00		**-356**	-215.56		**-276**	-171.11	-320.8	**-196**	-126.67
	-434	-258.89		**-354**	-214.44		**-274**	-170.00	-317.2	**-194**	-125.56
	-432	-257.78		**-352**	-213.33	-457.6	**-272**	-168.89	-313.6	**-192**	-124.44
	-430	-256.67		**-350**	-212.22	-454.0	**-270**	-167.78	-310.0	**-190**	-123.33
	-428	-255.56		**-348**	-211.11	-450.4	**-268**	-166.67	-306.4	**-188**	-122.22
	-426	-254.44		**-346**	-210.00	-446.8	**-266**	-165.56	-302.8	**-186**	-121.11
	-424	-253.33		**-344**	-208.89	-443.2	**-264**	-164.44	-299.2	**-184**	-120.00
	-422	-252.22		**-342**	-207.78	-439.6	**-262**	-163.33	-295.6	**-182**	-118.89
	-420	-251.11		**-340**	-206.67	-436.0	**-260**	-162.22	-292.0	**-180**	-117.78
	-418	-250.00		**-338**	-205.56	-432.4	**-258**	-161.11	-288.4	**-178**	-116.67
	-416	-248.89		**-336**	-204.44	-428.8	**-256**	-160.00	-284.8	**-176**	-115.56
	-414	-247.78		**-334**	-203.33	-425.2	**-254**	-158.89	-281.2	**-174**	-114.44
	-412	-246.67		**-332**	-202.22	-421.6	**-252**	-157.78	-277.6	**-172**	-113.33
	-410	-245.56		**-330**	-201.11	-418.0	**-250**	-156.67	-274.0	**-170**	-112.22
	-408	-244.44		**-328**	-200.00	-414.4	**-248**	-155.56	-270.4	**-168**	-111.11
	-406	-243.33		**-326**	-198.89	-410.8	**-246**	-154.44	-266.8	**-166**	-110.00
	-404	-242.22		**-324**	-197.78	-407.2	**-244**	-153.33	-263.2	**-164**	-108.89
	-402	-241.11		**-322**	-196.67	-403.6	**-242**	-152.22	-259.6	**-162**	-107.78
	-400	-240.00		**-320**	-195.56	-400.0	**-240**	-151.11	-256.0	**-160**	-106.67
	-398	-238.89		**-318**	-194.44	-396.4	**-238**	-150.00	-252.4	**-158**	-105.56
	-396	-237.78		**-316**	-193.33	-392.8	**-236**	-148.89	-248.8	**-156**	-104.44
	-394	-236.67		**-314**	-192.22	-389.2	**-234**	-147.78	-245.2	**-154**	-103.33
	-392	-235.56		**-312**	-191.11	-385.6	**-232**	-146.67	-241.6	**-152**	-102.22
	-390	-234.44		**-310**	-190.00	-382.0	**-230**	-145.56	-238.0	**-150**	-101.11
	-388	-233.33		**-308**	-188.89	-378.4	**-228**	-144.44	-234.4	**-148**	-100.00
	-386	-232.22		**-306**	-187.78	-374.8	**-226**	-143.33	-230.8	**-146**	-98.89
	-384	-231.11		**-304**	-186.67	-371.2	**-224**	-142.22	-227.2	**-144**	-97.78
	-382	-230.00		**-302**	-185.56	-367.6	**-222**	-141.11	-223.6	**-142**	-96.67
	-380	-228.89		**-300**	-184.44	-364.0	**-220**	-140.00	-220.0	**-140**	-95.56

(continued on the next page)

Table A3.7. Temperature Conversions (contd.)

F		C	F		C	F		C	F		C
-216.4	-138	-94.44	+35.6	+2	-16.67	287.6	142	61.11	539.6	282	138.89
-212.8	-136	-93.33	+39.2	+4	-15.56	291.2	144	62.22	543.2	284	140.00
-209.2	-134	-92.22	+42.8	+6	-14.44	294.8	146	63.33	546.8	286	141.11
-205.6	-132	-91.11	+46.4	+8	-13.33	298.4	148	64.44	550.4	288	142.22
-202.0	-130	-90.00	+50.0	+10	-12.22	302.0	150	65.56	554.0	290	143.33
-198.4	-128	-88.89	+53.6	+12	-11.11	305.6	152	66.67	557.6	292	144.44
-194.8	-126	-87.78	+57.2	+14	-10.00	309.2	154	67.78	561.2	294	145.56
-191.2	-124	-86.67	+60.8	+16	-8.89	312.8	156	68.89	564.8	296	146.67
-187.6	-122	-85.56	+64.4	+18	-7.78	316.4	158	70.00	568.4	298	147.78
-184.0	-120	-84.44	+68.0	+20	-6.67	320.0	160	71.11	572.0	300	148.89
-180.4	-118	-83.33	+71.6	+22	-5.56	323.6	162	72.22	575.6	302	150.00
-176.8	-116	-82.22	+75.2	+24	-4.44	327.2	164	73.33	579.2	304	151.11
-173.2	-114	-81.11	+78.8	+26	-3.33	330.8	166	74.44	582.8	306	152.22
-169.6	-112	-80.00	+82.4	+28	-2.22	334.4	168	75.56	586.4	308	153.33
-166.0	-110	-78.89	+86.0	+30	-1.11	338.0	170	76.67	590.0	310	154.44
-162.4	-108	-77.78	+89.6	+32	±0.00	341.6	172	77.78	593.6	312	155.56
-158.8	-106	-76.67	+93.2	+34	+1.11	345.2	174	78.89	597.2	314	156.67
-155.2	-104	-75.56	+96.8	+36	+2.22	348.8	176	80.00	600.8	316	157.78
-151.6	-102	-74.44	+100.4	+38	+3.33	352.4	178	81.11	604.4	318	158.89
-148.0	-100	-73.33	+104.0	+40	+4.44	356.0	180	82.22	608.0	320	160.00
-144.4	-98	-72.22	107.6	42	5.56	359.6	182	83.33	611.6	322	161.11
-140.8	-96	-71.11	111.2	44	6.67	363.2	184	84.44	615.2	324	162.22
-137.2	-94	-70.00	114.8	46	7.78	366.8	186	85.56	618.8	326	163.33
-133.6	-92	-68.89	118.4	48	8.89	370.4	188	86.67	622.4	328	164.44
-130.0	-90	-67.78	122.0	50	10.00	374.0	190	87.78	626.0	330	165.56
-126.4	-88	-66.67	125.6	52	11.11	377.6	192	88.89	629.6	332	166.67
-122.8	-86	-65.56	129.2	54	12.22	381.2	194	90.00	633.2	334	167.78
-119.2	-84	-64.44	132.8	56	13.33	384.8	196	91.11	636.8	336	168.89
-115.6	-82	-63.33	136.4	58	14.44	388.4	198	92.22	640.4	338	170.00
-112.0	-80	-62.22	140.0	60	15.56	392.0	200	93.33	644.0	340	171.11
-108.4	-78	-61.11	143.6	62	16.67	395.6	202	94.44	647.6	342	172.22
-104.8	-76	-60.00	147.2	64	17.78	399.2	204	95.56	651.2	344	173.33
-101.2	-74	-58.89	150.8	66	18.89	402.8	206	96.67	654.8	346	174.44
-97.6	-72	-57.78	154.4	68	20.00	406.4	208	97.78	658.4	348	175.56
-94.0	-70	-56.67	158.0	70	21.11	410.0	210	98.89	662.0	350	176.67
-90.4	-68	-55.56	161.6	72	22.22	413.6	212	100.00	665.6	352	177.78
-86.8	-66	-54.44	165.2	74	23.33	417.2	214	101.11	669.2	354	178.89
-83.2	-64	-53.33	168.8	76	24.44	420.8	216	102.22	672.8	356	180.00
-79.6	-62	-52.22	172.4	78	25.56	424.4	218	103.33	676.4	358	181.11
-76.0	-60	-51.11	176.0	80	26.67	428.0	220	104.44	680.0	360	182.22
-72.4	-58	-50.00	179.6	82	27.78	431.6	222	105.56	683.6	362	183.33
-68.8	-56	-48.89	183.2	84	28.89	435.2	224	106.67	687.2	364	184.44
-65.2	-54	-47.78	186.8	86	30.00	438.8	226	107.78	690.8	366	185.56
-61.6	-52	-46.67	190.4	88	31.11	442.4	228	108.89	694.4	368	186.67
-58.0	-50	-45.56	194.0	90	32.22	446.0	230	110.00	698.0	370	187.78
-54.4	-48	-44.44	197.6	92	33.33	449.6	232	111.11	701.6	372	188.89
-50.8	-46	-43.33	201.2	94	34.44	453.2	234	112.22	705.2	374	190.00
-47.2	-44	-42.22	204.8	96	35.56	456.8	236	113.33	708.8	376	191.11
-43.6	-42	-41.11	208.4	98	36.67	460.4	238	114.44	712.4	378	192.22
-40.0	-40	-40.00	212.0	100	37.78	464.0	240	115.56	716.0	380	193.33
-36.4	-38	-38.89	215.6	102	38.89	467.6	242	116.67	719.6	382	194.44
-32.8	-36	-37.78	219.2	104	40.00	471.2	244	117.78	723.2	384	195.56
-29.2	-34	-36.67	222.8	106	41.11	474.8	246	118.89	726.8	386	196.67
-25.6	-32	-35.56	226.4	108	42.22	478.4	248	120.00	730.4	388	197.78
-22.0	-30	-34.44	230.0	110	43.33	482.0	250	121.11	734.0	390	198.89
-18.4	-28	-33.33	233.6	112	44.44	485.6	252	122.22	737.6	392	200.00
-14.8	-26	-32.22	237.2	114	45.56	489.2	254	123.33	741.2	394	201.11
-11.2	-24	-31.11	240.8	116	46.67	492.8	256	124.44	744.8	396	202.22
-7.6	-22	-30.00	244.4	118	47.78	496.4	258	125.56	748.4	398	203.33
-4.0	-20	-28.89	248.0	120	48.89	500.0	260	126.67	752.0	400	204.44
-0.4	-18	-27.78	251.6	122	50.00	503.6	262	127.78	755.6	402	205.56
+3.2	-16	-26.67	255.2	124	51.11	507.2	264	128.89	759.2	404	206.67
+6.8	-14	-25.56	258.8	126	52.22	510.8	266	130.00	762.8	406	207.78
+10.4	-12	-24.44	262.4	128	53.33	514.4	268	131.11	766.4	408	208.89
+14.0	-10	-23.33	266.0	130	54.44	518.0	270	132.22	770.0	410	210.00
+17.6	-8	-22.22	269.6	132	55.56	521.6	272	133.33	773.6	412	211.11
+21.2	-6	-21.11	273.2	134	56.67	525.2	274	134.44	777.2	414	212.22
+24.8	-4	-20.00	276.8	136	57.78	528.8	276	135.56	780.8	416	213.33
+28.4	-2	-18.89	280.4	138	58.89	532.4	278	136.67	784.4	418	214.44
+32.0	±0	-17.78	284.0	140	60.00	536.0	280	137.78	788.0	420	215.56

Table A3.7. Temperature Conversions (contd.)

F		C	F		C	F		C	F		C
791.6	422	216.67	1130.0	610	321.11	2390.0	1310	710.00	3650.0	2010	1098.9
795.2	424	217.78	1148.0	620	326.67	2408.0	1320	715.56	3668.0	2020	1104.4
798.8	426	218.89	1166.0	630	332.22	2426.0	1330	721.11	3686.0	2030	1110.0
802.4	428	220.00	1184.0	640	337.78	2444.0	1340	726.67	3704.0	2040	1115.6
806.0	430	221.11	1202.0	650	343.33	2462.0	1350	732.22	3722.0	2050	1121.1
809.6	432	222.22	1220.0	660	348.89	2480.0	1360	737.78	3740.0	2060	1126.7
813.2	434	223.33	1238.0	670	354.44	2498.0	1370	743.33	3758.0	2070	1132.2
816.8	436	224.44	1256.0	680	360.00	2516.0	1380	748.89	3776.0	2080	1137.8
820.4	438	225.56	1274.0	690	365.56	2534.0	1390	754.44	3794.0	2090	1143.3
824.0	440	226.67	1292.0	700	371.11	2552.0	1400	760.00	3812.0	2100	1148.9
827.6	442	227.78	1310.0	710	376.67	2570.0	1410	765.56	3830.0	2110	1154.4
831.2	444	228.89	1328.0	720	382.22	2588.0	1420	771.11	3848.0	2120	1160.0
834.8	446	230.00	1346.0	730	387.78	2606.0	1430	776.67	3866.0	2130	1165.6
838.4	448	231.11	1364.0	740	393.33	2624.0	1440	782.22	3884.0	2140	1171.1
842.0	450	232.22	1382.0	750	398.89	2642.0	1450	787.78	3902.0	2150	1176.7
845.6	452	233.33	1400.0	760	404.44	2660.0	1460	793.33	3920.0	2160	1182.2
849.2	454	234.44	1418.0	770	410.00	2678.0	1470	798.89	3938.0	2170	1187.8
852.8	456	235.56	1436.0	780	415.56	2696.0	1480	804.44	3956.0	2180	1193.3
856.4	458	236.67	1454.0	790	421.11	2714.0	1490	810.00	3974.0	2190	1198.9
860.0	460	237.78	1472.0	800	426.67	2732.0	1500	815.56	3992.0	2200	1204.4
863.6	462	238.89	1490.0	810	432.22	2750.0	1510	821.11	4010.0	2210	1210.0
867.2	464	240.00	1508.0	820	437.78	2768.0	1520	826.67	4028.0	2220	1215.6
870.8	466	241.11	1526.0	830	443.33	2786.0	1530	832.22	4046.0	2230	1221.1
874.4	468	242.22	1544.0	840	448.89	2804.0	1540	837.78	4064.0	2240	1226.7
878.0	470	243.33	1562.0	850	454.44	2822.0	1550	843.33	4082.0	2250	1232.2
881.6	472	244.44	1580.0	860	460.00	2840.0	1560	848.89	4100.0	2260	1237.8
885.2	474	245.56	1598.0	870	465.56	2858.0	1570	854.44	4118.0	2270	1243.3
888.8	476	246.67	1616.0	880	471.11	2876.0	1580	860.00	4136.0	2280	1248.9
892.4	478	247.78	1634.0	890	476.67	2894.0	1590	865.56	4154.0	2290	1254.4
896.0	480	248.89	1652.0	900	482.22	2912.0	1600	871.11	4172.0	2300	1260.0
899.6	482	250.00	1670.0	910	487.78	2930.0	1610	876.67	4190.0	2310	1265.6
903.2	484	251.11	1688.0	920	493.33	2948.0	1620	882.22	4208.0	2320	1271.1
906.8	486	252.22	1706.0	930	498.89	2966.0	1630	887.78	4226.0	2330	1276.7
910.4	488	253.33	1724.0	940	504.44	2984.0	1640	893.33	4244.0	2340	1282.2
914.0	490	254.44	1742.0	950	510.00	3002.0	1650	898.89	4262.0	2350	1287.8
917.6	492	255.56	1760.0	960	515.56	3020.0	1660	904.44	4280.0	2360	1293.3
921.2	494	256.67	1778.0	970	521.11	3038.0	1670	910.00	4298.0	2370	1298.9
924.8	496	257.78	1796.0	980	526.67	3056.0	1680	915.56	4316.0	2380	1304.4
928.4	498	258.89	1814.0	990	532.22	3074.0	1690	921.11	4334.0	2390	1310.0
932.0	500	260.00	1832.0	1000	337.78	3092.0	1700	926.67	4352.0	2400	1315.6
935.6	502	261.11	1850.0	1010	543.33	3110.0	1710	932.22	4370.0	2410	1321.1
939.2	504	262.22	1868.0	1020	548.89	3128.0	1720	937.78	4388.0	2420	1326.7
942.8	506	263.33	1886.0	1030	554.44	3146.0	1730	943.33	4406.0	2430	1332.2
946.4	508	264.44	1904.0	1040	560.00	3164.0	1740	948.89	4424.0	2440	1337.8
950.0	510	265.56	1922.0	1050	565.56	3182.0	1750	954.44	4442.0	2450	1343.3
953.6	512	266.67	1940.0	1060	571.11	3200.0	1760	960.00	4460.0	2460	1348.9
957.2	514	267.78	1958.0	1070	576.67	3218.0	1770	965.56	4478.0	2470	1354.4
960.8	516	268.89	1976.0	1080	582.22	3236.0	1780	971.11	4496.0	2480	1360.0
964.4	518	270.00	1994.0	1090	587.78	3254.0	1790	976.67	4514.0	2490	1365.6
968.0	520	271.11	2012.0	1100	593.33	3272.0	1800	982.22	4532.0	2500	1371.1
971.6	522	272.22	2030.0	1110	598.89	3290.0	1810	987.78	4550.0	2510	1376.7
975.2	524	273.33	2048.0	1120	604.44	3308.0	1820	993.33	4568.0	2520	1382.2
978.8	526	274.44	2066.0	1130	610.00	3326.0	1830	998.89	4586.0	2530	1387.8
982.4	528	275.56	2084.0	1140	615.56	3344.0	1840	1004.4	4604.0	2540	1393.3
986.0	530	276.67	2102.0	1150	621.11	3362.0	1850	1010.0	4622.0	2550	1398.9
989.6	532	277.78	2120.0	1160	626.67	3380.0	1860	1015.6	4640.0	2560	1404.4
993.2	534	278.89	2138.0	1170	632.22	3398.0	1870	1021.1	4658.0	2570	1410.0
996.8	536	280.00	2156.0	1180	637.78	3416.0	1880	1026.7	4676.0	2580	1415.6
1000.4	538	281.11	2174.0	1190	643.33	3434.0	1890	1032.2	4694.0	2590	1421.1
1004.0	540	282.22	2192.0	1200	648.89	3452.0	1900	1037.8	4712.0	2600	1426.7
1007.6	542	283.33	2210.0	1210	654.44	3470.0	1910	1043.3	4730.0	2610	1432.2
1011.2	544	284.44	2228.0	1220	660.00	3488.0	1920	1048.9	4748.0	2620	1437.8
1014.8	546	285.56	2246.0	1230	665.56	3506.0	1930	1054.4	4766.0	2630	1443.3
1018.4	548	286.67	2264.0	1240	671.11	3524.0	1940	1060.0	4784.0	2640	1448.9
1022.0	550	287.78	2282.0	1250	676.67	3542.0	1950	1065.6	4802.0	2650	1454.4
1040.0	560	293.33	2300.0	1260	682.22	3560.0	1960	1071.1	4820.0	2660	1460.0
1058.0	570	298.89	2318.0	1270	687.78	3578.0	1970	1076.7	4838.0	2670	1465.6
1076.0	580	304.44	2336.0	1280	693.33	3596.0	1980	1082.2	4856.0	2680	1471.1
1094.0	590	310.00	2354.0	1290	698.89	3614.0	1990	1087.8	4874.0	2690	1476.7
1112.0	600	315.56	2372.0	1300	704.44	3632.0	2000	1093.3	4892.0	2700	1482.2

Index